Industrial Biotechnology and Microbiology

Industrial Biotechnology and Microbiology

Editor: Wendell Carter

Callisto Reference
www.callistoreference.com

Callisto Reference,
118-35 Queens Blvd., Suite 400,
Forest Hills, NY 11375, USA

Visit us on the World Wide Web at:
www.callistoreference.com

ISBN: 978-1-63239-857-4 (Hardback)

The publisher's policy is to use permanent paper from mills that operate a sustainable forestry policy. Furthermore, the publisher ensures that the text paper and cover boards used have met acceptable environmental accreditation standards.

Printed in China.

Cataloging-in-publication Data

Industrial biotechnology and microbiology / edited by Wendell Carter.
p. cm.
Includes bibliographical references and index.
ISBN 978-1-63239-857-4
1. Biotechnology. 2. Industrial microbiology. 3. Biotechnology industries. 4. Genetic engineering. I. Carter, Wendell.
TP248.2 .I53 2017
660.6--dc23

Table of Contents

Preface

The study of microorganisms for their application in the industrial process is known as industrial biotechnology. Microbiology, the study of cellular and acellular cultures as they are found in nature, is a vital area of study for agriculture related practices. This book on industrial biotechnology and microbiology discusses topics related to food fermentation and food processing and the various techniques related to the culturing of microbiota. This book is a compilation of the latest researches and emerging trends that have been seen in these fields. It covers topics that are relevant for the theory as well as practice of industrial biotechnology and microbiology. This book elucidates new techniques and their applications in a multidisciplinary approach. It presents researches and studies performed by experts across the globe. This book aims to assist those with a good goal of delving into this field. It will serve as a vital source of reference to students and scholars alike.

After months of intensive research and writing, this book is the end result of all who devoted their time and efforts in the initiation and progress of this book. It will surely be a source of reference in enhancing the required knowledge of the new developments in the area. During the course of developing this book, certain measures such as accuracy, authenticity and research focused analytical studies were given preference in order to produce a comprehensive book in the area of study.

This book would not have been possible without the efforts of the authors and the publisher. I extend my sincere thanks to them. Secondly, I express my gratitude to my family and well-wishers. And most importantly, I thank my students for constantly expressing their willingness and curiosity in enhancing their knowledge in the field, which encourages me to take up further research projects for the advancement of the area.

Editor

Efficient Conversion of Phenylpyruvic Acid to Phenyllactic Acid by Using Whole Cells of *Bacillus coagulans* SDM

Zhaojuan Zheng[1], Cuiqing Ma[1]*, Chao Gao[2], Fengsong Li[1], Jiayang Qin[1], Haiwei Zhang[1], Kai Wang[1], Ping Xu[1,2]*

1 State Key Laboratory of Microbial Technology, Shandong University, Jinan, People's Republic of China, **2** MOE Key Laboratory of Microbial Metabolism and School of Life Sciences and Biotechnology, Shanghai Jiao Tong University, Shanghai, People's Republic of China

Abstract

Background: Phenyllactic acid (PLA), a novel antimicrobial compound with broad and effective antimicrobial activity against both bacteria and fungi, can be produced by many microorganisms, especially lactic acid bacteria. However, the concentration and productivity of PLA have been low in previous studies. The enzymes responsible for conversion of phenylpyruvic acid (PPA) into PLA are equivocal.

Methodology/Principal Findings: A novel thermophilic strain, *Bacillus coagulans* SDM, was isolated for production of PLA. When the solubility and dissolution rate of PPA were enhanced at a high temperature, whole cells of *B. coagulans* SDM could effectively convert PPA into PLA at a high concentration (37.3 g l^{-1}) and high productivity (2.3 g l^{-1} h^{-1}) under optimal conditions. Enzyme activity staining and kinetic studies identified NAD-dependent lactate dehydrogenases as the key enzymes that reduced PPA to PLA.

Conclusions/Significance: Taking advantage of the thermophilic character of *B. coagulans* SDM, a high yield and productivity of PLA were obtained. The enzymes involved in PLA production were identified and characterized, which makes possible the rational design and construction of microorganisms suitable for PLA production with metabolic engineering.

Editor: Shuang-yong Xu, New England Biolabs, Inc., United States of America

Funding: This work was supported by National Basic Research Program of China (2007CB707803) and Chinese National Program for High Technology Research and Development (2006AA020102). The funders had no role in study design, data collection and analysis, decision to publish, or preparation of the manuscript.

Competing Interests: The authors have declared that no competing interests exist.

* E-mail: macq@sdu.edu.cn (CM); pingxu@sjtu.edu.cn (PX)

Introduction

Phenyllactic acid (PLA) has broad and effective antimicrobial activity against both bacteria and fungi and can therefore be employed and developed as a new type of natural antiseptic agent to extend the shelf life of food and feed [1–3]. It is also a useful precursor for the synthesis of many important drugs, including Danshensu (3,4-dihydroxyphenyllactic acid) which can inhibit platelet aggregation and coronary artery disease, hypoglycemic reagents, protease inhibitors, and anti-HIV reagents [4–8]. Because of its wide use in food and pharmaceutical industries, PLA production has attracted the attention of biotechnologists.

Chemical and biotechnological routes have been developed for PLA production. The chemical transformation strategy has some disadvantages, including a complex technology route, excessive by-products, and environmental pollution [9,10]. Regarding the requirement for environmental protection and sustainable development, biotransformation has emerged as a powerful strategy for the production of this valuable compound [11]. Recently, several microorganisms, including *Geotrichum candidum*, propionibacteria, and lactic acid bacteria (LAB) were found to be PLA producers [12–18].

Many studies have focused on the ability of LAB to produce PLA because LAB have GRAS (generally recognized as safe) status. LAB, especially *Lactobacillus* strains, yielded PLA at the low level of 0.05–0.57 mM via the phenylalanine (Phe) metabolic pathway [19,20]. The transamination reaction in the Phe pathway was the bottleneck for PLA formation [20], and the use of phenylpyruvic acid (PPA) as a substitute substrate led to a 14-fold increase in PLA production [21]. PPA showed obvious inhibitory effects in the biotransformation process, and therefore, fed-batch fermentation could be conducted for producing a high amount of PLA. It should be noted that PPA powder must be dissolved beforehand because of its slow dissolution rate at the biotransformation temperature [21]. PLA concentration was significantly increased with the application of this strategy [22].

PLA is the acknowledged reduction product of PPA, but the enzymes responsible for this reaction remain unclear. In previous studies, hydroxyisocaproate dehydrogenase (HicDH), phenyllactic acid dehydrogenase (PLDHase), and D-lactate dehydrogenase (D-LDH) of *Lactobacillus* were assumed to be involved in PLA production from PPA [19,23–26].

In this study, a thermophilic strain, *Bacillus coagulans* SDM, was isolated and its ability to produce PLA from PPA was confirmed.

The enzymes involved in the production of PLA were identified and characterized. In addition, with the enhanced solubility and dissolution rate of PPA at the high reaction temperature, a high concentration, yield and productivity of PLA were obtained in fed-batch bioconversion.

Results

Isolation of thermophilic bacteria for PLA production

Using the procedures described in Materials and Methods, 560 bacterial strains that grew at 50°C were tested for PLA production. Strain SDM showed the highest PLA production rate among the isolates and was selected for a detailed study. Identification of the strain was confirmed by Deutsche Sammlung von Mikroorganismen und Zellkulturen GmbH (DSMZ). The pattern of the fatty acid analysis was typical for *Bacillus* and suggested *B. coagulans*. Identification of the strain SDM as *B. coagulans* was also supported by the partial 16S rRNA sequence (GenBank accession number: HQ171055). The strain was deposited at the China Center for Type Culture Collection (CCTCC NO: M 2010012).

Optimization of biotransformation by whole cells of *B. coagulans* SDM

By using the thermophilic property of *B. coagulans* SDM, bioconversion of PLA from PPA was studied at a relatively high temperature. The reaction was carried out at 100 rpm for 0.5 h in phosphate buffer solution (PBS, 1/15 M [pH 7.4]) containing whole cells of *B. coagulans* SDM, PPA, and glucose. The optimal pH value was determined to be 6.5 (Figure 1a) and the optimal temperature was determined to be 50°C (Figure 1b). The optimal PPA concentration was determined to be 40 mM (Figure 1c) and substrate inhibition was observed at higher PPA concentrations. The effect of biomass concentration on PLA production was investigated and the highest specific bioconversion rate of PLA was obtained with 28 grams dry cell weight (DCW) per liter ($OD_{620\ nm} = 50$) (Figure 1d). The effect of the cell growth phase on PLA production was also optimized and the middle exponential phase showed the highest PLA production rate (Table 1).

Production of PLA from PPA with substrate feeding

Fed-batch bioconversion was conducted with intermittent substrate feeding to avoid substrate inhibition. The bioconversion

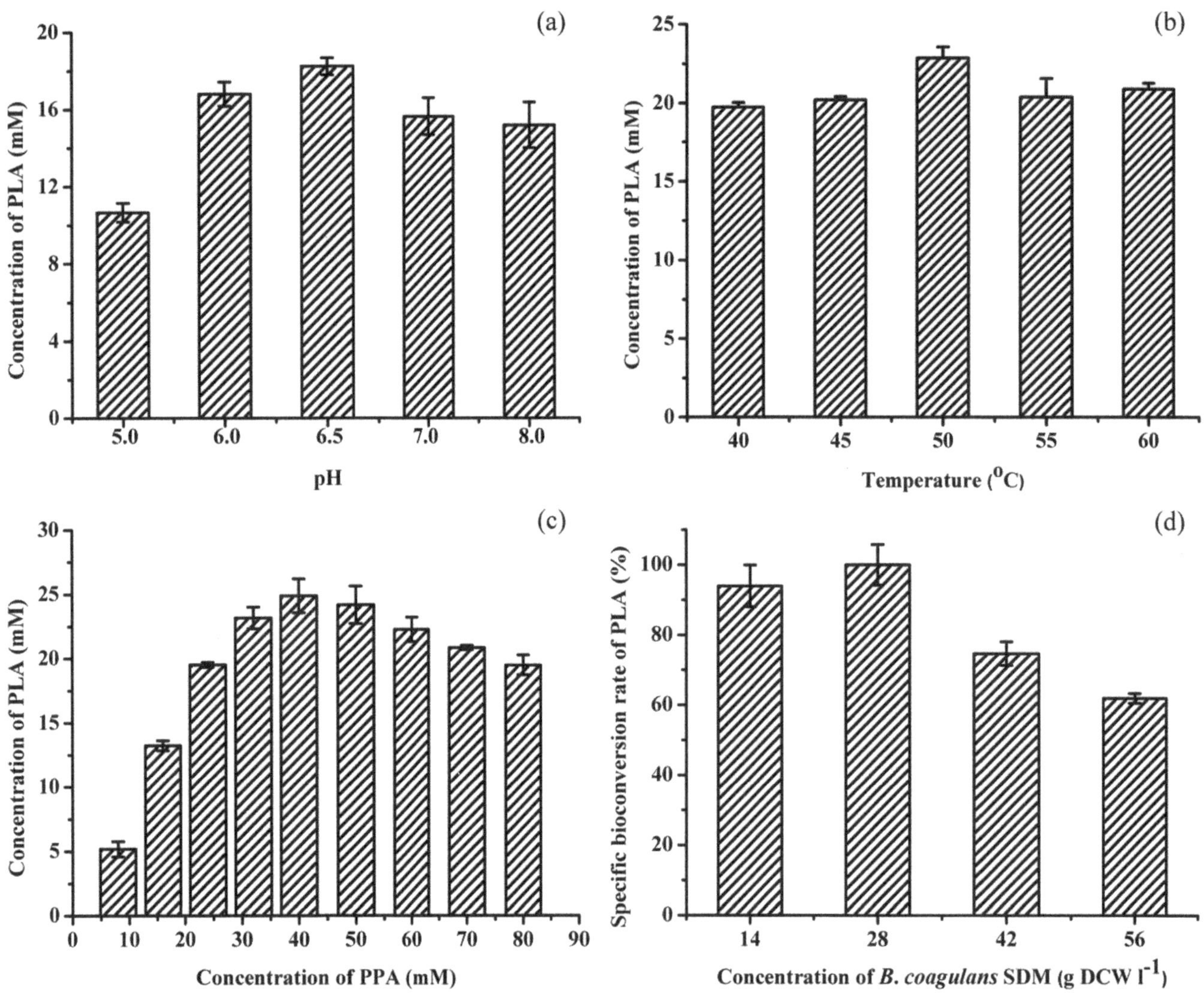

Figure 1. Optimization of bioconversion conditions. (a) Effect of pH on PLA production. (b) Effect of temperature on PLA production. (c) Effect of PPA concentration on PLA production. (d) Effect of biomass concentration on the specific bioconversion rate of PLA.

Table 1. Effects of cell growth phase on PLA production[a].

Growth phase	Volume of collected broth (ml)	Reaction time (h)	Production of PLA (mg)	Bioconversion ability (mg h^{-1} per ml broth)
Early exponential phase	25	0.5	26.4±1.2	1.27±0.06
Middle exponential phase	25	0.5	47.2±1.8	2.27±0.09
Late exponential phase	25	0.5	19.4±0.2	0.93±0.01
Stationary phase	25	0.5	17.4±0.8	0.84±0.04

[a]Cells were cultivated to different growth phases and 25 ml broth of different phases were harvested, washed, and resuspended to form 10 ml cell suspensions, respectively. Reactions were carried out in 50-ml flasks containing 10 ml of the reaction mixtures for 0.5 h, and then the concentrations of PLA were determined.

process was performed in a 5-l bioreactor (BIOSTAT B, B. Braun Biotech International GmbH, Germany) containing 2-l reaction mixtures. PPA was added in the solid form because of the observed enhanced dissolution of PPA at the high temperature. The initial concentrations of PPA and glucose were 6.6 g l^{-1} (40 mM) and 36 g l^{-1}, respectively. The pH was maintained at 6.5 by addition of NaOH solution. PPA and glucose powders were supplemented to maintain the initial concentrations by a pulse-feeding strategy.

The time course of PLA production from PPA was shown in Figure 2. A total of 105 g PPA and 252 g glucose were added to the 2-l reaction mixtures in this experiment. In addition to PLA, lactic acid was co-produced throughout the bioconversion process. The final PLA and lactic acid concentrations were 37.3 g l^{-1} and 66 g l^{-1}, respectively. The average productivity of PLA was 2.3 g l^{-1} h^{-1} and the yield was 70%.

Identification of enzymes responsible for conversion of PPA into PLA

The conversion between PPA and PLA is a redox reaction. To identify the enzymes responsible for the production of PLA, activity staining of the enzymes that catalyzed PLA oxidation was carried out. As shown in Figure 3 (lane 1), no PLA oxidation activity was detected without NAD addition. In the presence of NAD, two bands with distinct mobilities were presented when DL-PLA was used as the substrate (Figure 3, lane 2). The activities responsible for L- and D-lactate oxidation were also studied because of the co-production of lactic acid throughout the bioconversion process. The two bands exhibiting L- and D-lactate oxidation activities were located at the same positions as the two enzymes involved in L- and D-PLA oxidation, respectively (Figure 3). Since the biotransformation of lactic acid from pyruvic acid is catalyzed by the NAD-dependent lactate dehydrogenases (nLDHs), the results of activity staining implied that the enzymes responsible for PPA reduction present in *B. coagulans* SDM are nLDHs.

Activities of L-nLDH and D-nLDH in PPA reduction

To further identify the function of nLDHs in PLA production activities in strain SDM, the nLDHs were purified and further characterized. At present, only one genome of the *B. coagulans* strain, *B. coagulans* 36D1, has been released. Analysis of the genome of *B. coagulans* 36D1 identified one L-nLDH (accession number at GenBank, ZP_04430752) and one D-nLDH (accession number at GenBank, ZP_04430367). By using these two genes as guides for designing primers, the *ldh*L gene (GenBank accession number: HQ148709) and *ldh*D gene (GenBank accession number: HQ148710) of *B. coagulans* SDM were cloned.

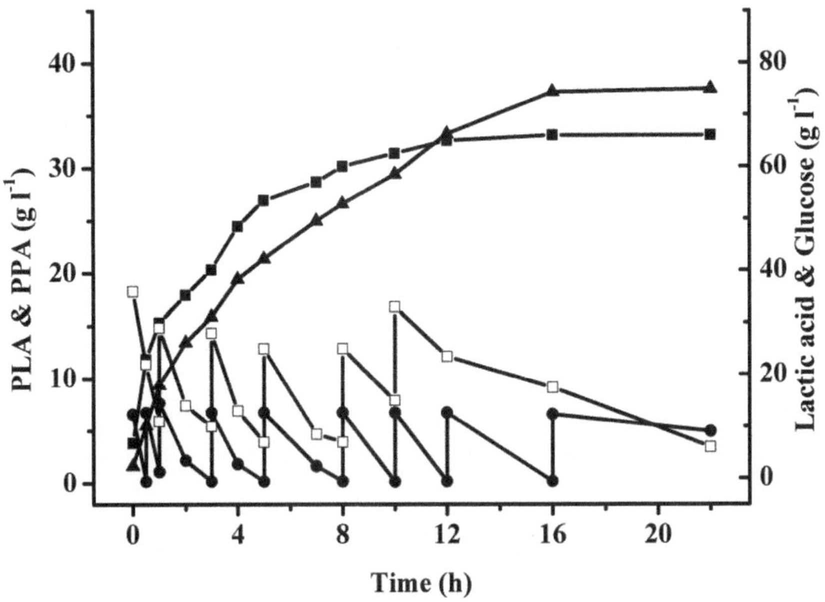

Figure 2. Time course of production of PLA from PPA under optimum conditions. Symbols represent: ▲, PLA; ●, PPA; ■, lactic acid; □, glucose.

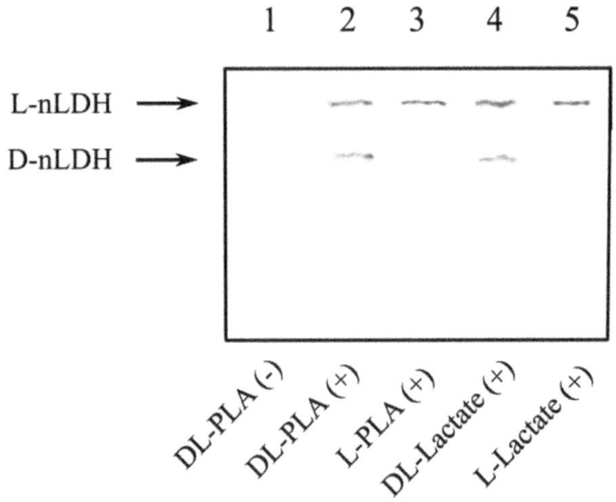

Figure 3. Native polyacrylamide gel electrophoresis staining for PLA and lactate oxidation activities. Symbol (−) represents solutions without NAD addition and symbol (+) represents solutions with 5 mM NAD. DL-PLA was used as substrate for active staining in lane 1 and 2, whereas L-PLA, DL-lactate, and L-lactate were used as substrates in lane 3, 4, and 5, respectively. All substrates were used at a concentration of 50 mM.

The amino acid sequences of the cloned nLDHs were deduced from the nucleotide sequences, and L-nLDH and D-nLDH exhibited considerable homology with other reported L-nLDHs and D-nLDHs of LAB, respectively (Figure S1 and Figure S2). Multiple alignments with other reported L-nLDHs showed that all catalytically important residues (Arg 109, Asp 168, Arg 171, and His 195) were conserved in the *B. coagulans* SDM L-nLDH (Figure S1) [27]. Moreover, the amino acids involved in activation by fructose 1,6-bisphosphate (FDP) (Arg 173 and His 188) were also found in L-nLDH (Figure S1) [28]. The essential residues of D-nLDH (Arg 235, Glu 264, and His 296) were also conserved in the *B. coagulans* SDM D-nLDH (Figure S2) [29,30].

The catalytic efficiencies of the recombinant nLDHs were determined by the reduction of pyruvate or PPA. Although notable differences in the catalytic activities were observed, both pyruvate and PPA could be catalyzed by the L-nLDH and D-nLDH of *B. coagulans* SDM (Table 2). The catalytic efficiencies of L-nLDH and D-nLDH on pyruvate were 40-fold and 3-fold higher than on PPA, respectively. A possible explanation is that PPA has a larger group at the C3-position, which results in a substrate that is unfavorable compared with pyruvate [31–33].

Discussion

In previous studies, PPA as the direct precursor of PLA was used as a substitute for Phe to enhance PLA production [21]. PPA is difficult to dissolve at a low temperature or under acidic and neutral conditions. This problem was solved by dissolving PPA with 2 M NaOH beforehand and adjusting the pH of the media by acid after the PPA solution was added [21]. In this study, a *B. coagulans* strain that transformed PPA to PLA efficiently at 50°C was isolated. PPA powder was directly added to the reaction mixture by using a pulse-feeding strategy in the fed-batch bioconversion. The rapid dissolution of PPA and the high metabolic activity at higher temperature resulted in a higher concentration and yield of PLA in *B. coagulans* SDM compared with other reported producers (Table 3).

The enzymes responsible for PPA reduction were uncertain in previous studies. HicDH was shown to catalyze the reduction of aromatic α-keto acids to the corresponding hydroxyl acids in the amino acid catabolism of many LAB [26]. In some studies, it was assumed that PLDHase was responsible for reducing PPA to PLA during Phe catabolism in LAB [23,25]. Recently, a D-LDH from the PLA producing strain, *Lactobacillus* sp. SK007, was purified and its reducing activity on PPA was detected [19]. In this study, the enzymes responsible for PPA reduction were identified as L-nLDH and D-nLDH through enzyme activity staining and kinetic analysis.

A cosubstrate was necessary to supply NADH because conversion of PPA to PLA was accompanied by the oxidation of NADH to NAD. In general, glucose is a good carbon source for cofactor regeneration because of its commercial availability, low price, and excellent bioavailability. Therefore, glucose was used for cofactor regeneration in this study. The biotransformation was conducted as follows: NADH was produced from glycolysis and PPA was reduced to PLA via nLDH-catalyzed reduction (Figure 4). Because both pyruvate and PPA are the substrates of nLDHs (Figure 4), the pyruvate derived from glucose would be expected to compete against PPA for binding to nLDHs and produce a mass of lactic acid (Figure 2). For a substrate such as PPA with larger groups at the C3-position, the catalytic efficiencies of nLDHs were markedly decreased (Table 2) [31–33]. Overcoming this bottleneck by redesign of the wild-type enzymes would significantly improve PLA production.

In conclusion, PLA is efficiently produced from PPA by the novel thermophilic strain *B. coagulans* SDM. In fed-batch bioconversion, 37.3 g l^{-1} PLA was obtained from 52.5 g l^{-1} PPA in 16 h at 50°C. The enzymes responsible for PPA reduction were further examined and identified as L-nLDH and D-nLDH. Illustrating this problem makes it possible to rationally modify microorganisms that are suitable for PLA production through metabolic engineering.

Materials and Methods

Isolation of thermotolerant bacteria

Soil samples were collected from chemical plants, farmlands, gardens, and dust heaps. Approximately 1 g of each sample was enriched in 50 ml of a liquid medium (GYSC) containing (g l^{-1}): glucose (50), yeast extract (10), soybean peptone (5), and $CaCO_3$

Table 2. Kinetic parameters of L-nLDH and D-nLDH for pyruvate and PPA.

Substrate	L-nLDH				D-nLDH			
	K_m (mM)	V_{max} (U/mg)	k_{cat} (s^{-1})	k_{cat}/K_m (M^{-1} s^{-1})	K_m (mM)	V_{max} (U/mg)	k_{cat} (s^{-1})	k_{cat}/K_m (M^{-1} s^{-1})
Pyruvate	2.6	1800.5	1172.2	4.5×10^5	2.2	33.7	23.6	1.1×10^4
PPA	4.3	72.6	47.2	1.1×10^4	4.4	23.5	16.5	3.9×10^3

Table 3. Comparison of PLA production by bacteria.

Organism	Substrate	Temperature (°C)	Time (h)	Production process	Concentration (g l^{-1})	Yield	Reference
Lactobacillus plantarum ITM21B	Phe	30	72	-	0.055	0.14	[18]
Lactobacillus plantarum TMW1.468	Phe	30	48	-	0.12	0.14	[20]
Lactobacillus sanfranciscensis DSM20451^T	Phe	30	48	-	0.042	0.05	[20]
Lactobacillus sp. SK007	Phe	30	72	Batch	0.061	0.06	[21]
Lactobacillus sp. SK007	PPA	30	24	Batch	0.86	0.85	[21]
Lactobacillus sp. SK007	PPA	30	72	Fed-batch	17.38	0.511	[22]
B. coagulans SDM	PPA	50	16	Fed-batch	37.3	0.70	This study

(20). The samples were incubated at 50°C and 100 rpm for 8 h, and an aliquot of the suspension was plated on GYSC solid medium and incubated at 50°C until multiple colonies emerged. Representative colonies were selected and incubated in GYSC liquid medium at 50°C and 100 rpm. After 8 h, 1% PPA (w/v) was added to the medium. The reaction mixtures were incubated for another 8 h, centrifuged at 4,500 rpm for 15 min, and the supernatant was analyzed for PPA and PLA. The strain that produced the highest amounts of PLA was designated SDM.

Biotransformation by whole cells of strain SDM

For whole cell preparation, overnight precultures of strain SDM were inoculated into fresh GYSC liquid medium with 10% inoculum and the cultures were incubated at 50°C and 100 rpm. Cells were harvested by centrifugation at 4,500 rpm for 15 min, washed with PBS, and resuspended in the same buffer to form cell suspensions. The biotransformation reactions were performed in 50-ml flasks containing 10 ml of the reaction mixtures. Parameters varied as follows for optimization of the reaction conditions: pH values were 5–8. Temperatures were 40–60°C. PPA concentrations were 8–80 mM, and cell concentrations were 14–56 g DCW l^{-1}. The time of harvesting cells was set at early exponential phase, middle exponential phase, late exponential phase, and stationary phase. After completion of the reactions, the samples were heated to 100°C and centrifuged. The concentrations of PPA and PLA in the resulting supernatants were quantitatively analyzed by high-performance liquid chromatography (HPLC).

Activity staining after native polyacrylamide gel electrophoresis (PAGE)

To identify the enzyme activities for PLA production in *B. coagulans* SDM, cells were harvested, washed, resuspended in PBS, and disrupted by sonication in an ice bath. The disrupted cells were centrifuged for 40 min at 13,000 rpm and the supernatant was used as the crude cell extract for native PAGE. Native PAGE was carried out at 4°C on a 10% native polyacrylamide gel. Active staining was performed after native PAGE according to a previous method with some modifications [34]. The gel was cut into five parts after electrophoresis and soaked in different solutions containing 100 mM Tris-HCl buffer (pH 8.3), 0.15 mM phenazine methanosulfate (PMS), 0.15 mM nitro blue tetrazolium (NBT), and corresponding substrates. Chemically reduced NBT dye by enzymatically produced NADH was precipitated as a dark blue band that was easily seen in the stained gels.

Bacterial strains, plasmids, and growth conditions

Escherichia coli strains and plasmids used in this study are listed in Table 4. *E. coli* was grown at 37°C in Luria-Bertani (LB) medium and kanamycin was added at a concentration of 50 μg ml^{-1}, if necessary.

Recombinant protein production and purification

The *ldh*L and *ldh*D genes were amplified by PCR using genomic DNA of *B. coagulans* SDM as the template. Primers were designed

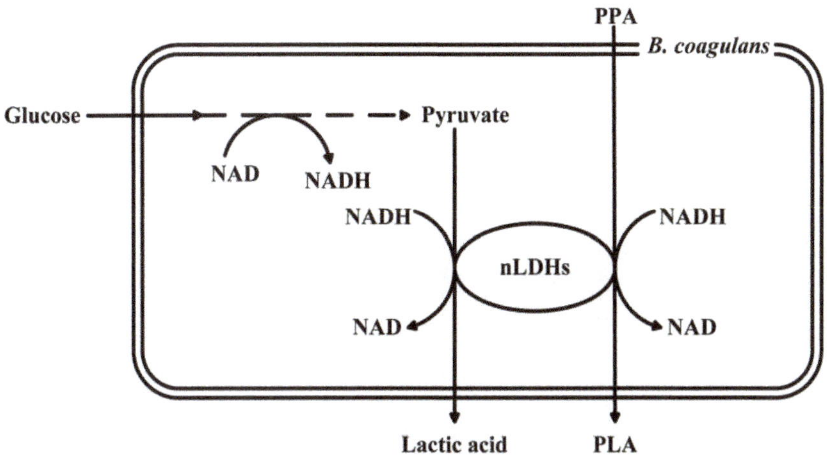

Figure 4. Schematic view of PLA production system of *B. coagulans* SDM.

Table 4. Strains, plasmids, and primers used in this work.

Strain, plasmid, or primer	Relevant characteristics	Source or reference
Strains		
E. coli DH5α	ϕ80 *lacZ*ΔM15 Δ(*lacZYA-argF*) U169 *recA*1 *endA*1 *hsdR*17 *supE*44λ- *thi*-1	Novagen
E. coli BL21(DE3)	F^- *ompT gal dcm lon hsdS*$_B$($r_B^- m_B^-$) λ(*DE3*)	Novagen
Plasmids		
pET-28a	Expression vector, Kan^r	Novagen
pET28a-*ldh*L	N and C-terminal His-tagged *ldh*L in pET-28a	This study
pET28a-*ldh*D	N and C-terminal His-tagged *ldh*D in pET-28a	This study
Primers		
P1	5′-CCGC*AAGCTT*CGATGAAAAAGGTCAATCGTATTGCAG-3′	This study
P2	5′-TTAA*CTCGAG*CAATACAGGTGCCATCGTTTCT-3′	This study
P3	5′-GCGC*AAGCTT*CGATGAGAAAAGTTGTTGCCTA-3′	This study
P4	5′-CTAA*CTCGAG*TGATTTTATCTCCCACCTGC-3′	This study

The italic face indicates the introduction of restriction sites.

based on the genome sequences of *B. coagulans* 36D1. Primers P1 and P2 were used for *ldh*L amplification and primers P3 and P4 were used for *ldh*D amplification (Table 4). The resulting PCR products were digested with *Hin*dIII-*Xho*I and cloned into the *Hin*dIII-*Xho*I sites of pET-28a to construct plasmids pET28a-*ldh*L and pET28a-*ldh*D. For protein expression, *E. coli* BL21(DE3) harboring recombinant plasmid was incubated in LB media (containing 50 μg ml^{-1} kanamycin) at 37°C with shaking. When the culture reached an optical density of 0.6 at 600 nm, 1 mM isopropyl-β-D-thiogalactopyranoside (IPTG) was added to induce gene expression. After induction at 25°C for 5 h, cells were harvested and suspended in the binding buffer (200 mM sodium phosphate, 500 mM sodium chloride, and 20 mM imidazole [pH 7.4]) and disrupted by sonication. The suspension was centrifuged to remove cell debris. The resultant supernatant was filtered and loaded onto a HisTrap HP 5 ml column (GE Healthcare) and the enzyme was eluted with 60% binding buffer and 40% elution buffer (200 mM sodium phosphate, 500 mM sodium chloride, and 500 mM imidazole [pH 7.4]) at a flow rate of 5 ml min^{-1}. The fractions containing the enzyme were detected by sodium dodecyl sulfate-polyacrylamide gel electrophoresis (SDS-PAGE). The protein concentration was determined by the Bradford method using bovine serum albumin for calibration [35].

Lactate dehydrogenase assay

Reduction activities of nLDHs on pyruvate or PPA were assayed at 50°C with freshly purified recombinant L/D-nLDHs. For the D-nLDH assay, the reaction mixture contained 50 mM Tris-HCl buffer (pH 7.0), 0.2 mM NADH, and 10 mM substrate. For the L-nLDH assay, the catalytic reaction mixtures contained all reagents mentioned above in addition to 5 mM FDP. One unit of L/D-nLDHs was defined as the amount that catalyzed the oxidation of 1 μmol of NADH per minute.

Analytical procedures

PPA and PLA were measured by HPLC (Agilent 1100 series, Hewlett-Packard, USA) equipped with an Agilent Zorbax SB-C18 column (150×4.6 mm, 5 μm) and a variable-wavelength detector at 210 nm. The mobile phase consisted of 1 mM H_2SO_4 and acetonitrile with a ratio of 85:15 (v/v) at a flow rate of 0.7 ml min^{-1} at 30°C.

Glucose and lactic acid concentrations were measured by HPLC equipped with a Bio-Rad Aminex HPX-87H column (300×7.8 mm) and a refractive index detector. Analysis was performed with a mobile phase of 10 mM H_2SO_4 at a flow rate of 0.4 ml min^{-1} at 55°C.

DCW was calculated from the optical density ($OD_{620\ nm}$) with a linear correlation factor (1 $OD_{620\ nm}$ = 0.56 g DCW l^{-1}).

Supporting Information

Figure S1 Multiple alignment of the amino acid sequences of L-nLDHs from selected species. Species and accession numbers of sequences are as follows: *Lactobacillus casei* ATCC 393 (Lcas), YP_001988625; *Lactobacillus rhamnosus* GG (Lrha), YP_003172269; *Lactococcus lactis* (Llac), AAB51674; *Lactobacillus helveticus* DPC 4571 (Lhel), YP_001576807; and *B. coagulans* SDM (Bcoa), this study.

Figure S2 Multiple alignment of the amino acid sequences of D-nLDHs from selected species. Species and accession numbers of sequences are as follows: *Lactobacillus johnsonii* NCC 533 (Ljoh), NP_964061; *Lactobacillus helveticus* DSM 20075 (Lhel), ZP_05752035; *Leuconostoc mesenteroides* (Lmes), AAA99506; *Lactobacillus pentosus* ATCC 8041 (Lpen), BAA14352; and *B. coagulans* SDM (Bcoa), this study.

Author Contributions

Conceived and designed the experiments: ZZ CM CG PX. Performed the experiments: ZZ CG FL HZ KW. Analyzed the data: ZZ CM CG FL JQ PX. Contributed reagents/materials/analysis tools: JQ CM PX. Wrote the paper: ZZ CM PX.

References

1. Dieuleveux V, Lemarinier S, Guéguen M (1998) Antimicrobial spectrum and target site of D-3-phenyllactic acid. Int J Food Microbiol 40: 177–183.

2. Lavermicocca P, Valerio F, Visconti A (2003) Antifungal activity of phenyllactic acid against molds isolated from bakery products. Appl Environ Microbiol 69: 634–640.

3. Schnürer J, Magnusson J (2005) Antifungal lactic acid bacteria as biopreservatives. Trends Food Sci Technol 16: 70–78.
4. Zhou LM, Zuo Z, Chow MSS (2005) Danshen: an overview of its chemistry, pharmacology, pharmacokinetics, and clinical use. J Clin Pharmacol 45: 1345–1359.
5. Yang RX, Huang SY, Yan FF, Lu XT, Xing YF, et al. (2010) Danshensu protects vascular endothelia in a rat model of hyperhomocysteinemia. Acta Pharmacol Sin 31: 1395–1400.
6. Budt KH, Stowasser B, Knolle J, Ruppert D, Meichsner C, et al. (1991) Inhibitors of retroviral proteases. DE Patent 4,030, 350.
7. Kano S, Yuasa Y, Yokomatsu T, Shibuya S (1998) Highly stereocontrolled synthesis of the four individual stereoisomers of statine. J Org Chem 53: 3865–3868.
8. Urban FJ, Moore BS (1992) Synthesis of optically-active 2-benzyldihydrobenzopyrans for the hypoglycemic agent englitazone. J Heterocycl Chem 29: 431–438.
9. Morita H, Mori H (1996) Process for production of phenyllactic acid derivative. US Patent 5,684,186.
10. Wong HNC, Xu ZL, Chang HM, Lee CM (1992) A modified synthesis of (±)-β-aryllactic acids. Synthesis 8: 793–797.
11. Xiao ZJ, Lv CJ, Gao C, Qin JY, Ma CQ, et al. (2010) A novel whole-cell biocatalyst with NAD^+ regeneration for production of chiral chemicals. PLoS ONE 5: e8860.
12. Dieuleveux V, Van Der Pyl D, Chataud J, Guéguen M (1998) Purification and characterization of anti-*Listeria* compounds produced by *Geotrichum candidum*. Appl Environ Microbiol 64: 800–803.
13. Lavermicocca P, Valerio F, Evidente A, Lazzaroni S, Corsetti A, et al. (2000) Purification and characterization of novel antifungal compounds from the sourdough *Lactobacillus plantarum* strain 21B. Appl Environ Microbiol 66: 4084–4090.
14. Lind H, Jonsson H, Schnürer J (2005) Antifungal effect of dairy propionibacteria-contribution of organic acids. Int J Food Microbiol 98: 157–165.
15. Magnusson J, Ström K, Roos S, Sjögren J, Schnürer J (2003) Broad and complex antifungal activity among environmental isolates of lactic acid bacteria. FEMS Microbiol Lett 219: 129–135.
16. Ström K, Sjögren J, Broberg A, Schnürer J (2002) *Lactobacillus plantarum* MiLAB 393 produces the antifungal cyclic dipeptides cyclo(L-Phe-L-Pro) and cyclo(L-Phe-*trans*-4-OH-L-Pro) and 3-phenyllactic acid. Appl Environ Microbiol 68: 4322–4327.
17. Thierry A, Maillard MB (2002) Production of cheese flavour compounds derived from amino acid catabolism by *Propionibacterium freudenreichii*. Lait 82: 17–32.
18. Valerio F, Lavermicocca P, Pascale M, Visconti A (2004) Production of phenyllactic acid by lactic acid bacteria: an approach to the selection of strains contributing to food quality and preservation. FEMS Microbiol Lett 233: 289–295.
19. Li XF, Jiang B, Pan BL, Mu WM, Zhang T (2008) Purification and partial characterization of *Lactobacillus* species SK007 lactate dehydrogenase (LDH) catalyzing phenylpyruvic acid (PPA) conversion into phenyllactic acid (PLA). J Agric Food Chem 56: 2392–2399.
20. Vermeulen N, Gänzle MG, Vogel RF (2006) Influence of peptide supply and cosubstrates on phenylalanine metabolism of *Lactobacillus sanfranciscensis* DSM20451^T and *Lactobacillus plantarum* TMW1.468. J Agric Food Chem 54: 3832–3839.
21. Li XF, Jiang B, Pan BL (2007) Biotransformation of phenylpyruvic acid to phenyllactic acid by growing and resting cells of a *Lactobacillus* sp. Biotechnol Lett 29: 593–597.
22. Mu WM, Liu FL, Jia JH, Chen C, Zhang T, et al. (2009) 3-Phenyllactic acid production by substrate feeding and pH-control in fed-batch fermentation of *Lactobacillus* sp. SK007. Bioresour Technol 100: 5226–5229.
23. Gummalla S, Broadbent JR (2001) Tyrosine and phenylalanine catabolism by *Lactobacillus* cheese flavor adjuncts. J Dairy Sci 84: 1011–1019.
24. Hummel W, Schütte H, Kula MR (1983) Large scale production of D-lactate dehydrogenase for the stereospecific reduction of pyruvate and phenylpyruvate. Appl Microbiol Biotechnol 18: 75–85.
25. Marilley L, Casey MG (2004) Flavours of cheese products: metabolic pathways, analytical tools and identification of producing strains. Int J Food Microbiol 90: 139–159.
26. Yvon M, Rijnen L (2001) Cheese flavour formation by amino acid catabolism. Int Dairy J 11: 185–201.
27. Dunn CR, Wilks HM, Halsall DJ, Atkinson T, Clarke AR, et al. (1991) Design and synthesis of new enzymes based on the lactate dehydrogenase framework. Philos Trans R Soc London, Ser B 332: 177–184.
28. Arai K, Hishida A, Ishiyama M, Kamata T, Uchikoba H, et al. (2002) An absolute requirement of fructose 1,6-bisphosphate for the *Lactobacillus casei* L-lactate dehydrogenase activity induced by a single amino acid substitution. Protein Eng 15: 35–41.
29. Stoll VS, Kimber MS, Pai EF (1996) Insights into substrate binding by D-2-ketoacid dehydrogenases from the structure of *Lactobacillus pentosus* D-lactate dehydrogenase. Structure 4: 437–447.
30. Taguchi H, Ohta T (1993) Histidine 296 is essential for the catalysis in *Lactobacillus plantarum* D-lactate dehydrogenase. J Biol Chem 268: 18030–18034.
31. Arai K, Kamata T, Uchikoba H, Fushinobu S, Matsuzawa H, et al. (2001) Some *Lactobacillus* L-lactate dehydrogenases exhibit comparable catalytic activities for pyruvate and oxaloacetate. J Bacteriol 183: 397–400.
32. Razeto A, Kochhar S, Hottinger H, Dauter M, Wilson KS, et al. (2002) Domain closure, substrate specificity and catalysis of D-lactate dehydrogenase from *Lactobacillus bulgaricus*. J Mol Biol 318: 109–119.
33. Tokuda C, Ishikura Y, Shigematsu M, Mutoh H, Tsuzuki S, et al. (2003) Conversion of *Lactobacillus pentosus* D-lactate dehydrogenase to a D-hydroxyisocaproate dehydrogenase through a single amino acid replacement. J Bacteriol 185: 5023–5026.
34. Ma CQ, Gao C, Qiu JH, Hao JR, Liu WW, et al. (2007) Membrane-bound L- and D-lactate dehydrogenase activities of a newly isolated *Pseudomonas stutzeri* strain. Appl Microbiol Biotechnol 77: 91–98.
35. Bradford MM (1976) A rapid and sensitive method for the quantitation of microgram quantities of protein utilizing the principle of protein-dye binding. Anal Biochem 72: 248–254.

2

Efficient Whole-Cell Biocatalyst for Acetoin Production with NAD+ Regeneration System through Homologous Co-Expression of 2,3-Butanediol Dehydrogenase and NADH Oxidase in Engineered *Bacillus subtilis*

Teng Bao[1]◔, Xian Zhang[1]◔, Zhiming Rao[1]*, Xiaojing Zhao[1], Rongzhen Zhang[2], Taowei Yang[1], Zhenghong Xu[3], Shangtian Yang[4]

1 The Key Laboratory of Industrial Biotechnology of Ministry of Education, School of Biotechnology, Jiangnan University, Wuxi, Jiangsu, People's Republic of China, **2** School of Biotechnology, Jiangnan University, Wuxi, Jiangsu, People's Republic of China, **3** School of Medicine and Pharmaceuticals, Jiangnan University, Wuxi, Jiangsu, People's Republic of China, **4** Department of Chemical Engineering, Ohio State University, Columbus, Ohio, United States of America

Abstract

Acetoin (3-hydroxy-2-butanone), an extensively-used food spice and bio-based platform chemical, is usually produced by chemical synthesis methods. With increasingly requirement of food security and environmental protection, bio-fermentation of acetoin by microorganisms has a great promising market. However, through metabolic engineering strategies, the mixed acid-butanediol fermentation metabolizes a certain portion of substrate to the by-products of organic acids such as lactic acid and acetic acid, which causes energy cost and increases the difficulty of product purification in downstream processes. In this work, due to the high efficiency of enzymatic reaction and excellent selectivity, a strategy for efficiently converting 2,3-butandiol to acetoin using whole-cell biocatalyst by engineered *Bacillus subtilis* is proposed. In this process, NAD^+ plays a significant role on 2,3-butanediol and acetoin distribution, so the NADH oxidase and 2,3-butanediol dehydrogenase both from *B. subtilis* are co-expressed in *B. subtilis* 168 to construct an NAD^+ regeneration system, which forces dramatic decrease of the intracellular NADH concentration (1.6 fold) and NADH/NAD^+ ratio (2.2 fold). By optimization of the enzymatic reaction and applying repeated batch conversion, the whole-cell biocatalyst efficiently produced 91.8 g/L acetoin with a productivity of 2.30 g/(L·h), which was the highest record ever reported by biocatalysis. This work indicated that manipulation of the intracellular cofactor levels was more effective than the strategy of enhancing enzyme activity, and the bioprocess for NAD^+ regeneration may also be a useful way for improving the productivity of NAD^+-dependent chemistry-based products.

Editor: Y-H Percival Zhang, Virginia Tech, United States of America

Funding: This work was supported by the Program for New Century Excellent Talents in University (NCET-10-0459), the National Basic Research Program of China (973 Program) (2012CB725202), the National Natural Science Foundation of China (21276110), the Research Project of Chinese Ministry of Education. (No. 113033A), the Fundamental Research Funds for the Central Universities (JUSRP51306A and JUSRP21121), the 111 Project (111-2-06) and a Project Funded by the Priority Academic Program Development of Jiangsu Higher Education Institution. The funders had no role in study design, data collection and analysis, decision to publish, or preparation of the manuscript.

Competing Interests: The authors have declared that no competing interests exist.

* Email: raozhm@jiangnan.edu.cn

◔ These authors contributed equally to this work.

Introduction

Acetoin (3-hydroxy-2-butanone, AC) is an extensively-used spice that naturally exists in corn, grape, cocoa, apple, butter, coffee, etc. Widely used in food and beverage industry, AC also serves as a platform compound in many other industries [1]. It is one of the 30 platform chemicals that are given priority to their development and utilization by the US Department of Energy [2]. Although there are many chemical synthetic methods for AC preparation [3], its market is limited by the disadvantages of traditional chemical synthesis. On the other hand, with the further development of green chemical technology and the constant improvement of the environmental protection consciousness, non-toxic and non-pollution biological technology inevitably become the main direction of industrial development and consumers prefer security natural products even though they are generally more expensive than the corresponding chemical compounds.

Nowadays, a lot of efforts have been made to develop natural AC production using fermentative [4], enzymatic [5] or biocatalytic technologies [6]. A number of bacteria have abilities to produce AC, including the genera *Klebsiella*, *Paenibacillus*, *Gluconobacter*, *Bacillus*, *Serratia*, etc. [7–11]. However, in most of these species, AC just plays a part as by-product of 2,3-butanediol (2,3-BD), which is another important bio-based platform chemical [12]. *Bacillus* species, which can produce various of industrial products [13], have been proved with AC as its major fermentation product under specific conditions [14]. Many efforts have been made to improve the production of AC from *Bacillus* strains. Liu et al. isolated a *B. licheniformis* strain that could produce 41.3 g/L of AC [4]. Zhang et al. isolated the *B. subtilis* JNA-3-10 and produced 42.2 g/L of AC [15]. Fermentation optimization strategies have been used to improve AC production, such as optimizing the medium components [16], controlling the level of dissolved oxygen and controlling the fermentation pH [17]. Metabolic engineering strategies were also applied to improve AC production through modifying metabolic branch-points in the network [14,18,19]. However, so much long fermentation duration lead to a low AC productivity. To our knowledge, the highest productivity of AC by *Bacillus* strains is just 1.42 g/(L·h) [4]. In addition, the mixed acid-butanediol fermentation of *Bacillus* strains will metabolize a certain portion of sugars to the by-products of organic acids such as lactic acid and acetic acid, which causes energy cost and increases the difficulty of product purification in downstream processes [20].

Recently, the introduction of NAD^+ regeneration system could dramatically improve AC production and decrease the yield of NADH-dependent by-products [6,11]. Sun et al. obtained 75.2 g/L AC with a productivity of 1.88 g/(L·h) by *S. marcescens* H32 with over-expression of a water-forming NADH oxidase [11]. Xiao et al. developed a co-expression system with 2,3-butanediol dehydrogenase and NADH oxidase in *E. coli* produced AC at a high productivity of 3.06 g/(L·h) [6]. Although have relative high AC productivities, AC yield of this biocatalyst was still far away from the highest report of 89.2 g/L achieved by Wang et al. using fermentation method with 2,3-BD as substrate by *Gluconobacter oxydans* DSM 2003 [10]. Therefore, combining both advantages of fermentation and cofactor regeneration, a potential strategy of introducing a biocatalytic process with NAD^+ regeneration system for efficient natural AC production in *B. subtilis* is proposed by us. Whole-cell biocatalyst has been intensively explored for the production of valuable compounds because excellent selectivity and NAD^+ reserves that provides a continuous source of cofactors [21]. Trough NAD^+ regeneration system, the cellular cofactor level, redox state and the corresponding enzymatic activity are expected to have major effects on the performance of the whole-cell biocatalysts. In this whole-cell biocatalyst, 2,3-BD is used as substrate and only AC can be obtained in the short biocatalyst period. This is also a good solution to develop derivative process for industrially produced 2,3-BD utilization, which can not be commercially utilized so far.

In previous work of our lab, when *B. subtilis* was fermented with glucose as substrate, 2,3-BD was the major product in prophase of fermentation, whereas in the anaphase of fermentation, 2,3-BD was reversely transformed to AC [15]. The reversible conversion between AC and 2,3-BD was responsible by the enzyme AC reductase/2,3-BD dehydrogenase (AR/BDH EC 1.1.1.4). Generally, AR/BDHs have the similar optimum pH-values for oxidation and reduction [22–24]. However, very special property of this enzyme that has different optimum pH-values for oxidation and reduction was proved [25,26]. Intrigued by this point, our group experimentally demonstrated that AR/BDH from *B. subtilis* also has very different optimum pH-values [27]. The results indicated that this enzyme preferentially catalyzes the reduction/oxidation reaction in the acidic/alkaline condition. Similar to other reductases and dehydrogenases, AR/BDH requires NAD^+ and NADH as cofactors [28,29]. As one pair of key cofactors, NADH and NAD^+ play a critical role in over 300 biochemical reactions including oxidation and reduction [30,31], which helps to maintain the elementary requirement for metabolism and growth in cells [32]. During the catalytic process of BDH, NAD^+ is reduced to NADH accompanied by the oxidization of 2,3-BD [6]. Therefore, an driving force performed by NADH oxidase (NOX EC 1.6.99.3) was required for sweeping away the obstruction proposed on NAD^+ regeneration. As a flavoprotein, NOX uses oxygen to produce either water or hydrogen peroxide [33], which can partially or completely inhibit cell growth. In this work, the homologous NOX from *B. subtilis*, similar to the NOX from Lactobacillus brevis [33], generates NAD^+ from NADH by reducing O_2 to H_2O was used as the core enzyme for NAD^+ regeneration [34], because construction of a homologous NAD^+ regeneration system can be more efficient and safe for AC production. The constructed biocatalyst of *B. subtilis*, in which AR/BDH and NOX were co-expressed (shown in Figure 1), adjusted the intracellular $NADH/NAD^+$ ratio and NAD(H) level and strongly pumped the catalytic reaction. After optimization of the biocatalytic conditions, AC production and productivity was highly improved.

Materials and Methods

Chemicals, media and cultivation

Ampicillin, kanamycin, NAD^+, NADH, FAD, dithiothreitol (DTT) and phenylmethanesulfonyl fluoride (PMSF) were obtained from Sangon Biotech (Shanghai, China). PCR primers were synthesized by Sangon Biotech (Shanghai, China). The restriction endonucleases and T4 DNA ligase were obtained from Takara (Dalian, China). All other chemicals were commercially available reagents of analytical grade.

B. subtilis and *E. coli* were cultivated in Luria-Broth (LB) medium routinely and were used as the cloning and expression hosts, respectively. When necessary, antibiotics (100 μg/mL ampicillin or 50 μg/mL kanamycin) were supplemented to the medium to maintain the plasmids.

Strains, plasmids and primers

The strains, plasmids and primers used in this study were given in Table 1.

Construction of recombinant plasmids

The *bdhA* and *yodC* genes were amplified using forward primer P1/P3 and reverse primer P2/P4 and *B. subtilis* 168 as template. The plasmid pMA5-*bdhA*/pMA5-*yodC* were constructed by inserting *bdhA*/*yodC* between the *Bam*H I and *Mlu* I sites of pMA5. The *yodC* gene with with the upstream region including *Hpa*II promoter and RBS was then amplified using forward primer P5 and reverse primer P6 and pMA5-*yodC* as template. The plasmid pMA5-*bdhA*-*yodC* was constructed by inserting *Hpa*II-*yodC* between the *Sph* I and *Hind* III sites of pMA5-*bdhA*.

Transformation and selection of recombinant strains

The ligated DNAs were used to transform *E. coli* JM109. Positive colonies were selected on agar plates containing ampicillin, and the plasmids were confirmed using restriction enzyme analysis and DNA sequencing. The recombinant plasmids were then used to transform *B. subtilis* 168 [35]. The transformants were screened for their ability to grow on LB agar plates containing kanamycin.

Assays of AR/BDH and NOX activities and SDS-PAGE analysis

The cells were harvested by centrifugation at 10000 rpm for 10 min at 4°C and washed twice with 50 mM sodium phosphate buffer (pH 6.5). Then the harvested cells were resuspended in

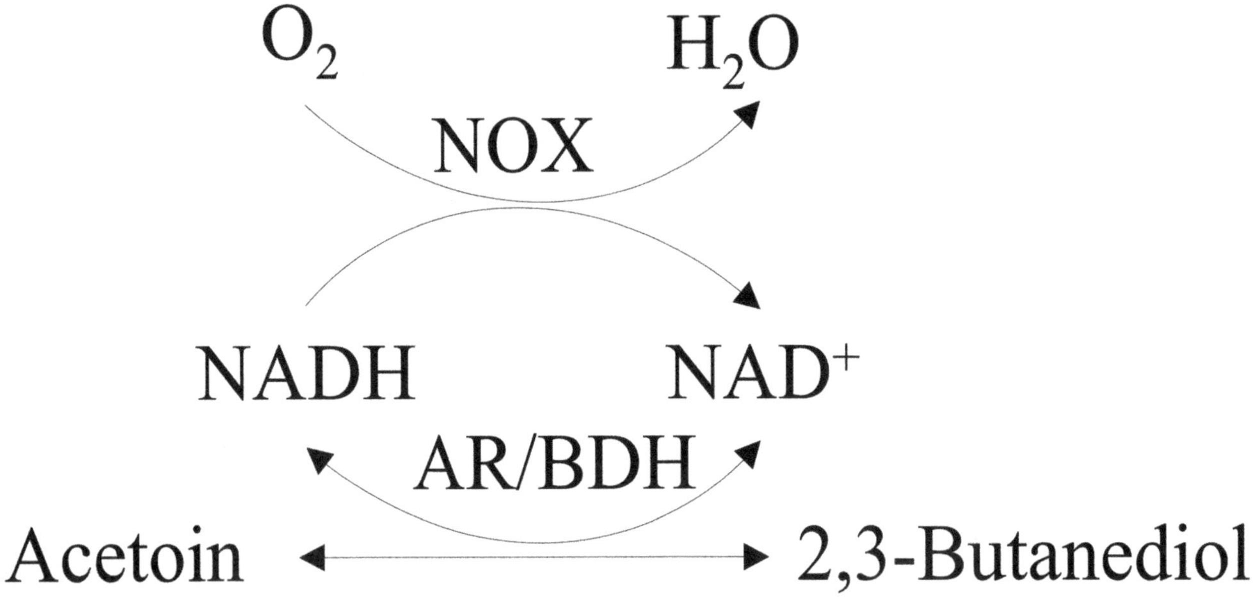

Figure 1. Whole-cell biocatalyst with NAD^+ regeneration system for the production of acetoin. AR/BDH: acetoin reductase/2,3-butanediol dehydrogenase; NOX: NADH oxidase.

Table 1. Bacterial strains, plasmids and primers used.

Strains/plasmids/primers	Characteristics	Sources
Strains		
Escherichia coli		
E. coli JM109	*recA1, endA1, gyrA96, thi-1, hsd R17*$(r_k^- m_k^+)$*supE44*	Invitrogen
E. coli JM109/ pMA5-*bdhA*	*E. coli* JM109 containing pMA5-*bdhA* (Amp^R)	This study
E. coli JM109/ pMA5-*yodC*	*E. coli* JM109 containing pMA5-*yodC* (Amp^R)	This study
E. coli JM109/ pMA5-*bdhA-yodC*	*E. coli* JM109 containing pMA5-*bdhA-yodC* (Amp^R)	This study
Bacillus subtilis		
B. subtilis 168	*trpC2*	Laboratory stock
B. subtilis 168/ pMA5	*B. subtilis* 168 containing pMA5 (Km^R)	Laboratory stock
B. subtilis 168/ pMA5-*bdhA*	*B. subtilis* 168 containing pMA5- *bdhA* (Km^R)	This study
B. subtilis 168/ pMA5-*yodC*	*B. subtilis* 168 containing pMA5-*yodC* (Km^R)	This study
B. subtilis 168/ pMA5-*bdhA-yodC*	*B. subtilis* 168 containing pMA5-*bdhA-yodC* (Km^R)	This study
Plasmids		
pMA5	*Hpa*II promoter, colE1 ori, *repB*, replicates in *E. coli* (Amp^R) or *B. subtilis* (Km^R)	Laboratory stock
pMA5-*bdhA*	pMA5 containing *bdhA*-His	This study
pMA5-*yodC*	pMA5 containing *yodC*-His	This study
pMA5-*bdhA-yodC*	pMA5 containing *bdhA*-His and *yodC*-His	This study
Primers 5′-3′		
P1	ACCGGGATCCATGAAGGCAGCAAGATGG (*BamH* I)	
P2	ACCGACGCGTTTAGTGGTGGTGGTGGTGGTGGTTAGGTCTAACAAGG (*Mlu* I)	
P3	ACCGGGATCCATGACGAATA CTCTGGAT (*BamH* I)	
P4	ACCGACGCGTTTAGTGGTGGTGGTGGTGGTGCAGCCAA GTTGATAC (*Mlu* I)	
P5	ACCGGCATGCGTAAGCTAGACAAAACGGAC (*Sph* I)	
P6	ACCGAAGCTTTTAGTGGTGGTGGTGGTGGTGCAGCCAAGTTGATAC (*Hind* III)	

Amp^R ampicillin-resistant, Km^R kanamycin-resistant, the restriction enzyme sites were bold typed, and the His-Taq coding region were underlined.

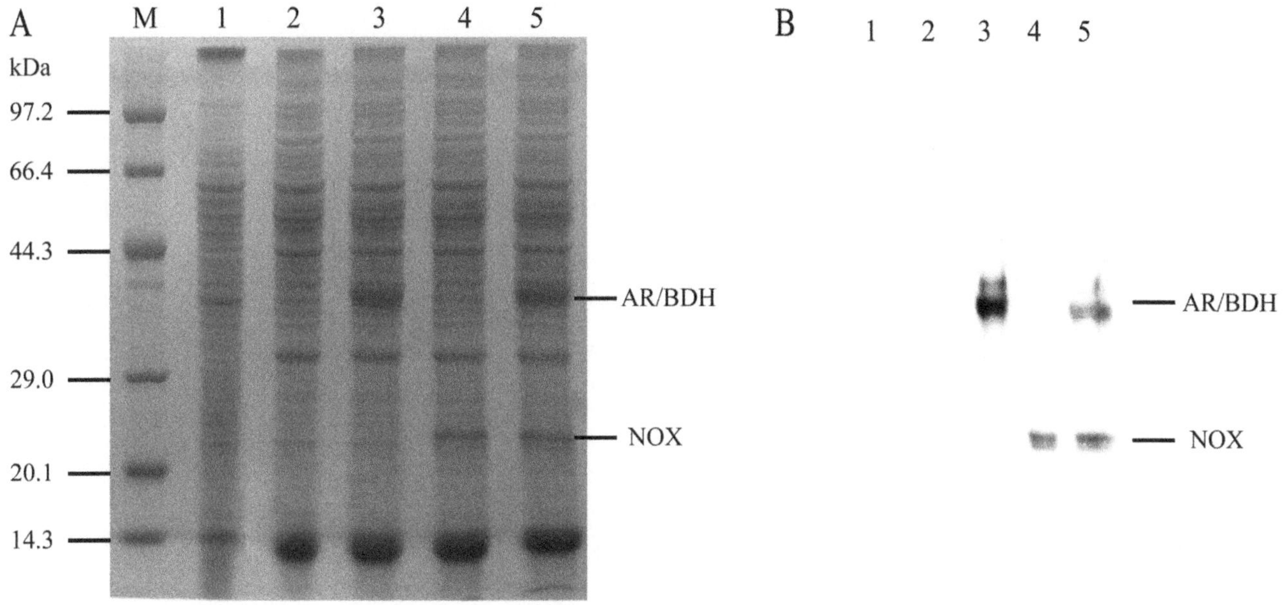

Figure 2. Expression analysis of recombinant *B. subtilis*. (A) SDS-PAGE result of AR/BDH and NOX expressed in *B. subtilis*. Lane M, protein marker; lane 1, *B. subtilis* 168; lane 2, *B. subtilis* 168/pMA5; lane 3, *B. subtilis* 168/pMA5-*bdhA*; lane 4, *B. subtilis* 168/pMA5-*yodC*; lane 5, *B. subtilis* 168/pMA5-*bdhA-yodC*. (B) western blot result of AR/BDH and NOX expressed in *B. subtilis*. lane 1, *B. subtilis* 168; lane 2, *B. subtilis* 168/pMA5; lane 3, *B. subtilis* 168/pMA5-*bdhA*; lane 4, *B. subtilis* 168/pMA5-*yodC*; lane 5, *B. subtilis* 168/pMA5-*bdhA-yodC*.

50 mM sodium phosphate buffer (pH 6.5) containing 0.1 mM PMSF, and 1 mM DTT. After ultrasonic disruption, cell debris was removed by centrifugation at 15000 rpm at 4°C for 30 min to obtain a crude enzyme solution for the enzyme assay.

The AR/BDH and NOX activities were determined spectrophotometrically by measuring the change of absorbance at 340 nm and 25°C corresponding to the oxidation of NADH ($\varepsilon_{340} = 6220/M{\cdot}cm$) or the reduction of NAD^+. AR activity was determined using 50 mM sodium phosphate buffer (pH 6.5) containing 50 mM acetoin and 5 mM NAD^+. BDH activity was determined using 50 mM glycine-NaOH buffer (pH 8.5) containing 100 mM 2,3-butanediol and 5 mM NADH. One unit of activity (U) corresponds to 1 μmol NAD(H) formed per minute [23].

NOX activity was determined using 50 mM sodium phosphate (pH 7.0) containing 0.3 mM EDTA, 50 μM FAD and 0.3 mM NADH. One unit of activity (U) corresponds to 1 μmol NAD^+ formed per minute [36].

The enzymes were assayed by sodium dodecyl sulfate-polyacrylamide gel electrophoresis (SDS-PAGE) according to Laemmli method [37]. AR/BDH and NOX were also detected by Western blotting by using a mouse monoclonal anti-His_6 antibody. The extracts were loaded in a 12% SDS-PAGE gel that was blotted onto a polyvinylidene difluoride (PVDF) membrane (Millipore) and treated with the primary anti-His_6 antibody. To visualize the relevant bands, the blot was treated with a secondary goat anti-mouse antibody (Bio-Rad) conjugated to horseradish peroxidase (HRP), and the bands were detected by chemiluminescence with luminol and peroxide with the aid of a Bio-Rad Chemidoc XRS. The molecular weight of the subunit of AR/BDH and NOX were determined by comparing the relative mobility of perfect protein Marker 14.3–97.2 kDa (Takara). The protein concentration was determined by Bradford method [38] using BSA as the standard protein.

Biocatalyst preparation and biocatalysis conditions

B. subtilis cells were grown at 37°C with fermentation medium (soya peptone 1 g/L, glucose 4 g/L, corn steep liquor 1.5 g/L, urea 0.3 g/L, $K_2HPO_4{\cdot}3H_2O$ 0.17 g/L, KH_2PO_4 0.23 g/L, $MnSO_4{\cdot}H_2O$ 0.075 g/L, NaCl 5 g/L, pH 6.8–7.0) in shake flasks. After cultivation for 36 h ($OD_{600} = 13.0$–16.0), the cells were harvested by centrifugation at 10000 rpm for 10 min at 4°C

Table 2. Enzyme activity and the whole-cell biocatalytic activity of different strains.

Strains	AR activity (mU/mg)	BDH activity (mU/mg)	NOX activity (mU/mg)	Whole-cell biocatalytic ability (U/L)
B. subtilis 168	40.2±0.96	1.9±0.03	33.4±1.03	14.10±0.41
B. subtilis 168/pMA5	38.3±0.84	1.5±0.04	31.6±0.73	15.21±0.44
B. subtilis 168/pMA5-*bdhA*	179.7±5.21	124.5±3.11	34.9±0.94	18.24±0.42
B. subtilis 168/pMA5-*yodC*	63.5±1.71	4.8±0.11	570.3±18.25	16.55±0.51
B. subtilis 168/pMA5-*bdhA-yodC*	171.8±5.32	152.8±4.12	346.4±9.70	24.12±0.53

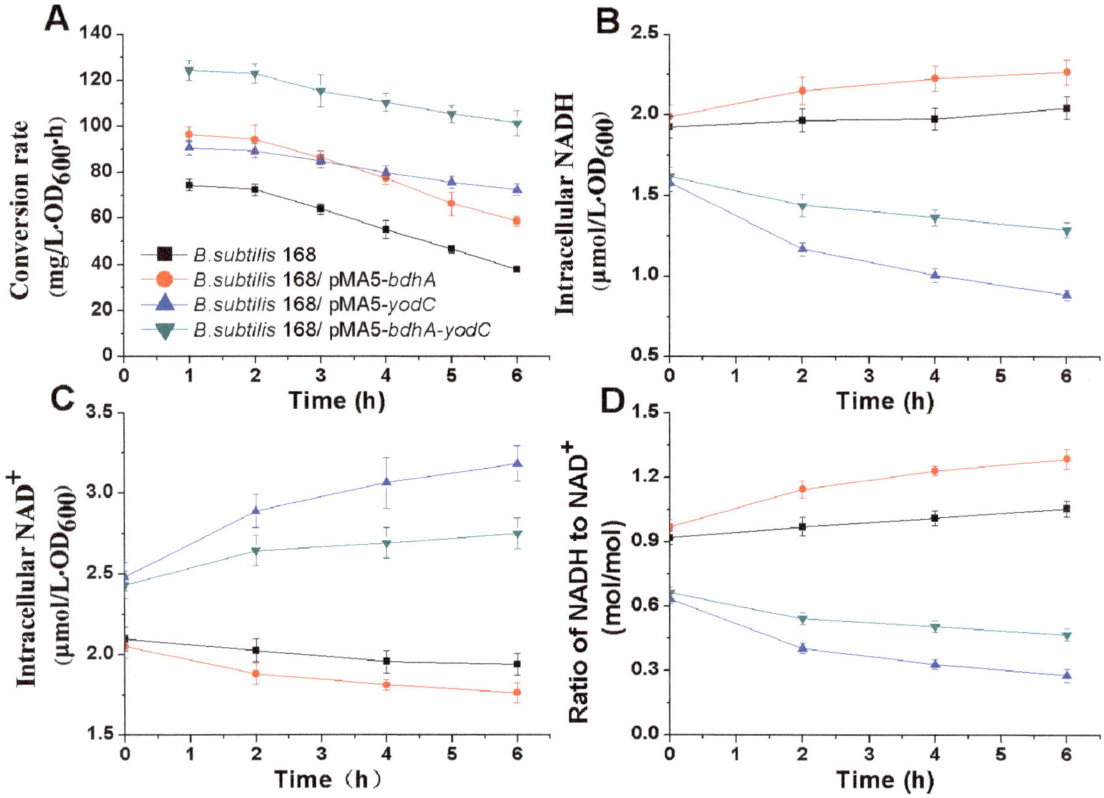

Figure 3. The intracellular NADH, NAD^+ concentrations in different biocatalysts and the effect of NADH/NAD^+ ratio on conversion rate.

and washed twice with 50 mM sodium phosphate buffer (pH 6.5). Then 500 mL cell cultures were resuspended into 50 mL 50 mM sodium phosphate buffer (pH 8.0) containing 40 g/L 2,3-BD. The whole-cell biocatalysis was performed on a rotary shaker (200 rpm) at 37°C.

When whole-cell biocatalysis carried out in a 5-L fermentor (Baoxing Co., Shanghai, China), 20 L of recombinant *B. subtilis* cultures was harvested and resuspended into 2 L of 50 mM sodium phosphate buffer (pH 8.5) containing 40 g/L 2,3-BD and 5 mM $MnCl_2$. The agitation speed of the whole-cell biocatalysis was 200 rpm.

The whole-cell biocatalytic activity was assayed by measuring the formation of AC, and one unit was defined as 1.0 mM AC produced per hour per OD_{600} of culture at 37°C.

Determination of NADH and NAD^+ concentrations

The intracellular concentrations of NADH and NAD^+ were determined using the Amplite Fluorimetric NAD/NADH Ratio Assay Kit (15263) from AAT Bioquest (Sunnyvale, CA, USA) according to the manufacturers' instructions. The assay kit provides a convenient method for sensitive detection of NAD^+, NADH and their ratio. The signal can be easily read by either a fluorescence microplate reader at Ex/Em = 530–570/590–600 nm (maximum Ex/Em = 540/590 nm) or an absorbance microplate reader at ~576 nm [39].

Optimization of the whole-cell biocatalyst

The following buffer systems were used to investigate the optimal pH of the biocatalyst: 50 mM acetate-sodium acetate buffer (pH 4.5–5.5), 50 mM sodium phosphate buffer (pH 5.5–8.5) and 50 mM glycine-NaOH buffer (pH 8.5–10.5). Conversion temperature (20–50°C), concentration of substrate (10–60 g/L) and the concentration of metal ion stimulaters ($MgCl_2$, $MnCl_2$, $CaCl_2$, $FeCl_2$ and $FeCl_3$) were also optimized.

Analysis methods

The bio-profiles were monitored periodically over the experiment period. Samples were centrifuged at 10000 rpm for 10 min, and the supernatant was used for further analysis. Concentrations of AC and 2,3-BD were monitored by gas chromatography (Jie Dao GC1600, China, FID detector, N_2 flow rate of 50 ml/min, detector temperature of 250°C, and a column temperature of 160°C). Biomass was measured at OD_{600} (UNICO UV-2000) of the culture broth after appropriate dilution with water. All assays were performed by triplicate cultures. The samples were determined at least twice for three extracts derived from independent cultures, and standard deviations of the biological replicates were represented by error bars.

Results and Discussion

Construction of the whole-cell biocatalyst

To construct the whole-cell biocatalyst, the following strains of *B. subtilis* 168, *B. subtilis* 168/pMA5, *B. subtilis* 168/pMA5-*bdhA*, *B. subtilis* 168/pMA5-*yodC* and *B. subtilis* 168/pMA5-*bdhA-yodC* were detected to their enzyme activities and whole-cell biocatalytic activities. The results were shown in Table 2. *B. subtilis* 168/pMA5-*bdhA* and *B. subtilis* 168/pMA5-*yodC* showed the BDH and NOX activity of 124.5 mU/mg and 570.3 mU/mg, respectively. As expected, the co-expression system of *B. subtilis*

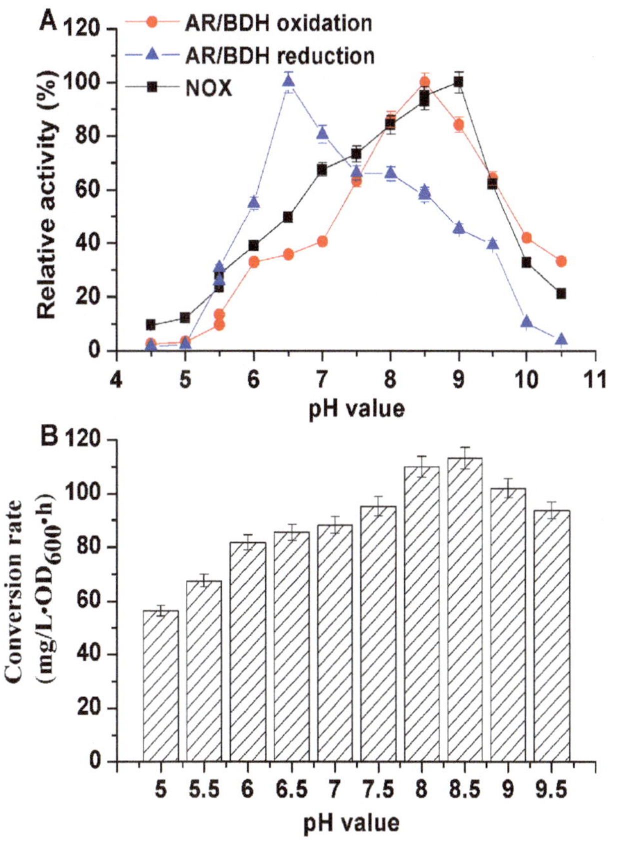

Figure 4. Effect of pH on the conversion rate of whole-cell biocatalyst. (A) Enzyme activities in different pH-values; (B) whole-cell biocatalyst conversion rate.

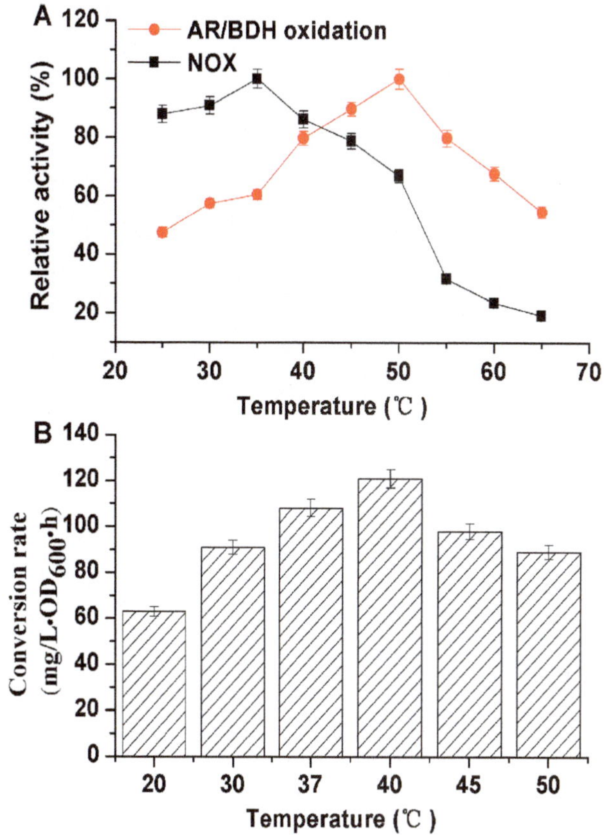

Figure 5. Effect of temperature on the conversion rate of whole-cell biocatalyst. (A) Enzyme activities under different temperature; (B) whole-cell biocatalyst conversion rate.

168/pMA5-*bdhA-yodC* showed a BDH activity of 152.8 mU/mg and a NOX activity of 346.4 mU/mg. SDS-PAGE analyze of the recombinant protein expression was shown in Figure 2. AR/BDH and NOX have the specific molecular weights of 37.3 kDa and 25.7 kDa, respectively. The results of SDS-PAGE and enzyme activity assay indicated that *bdhA* were successfully co-expressed with *yodC*.

In comparison with free enzyme activities, the whole-cell biocatalytic activities prepared from the recombinants were defined as the formation of 1.0 mmol AC per hour per OD_{600} of culture at 37°C, which were much more direct to prove the biocatalytic abilities. As shown in Table 1, *B. subtilis* 168 could not effectively catalyze 2,3-BD to AC because of low AR/BDH activity and the restriction of NAD^+ pool. By over-expression of AR/BDH or NOX, the whole-cell biocatalytic activities of *B. subtilis* 168/pMA5-*bdhA* and *B. subtilis* 168/pMA5-*yodC* were both increased, but no more than 30 % compared to *B. subtilis* 168. However, the whole-cell biocatalytic activity of *B. subtilis* 168/pMA5-*bdhA-yodC* was nearly twice that of *B. subtilis* 168, indicating the NAD^+ regeneration system worked successfully.

Although AR/BDH activity of *B. subtilis* 168/pMA5-*yodC* was lower than *B. subtilis* 168/pMA5-*bdhA*, the over-expressed NOX regenerated sufficient NAD^+ to catalyze 2,3-BD to AC. So both *B. subtilis* 168/pMA5-*yodC* and *B. subtilis* 168/pMA5-*bdhA-yodC* showed obviously increased whole-cell biocatalytic activities by over-expressing of an NAD^+ regeneration enzyme, suggesting cofactor regeneration is more significant than improving the catalytic enzyme activity.

Effects of intracellular NADH and NAD^+ concentrations on AC biosynthesis

The recombinant strains and the parent strain were cultured in the fermentation medium. Homologous over-expression of NOX in *B. subtilis* was expected to regenerate the overall intracellular NAD^+ pool for continuously converting AC to 2,3-BD. As shown in Figure 3A, *B. subtilis* 168/pMA5-*bdhA-yodC* achieved the highest conversion rate of 2,3-BD to AC comparing to the other biocatalysts. Both *B. subtilis* 168/pMA5-*yodC* and *B. subtilis* 168/pMA5-*bdhA-yodC* could maintain efficient productivities to catalyze 2,3-BD to AC, indicating NAD^+ was regenerated by NOX.

The significant role of cofactors played in these biocatalysts was further proved by comparing the intracellular concentrations of NAD^+ and NADH in the recombinants, in which the overall amount of intracellular NAD^+ and NADH were almost constant within 6 hours. The intracellular NAD^+ pools in recombinant *B. subtilis* 168/pMA5-*bdhA-yodC* and *B. subtilis* 168/pMA5-*yodC* were greatly improved by over-expression of NOX (Figure 3B and 3C), and NAD^+ was continuously regenerated to keep a persistent AC productivity. Although over-expression of AR/BDH by *B. subtilis* 168/pMA5-*bdhA* effectively converted 2,3-BD to AC in the first 2 hours, the reduced NAD^+ pool then restricted its process for the bioconversion.

By comparing of the intracellular NAD^+ and NADH levels and the ratio of NADH to NAD^+ (Figure 3D), NAD^+ was proved be regenerated by over-expression of NOX. After 6 h, the intracellular NAD^+ concentration of *B. subtilis* 168/pMA5-*bdhA-yodC*

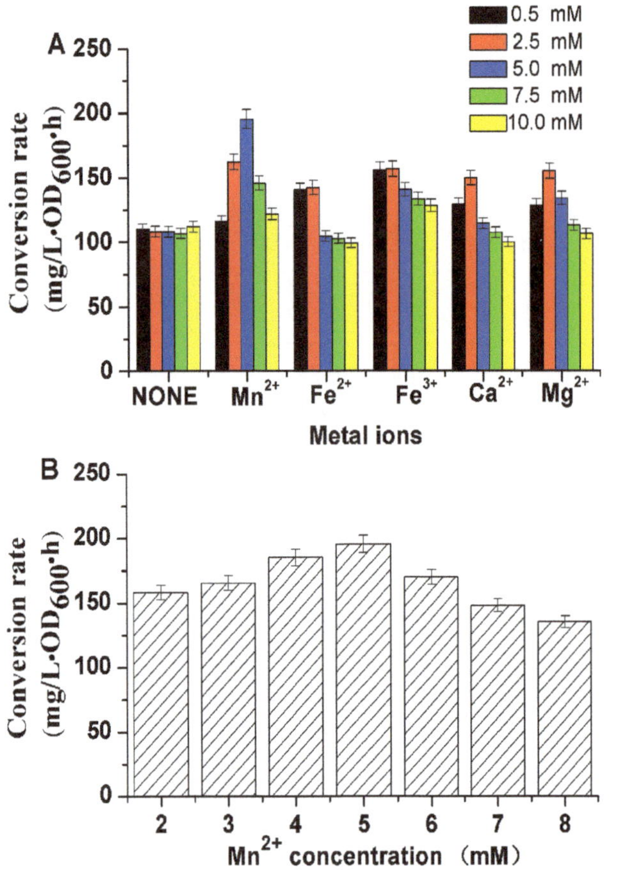

Figure 6. Effect of metal ion stimulators on the conversion rate of whole-cell biocatalyst. (A) Effect of different metal ions on the conversion rate of the biocatalyst; (B) Effect of Mn^{2+} concentration on the conversion rate of the biocatalyst.

increased to 2.75 μmol/(L·OD_{600}), higher than *B. subtilis* 168 and *B. subtilis* 168/pMA5-*bdhA*. Corresponding to above results, the intracellular NADH concentration of *B. subtilis* 168/pMA5-*bdhA-yodC* decreased to 1.29 μmol/(L·OD_{600}), lower than *B. subtilis* 168 and *B. subtilis* 168/pMA5-*bdhA*. Although *B. subtilis* 168/pMA5-*yodC* achieved the highest NOX activity, which resulted in the lowest NADH/NAD^+ ratio, it did not showed the highest conversion rate from 2,3-BD to AC. The results suggested that a relatively stable dynamic redox balance of low NADH/NAD^+ ratio as well as a high AR/BDH activity were necessary for efficiently and persistently converting 2,3-BD to AC. So that, the biocatalyst of *B. subtilis* 168/pMA5-*bdhA-yodC*, which co-expressed AR/BDH and NOX was chosen for later experiments.

Optimization of the biocatalytic reaction

To achieve higher productivity, the whole-cell biocatalytic reaction was optimized. As an important parameter that often limit enzyme activity and stability in technical applications [40], the effect of pH on the conversion rate of the biocatalyst was investigated. As shown in Figure 4A, AR/BDH had different optimum pH-values of 8.5 and 6.5 for oxidation and reduction, respectively. Meanwhile, the optimum pH value for NOX was 9.0, and this enzyme could maintain 80 % relative activity under pH 8.0–9.0. So the efficiency of the whole-cell biocatalyst could be affected by adjusting the pH-values of the conversion solution. As shown in Figure 4B, reactions with 40.0 g/L of 2,3-BD as substrate in the biocatalyst *B. subtilis* 168/pMA5-*bdhA-yodC* were conducted under different pH-values for 6 h. The conversion rate was measured by monitoring the yield of AC. The results indicated that the conversion rate of this biocatalyst was gradually increased by improving the pH levers in the alkaline condition and the highest conversion rate was acquired at pH 8.5. Thus, we use pH 8.5 as the optimum conversion pH-value for the following experiments.

Temperature can also affect the efficiency of the whole-cell catalytic processes, including enzyme activity and cellular maintenance [10]. Thus, we studied the effect of temperature on AC conversion rate using this biocatalyst. As shown in Figure 5A, AR/BDH and NOX had different optimum temperature. The optimum oxidation temperature for AR/BDH activity was 50°C (similar to the optimum reduction temperature, data was not shown), but the optimum temperature for NOX activity was 35°C. However, both enzymes could maintain 80 % relative activities under the temperature 40–45°C. So the whole-cell bioconversion was assayed under the following temperatures of 20°C, 30°C, 37°C (the best growing temperature), 40°C, 45°C and 50°C, respectively (Figure 5B). The results indicated that the highest conversion rate of this biocatalyst was obtained at a relatively high temperature 40°C, closed to the best growing temperature. Finally, a temperature of 40°C was chosen for the optimum temperature.

Chemical stimulators such as Mg^{2+}, Mn^{2+} and Ca^{2+} can improve the activity of AR/BDH [27]. However, another enzyme in this biocatalyst, NOX, had been proved that mental ions have neither stimulation effect nor inhibition effect on its activity [34]. Thus, the following chemicals of $MgCl_2$, $MnCl_2$, $CaCl_2$, $FeCl_2$ and $FeCl_3$ were studied of their effect on the biocatalyst, and they were added to the conversion solution at the final concentrations of 0.5 mM, 2.5 mM, 5.0 mM, 7.5 mM and 10.0 mM, respectively. Of all the chemicals listed in Figure 6A, 2.5–5.0 mM Mn^{2+} showed remarkable stimulation effect on AC conversion rate. Other metal ions just slightly stimulated the convention rate of 2,3-BD to AC. To find the optimum Mn^{2+} concentration on the efficiency of this biocatalyst, the final concentration of Mn^{2+} was adjusted from 2.0 mM to 8.0 mM (Figure 6B). The results indicated that 5.0 mM Mn^{2+} was favorable for this whole-cell biocatalyst.

Substrate concentration is another important factor in whole-cell catalytic processes, which may result in substrate inhibition. The effect of 2,3-BD concentration on the conversion rate using the biocatalyst was tested (Figure 7). The results showed that the biocatalyst achieved the highest conversion rate with 40 g/L 2,3-BD as substrate. While 2,3-BD concentration kept on increasing, the conversion rate could be inhibited by high concentration of substrate [6]. AR/BDH can catalyze the stereospecific oxidation of (2R,3R)-2,3-BD and meso-2,3-BD to (3R)-AC and (3S)-AC, respectively [41]. However, it cannot catalyze the conversion of (2S,3S)-2,3-BD. Thus, while using 40.0 g/L of mixed stereospecific 2,3-BD as substrate, the highest AC yield was 31.5 g/L after 12 h. Although 2,3-BD could not be totally converted into AC, it can be easily separated from AC which has a low boiling point. Therefore, 40.0 g/L of 2,3-BD was chosen as the optimum substrate concentration for this biocatalyst.

Whole-cell biocatalyst using repeated batch strategy under optimized conditions

The whole-cell biocatalyst was conducted under optimized conditions as described above in a 5-L fermentor. As shown in Figure 8A, 31.5 g/L AC was acquired from 40.0 g/L 2,3-BD after 12 h with a productivity 2.62 g/(L·h). No other products were

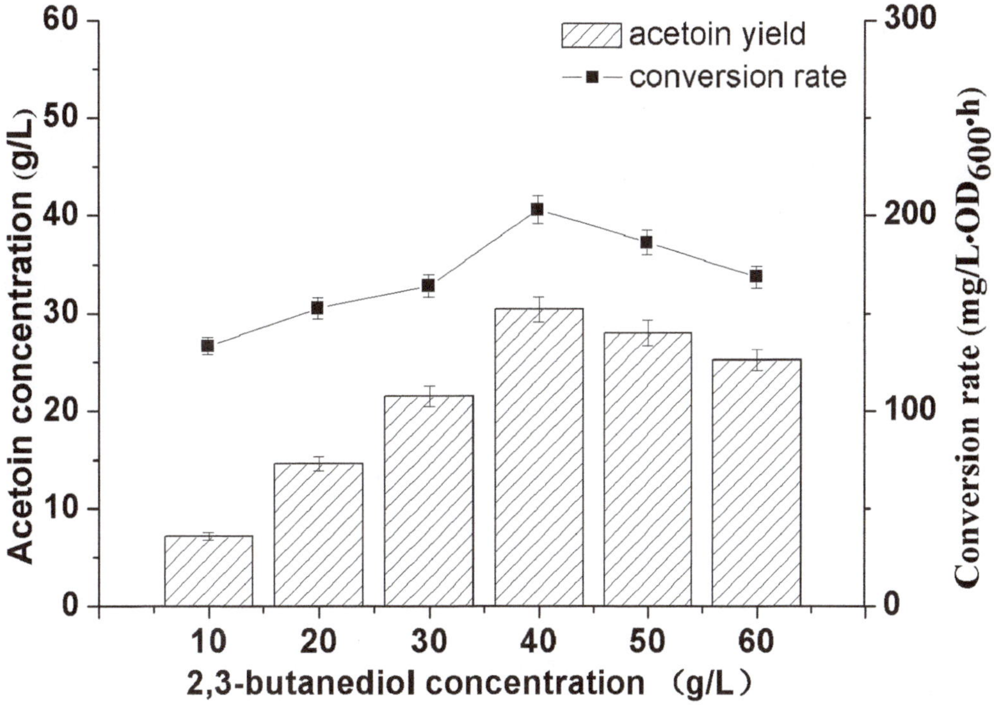

Figure 7. Effect of 2,3-butanediol concentration on the conversion rate of whole-cell biocatalyst.

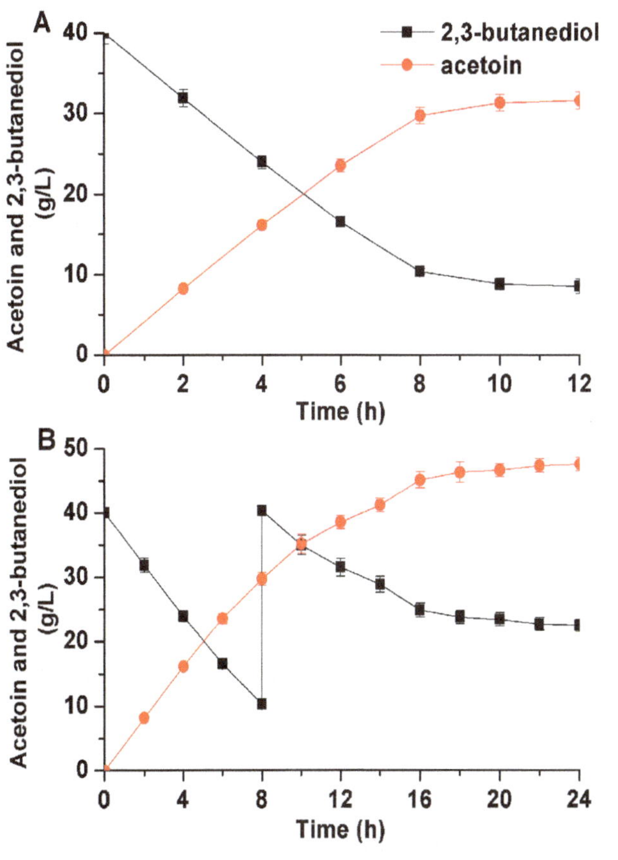

Figure 8. Time course of batch and fed-batch bioconversion of acetoin from 2,3-butanediol. (A) Batch bioconversion; (B) fed-batch bioconversion.

detected during the process. Since AR/BDH was a reversible enzyme, the whole-cell biocatalyst could not completely transform 2,3-BD to AC, high concentration of AC would restrain the conversion. As shown in Figure 8B, The initial 2,3-BD concentration was 40.0 g/L, and 30.2 g/L of 2,3-BD was added at 8 h. However, due to the limitation of high concentration of product, the yield of AC could not be effectively increased.

To achieve a higher product yield, repeated batch strategies could efficiently enhance the concentrations of the aim products. Due to the NAD^+ regeneration system, high AC yield could be obtained through more than once whole-cell biocatalyst by using the same batch bacterium without any other additional NADH/NAD^+. The initial 2,3-BD concentration was 40.0 g/L, after each conversion cycle of 8 h, the cells were harvested and then added by 40.0 g/L of fresh 2,3-BD solution to continue the enzymatic reaction. As shown in Figure 9, there was a stable and efficient increase on AC yield in the first two batch conversion. However, AC productivity gradually decreased in the third batch. Thus, after total 40 h of bioconversion, 91.8 g/L of AC was produced from 120.0 g/L 2,3-BD with the productivity of 2.30 g/(L·h), which was the highest production by biocatalyst (Table 3).

This is the first report of applying NAD^+ regeneration system to produce acetoin in *B. subtilis*. By repeated batch strategy, the whole-cell biocatalyst achieves the purpose of efficient and sustainable producing of AC, and finally reaches the highest AC production record by biocatalysis. In addition, compared to the difficulty of the downstream purification processes of mixed acid-butanediol fermentation, this bioprocess is relatively simple and security, and residual 2,3-BD can be easily separated from AC.

Conclusions

A biocatalyst for AC production was successfully constructed by introducing the NAD^+ regeneration system into *B. subtilis*, in which AR/BDH and NOX were co-expressed. After optimization

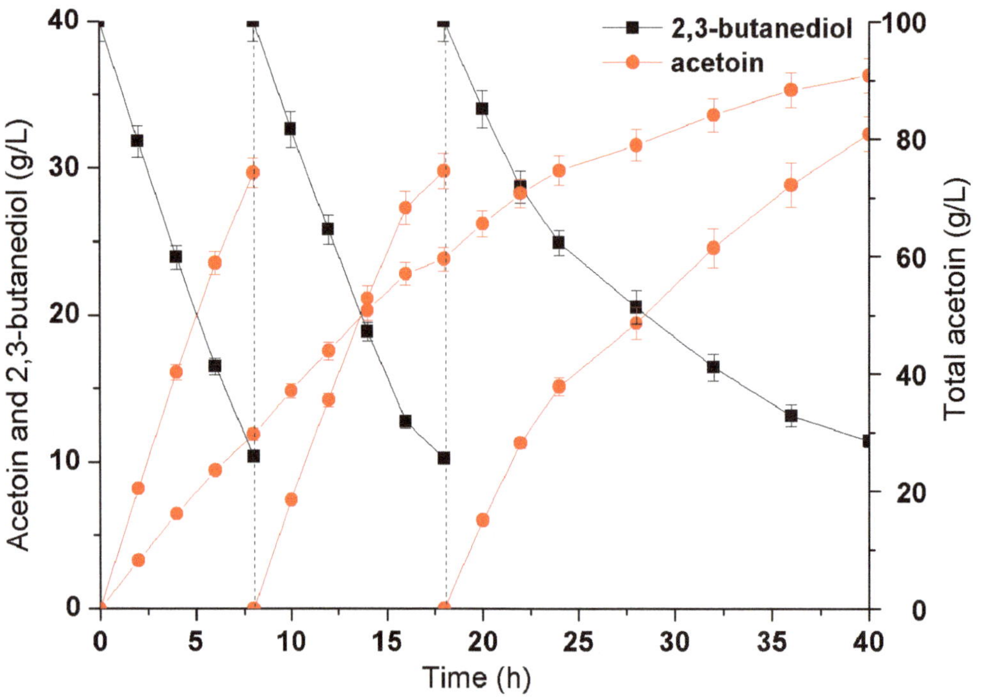

Figure 9. Time course of repeated-batch bioconversion of acetoin from 2,3-butanediol.

Table 3. Comparing of microbial production of acetoin using different fermentative strains or biocatalysts.

Strains	Productivity (g/L · h)	Concentration (g/L)	Yield (mol/mol)	References
Fermentation				
Klebsiella pneumoniae XZF-308	0.32	25.9	0.16	[42]
Gluconobacter oxydans DSM 2003	1.24	89.2	0.91	[10]
Serratia marcescens H32	1.88	75.2	0.78	[11]
Paenibacillus polymyxa CS107	1.32	55.3	0.76	[8]
Lactococcus lactis subsp. *lactis* 3022	0.19	9.28	0.20	[43]
Bacillus				
Bacillus licheniformis MEL09	1.15	41.3	0.84	[4]
Bacillus subtilis CICC10025	0.63	35.4	0.83	[16]
Bacillus subtilis 168	0.09	5.5	0.70	[19]
Bacillus subtilis 168	0.47	19.8	0.79	[1]
Bacillus amyloliquefaciens	1.42	51.2	0.43	[17]
Bacillus subtilis JNA-3-10	0.32	42.2	0.57	This lab[15]
Bacillus subtilis JNA-3-10-PAR	0.43	41.5	0.71	This lab[18]
Bacillus subtilis JNA-UD-6	0.37	53.9	0.74	This lab[14]
Biocatalyst				
Escherichia coli BL21 (DE3)	3.06	36.7	0.85	[6]
Purified NADPH-dependent carbonyl reductase and glucose dehydrogenase	9.76	12.2	0.85	[5]
Escherichia coli Rosetta (DE3)	2.25	13.5	0.91	[44]
Bacillus subtilis 168 and *Klebsiella pneumoniae* CICC 10011	1.89	56.7	0.62	[41]
Bacillus subtilis 168	2.30	91.8	0.78	This study

of this converting reaction, repeated batch strategy was further applied on this biocatalyst, and 120.0 g/L 2,3-BD was converted into 91.8 g/L AC with the productivity of 2.30 g/(L·h). To our knowledge, this is the highest report of AC production by biocatalyst. However, AR/BDH stability and the inhibitory effect of acetoin/2,3-butanediol to this enzyme should be further studied, and modeling of this enzyme and site directed mutation is now undergoing. This work proposed an efficient approach for nature AC production and microbial-based biofuel utilization of 2,3-BD.

Author Contributions

Conceived and designed the experiments: TB X. Zhang ZR. Performed the experiments: TB X. Zhang X. Zhao. Analyzed the data: TB X. Zhang RZ TY ZX SY. Contributed to the writing of the manuscript: TB X. Zhang ZR.

References

1. Wang M, Fu J, Zhang X, Chen T (2012) Metabolic engineering of *Bacillus subtilis* for enhanced production of acetoin. Biotechnol Lett 34: 1877–1885.
2. Sun JN, Zhang LY, Rao B, Han YB, Chu J, et al. (2012) Enhanced acetoin production by *Serratia marcescens* H32 using statistical optimization and a two-stage agitation speed control strategy. Biotechnology and Bioprocess Engineering 17: 598–605.
3. Toda F, Tanaka K, Tange H (1989) New Reduction Method of Alpha-Diketones, Oxo Amides, and Quinones with Zn-Etoh in the Presence of a Salt. Journal of the Chemical Society-Perkin Transactions 1: 1555–1556.
4. Liu YF, Zhang SL, Yong YC, Ji ZX, Ma X, et al. (2011) Efficient production of acetoin by the newly isolated *Bacillus licheniformis* strain MEL09. Process Biochemistry 46: 390–394.
5. Gao C, Zhang LJ, Xie YJ, Hu CH, Zhang Y, et al. (2013) Production of (3S)-acetoin from diacetyl by using stereoselective NADPH-dependent carbonyl reductase and glucose dehydrogenase. Bioresource Technology 137: 111–115.
6. Xiao ZJ, Lv CJ, Gao C, Qin JY, Ma CQ, et al. (2010) A Novel Whole-Cell Biocatalyst with NAD(+) Regeneration for Production of Chiral Chemicals. Plos One 5.
7. Xiao Z, Xu P (2007) Acetoin metabolism in bacteria. Crit Rev Microbiol 33: 127–140.
8. Zhang LY, Chen S, Xie HB, Tian YT, Hu KH (2012) Efficient acetoin production by optimization of medium components and oxygen supply control using a newly isolated *Paenibacillus polymyxa* CS107. Journal of Chemical Technology and Biotechnology 87: 1551–1557.
9. Cho S, Kim KD, Ahn JH, Lee J, Kim SW, et al. (2013) Selective production of 2,3-butanediol and acetoin by a newly isolated bacterium *Klebsiella oxytoca* M1. Appl Biochem Biotechnol 170: 1922–1933.
10. Wang X, Lv M, Zhang L, Li K, Gao C, et al. (2013) Efficient bioconversion of 2,3-butanediol into acetoin using *Gluconobacter oxydans* DSM 2003. Biotechnology for Biofuels 6: 155.
11. Sun JA, Zhang LY, Rao B, Shen YL, Wei DZ (2012) Enhanced acetoin production by *Serratia marcescens* H32 with expression of a water-forming NADH oxidase. Bioresour Technol 119: 94–98.
12. Ji XJ, Huang H, Ouyang PK (2011) Microbial 2,3-butanediol production: a state-of-the-art review. Biotechnol Adv 29: 351–364.
13. Perkins JB, Sloma A, Hermann T, Theriault K, Zachgo E, et al. (1999) Genetic engineering of *Bacillus subtilis* for the commercial production of riboflavin. Journal of Industrial Microbiology & Biotechnology 22: 8–18.
14. Zhang X, Zhang R, Yang T, Zhang J, Xu M, et al. (2013) Mutation breeding of acetoin high producing *Bacillus subtilis* blocked in 2,3-butanediol dehydrogenase. World J Microbiol Biotechnol 29: 1783–1789.
15. Zhang X, Yang TW, Lin Q, Xu MJ, Xia HF, et al. (2011) Isolation and identification of an acetoin high production bacterium that can reverse transform 2,3-butanediol to acetoin at the decline phase of fermentation. World Journal of Microbiology & Biotechnology 27: 2785–2790.
16. Xiao ZJ, Liu PH, Qin JY, Xu P (2007) Statistical optimization of medium components for enhanced acetoin production from molasses and soybean meal hydrolysate. Applied Microbiology and Biotechnology 74: 61–68.
17. Zhang Y, Li S, Liu L, Wu J (2013) Acetoin production enhanced by manipulating carbon flux in a newly isolated *Bacillus amyloliquefaciens*. Bioresour Technol 130: 256–260.
18. Zhang X, Zhang R, Bao T, Yang T, Xu M, et al. (2013) Moderate expression of the transcriptional regulator ALsR enhances acetoin production by *Bacillus subtilis*. J Ind Microbiol Biotechnol 40: 1067–1076.
19. Chen T, Liu WX, Fu J, Zhang B, Tang YJ (2013) Engineering *Bacillus subtilis* for acetoin production from glucose and xylose mixtures. J Biotechnol 168: 499–505.
20. Gong Y, Tang Y, Wang XL, Yu LX, Liu DH (2004) The possibility of the desalination of actual 1,3-propanediol fermentation broth by electrodialysis. Desalination 161: 169–178.
21. Zhou YJJ, Yang W, Wang L, Zhu ZW, Zhang SF, et al. (2013) Engineering NAD(+) availability for *Escherichia coli* whole-cell biocatalysis: a case study for dihydroxyacetone production. Microbial Cell Factories 12.
22. Crow VL (1990) Properties of 2,3-Butanediol Dehydrogenases from *Lactococcus lactis subsp. lactis* in Relation to Citrate Fermentation. Appl Environ Microbiol 56: 1656–1665.
23. Gonzalez E, Fernandez MR (2001) Characterization and functional role of *Saccharomyces cerevisiae* 2,3-butanediol dehydrogenase. Chemico-Biological Interactions 130–132: 425–434.
24. Heidlas J, Tressl R (1990) Purification and characterization of a (R)-2,3-butanediol dehydrogenase from *Saccharomyces cerevisiae*. Arch Microbiol 154: 267–273.
25. Machielsen R, Uria AR, Kengen SWM, van der Oost J (2006) Production and characterization of a thermostable alcohol dehydrogenase that belongs to the aldo-keto reductase superfamily. Applied and Environmental Microbiology 72: 233–238.
26. Wang Z, Song Q, Yu M, Wang Y, Xiong B, et al. (2013) Characterization of a stereospecific acetoin(diacetyl) reductase from *Rhodococcus erythropolis* WZ010 and its application for the synthesis of (2S,3S)-2,3-butanediol. Appl Microbiol Biotechnol.
27. Zhang X, Bao T, Rao Z, Yang T, Xu Z, et al. (2014) Two-Stage pH Control Strategy Based on the pH Preference of Acetoin Reductase Regulates Acetoin and 2,3-Butanediol Distribution in *Bacillus subtilis*. PLoS One 9: e91187.
28. Gonzalez E, Fernandez MR, Larroy C, Sola L, Pericas MA, et al. (2000) Characterization of a (2R,3R)-2,3-butanediol dehydrogenase as the *Saccharomyces cerevisiae* YAL060W gene product. Disruption and induction of the gene. J Biol Chem 275: 35876–35885.
29. HÖhn-Bentz H RF (1978) Bacterial 2,3-Butanediol Dehydrogenases. Arch Microbiol 116: 197–203.
30. San KY, Bennett GN, Berrios-Rivera SJ, Vadali RV, Yang YT, et al. (2002) Metabolic engineering through cofactor manipulation and its effects on metabolic flux redistribution in *Escherichia coli*. Metabolic Engineering 4: 182–192.
31. Berrios-Rivera SJ, Bennett GN, San KY (2002) Metabolic engineering of *Escherichia coli*: Increase of NADH availability by overexpressing an NAD(+)-dependent formate dehydrogenase. Metabolic Engineering 4: 217–229.
32. Heux S, Cachon R, Dequin S (2006) Cofactor engineering in *Saccharomyces cerevisiae*: Expression of a H2O-forming NADH oxidase and impact on redox metabolism. Metabolic Engineering 8: 303–314.
33. Geueke B, Riebel B, Hummel W (2003) NADH oxidase from *Lactobacillus brevis*: a new catalyst for the regeneration of NAD. Enzyme and Microbial Technology 32: 205–211.
34. Zhang X, Zhang R, Bao T, Rao Z, Yang T, et al. (2014) The rebalanced pathway significantly enhances acetoin production by disruption of acetoin reductase gene and moderate-expression of a new water-forming NADH oxidase in *Bacillus subtilis*. Metab Eng 23C: 34–41.
35. Dartois V, Coppee JY, Colson C, Baulard A (1994) Genetic analysis and overexpression of lipolytic activity in *Bacillus subtilis*. Appl Environ Microbiol 60: 1670–1673.
36. de Felipe FL, Hugenholtz J (2001) Purification and characterisation of the water forming NADH-oxidase from *Lactococcus lactis*. International Dairy Journal 11: 37–44.
37. Laemmli UK (1970) Cleavage of structural proteins during the assembly of the head of bacteriophage T4. Nature 227: 680–685.
38. Bradford MM (1976) A rapid and sensitive method for the quantitation of microgram quantities of protein utilizing the principle of protein-dye binding. Anal Biochem 72: 248–254.
39. O'Donnell JM, Kudej RK, LaNoue KF, Vatner SF, Lewandowski ED (2004) Limited transfer of cytosolic NADH into mitochondria at high cardiac workload. American Journal of Physiology-Heart and Circulatory Physiology 286: H2237–H2242.
40. Li LX, Wang Y, Zhang LJ, Ma CQ, Wang AL, et al. (2012) Biocatalytic production of (2S,3S)-2,3-butanediol from diacetyl using whole cells of engineered *Escherichia coli*. Bioresource Technology 115: 111–116.
41. Liu Z, Qin J, Gao C, Hua D, Ma C, et al. (2011) Production of (2S,3S)-2,3-butanediol and (3S)-acetoin from glucose using resting cells of *Klebsiella pneumonia* and *Bacillus subtilis*. Bioresour Technol 102: 10741–10744.
42. Ji XJ, Xia ZF, Fu NH, Nie ZK, Shen MQ, et al. (2013) Cofactor engineering through heterologous expression of an NADH oxidase and its impact on metabolic flux redistribution in *Klebsiella pneumoniae*. Biotechnol Biofuels 6.
43. Kaneko T, Takahashi M, Suzuki H (1990) Acetoin Fermentation by Citrate-Positive *Lactococcus lactis subsp. lactis* 3022 Grown Aerobically in the Presence of Hemin or Cu. Appl Environ Microbiol 56: 2644–2649.
44. Gao J, Xu YY, Li FW, Ding G (2013) Production of S-acetoin from diacetyl by *Escherichia coli* transformant cells that express the diacetyl reductase gene of *Paenibacillus polymyxa* ZJ-9. Lett Appl Microbiol 57: 274–281.

3

Investigation of Specificity Determinants in Bacterial tRNA-Guanine Transglycosylase Reveals Queuine, the Substrate of Its Eucaryotic Counterpart, as Inhibitor

Inna Biela[❡], **Naomi Tidten-Luksch**[❡], **Florian Immekus, Serghei Glinca, Tran Xuan Phong Nguyen, Hans-Dieter Gerber, Andreas Heine, Gerhard Klebe, Klaus Reuter***

Institut für Pharmazeutische Chemie, Philipps-Universität Marburg, Marburg, Germany

Abstract

Bacterial tRNA-guanine transglycosylase (Tgt) catalyses the exchange of the genetically encoded guanine at the wobble position of tRNAs^{His,Tyr,Asp,Asn} by the premodified base $preQ_1$, which is further converted to queuine at the tRNA level. As eucaryotes are not able to synthesise queuine *de novo* but acquire it through their diet, eucaryotic Tgt directly inserts the hypermodified base into the wobble position of the tRNAs mentioned above. Bacterial Tgt is required for the efficient pathogenicity of *Shigella* sp, the causative agent of bacillary dysentery and, hence, it constitutes a putative target for the rational design of *anti*-Shigellosis compounds. Since mammalian Tgt is known to be indirectly essential to the conversion of phenylalanine to tyrosine, it is necessary to create substances which only inhibit bacterial but not eucaryotic Tgt. Therefore, it seems of utmost importance to study selectivity-determining features within both types of proteins. Homology models of *Caenorhabditis elegans* Tgt and human Tgt suggest that the replacement of Cys158 and Val233 in bacterial Tgt (*Zymomonas mobilis* Tgt numbering) by valine and accordingly glycine in eucaryotic Tgt largely accounts for the different substrate specificities. In the present study we have created mutated variants of *Z. mobilis* Tgt in order to investigate the impact of a Cys158Val and a Val233Gly exchange on catalytic activity and substrate specificity. Using enzyme kinetics and X-ray crystallography, we gained evidence that the Cys158Val mutation reduces the affinity to $preQ_1$ while leaving the affinity to guanine unaffected. The Val233Gly exchange leads to an enlarged substrate binding pocket, that is necessary to accommodate queuine in a conformation compatible with the intermediately covalently bound tRNA molecule. Contrary to our expectations, we found that *a priori* queuine is recognised by the binding pocket of bacterial Tgt without, however, being used as a substrate.

Editor: Lennart Randau, Max-Planck-Institute for Terrestrial Microbiology, Germany

Funding: This work was supported by the Philipps-University of Marburg and the German Research Foundation (Graduiertenkolleg "Protein Function at the Atomic Level" and DFG grant KL1204/11-1). The travel costs to the BESSY II synchrotron in Berlin were covered by the Helmhotz-Zentrum Berlin. The funders had no role in study design, data collection and analysis, decision to publish, or preparation of the manuscript.

Competing Interests: The authors have declared that no competing interests exist.

* E-mail: reuterk@staff.uni-marburg.de

❡ These authors contributed equally to this work.

Introduction

Transfer RNA–guanine transglycosylase (Tgt, EC 2.4.2.29) catalyses the exchange of a specific guanine base in tRNA molecules by a substituted 7-deazaguanine. Although Tgt is found in all three domains of life, the specificity of this enzyme with respect to (i) the tRNA substrate, (ii) the position of the guanine base therein to be exchanged, and (iii) the 7-deazaguanine derivative to be inserted differs in archaea, bacteria and eucaryotes (for review see [1]). Thus, the archaeal Tgt inserts $preQ_0$ at position 15 of the majority of archaeal tRNAs where it is further converted to archaeine (for chemical formulae see Figure 1). Located at the "elbow" of the L-shaped tRNA the modification stabilises the overall tRNA structure by supporting the formation of the non-canonical Levitt base pair with cytosine 48 [2].

In contrast, both bacterial and eucaryotic Tgts are required for the introduction of the hypermodified base queuine into position 34 (the anticodon "wobble position") of tRNAs^{His,Tyr,Asp,Asn} all of which have a uracil 33 – guanine 34 – uracil 35 sequence in common [3,4]. Although no distinct function for queuine has been demonstrated yet, its presence within the tRNA anticodon suggests that it may be involved in coordinating translational fidelity and speed.

Bacterial Tgt replaces the genetically encoded guanine at the above-mentioned position by the queuine-precursor $preQ_1$. This premodified base is synthesised from guanosine 5′-triphosphate in a pathway employing GTP cyclohydrolase I (FolE) [5], 6-carboxy-5,6,7,8-tetrahydropterin synthase (QueD) [6], an *S*-adenosyl-L-methionine-dependent organic radical-generating enzyme (QueE), [7], $preQ_0$ synthetase (QueC = ToyM) [7] and a nitrile reductase (QueF) [8–10]. Once incorporated into tRNA, $preQ_1$ is further converted to the functional queuine in two consecutive steps catalysed by *S*-adenosylmethionine:tRNA ribosyltransferase-isomerase (QueA) [11–13] and the coenzyme B_{12}-dependent epoxyqueuosine reductase (QueG) [14,15] (Figure 2).

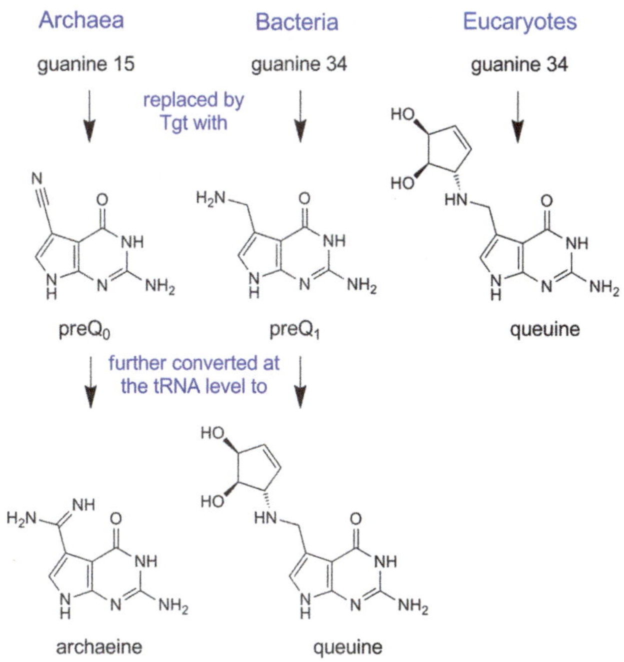

Figure 1. Substrate bases of the Tgt enzymes from the three domains of life. preQ$_0$, 7-cyano-7-deazaguanine; preQ$_1$, 7-aminomethyl-7-deazaguanine; queuine = 7-(((4,5-*cis*-dihydroxy-2-cyclopenten-1-yl) amino) methyl)-7-deazaguanine.

Bacterial Tgt functions as a homodimer in which the protomer consists of a $(\beta\alpha)_8$ barrel harbouring a Zn^{2+} binding subdomain close to the *C*-terminus [16]. For sterical reasons, this dimer is, although endowed with two active sites, able to bind and convert only one substrate tRNA molecule at a time [17]. The reaction catalysed by bacterial Tgt is accurately documented by numerous biochemical studies and crystal structures of Tgt from *Zymomonas mobilis* determined in its apo-form, in the presence of guanine, preQ$_1$ or small molecule inhibitors and in complex with an RNA substrate [16,18–24]. The reaction follows a ping-pong mechanism including a covalent Tgt·tRNA intermediate (Figure 3).

Since eucaryotes are, other than bacteria, not capable of synthesising queuine *de novo* but rather acquire this modified base from their diet or from the gut flora, eucaryotic Tgt has to directly recognise and insert queuine into position 34 of $tRNAs^{His,Tyr,Asp,Asn}$. This stands in contrast to bacterial Tgt which was reported to be unable to accept this base as a substrate [25,26]. Unlike its bacterial counterpart, eucaryotic Tgt constitutes a *hetero*dimer with both subunits, however, being homologous to the bacterial protomer [27,28]. Accordingly, both subunits contain the three cysteines as well as the histidine required for the coordination of the structural Zn^{2+} ion. However, only one subunit includes the residues known to be important for catalysis, among these is the active-site nucleophile, Asp280 (*Z. mobilis* Tgt numbering) [21,24]. Obviously, eucaryotic Tgt consists of a catalytic subunit and a second subunit which most probably supports the binding and the orientation of the tRNA substrate during catalysis.

It was shown that a functional Tgt is required for efficient pathogenicity of *Shigella* bacteria which are the causative agents of bacillary dysentery. A null-mutation in the *tgt* gene leads to a

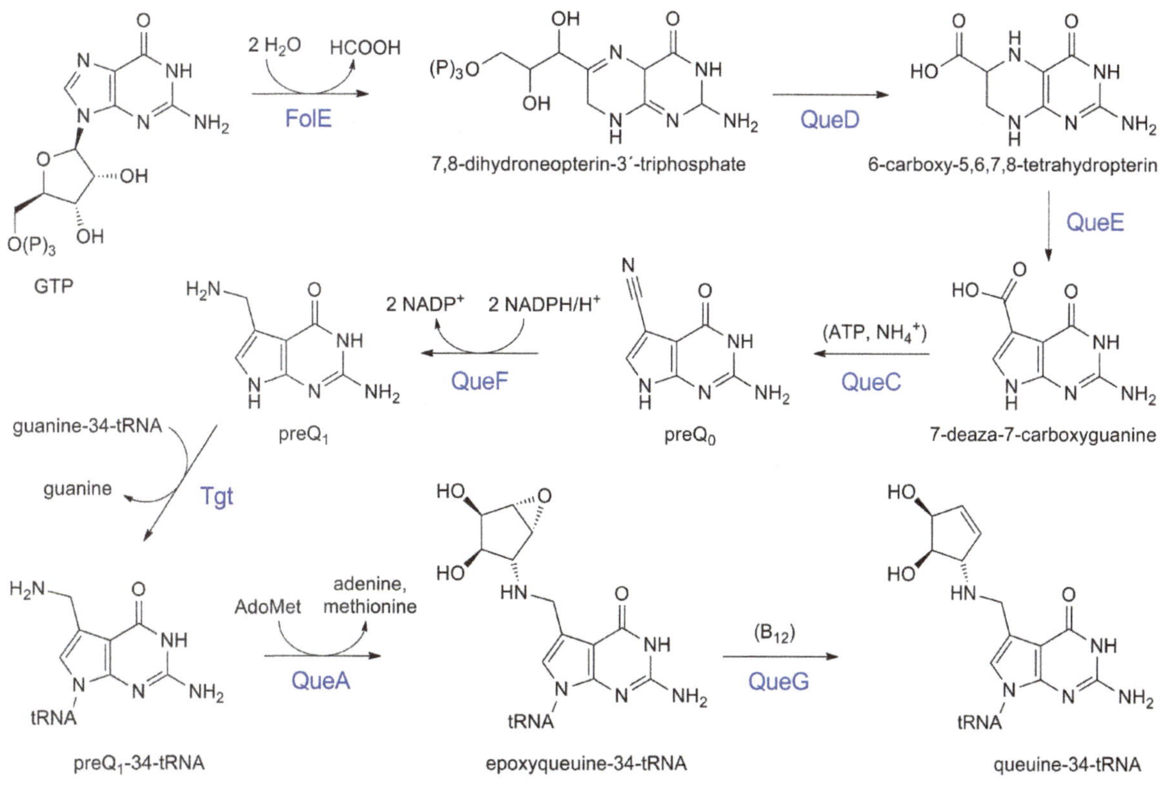

Figure 2. Biosynthesis of the modified tRNA base queuine. Substrates and cosubstrates: GTP, guanosine 5′-triphosphate; AdoMet, *S*-adenosylmethionine; B$_{12}$, coenzyme B$_{12}$; preQ$_0$, 7-cyano-7-deazaguanine; preQ$_1$, 7-aminomethyl-7-deazaguanine; enzymes: FolE, GTP cyclohydrolase I; QueD, 6-carboxy-5,6,7,8-tetrahydropterin synthase; QueE, 7-carboxy-7-deazaguanine synthase; QueC, preQ$_0$ synthetase; QueF, preQ$_0$ reductase; Tgt, tRNA-guanine transglycosylase; QueA, *S*-adenosylmethionine:tRNA ribosyltransferase-isomerase; QueG, epoxyqueuosine reductase.

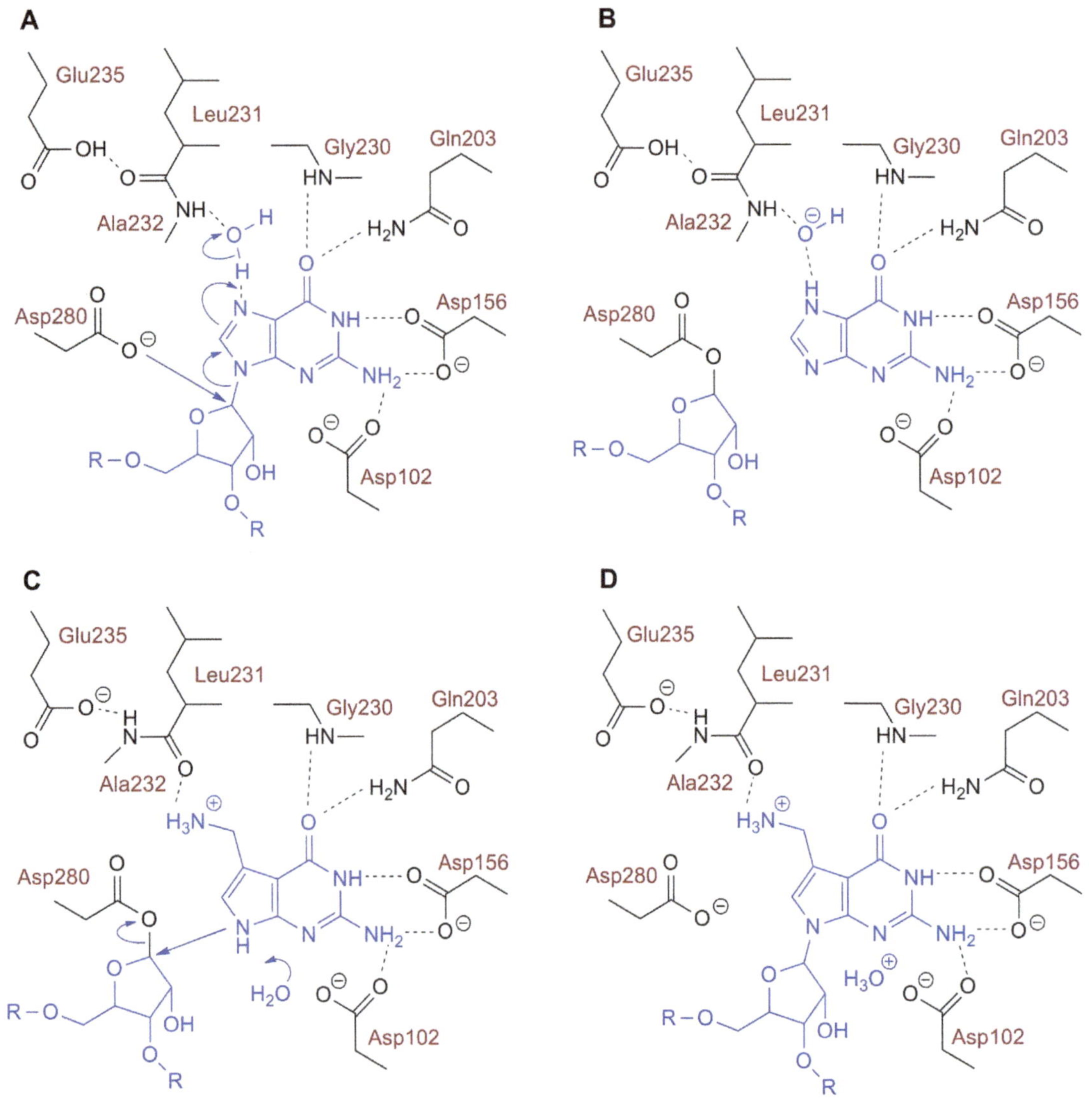

Figure 3. Assumed catalytic mechanism of bacterial Tgt. (A and B) The glycosidic bond of guanosine 34 is cleaved *via* nucleophilic attack by the Asp280 carboxylate resulting in the formation of a covalent Tgt·tRNA intermediate. (C and D) Guanine is replaced by $preQ_1$ which is incorporated into the tRNA *via* nucleophilic attack of the ribose 34 anomeric carbon by *N*9 of $preQ_1$. Notably the replacement of guanine by $preQ_1$ in the binding pocket of Tgt induces a flip of the Leu231/Ala232 peptide bond. The formation of a hydroxide and an oxonium ion as byproducts of the reaction is assumed to be responsible for its irreversibility as mutual neutralisation will efficiently detract these ions from equilibrium. *H*-bonds are indicated by dashed lines.

strongly reduced translation of *virF*-mRNA encoding a transcriptional activator required for the expression of a large number of *Shigella* pathogenicity genes [29]. So far, the exact reason for this phenomenon is unknown, though Hurt *et al.* showed that Tgt is able to replace guanine 421 of *virF* mRNA by $preQ_1$ [30]. This unusual modification of an mRNA molecule might possibly have a stimulating effect on its translation. In any case, the fact that full pathogenicity of *Shigella* sp. relies on Tgt activity encouraged us to use this enzyme as a target for the rational design of *anti*-Shigellosis compounds. Meanwhile, based on the high-resolution crystal structure of *Z. mobilis* Tgt, inhibitors which show an affinity to bacterial Tgt down to the single-digit nanomolar range have been synthesised [31–33].

Since a Tgt enzyme is present in humans as well, it is highly desirable to create inhibitors which preferably bind the bacterial enzyme while leaving the eucaryotic one unaffected. This seems of particular significance as recent studies performed on mice showed that a queuine deficiency or disruption of the *tgt* gene impairs the conversion of phenylalanine to tyrosine due to uncontrolled oxidation of the cofactor tetrahydrobiopterin [34]. Bearing this observation in mind, it is of utmost importance to study selectivity-determining features within both types of proteins.

Although no crystal structure of a eucaryotic Tgt has been determined yet, a homology model of the *Caenorhabditis elegans* Tgt catalytic subunit that is based on the structure of *Z. mobilis* Tgt as a template, suggests a large structural similarity of the active sites of both bacterial and eucaryotic Tgt [35]. However, in the eucaryotic enzyme, the recognition of the bulkier substrate is most probably enabled by a spatial expansion of the binding pocket. The enlargement of the binding pocket is caused by a single replacement of Val233 (*Z. mobilis* Tgt numbering) in bacterial Tgt by glycine in the eucaryotic catalytic subunit. Apart from that, the model suggests that only minor changes might be necessary in order to convert the binding pocket of bacterial Tgt such that it exhibits the specificity of the eucaryotic enzyme. Hence, in an attempt to generate a model system for the active site of human Tgt and to gain insights into the relevance of particular amino acids for substrate base selectivity, we adapted the *Z. mobilis* Tgt active site by site-directed mutagenesis in order to mimic that of the human enzyme. This paper describes the extensive analysis of mutated *Z. mobilis* Tgt variants created with this objective.

Results

Generation of an *in silico* homology model of the human Tgt catalytic subunit

In 2009 Boland *et al.* reported the first successful isolation and characterisation of an enzymatically active eucaryotic Tgt, which was the heterologously produced heterodimeric murine enzyme [27]. Shortly after that, Chen *et al.* published the production, purification and kinetic characterisation of recombinant human Tgt [28]. However, no crystal structure of either of these enzymes is available yet, and the only structural information concerning the active site of a eucaryotic Tgt originates from a homology model of the *Caenorhabditis elegans* catalytic subunit [35]. As the complete sequence of the human catalytic subunit has meanwhile become available [28,36] we created a model of this orthologue using the crystal structure of *Z. mobilis* Tgt as a template (for experimental details see the "Materials and Methods" section). The model indicates that most residues establishing the active centre and substrate binding pocket in the bacterial and human enzyme are identical. A significant difference, however, is the exchange of Val233 in *Z. mobilis* Tgt by a glycine (Gly232) in human Tgt (Figure 4). The exchange leads to an enlarged binding pocket of the human enzyme, being vital to accommodate the dihydroxy-cyclopentenyl side chain of queuine which is missing in $preQ_1$, the physiological substrate of the bacterial enzyme (Figure 1). A further conspicuous difference which might influence substrate selectivity is the exchange of Cys158 in *Z. mobilis* Tgt by a valine residue (Val161) in human Tgt. The side chain of Val161 in the model of human Tgt is at a position where it is likely to come in contact with the dihydroxy-cyclopentenyl moiety of a bound queuine molecule. Lastly, there is a conservative amino acid exchange within the binding pocket of the human compared to the *Z. mobilis* enzyme which will, however, most probably not affect substrate selectivity. Tyr106 within the active centre of *Z. mobilis* Tgt forms *via* its aromatic side chain a stacking interaction with the bound substrate base (see Figure 4). Not only in the human enzyme it is replaced by a phenylalanine (Phe109), but also in the majority of bacterial Tgts whose sequences are known so far. It must be noted that all of the mentioned amino acid exchanges can be observed in the model of the *C. elegans* Tgt catalytic subunit as well [35].

Point mutations aimed at enabling *Z. mobilis* Tgt to bind queuine and kinetic characterisation of the mutated variants

In order to investigate the importance of Gly232 and Val161 in human Tgt for queuine recognition, we created a doubly mutated variant of the *Z. mobilis* enzyme in which the corresponding Val233 and Cys158 were exchanged by glycine and valine, respectively. In addition, we created two further variants, each containing solely one of the above-mentioned amino acid exchanges. To resemble the situation in the active site of human Tgt, we did not introduce the mutations into wild type *Z. mobilis* Tgt but into the Tyr106Phe variant of *Z. mobilis* Tgt. This variant had been created in the course of a former study and it had been shown that the mutation of Tyr106 to phenylalanine has no measurable impact on any of the catalytic properties of the enzyme [37]. In the following, this mutation will not be mentioned explicitly anymore and Tgt(Tyr106Phe) will henceforth be referred to as "wild type" Tgt.

To analyse the influence the introduced Val233Gly and Cys158Val mutations have on the binding of the different substrate bases and on catalysis, we measured kinetic parameters for the mutated variants and "wild type" Tgt. At first, K_M and k_{cat} for guanine and $tRNA^{Tyr}$ were analysed *via* monitoring the incorporation of [8-^3H]-guanine into $tRNA^{Tyr}$. None of the mutations had, either alone or in combination, any significant influence on K_M(guanine) or K_M($tRNA^{Tyr}$), yet k_{cat} had, compared to "wild type" enzyme, decreased for all three variants (see Table 1). While for Tgt(Cys158Val) the turnover number was only

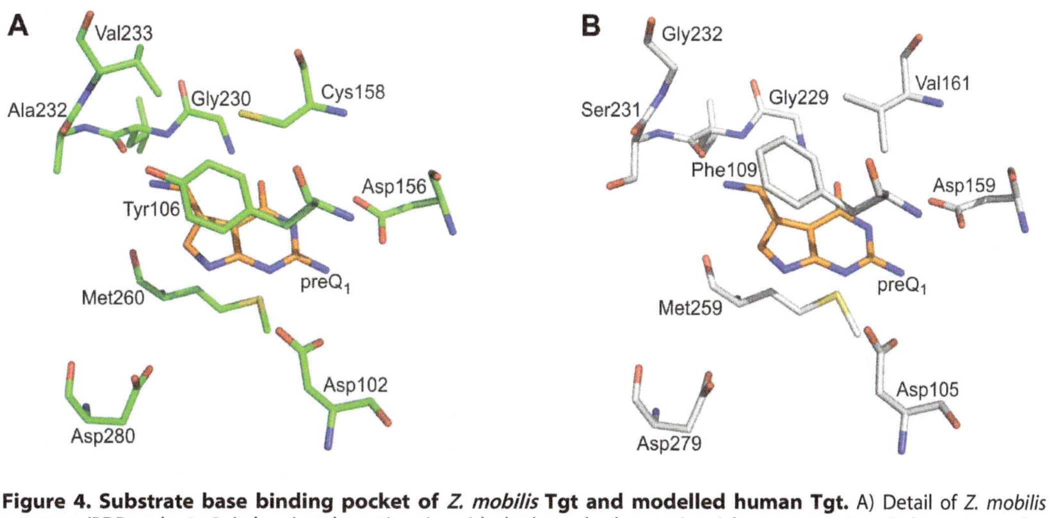

Figure 4. Substrate base binding pocket of *Z. mobilis* Tgt and modelled human Tgt. A) Detail of *Z. mobilis* Tgt·$preQ_1$ complex crystal structure (PDB-code: **1p0e**) showing the active site with the bound substrate in stick representation. Carbon atoms of protein residues are coloured in green, those of $preQ_1$ in orange. B) Homology model of human Tgt created with the *Z. mobilis* Tgt crystal structure as a template. The close up shows active site residues (carbon atoms in grey) superimposed with $preQ_1$ (carbon atoms in orange) as present in **1p0e**. The coordinates of the homology model are provided within the Supporting Information (Coordinates S1).

Table 1. Kinetic parameters for "wild type" Tgt and mutated Tgt variants.*

	$tRNA^{Tyr}$**	guanine	$preQ_1$
"Wild type" Tgt			
K_M (µmol·L^{-1})	0.9±0.2	1.2±0.2	0.9±0.1
k_{cat} (10^{-3} s^{-1})***	53.6±0.1	55.8±0.1	106.0±2.7
Tgt(Cys158Val)			
K_M (µmol·L^{-1})	2.1±0.2	1.4±0.2	16.0±2.4
k_{cat} (10^{-3} s^{-1})***	28.1±0.3	27.5±0.2	35.0±0.8
Tgt(Val233Gly)			
K_M (µmol·L^{-1})	2.0±0.3	1.0±0.2	0.6±0.1
k_{cat} (10^{-3} s^{-1})***	5.4±0.3	4.6±0.3	6.3±0.2
Tgt(Cys158Val/Val233Gly)			
K_M (µmol·L^{-1})	1.4±0.4	2.7±0.3	13.5±3.1
k_{cat} (10^{-3} s^{-1})***	0.8±0.1	0.6±0.1	1.0±0.3

*All Tgt variants considered in this table including "wild type" Tgt contain an additional Tyr106Phe mutation which in a previous study was shown to have no impact on any of the catalytic properties of the enzyme [37] (see main text).
**Using [8-^{3}*H*]-guanine as second substrate.
***Taking into account that Tgt is a homodimer able to bind and convert only one substrate tRNA molecule at a time [17].

marginally reduced by a factor of two it was lowered by one order of magnitude for Tgt(Val233Gly). In Tgt(Cys158Val/Val233Gly) the combination of both mutations led to a decrease in k_{cat} by two orders of magnitude.

Subsequently, kinetic parameters were determined for $preQ_1$ by measuring the incorporation of [^{3}*H*]-$preQ_1$ into $tRNA^{Tyr}$. For Tgt(Cys158Val) also $k_{cat}(preQ_1)$ was only slightly reduced by a factor of three compared to "wild type" Tgt. Remarkably, however, this mutation sparked an increase of $K_M(preQ_1)$ by one order of magnitude, a loss of affinity which had not been observed for guanine. In contrast, no decrease in $preQ_1$ affinity was detected for Tgt(Val233Gly) while, compared to "wild type" Tgt, $k_{cat}(preQ_1)$ had decreased by one order of magnitude. As observed during the use of guanine as a substrate, the combination of both mutations in Tgt(Cys158Val/Val233Gly) led to a decrease in $k_{cat}(preQ_1)$ by two orders of magnitude. Furthermore, $K_M(preQ_1)$ of Tgt(Cys158Val/Val233Gly) had increased by more than a factor of ten, confirming that the Cys158Val exchange results in a measurable loss of affinity to this substrate base.

Ultimately, we intended to determine kinetic parameters for the three mutated Tgt variants as well as for "wild type" Tgt with respect to queuine. Seemingly, low activity was observed with "wild type" Tgt, Tgt(Cys158Val) and Tgt(Val233Gly) by measuring the incorporation of [^{3}*H*]-queuine into $tRNA^{Tyr}$. Thorough analysis of the measured progress curves, however, revealed that the observed activities were obviously not due to the insertion of queuine into substrate tRNA. At queuine concentrations equal to or exceeding the concentration of $tRNA^{Tyr}$ (15 µmol·L^{-1}), the initially observed insertion of the radiolabelled base into tRNA already came to a halt when <10% of the tRNA was labelled (data not shown). This observation may be interpreted such that the used [^{3}*H*]-queuine preparation contained a [^{3}*H*]-labelled contaminant which was inserted with a measurable turnover (this contaminant might possibly be 7-deaza-7-hydroxymethylguanine, resulting from the reduction of the corresponding aldehyde which was used as starting material for queuine synthesis [26]). After the majority of this more potent impurity was incorporated into $tRNA^{Tyr}$, queuine was barely or not inserted at all. Accordingly, this unexpected restriction did not allow us to figure out whether "wild type" Tgt, Tgt(Cys158Val) and Tgt(Val233Gly) were indeed able to accept queuine as a substrate nor did it allow us to determine Michaelis-Menten parameters as an indicator for affinity and catalytic activity. Therefore, we applied an alternative assay tracing the decrease in radioactivity of $tRNA^{Tyr}$ labelled at position 34 with [8-^{3}*H*]-guanine due to the incorporation of unlabelled queuine. Compared to the commercially purchased [^{3}*H*]-queuine, the "cold" queuine used for this assay was produced *via* a completely different synthetic route with $preQ_1$ as a starting material. NMR analysis of this queuine preparation had revealed no detectable $preQ_1$ contamination implying that, if present, its portion amounted to <1‰ [38]. In fact, when queuine was used from this highly pure preparation at a concentration of 50 µmol·L^{-1} no insertion of queuine into $tRNA^{Tyr}$ was observed for "wild type" Tgt or any of the mutated variants during a time period of four hours. In contrast, when, as a positive control, 10 µmol·L^{-1} ("cold") $preQ_1$ was used in this assay, insertion of this base into 90% of the tRNA (in this assay present at a concentration of 8 µmol·L^{-1}) was achieved by "wild type" Tgt within <10 minutes (data not shown).

Affinity of $tRNA^{Tyr}$ to Tgt as determined by microscale thermophoresis

Since none of the Tgt variants investigated in this study showed turnover when queuine was used as a substrate it was not possible to determine K_M(queuine) as a measure of affinity. Therefore, in order to gain information about the ability of these variants to bind queuine we intended to use "microscale thermophoresis" to quantify dissociation constants (K_D) of the enzyme·queuine complexes. This method is based on a microscale temperature gradient which is produced in a solution within a capillary by an infrared LED laser. Proteins (as well as any other molecules) show a directional motion along the temperature gradient determined by their size, charge and hydration shell which in turn are influenced by the presence or absence of a bound ligand. Any change in the above-named parameters caused by ligand binding affects the thermophoretic movement of the investigated protein, allowing the determination of binding affinities [39]. To assess the applicability of this method to *Z. mobilis* Tgt we intended to determine K_D values for guanine, $preQ_1$, queuine and $tRNA^{Tyr}$ complexes of Tgt. However, no change in thermophoresis was observed upon addition of guanine, $preQ_1$ and queuine proving this method to be unsuitable to investigate the affinities of these bases to Tgt and its mutated variants.

In contrast, a strong influence on the thermophoretic behaviour of Tgt was noticed upon binding of the macromolecular substrate $tRNA^{Tyr}$ enabling the determination of $K_D(Tgt \cdot tRNA^{Tyr})$ which amounted to 36.2 nmol·L^{-1}±5.7 nmol·L^{-1} (Figure 5). This value is considerably lower than $K_M(tRNA^{Tyr})$ (Table 1) indicating an affinity of Tgt to this substrate which is clearly higher than so far assessed by enzyme kinetics studies. The reason for this discrepancy may be that the binding of guanine to apo-Tgt prevents the binding of $tRNA^{Tyr}$ to the enzyme. Thus, guanine not only acts as a substrate but also as a competitive inhibitor with respect to tRNA. As guanine is present in a saturating concentration in any assay that aims at determining $K_M(tRNA^{Tyr})$ of Tgt, the recorded value may not constitute a reliable measure but will rather lead to a significant underestimation of the enzyme's affinity to $tRNA^{Tyr}$.

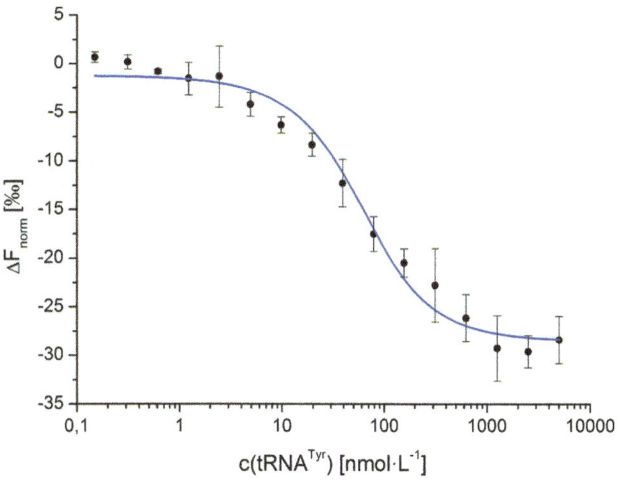

Figure 5. Binding of tRNA^{Tyr} to Tgt measured by Microscale Thermophoresis. Alexa Fluor®647 labelled Tgt at a concentration of 50 nmol·L^{-1} was incubated with varying amounts of tRNA^{Tyr}. The difference in normalised fluorescence as a measure of change in thermophoresis is plotted against the concentration of tRNA^{Tyr}.

Inhibition of Tgt and its mutated variants by queuine

Since queuine is not used as a substrate by "wild type" Tgt and our mutated variants, any specific binding of this base to the active centres of these enzymes should inevitably cause their inhibition. To verify an inhibitory effect of queuine on Tgt activity, we measured the activity of "wild type" Tgt, Tgt(Cys158Val), Tgt(Val233Gly) and Tgt(Cys158Val/Val233Gly) using guanine and tRNA^{Tyr} as substrates both in the presence and absence of a low µmol·L^{-1} concentration of queuine. Indeed, a considerable decrease in acitivity dependent on queuine was observed for all investigated Tgt variants (see below).

As inhibitors mimicking the nucleobases guanine or preQ$_1$ are basically able to inhibit Tgt in two different ways, we next determined the inhibition mode *via* that queuine acts on the investigated enzyme variants. If an inhibitor binds to the guanine 34/preQ$_1$ 34 subpocket of the apo-enzyme, it will inhibit the tRNA from accessing the enzyme and thus it will act as a competitive inhibitor with respect to the tRNA substrate. If, however, such an inhibitor occupies this subpocket after tRNA binding and the removal of guanine 34 it will prevent the substrate base preQ$_1$ from binding. Accordingly, any further reaction of the covalent Tgt·tRNA intermediate is disabled which corresponds to an uncompetitive inhibition with respect to the tRNA substrate. Nucleobase mimetics of comparable size as guanine or preQ$_1$ normally act *via* both inhibition modes. Nucleobase mimetics endowed with bulky exocyclic substituents, however, generally constitute purely competitive inhibitors with respect to tRNA as their binding to the covalent Tgt·tRNA complex would sterically interfere with the tRNA sugar phosphate backbone.

To figure out the inhibition mode *via* that queuine acts on the investigated Tgt variants we applied a trapping experiment which had been used for a similar purpose by Meyer *et al.* [23]. Incubation of Tgt with substrate tRNA in the presence of an uncompetitive inhibitor in excess leads to a considerable stabilisation of the covalent Tgt·tRNA intermediate which can be visualised in terms of two retarded bands on a Coomassie-stained gel after SDS-PAGE of the reaction mixture (the presence of two retarded bands representing this complex results from different tRNA conformers present under the applied PAGE conditions [40]). In the absence of any inhibitor, only a very small amount of the covalent intermediate is detectable under the used assay conditions. The presence of excess inhibitor which solely and efficiently acts as a competitive inhibitor with respect to substrate tRNA entirely suppresses the formation of the covalent intermediate.

Thus, we incubated "wild type" Tgt, Tgt(Cys158Val), Tgt(Val233Gly) and Tgt(Cys158Val/Val233Gly) with tRNA^{Tyr} in the presence of excess queuine. As a control, we incubated each of these Tgt variants solely with tRNA^{Tyr} as well as with tRNA^{Tyr} plus an excess of the preQ$_1$ mimetic 2,6-diamino-3*H*-quinazolin-4-one (DAQ) (Figure 6A) which constitutes a potent uncompetitive inhibitor of Tgt (K_{iu} = 0.6 µmol·L^{-1}) [23]. Subsequently, we subjected the reaction mixtures to SDS-PAGE. In the case of "wild type" Tgt and Tgt(Cys158Val) no retarded band representing the covalent Tgt·tRNA intermediate was detectable on the Coomassie-stained gel, when, in addition to the enzyme, only tRNA^{Tyr} was present in the reaction mixture (Figure 6B, lanes 2 and 8). This indicates that any excised guanine was quickly reinserted into tRNA keeping the amount of the covalent complex below the detection limit of the assay. In contrast, under the same conditions a faint band representing the covalent enzyme·tRNA complex was observed for Tgt(Val233Gly) and Tgt(Cys158Val/Val233Gly) indicating that the reinsertion rate of the excised guanine was significantly decreased with these Tgt variants (Figure 6B, lanes 5 and 11). When, in addition to tRNA^{Tyr} excess DAQ was added, in the case of all Tgt variants the majority of the enzyme was trapped in a covalent complex formed with tRNA^{Tyr}. On the Coomassie-stained gel this complex was visible as two clear retarded bands concomitant with considerable weakening of the band corresponding to the uncomplexed enzyme (Figure 6B, lanes 4, 7, 10 and 13). When queuine, instead of DAQ, was added to the reaction mixture, the formation of the covalent intermediate was completely suppressed in the case of "wild type" Tgt, Tgt(Cys158-Val) and Tgt(Val233Gly) (Figure 6B, lanes 3, 6 and 9). This suggests that for these Tgt variants, queuine predominantly acts as a competitive inhibitor with respect to substrate tRNA. Solely in the case of Tgt(Cys158Val/Val233Gly) a very faint band representing the covalent enzyme·tRNA complex was visible in the presence of excess queuine (Figure 6B, lane 12). This may indicate that, in the case of this mutated variant, queuine is to some extent able to bind to the guanine 34/preQ$_1$ subpocket of the covalent enzyme·tRNA complex resulting in an inhibition uncompetitive with tRNA binding. The fact, however, that the band representing the covalent enzyme·tRNA complex is clearly weaker than that observed in the absence of queuine suggests that queuine will also mainly bind to this mutated variant when present in the apo-form.

To gain information concerning the affinity of queuine to "wild type" Tgt and the mutated variants we determined, according to the method described in [23], the corresponding competitive inhibition constants, with respect to tRNA (K_{ic}), of queuine. K_{ic}(queuine) amounted to 3.8 µmol·L^{-1}±0.6 µmol·L^{-1} for "wild type" Tgt, 3.8 µmol·L^{-1}±0.5 µmol·L^{-1} for Tgt(Cys158Val), 3.1 µmol·L^{-1}±0.2 µmol·L^{-1} for Tgt(Val233Gly) and 8.9 µmol·L^{-1}±1.2 µmol·L^{-1} for Tgt(Cys158Val/Val233Gly). These results show that all studied Tgt variants including "wild type" Tgt are able to bind queuine with nearly identical affinity.

Crystal structure analyses of "wild type" Tgt and its mutated variants

To allow the interpretation of the obtained data at a structural level, we determined the crystal structures of the mutated Tgt variants created in this study in their apo-forms as well as in

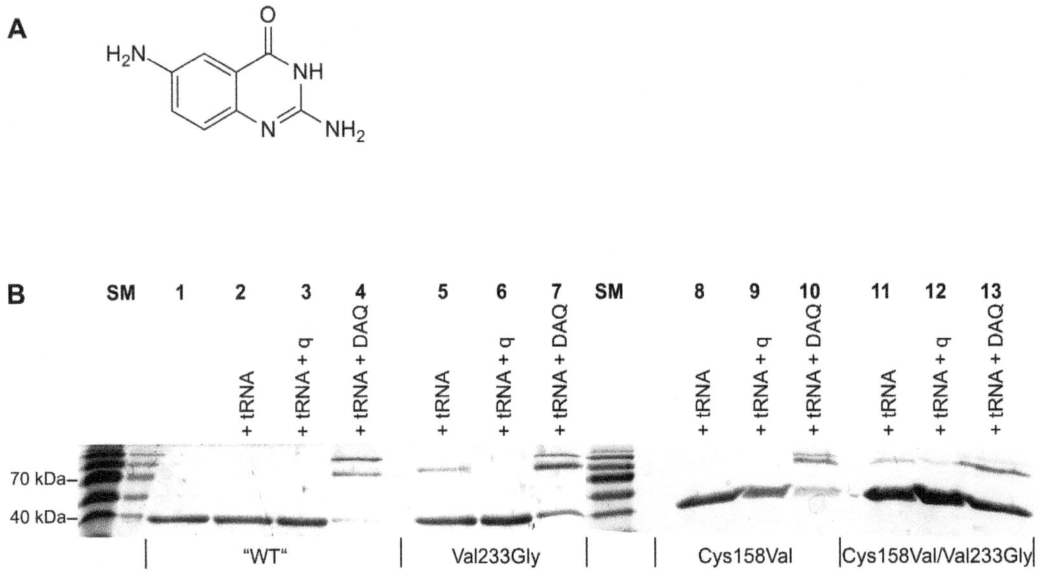

Figure 6. Trapping experiments performed with Tgt/tRNA mixtures in presence of queuine or 2,6-diamino-3*H*-quinazolin-4-one. A) Chemical structure of 2,6-diamino-3*H*-quinazolin-4-one (DAQ), an uncompetitive inhibitor of Tgt. B) SDS-PAGE analysis of reaction mixtures of Tgt or mutated variants thereof and $tRNA^{Tyr}$ under conditions indicated. SM, size marker; q, queuine. While DAQ causes retarded Tgt bands by stabilising the covalent Tgt·tRNA intermediate, queuine lacks this ability for most of the investigated Tgt variants. Solely in case of Tgt(Cys159Val/Val233Gly) a faint retarded band is visible indicating that queuine may to some extent be able to bind to the guanine 34/$preQ_1$ subpocket of the covalent enzyme·tRNA complex.

complex with $preQ_1$ and queuine. In addition, we determined the crystal structure of "wild type" Tgt with a bound queuine molecule. Structures of "wild type" Tgt in its apo-form and in complex with $preQ_1$ had already been published in a former study [37] and thus, were available in the protein data bank (PDB codes: **1ozm** and **1ozq**, respectively). An overview of the crystal structures considered in the present study is given in Table S1, while detailed data statistics are presented in Tables 2, 3 and 4.

The $2|F_o|-|F_c|$ electron density maps of all apo-structures clearly allow the identification of the introduced mutations, although a rather poorly defined density is present for Gly233 in both the structures of Tgt(Val233Gly) and Tgt(Cys158Val/Val233Gly). In the electron density maps of Tgt(Cys158Val) and of Tgt(Cys158Val/Val233Gly) the isopropyl side chain of Val158 is excellently defined. This is contrary to the sulfanylmethyl group of the original Cys158 whose side chain is, obviously due to its flexibility, poorly defined in the electron density maps of virtually all *Z. mobilis* Tgt crystal structures determined so far. Superimposition of the apo-structures of the mutated variants and "wild type" Tgt revealed that none of the amino acid exchanges implicates any significant structural change concerning the remaining part of the protein (mean rmsd 0.23 Å±0.02 Å).

Next, we analysed and compared the structures of the studied Tgt variants in complex with $preQ_1$. In all $preQ_1$-bound structures the ligand appears well-defined in the respective electron density map. Superimposition of the structural models reveals that $preQ_1$ is identically positioned in all investigated variants including "wild type" Tgt (mean rmsd = 0.20 Å±0.07 Å). Furthermore, no differences were observed with respect to the catalytic Asp280 and the conformations of amino acids known to form direct interactions with the bound $preQ_1$ ligand. The most conspicuous differences among the $preQ_1$-bound structures concern a small cavity separating the side chains of amino acid residues 158 and 233 (Figure 7). In "wild type" Tgt this cavity harbors two well-ordered water molecules (Figure 7A) which are displaced in the Tgt(Cys158Val)·$preQ_1$ complex structure by the isopropyl side chain of Val158 (Figure 7C). In the structure of $preQ_1$-bound Tgt(Val233Gly) this cavity is substantially widened but harbors only one well-ordered water molecule. In addition, the electron density assigned to the $preQ_1$ ligand does obviously not reflect the excellent resolution of this structure (1.19 Å) (Figures 7E and 8E). This is also indicated by the mean *B*-factor of $preQ_1$ which is, in this structure, considerably elevated compared to the mean *B*-factor of the protein (33.0 Å^2 *versus* 18.6 Å^2). In contrast, in the structures of all other $preQ_1$-bound Tgt variants the mean *B*-factor of $preQ_1$ deviates only slightly from the mean *B*-factor of the protein (see Tables 2 and 4). The structure of $preQ_1$-bound Tgt(Cys158Val/Val233Gly) reveals that, compared to Tgt(Val233Gly), the replacement of the hydrophilic and flexible sulfanylmethyl moiety of Cys158 by the hydrophobic isopropyl of valine leads to a regeneration of the original order in the substrate base binding pocket. The mean *B*-factor of the bound $preQ_1$, which is excellently defined in the electron density map, hardly differs from the mean *B*-factor of the protein (15.7 Å^2 *versus* 14.2 Å^2). Furthermore, an arrangement of three highly ordered water molecules is present within the spacious cavity between Val158 and Gly233 (Figure 7G).

A conspicuous structural change upon binding of $preQ_1$ is observed within the binding pockets of those Tgt variants which harbour the Cys158Val mutation. Both in Tgt(Cys158Val) and Tgt(Cys158Val/Val233Gly) the binding of $preQ_1$ provokes a shift of Val158 towards the ligand enabling five Van-der-Waals interactions of its isopropyl side chain with the 7-deazaguanine scaffold of $preQ_1$ (3.5 Å to 4.3 Å calculated by CONTACSYM [41]). In addition, a noticeable change of the proximate Thr159 whose side chain is rotated by about 90° is observed (Figure 9A). So far, in no other crystal structure of *Z. mobilis* Tgt or any variant thereof such a conformational change of Thr159 has been observed upon ligand binding. To obtain indication of whether the reduced affinity of Tgt(Cys158Val) and Tgt(Cys158Val/

Table 2. Crystallographic data collection and refinement statistics - Tgt(Cys158Val).

Crystal data*	Tgt(Cys158Val)	Tgt(Cys158Val)·guanine	Tgt(Cys158Val)·preQ$_1$	Tgt(Cys158Val)·queuine
PDB ID	**4gd0**	**4h7z**	**4e2v**	**4hvx**
A. Data Collection and Processing				
Beamline	BL 14.2	BL 14.1	BL 14.2	BL 14.2
λ (Å)	0.91841	0.91841	0.91841	0.91841
Space group	C2	C2	C2	C2
a, *b*, *c* (Å)	91.4, 65.0, 70.2	90.4, 64,8, 70,8	90.4, 64.9, 70.4	91.6, 64.5, 71.1
β (°)	96.1	95.6	95.8	96.2
Matthews coefficient (Å^3/Da)	2.4	2.4	2.4	2.4
Solvent content (%)	48.7	48.7	48.7	48.7
B. Diffraction Data				
Resolution range^a (Å)	50-1.29 (1.31-1.29)	30-1.68 (1.77-1.68)	21-1.18 (1.2-1.18)	50-1.82 (1.85-1.82)
No. of unique reflections	102,682 (4187)	45,941 (6706)	132,461 (5750)	35405 (1467)
Completeness^a (%)	97.1 (81.5)	99.2 (99.8)	95.0 (88.0)	94.9 (78.4)
Redundancy	2.7 (1.7)	2.9 (2.8)	2.5 (2.2)	3.0 (2.2)
$R(I)_{sym}{}^{a,b}$ (%)	4.7 (23.3)	7.0 (48.2)	4.2 (21.8)	4.5(36.3)
$I/\sigma(I)^a$	22.1 (3.0)	10.2 (2.1)	21.4 (4.1)	21.7 (2.2)
C. Refinement				
Program	Phenix	Phenix	Phenix	Phenix
$R_{work}{}^c/R_{free}{}^d$ (%)	16.2/18.5	15.7/19.1	13.7/15.2	16.6/19.7
Protein residues	373	371	364	358
Water molecules	524	318	464	165
Ligand atoms	—	11	13	13
Ramachandran plot				
Residues in most favored regions (%)	94.4	95.5	95.3	95.4
Residues in additionally allowed regions (%)	5.0	4.2	4.4	4.3
Residues in generously allowed regions (%)	0.6	0.3	0.3	0.3
Mean *B*-factors (Å^2)				
Protein	12.6	15.6	13.7	28.6
Water	29.1	29.4	29.2	35.8
Ligand	—	29.8	11.6	34.1
Rmsd from ideality				
rmsd angle (°)	1.1	1.8	1.2	1.2
rmsd bond (Å)	0.005	0.019	0.007	0.010

*All Tgt variants considered in this table contain an additional Tyr106Phe mutation which in a previous study was shown to have no impact on any of the catalytic properties of the enzyme [37] (see main text).

[a]) number in parentheses is for highest resolution shell.

[b]) $R_{sym} = \frac{\sum |I - \bar{I}|}{\sum I}$, with *I* representing the observed intensity and $\bar{I}$ representing the average intensities for multiple measurements.

[c]) $R_{work} = \frac{\sum_{hkl} |F_{obs} - F_{calc}|}{\sum_{hkl} |F_{obs}|}$.

[d]) R_{free} was calculated as R_{work} but on 5% of the data excluded from the refinement.

Val233Gly) towards preQ$_1$ arises from the described structural changes, we determined the crystal structure of Tgt(Cys158Val) in complex with guanine. If the observed conformational rearrangements actually account for the decrease in affinity to preQ$_1$ we expected them to not occur upon guanine binding since the Cys158Val exchange has no measurable impact on guanine recognition. The crystal structure of the Tgt(Cys158Val)·guanine complex reveals that upon guanine binding the side chain of Val158 becomes largely disordered as it is ill-defined in the electron density map. The position of the main chain atoms of Val158 indicates that there is hardly any shift of this residue towards the ligand. Strikingly, no change of the Thr159 side chain conformation which closely resembles that of the apo-structure is observed (Figure 9B). We would like to note that binding of guanine in Tgt(Cys158Val), as in wild type Tgt, leads to a Leu231/Ala232 peptide bond conformation which is flipped compared to the preQ$_1$-bound situation [22]. In the preQ$_1$-bound enzyme the carbonyl group of this peptide bond is exposed to the binding pocket while the backward-oriented amide forms an *H* bond to the deprotonated side chain carboxyl of Glu235. In

Table 3. Crystallographic data collection and refinement statistics – Tgt(Val233Gly).

Crystal data*	Tgt(Val233Gly)	Tgt(Val233Gly)·preQ$_1$	Tgt(Val233Gly)·queuine
PDB ID	3bl3	3bld	4hsh
A. Data Collection and Processing			
Beamline	In-house	BL 14.2	BL 14.2
λ (Å)	1.54178	0.91841	0.91841
Space group	C2	C2	C2
a, *b*, *c* (Å)	90.7, 64.8, 70.4	89.7, 64.9, 70.3	90.5, 64.8, 70.5
β (°)	95.9	95.6	95.9
Matthews coefficient (Å^3/Da)	2.4	2.4	2.4
Solvent content (%)	48.7	48.7	48.7
B. Diffraction Data			
Resolution range^a (Å)	50-2.25 (2.30-2.25)	50-1.19 (1.22 -1.19)	50-1.56 (1.59-1.56)
No. of unique reflections	49,829	116,963	57,783
Completeness^a (%)	93.4 (66.4)	91.6 (62.8)	100.0 (100.0)
Redundancy	2.8	3.7	4.2 (4.1)
$R(I)_{sym}{}^{a,b}$ (%)	11.0 (28.1)	4.1 (26.2)	8.8 (49.6)
$I/\sigma(I)^a$	9.0 (2.4)	21.0 (2.7)	15.4 (2.7)
C. Refinement			
Programm	ShelxL-97	ShelxL-97	Phenix
$R_{work}{}^c/R_{free}{}^d$ (%)	18.7/22.7	15.1/18.5	15.5/18.1
Protein residues	365	348	365
Water molecules	216	262	318
Ligand atoms	-	13/1	12/1
Ramachandran plot			
Residues in most favored regions (%)	94.9	95.6	95.5
Residues in additionally allowed regions (%)	4.8	4.1	4.2
Residues in generously allowed regions (%)	0.3	0.3	0.3
Mean *B*-factors (Å^2)			
Protein	19.9	18.6	16.8
Water	28.4	28.4	31.9
Ligand	—	33.0	36.0
RMSD from ideality			
rmsd angle (°)	1.3	2.1	1.6
rmsd bond (Å)	0.008	0.013	0.017

*All Tgt variants considered in this table contain an additional Tyr106Phe mutation which in a previous study was shown to have no impact on any of the catalytic properties of the enzyme [37] (see main text).

[a]) number in parentheses is for highest resolution shell.

[b]) $R_{sym} = \frac{\sum |I - \bar{I}|}{\sum I}$, with *I* representing the observed intensity and $\bar{I}$ representing the average intensities for multiple measurements.

[c]) $R_{work} = \frac{\sum_{hkl} |F_{obs} - F_{calc}|}{\sum_{hkl} |F_{obs}|}$.

[d]) R_{free} was calculated as R_{work} but on 5% of the data excluded from the refinement.

contrast, upon guanine binding the *NH*-group of this peptide bond is exposed, while the backward-oriented carbonyl oxygen interacts with the protonated Glu235 side chain carboxyl group (see also Figure 3).

Ultimately, we determined the crystal structures of the studied Tgt variants in complex with queuine. Although electron density which can be attributed to a bound ligand is clearly present in the binding pocket of the refined structure of "wild type" Tgt crystallised in the presence of queuine it turned out to be too ill-defined to allow a definite placement of queuine in the model (Figures 7B and 8B). In contrast, the electron density map of the Tgt(Cys158Val)·queuine complex structure allowed the placement of the 7-deazaguanine scaffold and the exocyclic aminomethyl group of queuine. Clearly, this part of the ligand shows, as expected, a binding mode which strongly resembles that of preQ$_1$ (Figures 7D and 8D). Only rudimentary density, however, is visible for the dihydroxy-cyclopentenyl moiety of queuine probably due to its enhanced residual mobility. Yet, the position of the aminomethyl group implicates the dihydroxy-cyclopentenyl pointing towards the uracil 33 nucleotide subpocket of the enzyme. This

Table 4. Crystallographic data collection and refinement statistics – Tgt(Cys158Val/Val233Gly), "WT".

Crystal data*	Tgt(Cys158Val/ Val233Gly)	Tgt(Cys158Val/ Val233Gly)·preQ$_1$	Tgt(Cys158Val/ Val233Gly)·queuine	"WT"-Tgt·queuine
PDB ID	**4h6e**	**4gcx**	**4hqv**	**
A. Data Collection and Processing				
Beamline	BL 14.2	BL 14.2	BL 14.2	BL 14.2
λ (Å)	0.91841	0.91841	0.91841	0.91841
Space group	C2	C2	C2	C2
a, b, c (Å)	91.0, 64.8, 70.5	91.2, 65.1, 70.6	90.9, 65.0, 70.6	90.8, 64.7, 70.6
β (°)	96.1	96.1	96.3	95.9
Matthews coefficient (Å^3/Da)	2.4	2.4	2.4	2.4
Solvent content (%)	48.7	48.7	48.7	48.7
B Diffraction Data				
Resolution range^a (Å)	30-1.42 (1.44-1.42)	50-1.42 (1.44-1.42)	50-1.66 (1.69-1.66)	50-1.22(1.24-1.22)
No. of unique reflections	76,896 (3,342)	77,486 (3,750)	48,267 (2,279)	117,888 (4,937)
Completeness^a (%)	96.7 (87.2)	99.0 (96.2)	97.3 (95.5)	97.3 (81.7)
Redundancy	2.8 (2.1)	3.1 (2.6)	3.0 (2.9)	4.0 (3.3)
$R(I)_{sym}{}^{a,b}$ (%)	4.3 (15.5)	6.7 (41.6)	9.7 (47.3)	4.6 (41,7)
$I/\sigma(I)^a$	22.9 (4.6)	16.5 (2.2)	10.7 (2.5)	26.1 (2.6)
C. Refinement				
Programm	Phenix	Phenix	Phenix	Phenix
$R_{work}{}^c/R_{free}{}^d$ (%)	14.9/17.4	13.1/16.1	17.5/20.7	13.6/15.7
Protein residues	272	369	273	371
Water molecules	421	397	397	334
Ligand atoms	—	13	13	—
Ramachandran plot				
Residues in most favored regions (%)	95.5	94.9	94.7	95.9
Residues in additionally allowed regions (%)	4.7	4.8	4.7	3.8
Residues in generously allowed regions (%)	0.3	0.3	0.6	0.3
Mean *B*-factors (Å^2)				
Protein	8.3	14.2	17.7	16.4
Water	22.8	32.5	29.7	30.0
Ligand	—	15.7	30.0	—
RMSD from ideality				
rmsd angle (°)	1.4	1.1	1.1	1.6
rmsd bond (Å)	0.010	0.006	0.007	0.014

*All Tgt variants considered in this table including "wild type" Tgt contain an additional Tyr106Phe mutation which in a previous study was shown to have no impact on any of the catalytic properties of the enzyme [37] (see main text).

**As the refined model contains no ligand the coordinates and structure factors were not deposited with the Protein Data Base but are provided within the Supporting Information (Coordinates S2; Structure factors S1).

^{a}) number in parentheses is for highest resolution shell.

^{b}) $R_{sym} = \frac{\sum |I-\bar{I}|}{\sum I}$, with *I* representing the observed intensity and $\bar{I}$ representing the average intensities for multiple measurements.

^{c}) $R_{work} = \frac{\sum_{hkl} |F_{obs} - F_{calc}|}{\sum_{hkl} |F_{obs}|}$.

^{d}) R_{free} was calculated as R_{work} but on 5% of the data excluded from the refinement.

leads, compared to the preQ$_1$-bound situation, to the displacement of a water molecule which forms an *H* bond to the exocyclic amino group of preQ$_1$. This water is strictly conserved in the preQ$_1$ complex structures of all investigated Tgt variants and designated "W" in Figures 8A, 8C, 8E and 8G. Superimposition of the queuine-bound Tgt(Cys158Val) structure with the structure of Tgt in complex with RNA [21] reveals that in the present conformation the dihydroxy-cyclopentenyl moiety of queuine sterically interferes with the ribose of the uracil 33 nucleotide and the phosphate at position 34. Hence, the binding of queuine in the observed conformation is obviously not compatible with the binding of a tRNA substrate. Notably, as observed for guanine, the presence of queuine in the substrate pocket of Tgt(Cys158Val) apparently completely disorders the Val158 side chain as there is no electron density assignable to this isopropyl moiety (Figure 8D). In contrast, this side chain is excellently defined in the electron

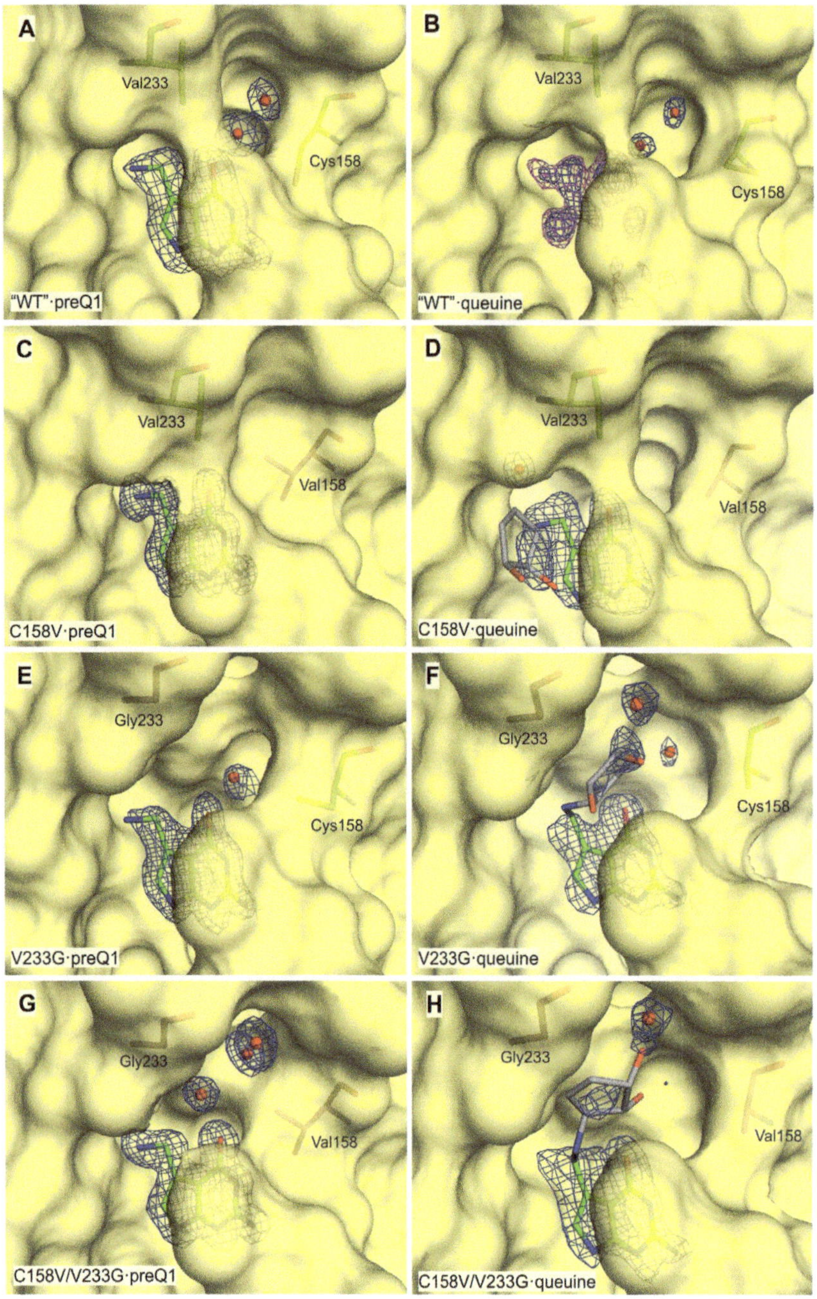

Figure 7. Crystal structures of Tgt variants in complex with preQ$_1$ or queuine: surface representation of substrate pockets. The solvent accessible surfaces of active sites from *Z. mobilis* Tgt variants are shown in bright yellow. The respective Tgt variant plus the bound ligand (shown in stick representation) are indicated in each sub-figure. Also amino acid residues at positions 158 and 233 are shown in stick representation. Carbon atoms of original amino acids and of the bound ligand are coloured green, those of mutated amino acids magenta. As the electron density assignable to the dihydroxy-cyclopentenyl moiety of queuine is poorly defined in all structures containing this ligand the coordinates of this moiety are not present in the respective structures deposited with the Protein Data Base. Accordingly, the conformations of the dihydroxy-cyclopentenyl shown in (D), (F) and (H) are tentative. To indicate this fact, the carbon atoms of this moiety are shown in grey. Selected water molecules are shown as red spheres. $2|F_o|-|F_c|$ (at σ 1.0) electron density is shown for the bound ligand and water molecules. $2|F_o|-|F_c|$ electron density contoured at a σ level of 1.0 is coloured blue, $|F_o|-|F_c|$ electron density contoured at a σ level of 2.5 is coloured magenta. An overview of the crystal structures analysed in this study including resolutions and PDB codes is given in Table S1.

density maps of apo- and preQ$_1$-bound Tgt(Cys158Val) and Tgt(Cys158Val/Val233Gly). The positioning of the Val158 main chain atoms, however, suggests a significant shift of this residue towards the bound ligand although less pronounced as observed in the preQ$_1$ complex structure of Tgt(Cys158Val). In accordance with the structure of the guanine-bound Tgt(Cys158Val), this shift is not concomitant with a rotation of the Thr159 side chain (Figure 9C).

Also the electron density maps of the Tgt(Val233Gly)·queuine and Tgt(Cys158Val/Val233Gly)·queuine complex structures enabled reliable placement of the 7-deazaguanine scaffold of the ligand which is, in both cases, similarly bound as in the

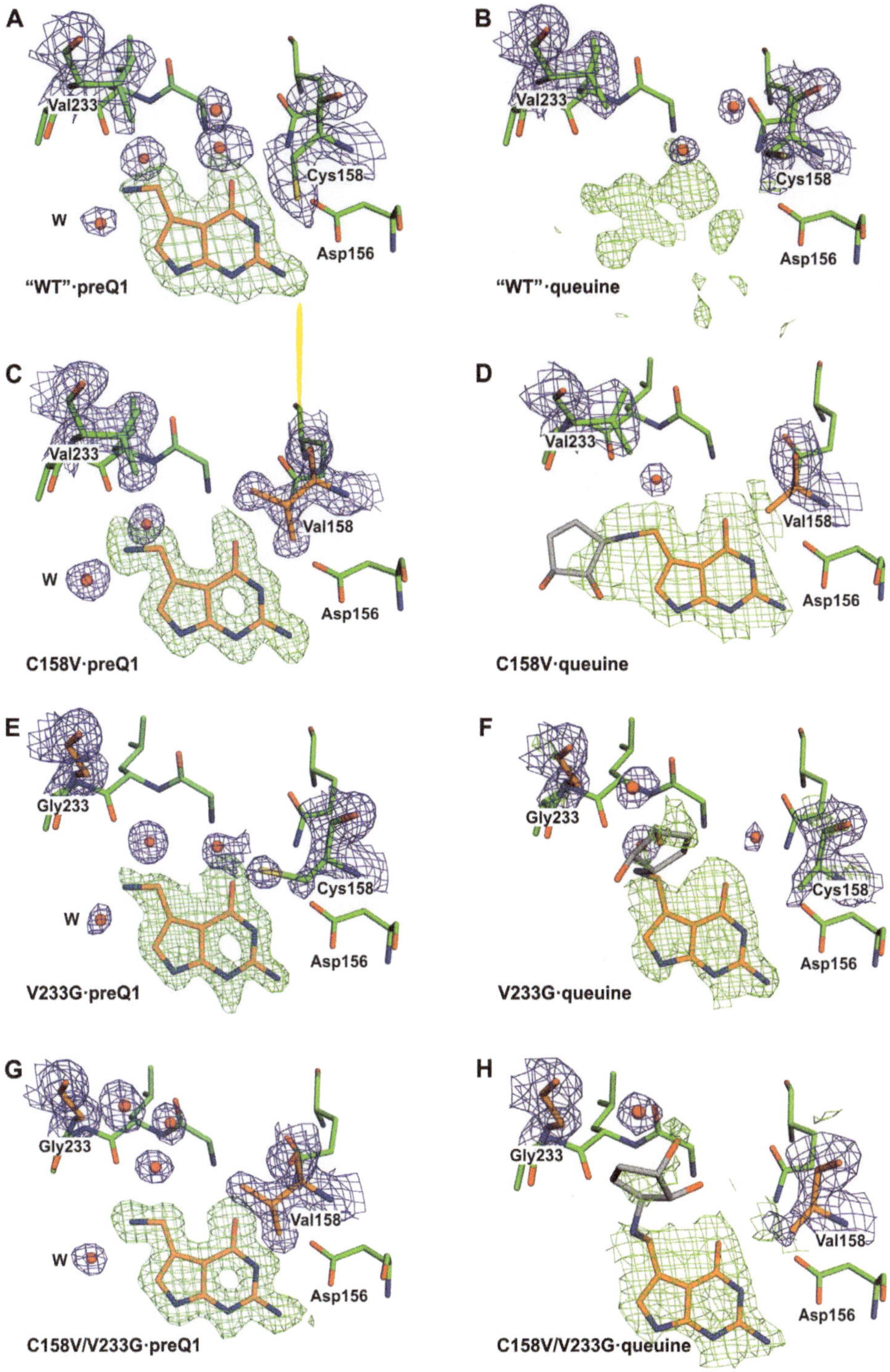

Figure 8. Crystal structures of Tgt variants in complex with preQ$_1$ or queuine: stick representation of substrate pockets. The respective Tgt variant plus the bound ligand are indicated in each sub-figure. Carbon atoms of original amino acids are coloured green, those of mutated amino acids as well as of the bound ligand orange. As the electron density assignable to the dihydroxy-cyclopentenyl moiety of queuine is poorly defined in all structures containing this ligand the coordinates of this moiety are not present in the respective structures deposited with the Protein Data Base. Accordingly, the conformations of the dihydroxy-cyclopentenyl shown in (D), (F) and (H) are tentative. To indicate this fact, the carbon atoms of this moiety are shown in grey. Selected water molecules are shown as red spheres. Electron density is shown for amino acid residues at positions 158 and 233 as well as for the ligand and for water molecules. The $2|F_o|-|F_c|$ electron density map contoured at a σ level of 1.0 is coloured blue. The green density represents an $|F_o|-|F_c|$ omit map contoured at 2.5 σ. An overview of the crystal structures analysed in this study including nominal resolutions and PDB codes is given in Table S1.

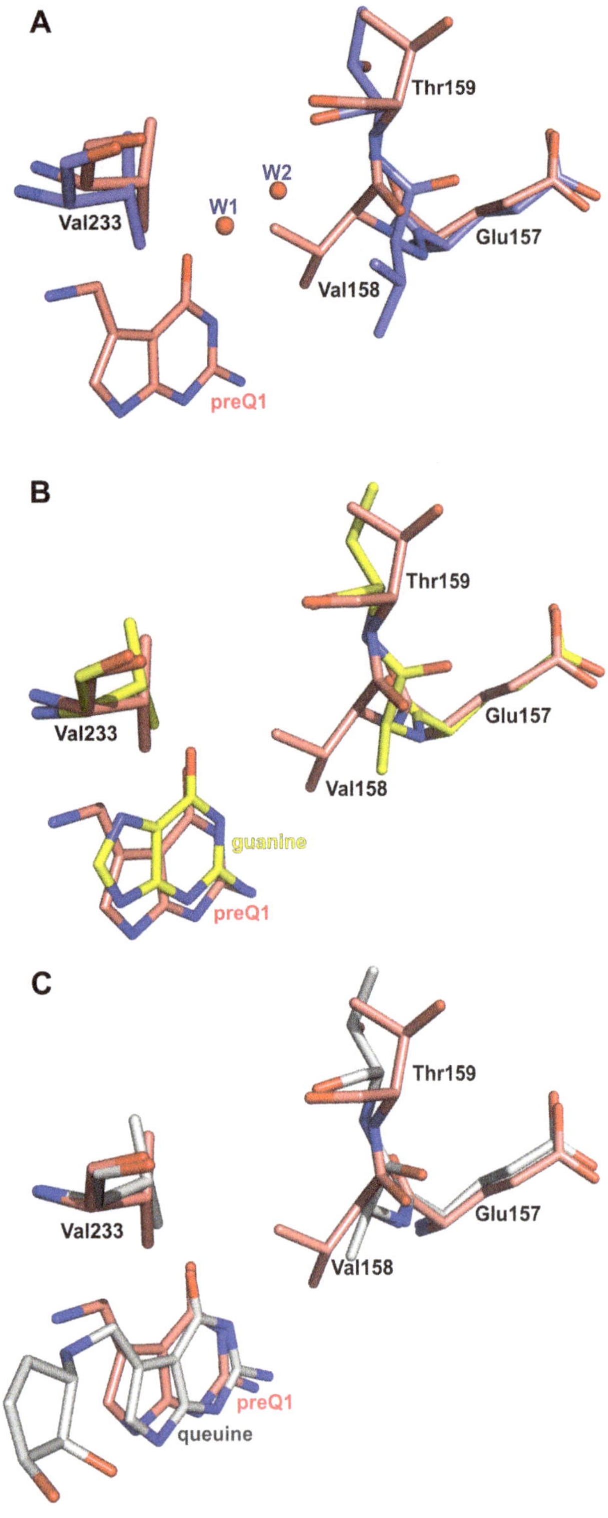

Figure 9. Superimposition of preQ$_1$-bound Tgt(Cys158Val) with its apo-, its guanine- and its queuine-bound form. Carbon atoms of the Tgt(Cys158Val)·preQ$_1$ complex are coloured pink in all sub-figures. (A) Superimposition (based on c$_\alpha$) of preQ$_1$-bound Tgt(Cys158Val) and apo-Tgt(Cys158Val). Carbon atoms of apo-Tgt(Cys158Val) are coloured blue. Binding of preQ$_1$ to Tgt(Cys158Val) provokes a shift of Val158 towards the ligand leading to the displacement of two water molecules (W1 and W2; shown as red spheres) which are present within this region in apo-Tgt(Cys158Val) and absent in the complex structure. In addition, the side chain of the proximate Thr159 rotates by about 90°. Exactly the same structural changes upon binding of preQ$_1$ are observed for Tgt(Cys158Val/Val233Gly). (B) Superimposition (based on c$_\alpha$) of preQ$_1$-bound Tgt(Cys158Val) and guanine-bound Tgt(Cys158Val). Carbon atoms of the Tgt(Cys158Val)·guanine complex are coloured yellow. In the Tgt(Cys158Val)·guanine complex, the side chain of Val158 becomes largely disordered. The Thr159 side chain adopts a similar conformation as observed in the apo-structure. (C) Superimposition (based on c$_\alpha$) of preQ$_1$-bound Tgt(Cys158Val) and queuine-bound Tgt(Cys158Val). Carbon atoms of the Tgt(Cys158Val)·queuine complex are shown in grey. Binding of queuine obviously leads to disordering of the Val158 side chain as no electron density attributable to this isopropyl moiety is present in the electron density map of the refined Tgt(Cys158Val)·queuine complex structure. Also upon binding of queuine no conformational change of the Thr159 side chain is observed. It adopts a similar conformation as in the apo- and guanine-bound structures.

Tgt(Cys158Val)·queuine complex. In both maps fragmentary density is visible for the dihydroxy-cyclopentenyl substituent (Figures 7F, 7H as well as 8F, 8H). Figures 7F and 7H show that the mutation of Val233 to glycine results in a considerable gain of space in the substrate binding pocket which allows the dihydroxy-cyclopentenyl of queuine to occupy the cavity located between Gly233 and the side chain of Cys/Val158. Within this cavity the dihydroxy-cyclopentenyl apparently has no specific interaction with the protein. As a consequence, it seems to be highly flexible which is reflected by the ill-defined electron density assigned to this moiety in both structures. As in the structure of queuine-bound Tgt(Cys158Val), nearly no electron density can be attributed to the side chain of Val158 in the Tgt(Cys158Val/Val233Gly)·queuine complex structure, indicating an increased mobility of this residue upon the binding of queuine. Also, in agreement with the structure of queuine-bound Tgt(Cys158Val), the positioning of the Val158 main chain atoms suggests a shift of this residue towards the ligand which is not concomitant with a rotation of the Thr159 side chain.

It must be noted that in the case of all queuine complex structures determined in this study the electron density assigned to the 7-deazaguanine scaffold of the ligand by far does not comply with the nominal resolution of the respective structure. Even in the case of the mutated variants where reliable placement of the scaffold into the binding pocket was possible the electron density within this area appears rather fuzzy or even patchy and lacks detail. It is likely that multiple conformations and the low occupancy (on average 60%) of queuine in the binding pocket provoked the poor electron density.

Discussion

While eucaryotic Tgt exchanges the genetically encoded guanine by queuine at position 34 of tRNAs^{His,Tyr,Asp,Asn}, bacterial Tgt is not able to insert the latter hypermodified base into these tRNAs. Instead, it uses the queuine-precursor preQ$_1$ as a substrate which is then further converted to the final queuine at the tRNA level. A homology model of the *C. elegans* Tgt catalytic subunit suggests that the exchange of Cys158 and Val233 in bacterial Tgt (*Z. mobilis* Tgt numbering) by valine and glycine, respectively, in eucaryotic Tgt largely accounts for the difference in substrate specificity between the Tgt enzymes of both domains of life [35]. This assumption was supported by a homology model of the human Tgt catalytic subunit created in the present study. The replacement of Val233 by glycine results in a considerable gain of space in the substrate binding pocket most probably required for the accommodation of the dihydroxy-cyclopentenyl moiety of queuine. The side chain of the valine residue corresponding to

Cys158 of *Z. mobilis* Tgt appears at a position where it might form an interaction with the dihydroxy-cyclopentenyl substituent. While Cys158 and Val233 are highly conserved in bacterial Tgt, the corresponding valine and glycine residues are invariant in all eucaryotic Tgt catalytic subunits whose sequences have been determined so far [42].

Based on the homology models of the *C. elegans* and human Tgt catalytic subunits, we replaced, *via* site-directed mutagenesis, Val233 and Cys158 in *Z. mobilis* Tgt by glycine and valine, respectively. In addition, to separately investigate the influence of these amino acid exchanges on substrate specificity and catalysis, we created two further variants, each containing solely one of the two mutations. However, the introduced changes which we desired for to convert substrate selectivity of bacterial towards eucaryotic Tgt did not result in an enzyme that is able to insert queuine into tRNA. Instead, the amino acid exchanges mainly interfere with catalysis by reducing k_{cat}(guanine) and k_{cat}(preQ$_1$). While the reduction is not of great significance for Tgt(Cys158Val) (2- to 3-fold), it accounts for one order of magnitude for Tgt(Val233Gly) and for two for Tgt(Cys158Val/Val233Gly). The crystal structures of these Tgt variants, determined in complex with preQ$_1$, clearly show that preQ$_1$ is, in each case, identically positioned as in the wild type, ruling out a substrate misorientation with respect to any catalytically relevant residue. Instead, the mutated residues may exert influence on the properties of a nearby catalytic water molecule acting as a proton donor during guanine excision (Figures 3A and 3B).

A notable consequence of the Cys158Val exchange is the significantly decreased affinity of the enzyme towards preQ$_1$ but not guanine. A similar phenomenon was reported by Chen *et al.* [26] who mutated the corresponding cysteine of *E. coli* Tgt to valine, resulting in a dramatical loss of affinity to preQ$_1$. Consistently, a measurable gain in preQ$_1$ affinity was observed by this group for the human enzyme upon mutation of the corresponding Val161 to cysteine. The study does, however, not provide any data demonstrating the impact of these mutations on guanine as a substrate. As already proposed by Chen *et al.* [26], it seems likely that, in eucaryotes, a valine at the mentioned position in the catalytic subunit of eucaryotic Tgt preferably prevents the (irreversible!) insertion of preQ$_1$ into tRNA. Our data provide evidence that this is achieved without the loss of affinity towards guanine which, most likely, would result in an attenuated affinity to the unmodified substrate tRNAs comprising guanine at position 34.

The crystal structures of Tgt(Cys158Val) and Tgt(Cys158Val/Val233Gly), each in complex with preQ$_1$, reveal, compared to the corresponding apo-forms, a shift of Val158 towards the ligand. This allows the isopropyl side chain of this residue to form Van-der-Waals contacts to the substrate. This structural change is concomitant with a ca. 90° rotation of the Thr159 side chain thus resulting in a novel conformation which has, until now, not been observed in any other crystal structure of *Z. mobilis* Tgt or a mutated variant thereof. The crystal structure of Tgt(Cys158Val) in complex with guanine reveals that there is no conformational change of Thr159 upon guanine binding. Obviously, the Thr159 conformational change is triggered by preQ$_1$ and not by guanine. This observation is of greater remarkability as the Thr159 side chain remains at large distance without any direct contact to the aminomethyl group of preQ$_1$. The *Penultimate Rotamer Library* published by Lovell *et al.* [43] reports three preferred rotamers for threonine observed with a relative frequency of 7%, 43%, and 49%. Notably, in the preQ$_1$ complexes of Tgt(Cys158Val) and Tgt(Cys158Val/Val233Gly) Thr159 adopts the least populated conformer, which probably corresponds to the rotamer that is least energetically favourable. In contrast, the Thr159 side chain conformation observed in all other crystal structures of *Z. mobilis* Tgt matches with the rotamer found with a frequency of 43%. Accordingly, the side chain conformation of Thr159 induced upon preQ$_1$ binding may well be correlated with the reduced preQ$_1$ affinity observed upon mutation of Cys158 to valine. It should be noted, that Thr159 is in most eucaryotic Tgt catalytic subunits replaced by a nearly isosteric valine residue [42].

As neither *Z. mobilis* Tgt nor any of our mutated variants are able to use queuine as a substrate we investigated whether or not queuine is actually recognised by these enzymes. For that purpose, we tested if queuine was able to act as inhibitor of bacterial Tgt since its binding to the active site should inevitably prevent radiolabelled guanine from being inserted into tRNA. Indeed, a considerable inhibitory effect of queuine on Tgt activity was observed for all investigated enzyme variants including "wild type" Tgt. The recognition of queuine in the substrate binding pocket of "wild type" Tgt and Tgt(Cys158Val) seems surprising as these variants do not exhibit the space-opening Val233Gly exchange believed to be prerequisite for the accommodation of the dihydroxy-cyclopentenyl side chain of queuine. Trapping experiments revealed that queuine predominantly acts as a competitive inhibitor with respect to tRNA as a substrate on all studied Tgt variants. This means queuine binds to the respective apo-enzymes but hardly to the covalent enzyme·tRNA complexes formed upon the release of guanine. Solely in the case of Tgt(Cys158Val/Val233Gly), the variant that should best approach the eucaryotic enzyme, the result of the trapping experiment indicates that the covalent enzyme·tRNA complex may, though to a low extent, be able to bind queuine.

The crystal structure of Tgt(Cys158Val) in complex with queuine provides an evident explanation why this variant is able to bind queuine exclusively in its apo-form. The bulky side chain of Val233 does not prevent queuine from binding to the active site. Yet, it forces its dihydroxy-cyclopentenyl moiety into a position which is not compatible with tRNA binding due to clashes with the tRNA sugar phosphate backbone. The same is likely to be true for "wild type" Tgt although ill-defined electron density in this crystal structure does not allow a reliable placement of queuine into the substrate binding pocket of the refined model.

The crystal structures of Tgt(Val233Gly) and Tgt(Cys158Val/Val233Gly) with queuine reveal that the Val233Gly exchange allows the dihydroxy-cyclopentenyl substituent to occupy the created cavity between position 233 and 158. The adopted binding mode of queuine will not interfere sterically with the sugar-phosphate backbone of bound tRNA. Yet, the results of the trapping experiment suggest that Tgt(Val233Gly) is also not able to bind queuine after the formation of the covalent enzyme·tRNA complex. Tgt(Cys158Val/Val233Gly) is, if at all, only to a minor extent able to bind queuine after the formation of this complex. The fuzzy electron density corresponding to the bound queuine suggests high residual ligand flexibility accompanied by a significant disorder induced in the active sites of the studied Tgt variants. The latter becomes evident from elevated *B*-factors and ill-defined electron density of residues lining the substrate binding pocket. Possibly, the enhanced mobility of the substrate binding pocket and of the bound queuine in these complexes is not compatible with the constrictions induced by the formation of the covalent enzyme·tRNA complex. Anyway, the fact that queuine is not able to bind efficiently to the covalent complex of any Tgt variant provides a plausible explanation why it does not constitute a substrate. None of the crystal structures determined in this study nor the *in silico* model of the catalytic subunit of human Tgt provide any clue on how this problem is overcome by eucaryotic

Tgt. Probably, further residues, that are more distant from the active centre but involved in positioning the tRNA substrate have to be changed as well to enable bacterial Tgt to insert queuine into tRNA. In addition, it has to be pointed out once more that eucaryotic Tgt not only consists of a catalytic subunit, which is closely related to the bacterial protomer but, in addition, also contains a second subunit, which is catalytically inactive. This second subunit might possibly be prerequisite to enable the enzyme to use queuine as a substrate by aligning the tRNA substrate in a manner that somehow compensates for the loss of activity observed in our mutated variants.

Comparison of the active centre of *Z. mobilis* Tgt with the modelled human Tgt catalytic subunit reveals that Ala232 adjacent to Val233 in the bacterial enzyme is replaced by a serine (Ser231) in human Tgt (Figure 4). This residue is involved in a peptide switch necessary to accommodate $preQ_1$ after the excised guanine 34 left the binding pocket [22] (Figure 3). Although the main chain amide of Ala232 binds indirectly, *via* a water molecule, to *N7* of guanine 34 its side chain methyl group forms no interaction with any substrate. As this residue is replaced by a serine in many bacterial Tgts too, it seems unlikely that it contributes to substrate specificity or catalytic activity. Nevertheless, we introduced the Ala232Ser mutation into Tgt(Cys158Val/Val233Gly) to investigate its impact. For comparison, the Ala232Ser mutation was also introduced into "wild type" Tgt. Kinetic measurements showed that the Ala232Ser conversion further reduced the catalytic activity of Tgt(Cys158Val/Ala232-Ser/Val233Gly). The residual turnover observed for this variant with both guanine and $preQ_1$ as substrates was too small to permit the determination of reliable Michaelis-Menten parameters, due to the low signal to noise ratio of the measured data. Interestingly, a pronounced 10^2-fold decrease in k_{cat} compared to "wild type" Tgt was also observed for Tgt(Ala232Ser) using guanine or $preQ_1$ as a substrate (data not shown). Therefore, the Ala232Ser variants of Tgt were not considered further in this study. In addition, crystal structures that were determined of Tgt(Cys158Val/Ala232Ser/Val233Gly) in its apo-form, in complex with $preQ_1$ and in complex with queuine do not show any significant differences to the corresponding ones of Tgt(Cys158Val/Val233Gly) and, due to redundance, will not be discussed in detail (data statistics and PDB codes of these structures are provided in Table S2 of the Supporting Information). Nevertheless, it is worth mentioning that, in the Tgt(Cys158Val/Ala232Ser/Val233Gly)·$preQ_1$ complex the above-mentioned side chain rotation of Thr159, thought to correlate with the Cys158Val mutation upon binding of $preQ_1$, is also observed. This conformational change is not visible in the structure of this variant in complex with queuine.

Materials and Methods

Homology modelling

The crystal structure of *Z. mobilis* Tgt (PDB code **1r5y** chain A) was used as a template to create a model of the human Tgt catalytic subunit (UniProtKB/Swiss-Prot [44] accession code Q9BXR0). Using ClustalW 1.83 [45] the corresponding sequences aligned with 42% identity. Based on the alignment, ten homology models for human Tgt were calculated with MODELLER 6a [46]. As no significant differences were apparent, the model with the lowest MODELLER target function value (11,429.19) was selected as representative. The coordinates of the homology model are provided within the Supporting Information (Coordinates S1).

Cloning and Tgt preparation

Site-directed mutagenesis was performed using the QuickChange™ kit (Stratagene) following the vendor's protocol. The oligonucleotides listed in Table 5 were used to introduce the mutations into the *Z. mobilis tgt*(Tyr106Phe) expression plasmid pET9d-ZM-Y106F [37]. Sequence analysis of the entire *tgt* gene (MWG Biotech, Ebersberg) confirmed in each case the presence of the desired mutation(s) as well as the absence of any further unwanted mutation. Subsequently, the mutated plasmids were transformed into *E. coli* BL21(DE3)/pLysS cells. These cells were used for the preparation of the mutated Tgt variants following the protocol of Romier *et al.* [47].

Preparation of *Escherichia coli* $tRNA^{Tyr}$

Preparation of unmodified *E. coli* $tRNA^{Tyr}$ (ECY2) [48] *via in vitro* transcription was done using the RiboMAX™ Large Scale RNA Production System-T7 (Promega) according to the vendor's protocol. The concentration of tRNA was determinded *via* UV photometry ($\lambda = 280$ nm).

Preparation of $preQ_1$ and queuine

Preparation of $preQ_1$ and queuine was carried out according to [38].

Determination of kinetic parameters

Progress curves of Tgt activity using guanine and $tRNA^{Tyr}$ as substrates were measured by monitoring incorporation of radiolabelled guanine into $tRNA^{Tyr}$. The reaction proceeded at 37°C in 75 μL solutions containing 150 $nmol \cdot L^{-1}$ (of monomer) *Z. mobilis* Tgt [300 $nmol \cdot L^{-1}$ in case of Tgt(Cys158Val/Val233Gly)] and variable concentrations of *E. coli* $tRNA^{Tyr}$ and guanine (7.5% [8-^3H]-guanine; 12 Ci/mmol, Hartmann Analytic). The assay was performed in 200 $mmol \cdot L^{-1}$ HEPES buffer pH 7.3, 20 $mmol \cdot L^{-1}$ $MgCl_2$, and 0.037% (v/v) Tween 20 (Roth). Reactions were started by adding tRNA and [8-^3H]-guanine to the protein solution. To allow the protein solution to adjust to assay temperature it was pre-incubated at 37°C for 10 min prior to substrate addition. After the reactions were started, 15 μL aliquots were taken at intervals of 1 to 4 min. Aliquots were immediately transferred to Whatman GC-F glass microfiber filters (GE Healthcare) and quenched with 10% (w/v) trichloroacetic acid at 0°C for 15 min. Unbound guanine was washed from the filters in 7 min intervals twice with 5% (w/v) trichloroacetic acid and twice with technical grade ethanol. Filters were dried at 60°C for 45 min. ^3H incorporated into tRNA was quantified using liquid scintillation counting. Results from the aliquots were used to calculate initial velocity using GraFit (V. 4.09; Erithacus Software Ltd., 1999).

Michaelis-Menten parameters for tRNA and guanine were determined separately in duplicate and average values were calculated. Kinetic parameters for guanine were measured using 15 $\mu mol \cdot L^{-1}$ $tRNA^{Tyr}$ and concentrations of [8-^3H]-guanine varying in the range of 0.5–20 $\mu mol \cdot L^{-1}$. Kinetic parameters for tRNA were measured using 20 $\mu mol \cdot L^{-1}$ [8-^3H]-guanine and $tRNA^{Tyr}$ concentrations varying in the range of 0.25–15 $\mu mol \cdot L^{-1}$. Initial velocities in "counts per min" were transferred to $\mu mol \cdot L^{-1} \cdot min^{-1}$ using a calibration constant derived from liquid scintillation counting of [8-^3H]-guanine solutions with variable concentrations. Kinetic parameters were determined *via* double-reciprocal linearisation and linear regression using GraFit.

Michaelis-Menten parameters for $preQ_1$ were determined in the same manner as for guanine but using radiolabelled $preQ_1$

Table 5. Oligonucleotides used in mutagenesis.

mutation/name of oligo	Sequence
C158V-f	5′-GTA ATG GCC TTT GAC GAA GTC ACG CCT TAT CCA GC-3′
C158V-b	5′-GC TGG ATA AGG CGT GAC TTC GTC AAA GGC CAT TAC-3′
V233G-f	5′-GCT GTT GGG GGA TTG GCT GGG GGT GAA GGA CAG GAT G-3′
V233G-b	5′-C ATC CTG TCC TTC ACC CCC AGC CAA TCC CCC AAC AGC-3′
A232S-f	5′-GCT GTT GGG GGA TTG TCT GTG GGT GAA GGA CAG GAT G-3′
A232S-b	5′-C ATC CTG TCC TTC ACC CAC AGA CAA TCC CCC AAC AGC-3′
A232S-V233G-f	5′-GCT GTT GGG GGA TTG TCT GGG GGT GAA GGA CAG GAT G-3′
A232S-V233G-b	5′-C ATC CTG TCC TTC ACC CCC AGA CAA TCC CCC AAC AGC-3′

Bases deviating from the original *tgt* sequence are underlined.

([methylene-^3H]-7-aminomethyl-7-deazaguanine = [^3H]-preQ_1; 7.0 Ci/mmol, Moravek Biochemicals Inc., California, USA) as substrate base. The concentration of preQ_1 was varied between 0.5–15 μmol·L^{-1} for "wild type" Tgt and Tgt(Val233Gly) or rather 1–50 μmol·L^{-1} for Tgt(Cys158Val) and Tgt(Cys158Val/Val233Gly).

In the same manner, enzymatic activity was analysed using [^3H]-queuine (7.0 Ci/mmol, Moravek Biochemicals Inc., California, USA) as substrate base at concentrations of up to 50 μmol·L^{-1}. Alternatively, enzymatic activity with queuine as substrate base was measured *via* monitoring the removal of [8-^3H]-guanine from tRNA^{Tyr} radiolabelled in position 34 as described in [22] but with the substrate base preQ_1 replaced by queuine. Radiolabelled tRNA^{Tyr} was used in a concentration of 8 μmol·L^{-1}. The concentration of queuine amounted to 50 μmol·L^{-1} for all inspected Tgt variants.

Determination of inhibition constants

Inhibition constants of queuine (competitive against tRNA) were determined for "wild type" Tgt and mutated variants as described in [23].

SDS-PAGE trapping experiment

5 μmol·L^{-1} *Z. mobilis* Tgt (monomer), 100 μmol·L^{-1} *E. coli* tRNA^{Tyr} (ECY2) and, if necessary, 1 mmol·L^{-1} of queuine or rather 2,6-diamino-3*H*-quinazolin-4-one in 10 μL of 100 mmol·L^{-1} HEPES buffer, pH 7.3, 20 mmol·L^{-1} $MgCl_2$, and 5 mmol·L^{-1} dithiothreitol were incubated for 1 h at 25°C. A total of 10 μL of SDS loading buffer [250 mmol·L^{-1} TrisHCl pH 6.8, 8% (w/v) SDS, 40% (v/v) Glycerol, 0.04% (w/v) bromophenol blue, 8% (v/v) 2-sulfanylethanol] was added and incubated for another hour at 25°C to allow proper unfolding. 10 μL of each sample were then loaded onto a 15% SDS polyacrylamide gel and electrophoresis was performed with SDS running buffer [25 mmol·L^{-1} TrisHCl pH 8.8, 200 mmol·L^{-1} glycine, 0.1% (w/v) SDS]. Gels were stained with Coomassie® brilliant blue R-250 (Bio-RAD).

Microscale thermophoresis measurements

Z. mobilis Tgt at a concentration of 10 μmol·L^{-1} was labelled with Alexa Fluor®647 succinimidyl ester at a concentration of 40 mg·L^{-1} at room temperature for 30 min in a 500 mmol·L^{-1} NaCl solution buffered with 50 mmol·L^{-1} HEPES pH 8.1 (molar dye : protein ratio≈3 : 1). Excessive unreacted Alexa Fluor®647 was removed with a NAP5 sephadex column (GE Healthcare) equilibrated with 2 mol·L^{-1} NaCl, 1 mmol·L^{-1} EDTA solution buffered with 10 mmol·L^{-1} TrisHCl pH 7.8. Photometry at 650 nm and 280 nm indicated a label : protein ratio of 0.8. The Alexa Fluor®647-Tgt solution was adjusted to 100 nmol·L^{-1} with 100 mmol·L^{-1} HEPES pH 7.3 buffer containing 20 mmol·L^{-1} $MgCl_2$ and 0.037% (v/v) (≡323 μmol·L^{-1}) Tween 20 (Roth). The final solution contained NaCl at a concentration of approximately 300 mmol·L^{-1}. The tRNA^{Tyr} ligand was dissolved in the same buffer (free of NaCl) at a concentration of 10 μmol·L^{-1}. A series of fifteen 1 : 1 dilutions of tRNA^{Tyr} solution : buffer solution was prepared producing tRNA^{Tyr} concentrations ranging from 305 pmol·L^{-1} to 10 μmol·L^{-1}. For thermophoresis, each of these solutions was mixed with one volume of Alexa Fluor®647-Tgt solution resulting in a constant concentration of fluorescence labelled Tgt of 50 nmol·L^{-1} and tRNA^{Tyr} concentrations ranging from 153 pmol·L^{-1} to 5 μmol·L^{-1}. After 10 min incubation followed by centrifugation at 10 000× g for 10 min, approximately 2 μL of each solution were filled into Monolith NT Standard treated capillaries (NanoTemper Technologies GmbH). Thermophoresis (including temperature jump) was measured at room temperature for 10 s by means of a Monolith NT.015 instrument (NanoTemper Technologies GmbH), using 100% LED power and 60% infrared laser power. Data of three independent runs were averaged and analysed using Origin 7 (Origin Lab). Curve fitting and K_D calculation was done using the program NanoTemper Analysis 1.2.009 (NanoTemper Technologies GmbH) based on the following equation [49]:

T = thermophoresis signal
U = minimal signal (unbound protein)
B = maximal signal (protein saturated with ligand)
c_p = concentration of labelled protein
c_i = concentration of ligand
K_D = dissociation constant

$$T = U + (B - U) \cdot \frac{(c_p + c_i + K_D) - \sqrt{(c_p + c_i + K_D)^2 - 4 \cdot c_p \cdot c_i}}{2 \cdot c_p}$$

Crystallisation, data collection, and processing

Crystals of the investigated Tgt variants suitable for ligand soaking were grown *via* the hanging-drop vapour diffusion method. Droplets were prepared by mixing 1 μL of concentrated protein solution (12 to 15 mg·mL^{-1} in 2 mol·L^{-1} NaCl, 10 mmol·L^{-1} Tris/HCl pH 7.8, 1 mmol·L^{-1} EDTA, 1 mmol·L^{-1} DTT) with 1 μL reservoir solution (11% (w/v)

PEG 8,000, 100 mmol·L^{-1} Tris/HCl, pH 8.5, 1 mmol·L^{-1} DTT, 10% (v/v) DMSO). Crystals grew within one week at 18°C in the presence of 1.0 mL of reservoir solution.

For cocrystallisation of Tgt(Cys158Val) with guanine, guanine·HCl was dissolved in DMSO and added to the crystallisation droplet to a final concentration of 5 mmol·L^{-1}. As guanine is, under the above-named conditions, not soluble up to this concentration a substantial portion of the compound precipitated. Crystals grew within the guanine precipitate in two weeks. To allow cocrystallisation of the Tgt variants with Q the compound was dissolved in DMSO and added to the droplet to a final concentration of 20 mmol·L^{-1}. Crystals of the Tgt variants complexed to preQ$_1$ were produced by a soaking procedure. The compound was dissolved in DMSO and added to 2 µL of reservoir solution to a final concentration of 10 to 20 mmol·L^{-1}. Finally, a single pre-grown Tgt crystal was transferred into the droplet which was sealed against 1.0 mL of reservoir solution and soaked for 30 min.

For data collection, crystals were cryoprotected using glycerol. The glycerol and PEG 8,000 concentrations of the reservoir buffers were increased stepwise by transferring the crystals to six different 2 µL cryodroplets each with 30-min incubation times while sealed against 1.0 mL of the same solution (glycerol concentrations (v/v): 5%→10%→15%→20%→25%→30%; and PEG 8,000 concentrations (w/v): 5.0%→6.3%→7.5%→8.0%→8.8%→9.8%, respectively). These droplets also contained the ligands at the same concentrations as the soaking and cocrystallisation solutions. The cryo-soaked crystals were flash-frozen in liquid nitrogen. Data sets (**3bld**, **4dg0**, **4e2v**, **4hvx**, **4h7z**, **4h6e**, **4hsh**, **4hqv**, **4gcx** and "WT"-Tgt·queuine) were collected at cryo conditions (−173°C≡100 K) at the BESSY II (Helmholz-Zentrum, Berlin, Germany) beamlines BL-14.1 (λ = 0.91841 Å) and BL-14.2 (λ = 0.91841 Å) and with CuKα radiation (λ = 1.5418 Å) using a Rigaku RU-300 rotating-anode generator at 50 kV and 90 mA equipped with Xenocs focussing optics and an R-AXIS IV detector. All crystals tested exhibited monoclinic symmetry in space group *C*2 containing one monomer per asymmetric unit with Matthews coefficients of 2.3–2.4. Data processing and scaling was performed using the HKL2000 package [50] except for structure **4h7z** which was processed using iMOSFLM 1.0.6 [51] and SCALA [52]. For all refined structures unit cell dimensions for the crystals, data collection and processing statistics are given in Tables 2, 3 and 4.

Coordinates of the apo-Tgt crystal structure (PDB-code: **1pud** in case of the structures **3bl3**, **3bld**, **3blo**) or rather coordinates of Tgt in complex with 2-[(thiophen-2-ylmethyl)amino]-1,7-dihydro-8H-imidazo[4,5-g]quinazolin-8-one (PDB-code: **3gev** in case of structures **4dg0**, **4e2v**, **4hvx**, **4h7z**, **4h6e**, **4hsh**, **4hqv**, **4gcx** and "WT"-Tgt·queuine) were slightly modified (deletion of water and ligand as well as coordinates at the sites of mutation) before applying for initial rigid-body refinement of the protein molecule followed by repeated cycles of conjugate gradient energy minimisation, simulated annealing and *B*-factor refinement using the CNS program package [53]. Refinement at the later stages was performed with SHELXL [54] for structures **3bl3**, **3bld**, and **3blo**. Here, up to 50 cycles of conjugate gradient minimisation were performed with default restraints on bonding geometry and *B*-values: 5% of all data were used for R_{free} calculation. For the structures **4e2v**, **4dg0**, **4h7z**, **4h6e**, **4hqv**, **4hsh**, **4hqv**, **4gcx** and "WT"-Tgt·queuine refinement was performed with PHENIX 1.8-1069 [55] with 5% to 10% of all data being used for R_{free} calculation and followed by repeated cycles of maximum likehood energy minimization. For the structures **4h7z**, **4hvx**, **4hqv**, **4hsh** an additional optimization of the weights between X-ray target and stereochemistry as well as between ADP restraints was performed by PHENIX 1.8-1069 [55]. Amino acid side chains were fitted to σA-weighted $2|F_o|$ - $|F_c|$ and $|F_o|$ - $|F_c|$ electron density maps using Coot [56]. Ligands were generated using SYBYL 8.0 (Tripos International: 1699 South Hanley Rd., St. Louis, Missouri, 63144, USA). Afterwards, geometric restraints of the ligands were calculated by Monomer Library Sketcher (ccp4) [57]. For structures **4e2v**, **4dg0**, **4h7z**, **4h6e**, **4hqv**, **4hsh**, **4hqv**, **4gcx** and "WT"-Tgt·queuine water was located using the refinement settings "update waters" implemented in the program PHENIX 1.8-1068 [55]. After further refinement cycles water was checked visually. Glycerol molecules and ligands were located in the difference electron density and added to the model for further refinement cycles. The temperature factors (individual *B*-factors) for structures **4e2v**, **4gcx**, **4gd0** and **3bld** were anisotropically refined, whereas for structures **4hvx**, **4h7z**, **4hsh**, **4hqv** and **4h6e** and "WT"-Tgt·queuine TLS refinement was applied. The definition of the TLS groups was done with the TLSMD server [58]. Restraints were applied to bond lengths and angles, planarity of aromatic rings, and Van-der-Waals contacts. Multiple side-chain conformations were built in case an appropriate electron density was observed and maintained during refinement and if the minor populated side chain showed at least 20% occupancy. All final models were validated using PROCHECK [59]. Analysis of temperature factors was done with Moleman [60].

Alignment and Figures

Alignment of structures based on c_α with similar or identical sequences was performed with the alignment function implemented in Pymol (http://www.pymol.org). Figures were prepared using ChemDraw Std 12.0 (PerkinElmer, Massachusetts, USA) and Pymol.

Protein Data Bank Accession Codes

The Protein Data Bank (PDB) accession codes allocated to the crystal structures determined in the course of this study are given in Tables 2, 3 and 4 and in Table S1.

Supporting Information

Coordinates S1 Coordinates of the homology model of the human Tgt catalytic subunit.

Coordinates S2 Coordinates of the *Z.mobilis* Tgt(Tyr106Phe) crystal structure determined in in the presence of queuine.

Structure Factors S1 Structure factors of the *Z.mobilis* Tgt(Tyr106Phe) crystal structure determined in in the presence of queuine.

Table S1 Overview of crystal structures analysed by Biela *et al.* (2013).

Table S2 Crystallographic data collection and refinement statistics of further structures related to the study of Biela *et al.* (2013).

Acknowledgments

We gratefully acknowledge the beamline staff at BESSY II (Helmholz-Zentrum Berlin) in Berlin, Germany, for providing us outstanding support.

We thank Christian Sohn for his support during "in house" X-ray data collection, François Diederich and Emmanuel Meyer for providing the uncompetitive Tgt inhibitor 2,6-diamino-3*H*-quinazolin-4-one, Steffi Dörr for exceedingly great technical assistance, Lena Siegling for experimental help during an internship and for editing the manuscript as well as Andreas Blum and Bernhard Stengl for beneficial discussions.

Author Contributions

Conceived and designed the experiments: GK KR. Performed the experiments: IB NT FI SG TXPN H-DG AH. Analyzed the data: IB NT AH GK KR. Wrote the paper: KR.

References

1. Stengl B, Reuter K, Klebe G (2005) Mechanism and substrate specificity of tRNA-guanine transglycosylases (TGTs): tRNA-modifying enzymes from the three different kingdoms of life share a common catalytic mechanism. ChemBioChem 6: 1926–1939.
2. Oliva R, Tramontano A, Cavallo L (2007) Mg^{2+} binding and archaeosine modification stabilise the G15-C48 Levitt base pair in tRNAs. RNA 13: 1427–1436.
3. Nakanishi S, Ueda T, Hori H, Yamazaki N, Okada N, et al. (1994) A UGU sequence in the anticodon loop is a minimum requirement for recognition by *Escherichia coli* tRNA-guanine transglycosylase. J Biol Chem 269: 32221–32227.
4. Curnow AW, Garcia GA (1995) tRNA-guanine transglycosylase from *Escherichia coli*. Minimal tRNA structure and sequence requirements for recognition. J Biol Chem 270: 17264–17267.
5. Phillips G, Yacoubi BE, Lyons B, Alvarez S, Iwata-Reuyl D, et al. (2008) Biosynthesis of 7-deazaguanosine-modified tRNA nucleosides: a new role for GTP cyclohydrolase I. J Bacteriol 190: 7876–7884.
6. McCarthy RM, Somogyi A, Bandarian V (2009) *Escherichia coli* QueD is a 6-carboxy-5,6,7,8-tetrahydropterin synthase. Biochemistry 48: 2301–2303.
7. McCarthy RM, Somogyi A, Lin G, Jacobsen NE, Bandarian V (2009) The deazapurine biosynthetic pathway revealed: *in vitro* enzymatic synthesis of $preQ_0$ from guanosine 5′-triphosphate in four steps. Biochemistry 48: 3847–3852.
8. Van Lanen SG, Reader JS, Swairjo MA, de Crécy-Lagard V, Lee B, et al. (2005) From cyclohydrolase to oxidoreductase: discovery of nitrile reductase activity in a common fold. Proc Natl Acad Sci USA 102: 4264–4269.
9. Lee BWK, Van Lanen SG, Iwata-Reuyl D (2007) Mechanistic studies of *Bacillus subtilis* QueF, the nitrile oxidoreductase involved in queuosine biosynthesis. Biochemistry 46: 12844–12854.
10. Chikwana VM, Stec B, Lee BWK, de Crécy-Lagard V, Iwata-Reuyl D, et al. (2012) Structural basis of biological nitrile reduction. J Biol Chem 287: 30560–30570.
11. Van Lanen SG, Kinzie GS, Matthieu S, Link T, Culp J, et al. (2003) tRNA modification by *S*-adenosylmethionine:tRNA ribosyltransferase-isomerase. Assay development and characterisation of the recombinant enzyme. J Biol Chem 278: 10491–10499.
12. Mathews I, Schwarzenbacher R, McMullan D, Abdubek P, Ambing E, et al. (2005) Crystal structure of *S*-adenosylmethionine:tRNA ribosyltransferase-isomerase (QueA) from *Thermotoga maritima* at 2.0 Å resolution reveals a new fold. Proteins 59: 869–874.
13. Grimm C, Ficner R, Sgraja T, Haebel P, Klebe G, et al. (2006) Crystal structure of *Bacillus subtilis S*-adenosylmethionine:tRNA ribosyltransferase-isomerase. Biochem Biophys Res Comm 351: 695–701.
14. Frey B, McCloskey J, Kersten W, Kersten H (1988) New function of vitamin B12: cobamide-dependent reduction of epoxyqueuosine to queuosine in tRNAs of *Escherichia coli* and *Salmonella typhimurium*. J Bacteriol 170: 2078–2082.
15. Miles ZD, McCarthy RM, Molnar G, Bandarian V (2011) Discovery of epoxyqueuosine (oQ) reductase reveals parallels between halorespiration and tRNA modification. Proc Natl Acad Sci USA 108: 7368–7372.
16. Romier C, Reuter K, Suck D, Ficner R (1996) Crystal structure of tRNA-guanine transglycosylase: RNA modification by base exchange. EMBO J 15: 2850–2857.
17. Ritschel T, Atmanene C, Reuter K, Van Dorsselaer A, Sanglier-Cianferani S, et al. (2009) An integrative approach combining noncovalent mass spectrometry, enzyme kinetics and X-ray crystallography to decipher Tgt protein-protein and protein-RNA interaction. J Mol Biol 393: 833–847.
18. Grädler U, Ficner R, Garcia GA, Stubbs MT, Klebe G, et al. (1999) Mutagenesis and crystallographic studies of *Zymomonas mobilis* tRNA-guanine transglycosylase to elucidate the role of serine 103 for enzymatic activity. FEBS Lett 454: 142–146.
19. Grädler U, Gerber H-D, Goodenough-Lashua DAM, Garcia GA, Ficner R, et al. (2001) A new target for Shigellosis: rational design and crystallographic studies of inhibitors of tRNA-guanine transglycosylase. J Mol Biol 306: 455–467.
20. Goodenough-Lashua DM, Garcia GA (2003) tRNA-guanine transglycosylase from *E. coli*: a ping-pong kinetic mechanism is consistent with nucleophilic catalysis. Bioorg Chem 31: 331–344.
21. Xie W, Liu X, Huang RH (2003) Chemical trapping and crystal structure analysis of a catalytic tRNA guanine transglycosylase covalent intermediate. Nature Struct Biol 10: 781–788.
22. Tidten N, Stengl B, Heine A, Garcia GA, Klebe G, et al. (2007) Glutamate versus glutamine exchange swaps substrate selectivity in tRNA-guanine transglycosylase: insight into the regulation of substrate selectivity by kinetic and crystallographic studies. J Mol Biol 374: 764–776.
23. Meyer EA, Donati N, Guillot M, Schweizer WB, Diederich F, et al. (2006) Synthesis, biological evaluation, and crystallographic studies of extended guanine-based (*lin*-benzoguanine) inhibitors for tRNA-guanine transglycosylase (TGT). Helv Chim Acta 89: 573–597.
24. Garcia GA, Chervin SM, Kittendorf JD (2009) Identification of the rate-limiting step of tRNA-guanine transglycosylase from *Escherichia coli*. Biochemistry 48: 11243–11251.
25. Okada N, Nishimura S (1979) Isolation and characterization of a guanine insertion enzyme, a specific tRNA transglycosylase, from *Escherichia coli*. J Biol Chem 254: 3061–3066.
26. Chen V-C, Brooks AF, Goodenough-Lashua DM, Kittendorf JD, Showwalter HD, et al. (2011) Evolution of eucaryal tRNA-guanine transglycosylase: insight gained from the heterocyclic substrate recognition by the wild-type and mutant human and *Escherichia coli* tRNA-guanine transglycosylases. Nucleic Acids Res 39: 2834–2844.
27. Boland C, Hayes P, Santa-Maria I, Nishimura S, Kelly VP (2009) Queuosine formation in eucaryotic tRNA occurs *via* a mitochondria-localized heteromeric transglycosylase. J Biol Chem 284: 18218–18227.
28. Chen V-C, Kelly VP, Stachura SV, Garcia GA (2010) Characterization of the human tRNA-guanine transglycosylase: confirmation of the heterodimeric subunit structure. RNA 16: 958–968.
29. Durand JM, Dagberg B, Uhlin BE, Björk GR (2000) Transfer RNA modification, temperature and DNA superhelicity have a common target in the regulatory network of the virulence of *Shigella flexneri*: the expression of the *virF* gene. Mol Microbiol 35: 924–935.
30. Hurt JK, Olgen S, Garcia GA (2007) Site-specific modification of *Shigella flexneri virF* mRNA by tRNA-guanine transglycosylase *in vitro*. Nucleic Acids Res 35: 4905–4913.
31. Ritschel T, Hörtner S, Heine A, Diederich F, Klebe G (2009) Crystal structure analysis and *in silico* pKa calculations suggest strong pKa shifts of ligands as driving force for high-affinity binding to TGT. ChemBioChem 10: 716–727.
32. Ritschel T, Kohler PC, Neudert G, Heine A, Diederich F, et al. (2009) How to replace the residual solvation shell of polar active site residues to achieve nanomolar inhibition of tRNA-guanine transglycosylase. ChemMedChem 4: 2012–2023.
33. Kohler PC, Ritschel T, Schweizer WB, Klebe G, Diederich F (2009) High-affinity inhibitors of tRNA-guanine transglycosylase replacing the function of a structural water cluster. Chem Eur J 15: 10809–10817.
34. Rakovich T, Boland C, Bernstein I, Chikwana VM, Iwata-Reuyl D, et al. (2011) Queuosine deficiency in eukaryotes compromises tyrosine production through increased tetrahydrobiopterin oxidation. J Biol Chem 286: 19354–19363.
35. Romier C, Meyer JE, Suck D (1997) Slight sequence variations of a common fold explain the substrate specificities of tRNA-guanine transglycosylases from the three kingdoms. FEBS Lett 416: 93–98.
36. Deshpande KL, Katze JR (2001) Characterization of cDNA encoding the human tRNA-guanine transglycosylase (TGT) catalytic subunit. Gene 265: 205–212.
37. Brenk R, Stubbs MT, Heine A, Reuter K, Klebe G (2003) Flexible adaptations in the structure of the tRNA-modifying enzyme tRNA-guanine transglycosylase and their implications for substrate selectivity, reaction mechanism and structure-based drug design. ChemBioChem 4: 1066–1077.
38. Gerber H-D, Klebe G (2012) Concise and efficient syntheses of $preQ_1$, queuine and *ent*-queuine. Org Biomol Chem 10: 8660–8668.
39. Jerabek-Willemsen M, Wienken CJ, Braun D, Baaske P, Duhr S (2011) Molecular interaction studies using microscale thermophoresis. ASSAY and Drug Development Technologies 9: 342–353.
40. Kung FL, Garcia GA (1998) tRNA-guanine transglycosylase from *Escherichia coli*: recognition of full-length 'queuine-cognate' tRNAs. FEBS Lett 431: 427–432.
41. Sheriff S, Hendrickson WA, Smith JL (1987) Structure of myohemerythrin in the azidomet state at 1.7/1.3 Å resolution. J Mol Biol 197: 273–296.
42. Thomas CE, Chen Y-C, Garcia GA (2011) Differential heterocyclic substrate recognition by, and pteridine inhibition of *E. coli* tRNA-guanine transglycosylase. Biochem Biophys Res Comm 410: 34–39.
43. Lovell SC, Word JM, Richardson JS, Richardson DC (2000) The penultimate rotamer library. Proteins 40: 389–408.
44. Boeckmann B, Bairoch A, Apweiler R, Blatter MC, Estreicher A, et al. (2003) The SWISS-PROT protein knowledgebase and its supplement TrEMBL in 2003. Nucleic Acids Res 31: 365–370.
45. Chenna R, Sugawara H, Koike T, Lopez R, Gibson TJ, et al. (2003) Multiple sequence alignment with the Clustal series of programs. Nucleic Acids Res 31: 3497–3500.
46. Eswar N, Marti-Renom MA, Webb B, Madhusudhan MS, Eranian D, et al. (2000) Comparative protein structure modelling with MODELLER. Current Protocols in Bioinformatics: Supplement 15: 5.6.1–5.6.30.

47. Romier C, Ficner R, Reuter K, Suck D (1996) Purification, crystallization, and preliminary X-ray diffraction studies of tRNA-guanine transglycosylase from *Zymomonas mobilis*. Proteins 24: 516–519.
48. Curnow AW, Kung FL, Koch KA, Garcia GA (1993) tRNA-guanine transglycosylase from *Escherichia coli*: gross tRNA structural requirements for recognition. Biochemistry 32: 5239–5246.
49. Wienken CJ, Baaske P, Rothbauer U, Braun D, Duhr S (2010) Protein-binding assays in biological liquids using microscale thermophoresis. Nat Commun 1: 100.
50. Otwinowski Z, Minor W (1997) Processing of X-ray diffraction data collected in oscillation mode. Methods Enzymol 276: 307–326.
51. Battye TGG, Kontogiannis L, Johnson O, Powell HR, Leslie AGW (2011) iMOSFLM: a new graphical interface for diffraction-image processing with MOSFLM. Acta Crystallogr Sect D 67: 271–281.
52. Evans PR (2011) An introduction to data reduction: space-group determination, scaling and intensity statistics. Acta Crystallogr Sect D 67: 282–292.
53. Brunger AT, Adams PD, Clore GM, DeLano WL, Gros P, et al. (1998) Crystallography and NMR system: A new software suite for macromolecular structure determination. Acta Crystallogr Sect D 54: 905–921.
54. Sheldrick GM, Schneider TR (1997) SHELXL: high-resolution refinement. Methods Enzymol 277b: 319–343.
55. Adams PD, Afonine PV, Bunkóczi G, Chen VB, Davis IW, et al. (2010) PHENIX: a comprehensive Python-based system for macromolecular structure solution. Acta Crystallogr Sect D 66: 213–221.
56. Emsley P, Cowtan K (2004) Coot: model-building tools for molecular graphics. Acta Crystallogr Sect D 60: 2126–2132.
57. Winn MD, Ballard CC, Cowtan KD, Dodson EJ, Emsley P, et al. (2011) Overview of the *CCP4* suite and current developments. Acta Crystallogr Sect D 67: 235–242.
58. Painter J, Merritt EA (2006) TLSMD web server for the generation of multi-group TLS models. J Appl Crystallogr 39: 109–111.
59. Laskowski RA, MacArthur MW, Moss DS, Thornton JM (1993) PROCHECK: a program to check the stereochemical quality of protein structures. J Appl Crystallogr 26: 283–291.
60. Kleywegt GJ, Zou JY, Kjeldgaard M, Jones TA, Around O (2001) In International Tables for Crystallography; Rossmann MG, Arnold E, Eds, Kluwer Academic Publishers: Dordrecht Vol F, 353–356.

Structural Variation among Wild and Industrial Strains of *Penicillium chrysogenum*

Valerie L. Wong[1]*, Christopher E. Ellison[1], Michael B. Eisen[2], Lior Pachter[3], Rachel B. Brem[2]

1 Department of Plant and Microbial Biology, University of California Berkeley, Berkeley, California, United States of America, **2** Department of Molecular and Cell Biology, University of California Berkeley, Berkeley, California, United States of America, **3** Departments of Mathematics, Molecular and Cell Biology, and Electrical Engineering and Computer Science, University of California Berkeley, Berkeley, California, United States of America

Abstract

Strain selection and strain improvement are the first, and arguably most important, steps in the industrial production of biological compounds by microorganisms. While traditional methods of mutagenesis and selection have been effective in improving production of compounds at a commercial scale, the genetic changes underpinning the altered phenotypes have remained largely unclear. We utilized high-throughput Illumina short read sequencing of a wild *Penicillium chrysogenum* strain in order to make whole genome comparisons to a sequenced improved strain (WIS 54–1255). We developed an assembly-free method of identifying chromosomal rearrangements and validated the *in silico* predictions with a PCR-based assay and Sanger sequencing. Despite many rounds of mutagen treatment and artificial selection, WIS 54–1255 differs from its wild progenitor at only one of the identified rearrangements. We suggest that natural variants predisposed for high penicillin production were instrumental in the success of WIS 54–1255 as an industrial strain. In addition to finding a previously published inversion in the penicillin biosynthesis cluster, we located several genes related to penicillin production associated with these rearrangements. By comparing the configuration of rearrangement events among several historically important strains known to be high penicillin producers to a collection of recently isolated wild strains, we suggest that wild strains with rearrangements similar to those in known high penicillin producers may be viable candidates for further improvement efforts.

Editor: Gabriel Moreno-Hagelsieb, Wilfrid Laurier University, Canada

Funding: The authors have no support or funding to report.

Competing Interests: MBE is a member of the PLOS Board of Directors. This does not alter the authors' adherence to all the PLoS ONE policies on sharing data and materials.

* E-mail: vwong@umn.edu

These authors contributed equally to this work.

Introduction

The discovery of penicillin and its antibiotic properties begun by Alexander Fleming and developed by Chain and Florey was a landmark in medicine and pharmacology [1,2]. However, Fleming's original strain produced only small quantities of penicillin. The efforts to make antibiotics more available, particularly in response to great demand during World War II, entailed both a search for wild strains with enhanced production of penicillin and improvement of strains already in culture [3]. Notably, Raper, Alexander, and Coghill [4] cultivated and tested isolates from a variety of food products, spoiled produce, and soils. Nearly all of their *Penicillium* strains produced detectable levels of penicillin, but very few were comparable to the best industrially important strains of the time [4]. The new isolates formed a bimodal distribution of penicillin production [4], indicative of natural variation in the wild population and suggesting that some wild strains may be predisposed to be high penicillin producers and to give rise to viable industrial strains.

Many commercial strains used by pharmaceutical companies such as Lilly Industries and Wyeth Lab trace their ancestry back to a single wild strain (*P. chrysogenum* NRRL 1951) isolated from a moldy cantaloupe found in Peoria, Illinois [3,5–7]. Compared to its improved progeny, NRRL 1951 is a relatively low penicillin producer [4]. Multiple rounds of non-directed mutagenesis and selection led to numerous sub-lineages, including the well-studied Wisconsin 54–1255, and later industrial strains with vastly higher production levels [8]. Despite decades of work on strain improvement, little is known about the indirect regulation of penicillin biosynthesis and how improvement occurred in this "Wisconsin family" of strains. With the recent availability of high-throughput DNA sequencing, it is now possible to compare whole genomes of wild and industrial strains in order to identify genomic differences that may be responsible for the improved phenotype [9].

The *P. chrysogenum* core genes for penicillin biosynthesis (*pcbAB*, *pcbC*, and *penDE*) are clustered together among other ORFs in a 56.8 kb region [10,11]. Tandem duplications of this cluster can be found in *P. chrysogenum* strains that are high penicillin producers [12]. Other enzymes outside the core cluster are required to activate the first step in the pathway as well as to activate the side chains [13]. The last two steps in the penicillin biosynthesis pathway take place in the microbody (peroxisome), and strains with more microbodies produce more penicillin [14]. Adding precursors such as penylacetic acid (PAA) to the culture medium pushes synthesis towards penicillin G, one of two main commercial penicillins [15]. The penicillin biosynthesis cluster appears devoid

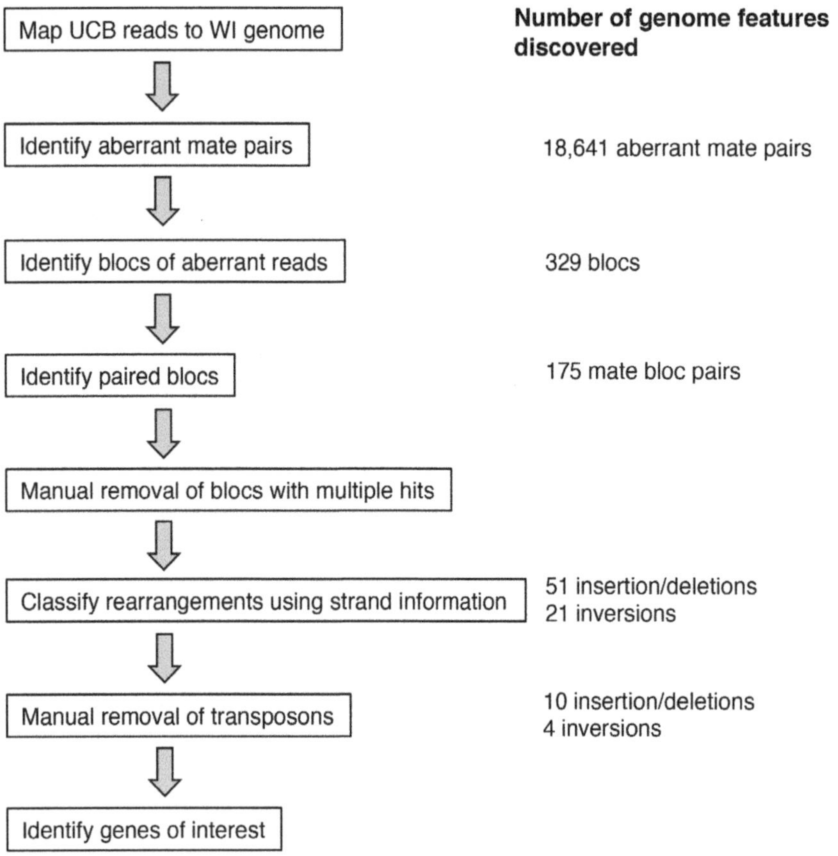

Figure 1. Flow chart for assembly-free rearrangement identification. Steps for identifying, classifying, and validating rearrangements. The number of genome features discovered at each stage are noted.

of regulators specific to penicillin production [11,16], and regulation of the process seems to be controlled by heterochromatin modification, nitrogen regulation, and pH-dependent carbon source regulation [7,13].

Improvement for industrial production required selection not only for β-lactam synthesis but also for growth in submerged culture. Relative to NRRL 1951, improved strains have an increased ability to deal with oxidative stress, a reduced range of secondary metabolite production concomitant with an increase in penicillin output, and a decrease in the expression of proteins associated with virulence and cell wall degradation [17]. Other methods of increasing penicillin production include modifying the growth conditions and reducing sporulation and growth, which occur at the expense of secondary metabolite production [18].

The sequencing of one improved Wisconsin family strain (WIS 54–1255) has produced insights into the genetics of penicillin biosynthesis [16], but this information alone is insufficient to elucidate the genomic changes between wild strains, improved strains, and wild strains with enhanced penicillin production. The WIS 54–1255 strain (hereafter referred to as WI) was produced via multiple rounds of selection and mutagenesis, including ultraviolet radiation, X rays, and nitrogen mustard, which may have led to chromosomal rearrangements [3,5,6]. There is a previously identified inversion in the biosynthesis cluster [11,12], presumably induced by strain improvement efforts.

We utilized Illumina short read sequencing of a wild *P. chrysogenum* strain (PC0184C, hereafter referred to as UCB) to develop an assembly-free computational pipeline using mate-pair information to identify chromosomal rearrangements between this wild strain and an improved strain (WIS 54–1255). We further validated our *in silico* predictions with a PCR-based assay and screened additional wild and industrially important *P. chrysogenum* strains to assess which, if any, genomic changes are specific to the industrially important strains and thus potentially contribute to their improved capacity for penicillin biosynthesis.

Results

Chromosomal rearrangements between the wild UCB strain of *P. chrysogenum* and the industrial WI strain

We carried out paired-end Illumina sequencing of genomic DNA of the wild UCB strain, mapped reads to the published WI genome, and used the mapping data to identify putative chromosomal rearrangements that differentiated this strain from the industrial Wisconsin strain (Fig. 1; see Methods for details). Briefly, we expected that rearrangement events would be detectable based on the mapping of mate pairs to locations in the WI genome assembly much further apart than the 300-500 bp fragment insert sizes used to prepare the library (Fig. 2). Reads from any rearrangement event should cluster together on opposite sides of the breakpoint positions, given the amplification using primers complementary to opposite strands of genomic DNA fragments during Illumina library construction [19]. An analysis pipeline based on this expectation (Fig. 1) located 51 candidate insertion/deletion events and 21 candidate inversion events. Manual inspection suggested that the origin of many of these

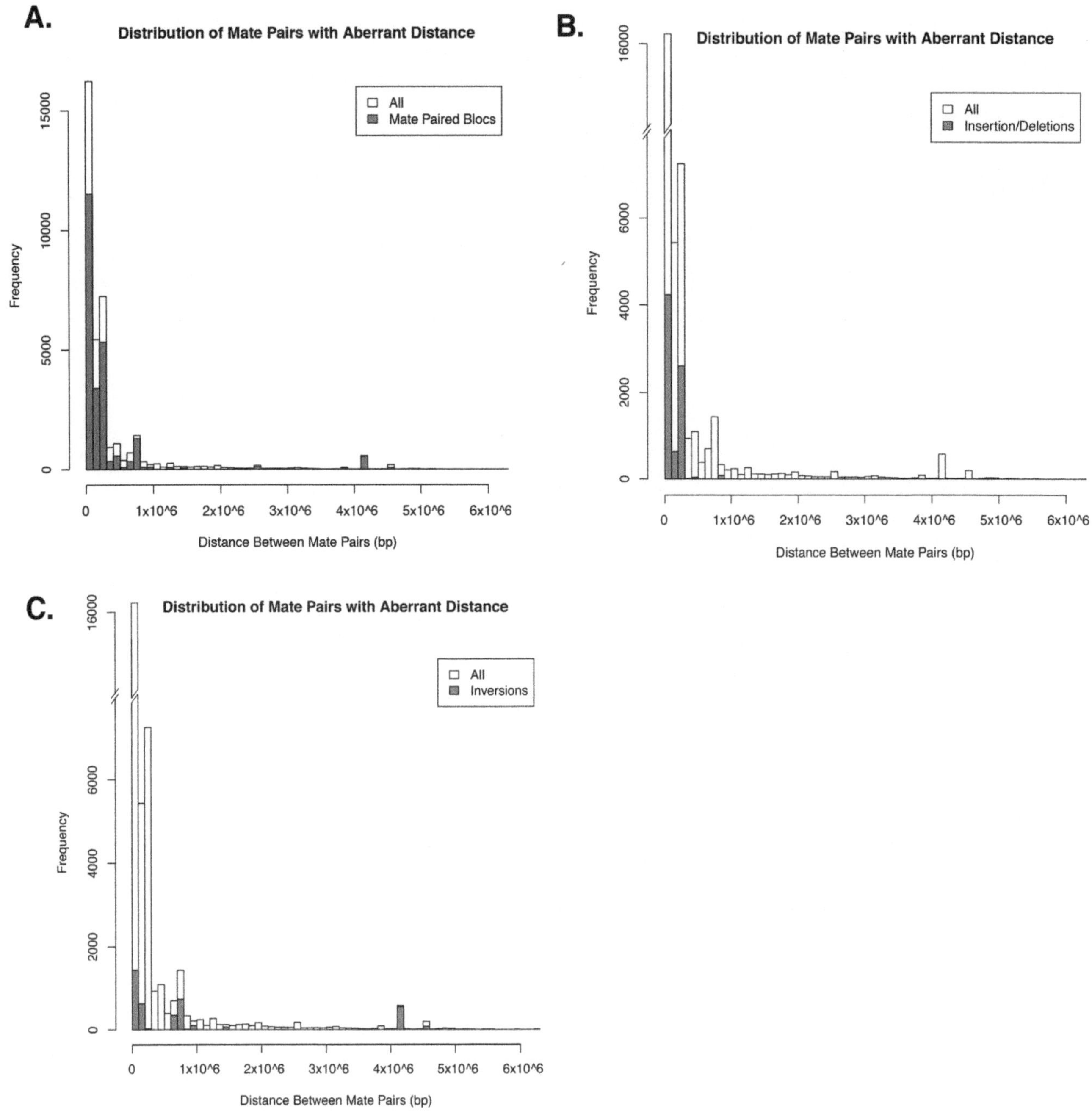

Figure 2. Distribution of mate pairs. Distribution of all aberrant mate pairs by distance (white) overlaid with mate pairs in all mated blocs (**A.**), mate pairs in blocs defining putative insertion/deletions (**B.**), and mate pairs in blocs defining putative inversions (**C.**).

candidate rearrangements likely lay in transposable element gains and losses (see Methods). Eliminating the latter from further consideration, we identified 10 large insertion/deletion events (> 1 kb) and 4 inversions (Fig. 3 and Table 1) in gene-rich regions. These inferred events included the single previously described structural difference between WI and other wild strains, which lay in the penicillin biosynthesis gene cluster [12] (event 326 in Table 1).

To validate each of these inferred rearrangements between the UCB and WI strains, we designed single-locus PCR-based assays (Fig. 4 and Table 1). Each amplification gave the product size expected from the event inferred *in silico*, with the exception of a likely complex rearrangement near the gene Pc22g09350 (Fig. 4). Sequencing of amplified products likewise confirmed the inferred events in each case (Table S1).

Rearrangements between UCB and WI are positioned preferentially near penicillin-related genes

We hypothesized that many of the chromosomal rearrangements we identified between the WI and UCB strains could act in *cis* on penicillin-related genes to contribute to the high penicillin production observed in WI. As an unbiased test of this notion, we used a set of 522 genes with functions related to penicillin synthesis based on expression profiles from Harris et al. [20]. Excluding the

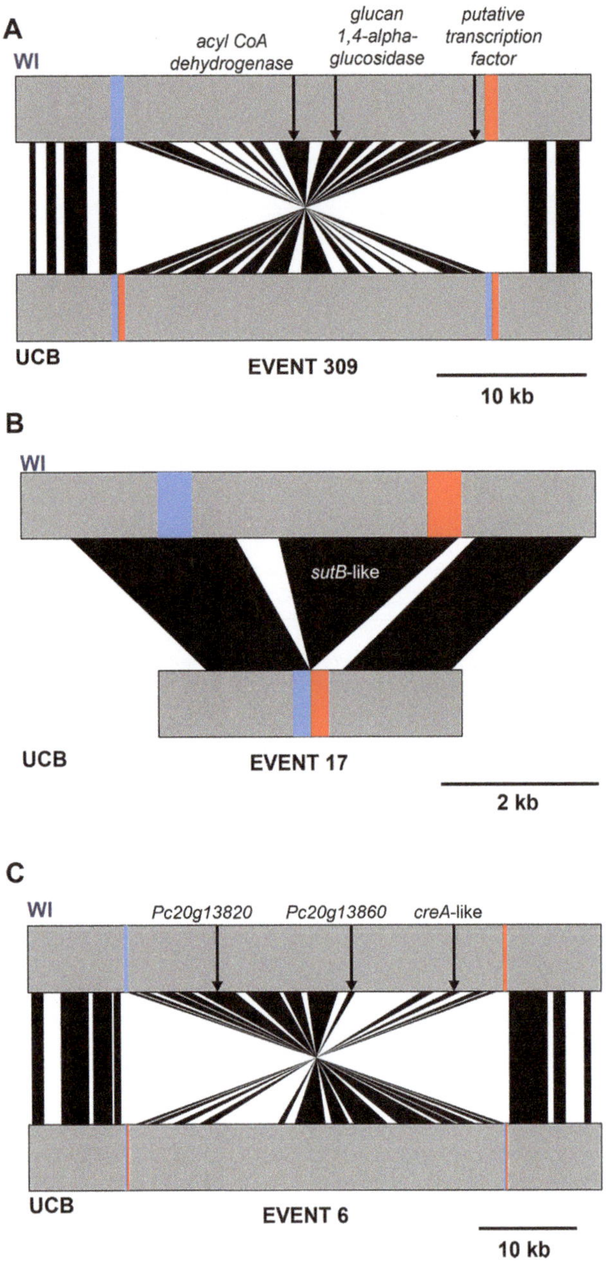

Figure 3. Architecture of rearrangements between the WI strain and the wild UCB strain. Rearrangement events containing genes with potential roles in penicillin biosynthesis are shown with their predicted gene contents (black spans). Flanking blocs of aberrantly mapping reads are marked in blue and red and penicillin-related genes are denoted by arrows. For those genes with known function, the gene name is listed, otherwise the gene ID is given. (A) Event 309 is an inversion and contains the predicted glucan 1,4-alpha-glucosidase Pc13g11940 and the gene of unknown function Pc13g11930, both of which show elevated expression in high penicillin G producing strains [16,20] as well as Pc13g11930, which is predicted to localize to the microbody where penicillin biosynthesis takes place. (B) Event 17 is an insertion of Pc12g01540, a gene with strong similarity to the sulfate transporter *sutB*, which also shows elevated expression in high penicillin-producing strains and may be involved in biosynthesis of amino acid β-lactam precursors [16,20]. (C) Event 6 is an inversion containing three genes that are repressed in high penicillin-producing strains: Pc20g13820, Pc20g13860, and Pc20g13880 [16], the latter of which is a homolog of the *Aspergillus niger creA* regulator of β-lactam biosynthesis.

previously characterized rearrangement at the penicillin biosynthesis cluster (event 326 in Table 1) to avoid potential bias towards known factors in high penicillin production, we found that 8 penicillin-related genes were also among the 96 genes residing within 5 kb of our PCR-verified rearrangement breakpoints, an enrichment beyond that expected by chance (Tables 3, S2, and S3) (hypergeometric $P = 0.038$).

Manual inspection of the genomic regions flanked by the rearrangement breakpoints between WI and UCB further supported the hypothesis that these structural changes influenced genes involved in penicillin biosynthesis. For example, a large inversion (event 309 in Table 1, Fig. 3, and Table S2) involved the predicted glucan 1,4-alpha-glucosidase Pc13g11940, highly expressed in a high penicillin G producer [16]; Pc13g11930, predicted to localize to the microbody where penicillin biosynthesis takes place [21]; and Pc13g11930, which is highly expressed in penicillin-producing strains [16,20]. An insertion (event 17 in Fig. 3 and Tables S2 and S4) involved Pc12g01540, a gene with strong similarity to the sulfate transporter *sutB*, which likewise is highly expressed in high penicillin-producing strains and may be involved in biosynthesis of amino acid β-lactam precursors of penicillin [16,20]. Another large inversion (event 6 in Table 1, Fig. 3, and Table S2) involved three genes repressed in high penicillin-producing strains, Pc20g13820, Pc20g13860, and Pc20g13880 [16], the latter of which is a homolog of the *Aspergillus niger creA* regulator of β-lactam biosynthesis. These findings establish a strong relationship between penicillin genes and structural rearrangements in the comparison of the WI industrial strain with the UCB wild isolate.

Most genomic rearrangements present in WI are segregating in wild *P. chrysogenum*

We expected that if chromosomal rearrangements occurred while the progenitor of the WI strain was subjected to mutagenesis and selection for increased penicillin production, such structural changes should be specific to the WI genome. To test this, we used our PCR assays to determine the orientation of these regions in the wild progenitor of the WI strain, NRRL 1951. Surprisingly, for nearly all the rearrangements that distinguished UCB from WI, the latter resembled its wild progenitor (Table 2). Only at the previously characterized rearrangement at the penicillin biosynthesis gene cluster (event 326 in Table 2 and Fig. 4) did the WI strain differ from NRRL 1951. These results strongly suggested that the majority of differences in chromosome structure between the UCB and WI strains had not arisen recently during artificial selection in the industrial setting. Instead, we hypothesized that these rearrangements were ancient alleles segregating in wild *P. chrysogenum* populations.

To assess structural variation across *P. chrysogenum* strains, we assayed the 14 rearrangements we had identified between UCB and WI (Table 1) in a panel of additional isolates: the original strain isolated by Fleming (NRRL 824); NRRL 832, identified as a high penicillin producer in submerged culture [6]; and two recently isolated wild strains (Henk PC08-3A and NRRL A3704). As predicted from our analyses of WI and UCB, almost all the rearrangements were polymorphic across this strain panel (Table 2), apart from the event at the penicillin gene cluster (event 326 in Table 2). We posit that the rearrangements are polymorphic within a single *P. chrysogenum* population, given the evidence for globally recombining populations [22]. Taken together, our results make clear that chromosomal rearrangements are widespread among *P. chrysogenum* strains, and they fall preferentially in regions proximal to genes involved in penicillin biosynthesis.

Table 1. Summary of PCR validations of rearrangement events.

Event Type	Event	Size (bp)	WI DNA		UCB DNA	
			WI Config	UCB Config	WI Config	UCB Config
Inversion	6	37,858	+	–	–	+
Insertion/Deletion	12	1,905	+	–	–	+
Insertion/Deletion	17	2,956	+	–	–	+
Insertion/Deletion	231	1,954	*	–	–	+
Insertion/Deletion	237	3,777	+	–	–	+
Insertion/Deletion	267	1,857	+	–	–	+
Insertion/Deletion	269	649	+	–	–	+
Insertion/Deletion	270	5,892	+	–	–	+
Insertion/Deletion	287	2,059	+	–	–	+
Inversion	309	23,748	+	–	–	+
Inversion	312	4,135,317	+	–	–	+
Insertion/Deletion	317	683	+	–	–	+
Insertion/Deletion	318	5,891	+	–	–	+
Inversion	326	2,275	+	–	–	+

PCR products for each primer configuration were scored as present (+), absent (–), or ambiguous (*) based on their migration on a 1.5% agarose gel (Fig. S2). Present scores indicate very bright bands of the expected size for both of the blocs spanning each breakpoint of an event. Absent scores indicate the lack of a bright band of the expected size for one or both blocs. The ambiguous score indicates unclear results. Event sizes are the number of base pairs in the WI genome internal to two mate blocs.

Discussion

How to maximize production of commercially relevant biomolecules is one of the central questions in industrial microbiology. For the vast majority of industrial strains producing any given molecule, the genetic basis of the production trait and its potential for further improvement remain unknown. In this work, we have identified structural rearrangements in a strain of *P. chrysogenum* used for decades in industrial penicillin production. We have shown that most of these genomic changes have the potential to affect penicillin biosynthesis yet are unlikely to be the product of artificial selection. The one exception, a rearrangement at the penicillin biosynthesis gene cluster, is known to distinguish the WI family of strains from wild isolates of *P. chrysogenum* [12], supporting a model in which this mutation was indeed a product of the strain improvement process. The fact that structural changes at this locus have not been observed in wild strains suggests the presence of strong purifying selection outside of the industrial setting.

Apart from structural changes at the penicillin biosynthesis gene cluster, our findings suggest that NRRL 1951, the progenitor of the WI strain and of many others used in industrial production, naturally inherited genomic attributes that predisposed it toward high penicillin yield. Of the many wild strains that have been screened for penicillin production, NRRL 1951 was among a very few chosen for improvement. Our findings raise the possibility that additional determinants of penicillin production are segregating among wild *P. chrysogenum* strains. We speculate that more extensive surveys of wild isolates for penicillin yield may enable industrial microbiologists to take full advantage of the variation already present in nature.

Materials and Methods

Ethics Statement

No human or animal subjects were used in the course of this work.

Cultures and Strains

The following *P. chrysogenum* strains were obtained from the ATCC culture collection: 28089 (designation WIS 54–1255 and hereafter referred to as WI), 9480 (NRRL 1951), 9179 (NRRL 832), and 9478 (NRRL 824, originally deposited as *P. notatum*). NRRL 1951 is the founding member of the industrially important Wisconsin family of improved *P. chrysogenum* strains [6]. The WI strain is descended from NRRL 1951. ATCC 9179 was identified as a strain particularly suited for penicillin production in submerged culture, which is preferable to surface culture for industrial production [23]. ATCC 9478 is Fleming's original strain [1].

The following wild strains were graciously provided by Daniel Henk (Imperial College, London): PC0184C, PC081B, PC083A, PC085A, PC088B, PC0820A, PC0897A, NRRL A1723, NRRL A2416, NRRL A24700, NRRL A3704, NRRL 1975, NRRL A3704, NRRL A3905, NRRL A20320, NRRL 32232, 924700, and PC0887A. We sequenced PC0814C, which was collected by Dr. Mark Enright (Imperial College) in January 2008 from Morzine, France. The strain was isolated from an adhesive film exposed to the air for 6 hours [22]. For brevity we refer to this strain as UCB. All *P. chrysogenum* strains were maintained on MEA (20 g/L malt extract, 1.5% agar) at room temperature.

UCB Strain (PC0814C) Sequencing

The UCB strain was sequenced with three lanes of paired-end Illumina (San Diego, CA) 36 bp reads, following standard Illumina protocols, at the Vincent J. Coates Genomic Sequencing Laboratory (UC Berkeley, Berkeley, CA). Insert sizes were 300 bp and 500 bp. Sequence reads were submitted to the NCBI Sequence Read Archive (accession number SRP040942).

Read Mapping

We mapped reads to the published *P. chrysogenum* WI genome assembly [3] using Bowtie [24], discarding all alignments for reads

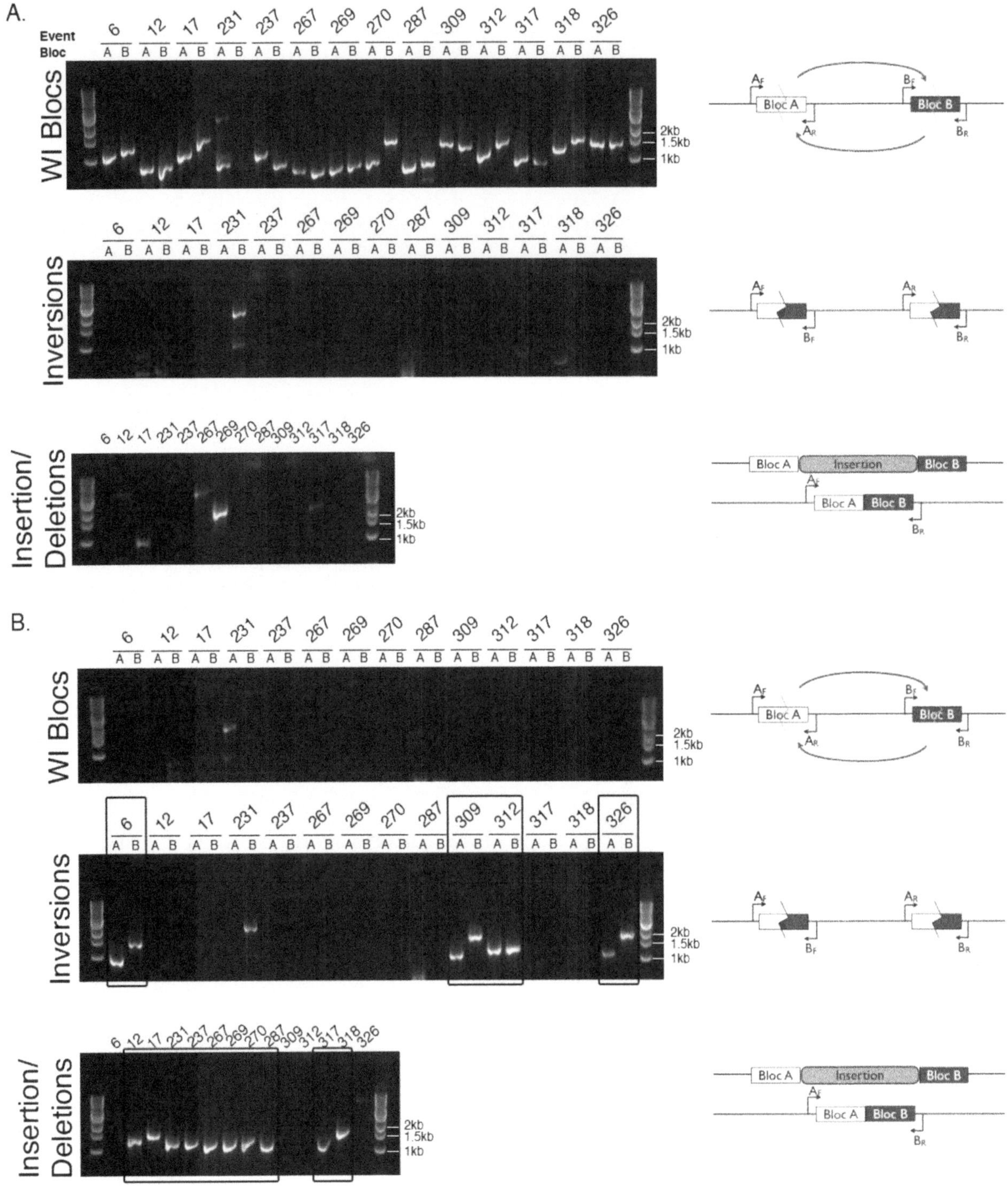

Figure 4. PCR validation of rearrangements identified *in silico*. PCR reactions used DNA from the **A.** WI strain and **B.** UCB strain. The top gel image in each figure used primers in the orientation to amplify across blocs in the WI strain. Middle images had primers to amplify across blocs in the UCB strain for a relative inversion. Bottom images show reactions priming across insertions in the WI strain relative to the UCB strain. Diagrams to the right of each gel image depict the primer configurations for PCR amplification across blocs. Small arrows indicate relative position and direction of forward (F) and reverse (R) primers for a set of paired mate blocs in the WI strain. Zigzag lines indicate breakpoints. Curved arrows indicate the direction of inversion. All PCR products for each strain were run on a single gel. Each PCR reaction is labeled by event and bloc (A or B), and boxes highlight successful amplification of rearrangement events in the UCB strain.

that mapped equally well to multiple places in the genome. To ensure that mate pairs with larger than expected mapping distances were not discarded during the alignment process, we disregarded mate pair information while mapping. We then identified the location of mate pairs in the WI genome assembly post-alignment using mate-pair information encoded in the Bowtie mapping output. After pairing mates together, the distance between their genomic locations was calculated, identifying

Table 2. PCR validation results for all rearrangement events across several strains.

Event	WI	NRRL 1951	NRRL 832	NRRL 824	Henk PC08-3A	NRRL A3704	UCB
6	WI	WI	WI	UCB	UCB	WI	UCB
12	WI	WI	WI	UCB	UCB	UCB	UCB
17	WI	WI	WI	UCB	UCB	UCB	UCB
231	WI	WI	WI	WI	WI	UCB	UCB
237	WI	WI	WI	WI	WI	WI	UCB
267	WI	WI	*	WI	WI	UCB	UCB
269	WI	WI	UCB	UCB	UCB	UCB	UCB
270	WI	WI	UCB	UCB	UCB	UCB	UCB
287	WI	WI	WI	UCB	*	UCB	UCB
309	WI	WI	WI	WI	WI	WI	UCB
312	WI	WI	UCB	UCB	UCB	UCB	UCB
317	WI	WI	WI	UCB	WI	WI	UCB
318	WI	WI	UCB	UCB	UCB	UCB	UCB
326	WI	UCB	*	UCB	UCB	UCB	UCB

NRRL 1951 differed from its improved descendant, the WI strain, in the state of only event 326. The WI strain consistently showed the same ambiguous banding for event 231 (Fig.S2), and strains labeled WI for that event were ambiguous in the same manner. Cells are designated for WI conformation (WI), UCB conformation (UCB), or ambiguous banding (*). The WI strain, NRRL 1951, NRRL 832, and NRRL 824 are known penicillin producers.

Table 3. Genes involved in penicillin biosynthesis and associated with rearrangement breakpoints.

Gene ID	Annotation (from Harris *et. al.* 2009)	Event #
Pc12g00460	strong similarity to multidrug resistance protein fnx1p – *S. pombe*	12
Pc12g01540	strong similarity to sulfate permease SutB – *P. chrysogenum*	17
Pc18g03010	strong similarity to choline permease Hnm1 – *S. cerevisiae*	267
Pc18g03120	strong similarity to hypothetical protein mg00375.1 – *M. grisea*	269
Pc21g09650	weak similarity to ecto-ATPase c-cam105 – *R. norvegicus*	317
Pc21g13940	strong similarity to hypothetical protein An03g01270 – *A. niger*	318
Pc21g13950	strong similarity to aspergillopepsin II – *A. niger*	318
Pc22g09350	strong similarity to vacuolar H(+) Ca(2+) exchanger Vcx1 – *S. cerevisiae*	231

These genes are located within 5 kb of rearrangement breakpoints and have putative functions related to penicillin biosynthesis based on an expression profiling experiment performed by Harris et al (2009).

18,641 aberrant mate pairs on the same contigs with distances over 1,000 bp. The distances between aberrant mates that mapped to the same contig were not evenly distributed, with most mate pairs being either less than 1 Mb apart or just over 4 Mb apart (Fig. 2). These aberrant mate pairs formed 329 blocs of at least 20 reads with start positions no more than 100 bp from each other. These blocs were further narrowed down to 175 pairs of mate blocs by taking the median mate pair distance for reads within a bloc and searching for other blocs within 1 kb of that distance. Mate paired blocs were manually curated to correct for instances where one bloc had multiple hits.

Support for Rearrangements from High Coverage

Because the UCB strain genome was sequenced at ~30X coverage, each rearrangement breakpoint in the WI genome should be spanned by ~30 different mate pairs from the UCB reads. These pairs (presumably located within 500 bp of each other in the UCB genome) should map to locations in the WI genome at distances equal to the size of the rearrangement event (Fig. 2). We required read clusters or "blocs" to be composed of at least 20 reads from aberrant mate pairs located no more than 100 bp from each other. We also identified "mate blocs" by pairing up blocs that had at least 90% of the reads mate paired together. A pair of mate blocs therefore represents the boundaries of a single putative rearrangement event. Mate blocs were further manually curated to filter out putative repetitive elements.

Using Strand Information to Classify Rearrangements

Strandedness of aberrant mate pairs that map to the same contig was used to classify rearrangements as either inversions or insertions/deletions. An insertion event in the WI strain or a deletion in the UCB strain would result in mate pairs mapping to opposite strands, preserving the mate pair directions. An inversion event would flip the strand of one read in each mate pair, resulting in mate pairs mapping to the same strand. Events classified as insertions were required to be supported by at least 90% of aberrant mate pairs mapping to opposite strands, while inversion events were required to be supported by at least 90% of aberrant mate pairs mapping to the same strand.

PCR Validation

In order to validate the rearrangement events predicted *in silico*, we designed primers to amplify across blocs as arranged in the WI strain (Fig. 3). If the predicted rearrangement were correct, using the primers in the same orientation with DNA from the UCB strain, no product should be detected. However, if PCR product amplified when we switched the primer pairs to be either both forward or both reverse primers from a mate bloc pair, this was consistent with an inversion event. Insertions were detected by priming across the insertion, using a forward primer for one bloc in a mate bloc pair and the reverse primer from its mate. PCR products for the WI and UCB strains were further validated by sequencing at the UC Berkeley Sequencing Facility.

DNA Extraction

Prior to extraction, all *P. chrysogenum* cultures were grown in flasks containing 100 ml liquid malt extract medium (20 g/L malt extract) for two days at 25°C with gentle shaking. Tissue was harvested by filtering through Miracloth (Calbiochem, Darmstadt, Germany) and rinsing with ~100 mL sterile distilled water. Samples were frozen in liquid nitrogen and stored at −80°C prior to lyophilization for 2 days. DNA was extracted by beadbeating with 0.3 g zirconia/silica beads (BioSpec Products, Bartlesville, OK) in a screwtop tube for 30 sec. Following addition of 0.5 mL lysis buffer (50 mM Tris-HCl, 50 mM EDTA, 3% SDS, and 1% β-mercaptoethanol), tubes were vortexed to resuspend all ground tissue and then incubated at 65°C for 45 min. Chloroform (0.5 mL) was added, and tubes were vortexed and then spun at 1,320 rpm for 5 min. The aqueous phase (~350 μl) was transferred to a new tube with 35 μl Proteinase K and 350 μl Buffer AL from the DNeasy Blood and Tissue kit (Qiagen, Valencia, CA), and manufacturer's directions were subsequently followed.

Primer Design and PCR

Primers were designed with the PrimerQuest^{SM} tool by Integrated DNA Technologies (IDT) to amplify across each bloc in the WI-54-1255 genome (Table S1). Various primer pairs were selected to identify blocs in the WI arrangement, blocs that were inverted relative to WI, and blocs with insertions relative to WI. Each 25 μl reaction was made according to the following recipe: 1 μl of DNA extract, 2.5 μl of 2 mM dNTPs (Fermentas, Glen Burnie, MD), 2.5 μl buffer (0.5 M KCl, 0.1 M TrisHCl pH 8.3, 25 mM $MgCl_2$, and 1 mg/mL gelatin), 0.25 μl of each primer (50 μM, IDT, San Diego, CA), 0.5 μl Taq DNA polymerase (New England Biolabs, Ipswitch, MA), and 18 μl water. Reactions were run with the following PCR program:

1. 94°C for 1 min
2. 94°C for 1 min
3. 68°C for 1 min

4. 72°C for 1.5 min
5. Repeat steps 2 through 4 30 times.
6. 72°C for 8 min

P. chrysogenum WI-54-1255 Genes Associated with Rearrangements

Genes of interest were identified as those whose start was either within a rearrangement event or less then 500 bp outside it. The very large rearrangements were ignored for this purpose due to the sheer number of genes involved.

Supporting Information

Table S1 Primer sequences designed to amplify across Wisconsin blocs. Each primer was named based on the target bloc and the forward (F) or reverse (R) direction of priming.

Table S2 Genes associated with validated rearrangement events and found in the literature. Genes of interest were identified as those whose start was either within a rearrangement event or less than 500 bp outside it. The very large rearrangements were ignored for this purpose due to the sheer number of genes involved. Genes were annotated by van den Berg et al.

Table S3 Genes associated with validated rearrangement events and annotated by van Den Berg et al. but otherwise undescribed. Genes of interest were identified as those whose start was either within a rearrangement event or less then 500 bp outside it. The very large rearrangements were ignored for this purpose due to the sheer number of genes involved. Interpro terms were not available (NA) for all annotations.

Table S4 Contents of insertion events.

Acknowledgments

Devin Scannell produced the genomic libraries for sequencing. Daniel Henk graciously provided strains. We thank two reviewers for their helpful comments.

Author Contributions

Conceived and designed the experiments: VLW CEE RBB. Performed the experiments: VLW CEE. Analyzed the data: VLW CEE. Contributed reagents/materials/analysis tools: VLW CEE MBE LP RBB. Wrote the paper: VLW CEE RBB. Supervised the research: MBE LP RBB.

References

1. Fleming A (1929) On the antibacterial action of *Penicillium*, with special reference to their use in the isolation of *B. influenzae*. Br J Exp Pathol 10: 185–194.
2. Chain E, Florey H, Gardner A, Heatley N, Jennings M, et al. (1940) Symptoms relating. Lancet: 226–228.
3. Elander RP (1967) Enhanced Penicillin Biosyntheis in Mutant and Recombinant Strains of *Penicillium chrysogenum*. In: Stubbe H., editor Induzierte Mutationen un ihre Nutzung Berlin: Akademie-Verlag. pp. 403–423.
4. Raper KB, Alexander DF, Coghill RD (1944) Penicillin: II. Natural variation and penicillin production in *Penicillium notatum* and allied species. J Bacteriol 48: 639–659.
5. Muñiz C, Zelaya T, Esquivel G (2007) Penicillin and cephalosporin production: A historical perspective. Rev Latinoam 49: 88–98.
6. Raper KB, Alexander DF (1945) Penicillin: V. Mycological aspects of penicillin production. J Mitchell Soc.
7. Demain AL, Elander RP (1999) The beta-lactam antibiotics: past, present, and future. Antonie Van Leeuwenhoek 75: 5–19.
8. Lein J (1986) The Panlabs penicillin strain improvement program. In: Zdenko V, Hosbllek Z, editors.Overproduction of Microbial Metabolites: Strain Improvement and Process Control Strategies.Boston: Butterworth Publishers. pp. 105–139.
9. Le Crom S, Schackwitz W, Pennacchio L, Magnuson JK, Culley DE, et al. (2009) Tracking the roots of cellulase hyperproduction by the fungus *Trichoderma reesei* using massively parallel DNA sequencing. Proc Natl Acad Sci U S A 106: 16151–16156.
10. Díez B, Gutiérrez S, Barredo JL, van Solingen P, van der Voort LH, et al. (1990) The cluster of penicillin biosynthetic genes. Identification and characterization of the pcbAB gene encoding the alpha-aminoadipyl-cysteinyl-valine synthetase and linkage to the pcbC and penDE genes. J Biol Chem 265: 16358–16365.
11. Fierro F, García-Estrada C, Castillo NI, Rodríguez R, Velasco-Conde T, et al. (2006) Transcriptional and bioinformatic analysis of the 56.8 kb DNA region amplified in tandem repeats containing the penicillin gene cluster in *Penicillium chrysogenum*. Fungal Genet Biol 43: 618–629.
12. Fierro F, Barredot JL, Dfezt B, Gutierrez S, Fernandez FJ, et al. (1995) The penicillin gene cluster is amplified in tandem repeats linked by conserved hexanucleotide sequences. Proc Natl Acad Sci U S A 92: 6200–6204.
13. Martín JF, Ullán RV, García-Estrada C (2010) Regulation and compartmentalization of β-lactam biosynthesis. Microb Biotechnol 3: 285–299.
14. Kiel JAKW, van den Berg MA, Fusetti F, Poolman B, Bovenberg RAL, et al. (2009) Matching the proteome to the genome: the microbody of penicillin-producing *Penicillium chrysogenum* cells. Funct Integr Genomics 9: 167–184.
15. Gordon M, Pan S, Virgona A, Numerof P (1953) Biosynthesis of penicillin. I. Role of phenylacetic acid. Science 118: 43.
16. Van Den Berg MA, Albang R, Albermann K, Badger JH, Daran J-M, et al. (2008) Genome sequencing and analysis of the filamentous fungus *Penicillium chrysogenum*. Nat Biotechnol 26: 1161–1168.
17. Jami M-S, Barreiro C, García-Estrada C, Martín J-F (2010) Proteome analysis of the penicillin producer *Penicillium chrysogenum*: characterization of protein changes during the industrial strain improvement. Mol Cell Proteomics 9: 1182–1198.
18. Elander RP, Espenshade MA (1976) The role of microbial genetics in industrial microbiology. In: Miller BM, Litsky W, editors.Industrial Microbiology.New York: McGraw-Hill. pp. 192–256.
19. Bentley DR, Balasubramanian S, Swerdlow HP, Smith GP, Milton J, et al. (2008) Accurate whole human genome sequencing using reversible terminator chemistry. Nature 456: 53–59.
20. Harris DM, van der Krogt ZA, Klaassen P, Raamsdonk LM, Hage S, et al. (2009) Exploring and dissecting genome-wide gene expression responses of *Penicillium chrysogenum* to phenylacetic acid consumption and penicillinG production. BMC Genomics 10: 75.
21. Müller W, van der Krift T, Krouwer A, Wosten H, van der Voort L, et al. (1991) Localization of the pathway of the penicillin biosynthesis in *Penicillium chrysogenum*. EMBO J 10: 489–495.
22. Henk DA, Eagle CE, Brown K, Van den Berg MA, Dyer PS, et al. (2011) Speciation despite globally overlapping distributions in *Penicillium chrysogenum*: the population genetics of Alexander Fleming's lucky fungus. Mol Ecol: 4288–4301.
23. Moyer AJ, Coghill RD (1945) Penicillin IX. The Laboratory scale production of penicillin in submerged cultures by *Penicillium notatum* Westling (NRRL 832). J Bacteriol 51: 79–93.
24. Langmead B, Trapnell C, Pop M, Salzberg SL (2009) Ultrafast and memory-efficient alignment of short DNA sequences to the human genome. Genome Biol 10: R25.

5

Enhanced Biocatalytic Esterification with Lipase-Immobilized Chitosan/Graphene Oxide Beads

Siaw Cheng Lau[1], Hong Ngee Lim[1,2]*, Mahiran Basri[1,3], Hamid Reza Fard Masoumi[1], Asilah Ahmad Tajudin[4], Nay Ming Huang[5], Alagarsamy Pandikumar[5], Chi Hua Chia[6], Yoshito Andou[7]

1 Department of Chemistry, Faculty of Science, Universiti Putra Malaysia, Serdang, Selangor, Malaysia, **2** Institute of Advanced Technology, University Putra Malaysia, Serdang, Selangor, Malaysia, **3** Laboratory of Molecular Biomedicine, Institute of Bioscience, Universiti Putra Malaysia, Serdang, Selangor, Malaysia, **4** Faculty of Biotechnology and Biomolecular Sciences, Universiti Putra Malaysia, Serdang, Selangor, Malaysia, **5** Low Dimensional Materials Research Centre, Department of Physics, Faculty of Science, University of Malaya, Kuala Lumpur, Malaysia, **6** School of Applied Physics, Faculty of Science and Technology, Universiti Kebangsaan Malaysia, Bangi, Selangor, Malaysia, **7** Eco-Town Collaborative Research and Development Center for the Environment and Recycling, Kyushu Institute of Technology, Kitakyushu-shi, Fukuoka, Japan

Abstract

In this work, lipase from *Candida rugosa* was immobilized onto chitosan/graphene oxide beads. This was to provide an enzyme-immobilizing carrier with excellent enzyme immobilization activity for an enzyme group requiring hydrophilicity on the immobilizing carrier. In addition, this work involved a process for the preparation of an enzymatically active product insoluble in a reaction medium consisting of lauric acid and oleyl alcohol as reactants and hexane as a solvent. This product enabled the stability of the enzyme under the working conditions and allowed the enzyme to be readily isolated from the support. In particular, this meant that an enzymatic reaction could be stopped by the simple mechanical separation of the "insoluble" enzyme from the reaction medium. Chitosan was incorporated with graphene oxide because the latter was able to enhance the physical strength of the chitosan beads by its superior mechanical integrity and low thermal conductivity. The X-ray diffraction pattern showed that the graphene oxide was successfully embedded within the structure of the chitosan. Further, the lipase incorporation on the beads was confirmed by a thermo-gravimetric analysis. The lipase immobilization on the beads involved the functionalization with coupling agents, N-hydroxysulfosuccinimide sodium (NHS) and 1-ethyl-(3-dimethylaminopropyl) carbodiimide (EDC), and it possessed a high enzyme activity of 64 U. The overall esterification conversion of the prepared product was 78% at 60°C, and it attained conversions of 98% and 88% with commercially available lipozyme and novozyme, respectively, under similar experimental conditions.

Editor: Eugene A. Permyakov, Russian Academy of Sciences, Institute for Biological Instrumentation, Russian Federation

Funding: This research work was supported by the Fundamental Research Grant Scheme (FRGS/1/2012/ST05/UPM/02/3) and a High Impact Research Grant (UM.C/625/1/HIR/MOHE/SC/21) from the Ministry of Higher Education. The funders had no role in study design, data collection and analysis, or preparation of the manuscript.

Competing Interests: The authors have declared that no competing interests exist.

* Email: janet_limhn@yahoo.com

Introduction

Esterification is widely used for the synthesis of raw materials as emulsifiers in foods, lubricants, paints, emollients in cosmetics, and in perfumes [1]. Lipase is one of the biocatalysts that catalyzes an esterification process such as the hydrolysis of fats or lipids [2]. Lipase-catalyzed esterification is a vital reaction because of its useful ester products. More specifically, lipase-catalyzed esterification reactions have attracted research interest due to an increased use of organic esters in biotechnology and the chemical industry [3]. Lipase consists of two different structures in a homogeneous solution. One is a flap structure with its active site separated from the reaction medium, and the other is an open structure with its active site exposed to the reaction medium [4,5]. Homogeneous biocatalysis has some drawbacks related to the recovery of the catalyst and disposal of the wastewater. However, these problems can be overcome by using a heterogeneous solid catalyst because it will be more stable against different pH and temperature ranges [5–8]. It can also be continuously reused in various reactors [9]. Moreover, lipase immobilization is used to prevent product contamination, ease recovery, enhance the operational lifetime and stability, prevent the solvent denaturation of the lipase, and expose the active sites of the lipase for more efficient bonding with a support [6,10–12].

Generally, two methods are commonly employed for the enzyme immobilization: chemical methods involving covalent bonds with the enzyme and physical methods involving weak interactions with the enzyme [13,14]. Immobilization using covalent binding is more favorable because it provides a strong bonding between the lipase and support. In addition, covalent binding eases the reuse of the lipase more efficiently than physical methods like adsorption and entrapment [10]. Various supports have been used for the immobilization of lipase [15]. These encompass inorganic materials such as diatomaceous earth, silica and porous glass, synthetic resins and resin ion exchangers, and natural polysaccharides such as cellulose and cross-linked dextrin

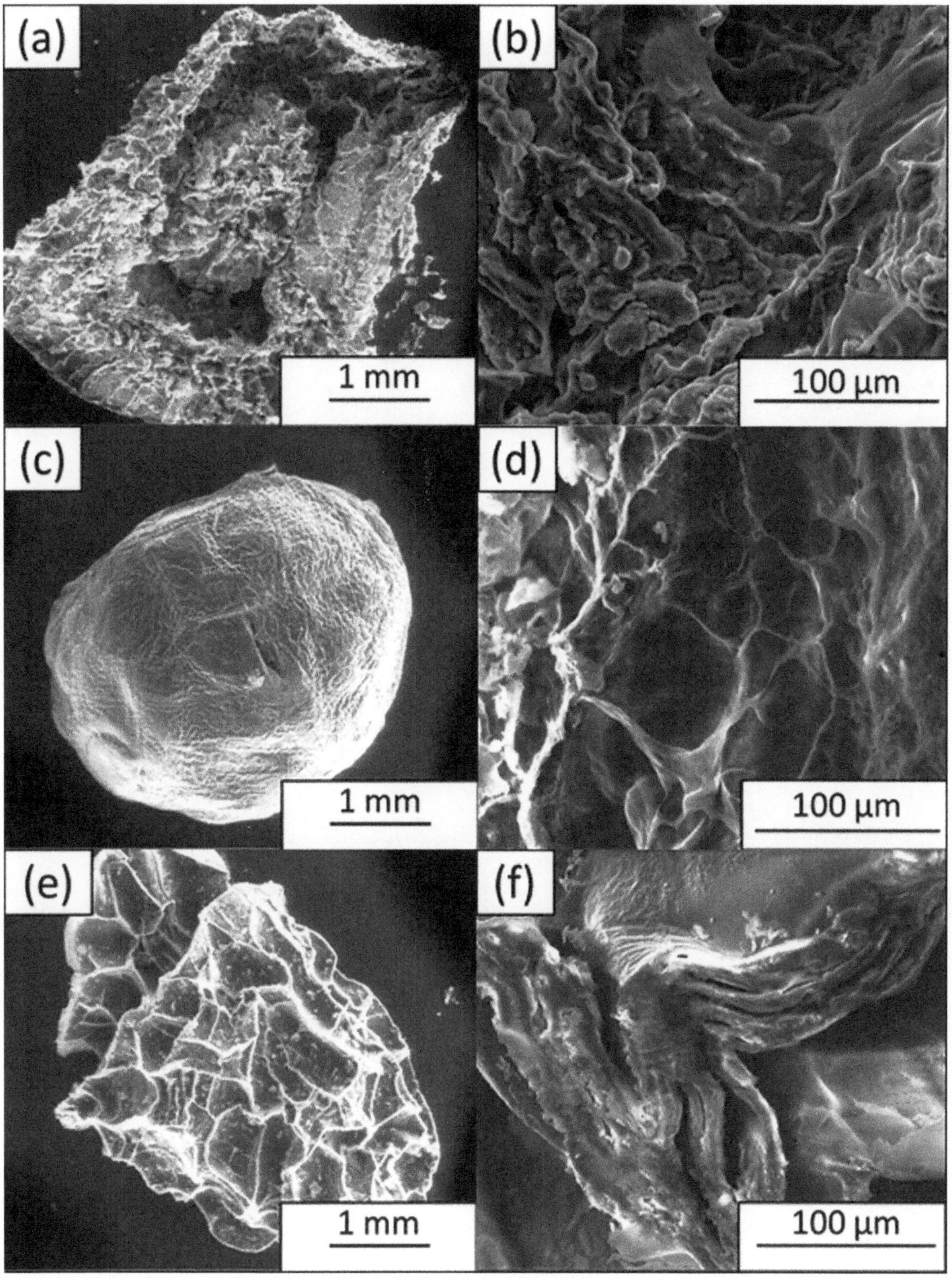

Figure 1. FESEM images of CS beads after coagulation time intervals of (a and b) 2 h, (c and d) 4 h, and (e and f) 20 h.

[15]. Chitosan is a natural polymer obtained by the deacetylation of chitin in a strongly basic medium at elevated temperature. It is insoluble in water [16] and can be easily reformed and reshaped from its solution by the addition of an alkali [17]. Chitosan is a more suitable support to immobilize the enzyme because it has a low cost and is biocompatible, nontoxic, available in various forms, physiological inert, and has high mechanical strength [11,18–20].

Immobilizing lipase on chitosan enhances the enzyme stability and eases the removal of the enzyme from the support without inactivating the enzyme or contaminating the support [16]. Chitosan beads can be prepared as a support for the immobilization of an enzyme by precipitating an acidic chitosan solution into a sodium hydroxide solution and then performing a solvent exchange with water [17]. Larger pores are more uniformly present from the surface to the inside of the carrier when chitosan is used as a carrier rather than synthetic resin carriers, thus providing greater support diffusibility [15]. Chitosan is a useful polysaccharide in protein conjugation via cross-linking agents such

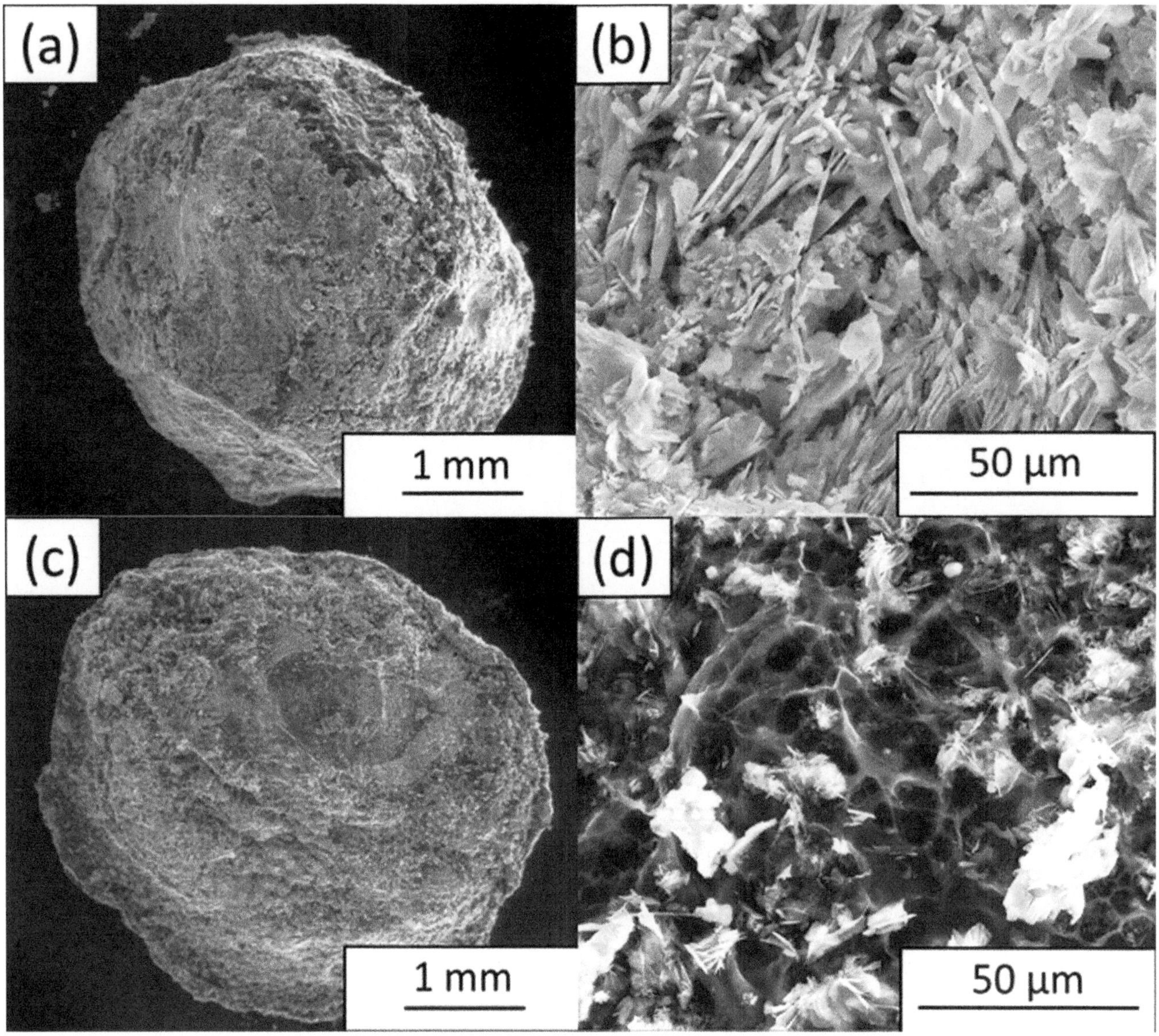

Figure 2. FESEM images of (a and b) CS/GO-S and (c and d) CS/GO beads coagulated for 4 h.

as N-(3-dimethylaminopropyl)-N-ethylcarbodiimide (EDC) [21]. One of the uses of EDC is to activate the hydroxyl groups of chitosan for the immobilization of lipase [22]. The highly reactive amino groups convenient for immobilizing enzymes via covalent bonding are disposed within the molecule [15]. In addition, chitosan has a higher affinity to enzymes because a higher amount of the enzymes may be immobilized onto the highly reactive amino groups of chitosan, compared to inorganic materials and synthetic resins [15].

An aerogel of chitosan beads embedded with graphene oxide (GO) obtained from freeze-drying is a type of matter that has an open cell foam structure with ultrafine pores and a cell size that results in a large surface area [23]. GO is able to enhance the physical strength of chitosan beads because of its superior mechanical integrity and low thermal conductivity, which are important properties for a catalyst support [23]. Moreover, the oxide functional groups of GO can provide additional active sites for the immobilization of lipase. Lipase is covalently immobilized onto chitosan beads embedded with GO (CS/GO) before they are used for esterification. Upon immobilization, the coiled structure of lipase will be expanded, hence exposing its active sites and promoting effective bonding with the support.

In the present study, the lipase from *Candida rugosa* was immobilized on chitosan/graphene oxide beads that were prepared and characterized using suitable techniques, which are field emission scanning electron microscopy, an X-ray diffraction technique, a thermogravimetric analysis, and a lipase assay [29]. The lipase immobilization on the beads involved functionalization with coupling agents, N-hydroxysulfosuccinimide sodium (NHS) and 1-ethyl-(3-dimethylaminopropyl) carbodiimide (EDC), and it possessed a high enzyme activity of 64 U. The commercially available lipases that were used for the reaction was a lipase of *Rhizomucor miehei* immobilized on a macroporous anion exchange resin (Lipozyme RM IM) and a lipase of *Candida antartica* Lipase B immobilized on acrylic resin (Novozyme 435). They are both currently being used in biodiesel productions [24,25]. The overall esterification conversion of the prepared product was 78% at 60°C, and it reached 98% and 88% with

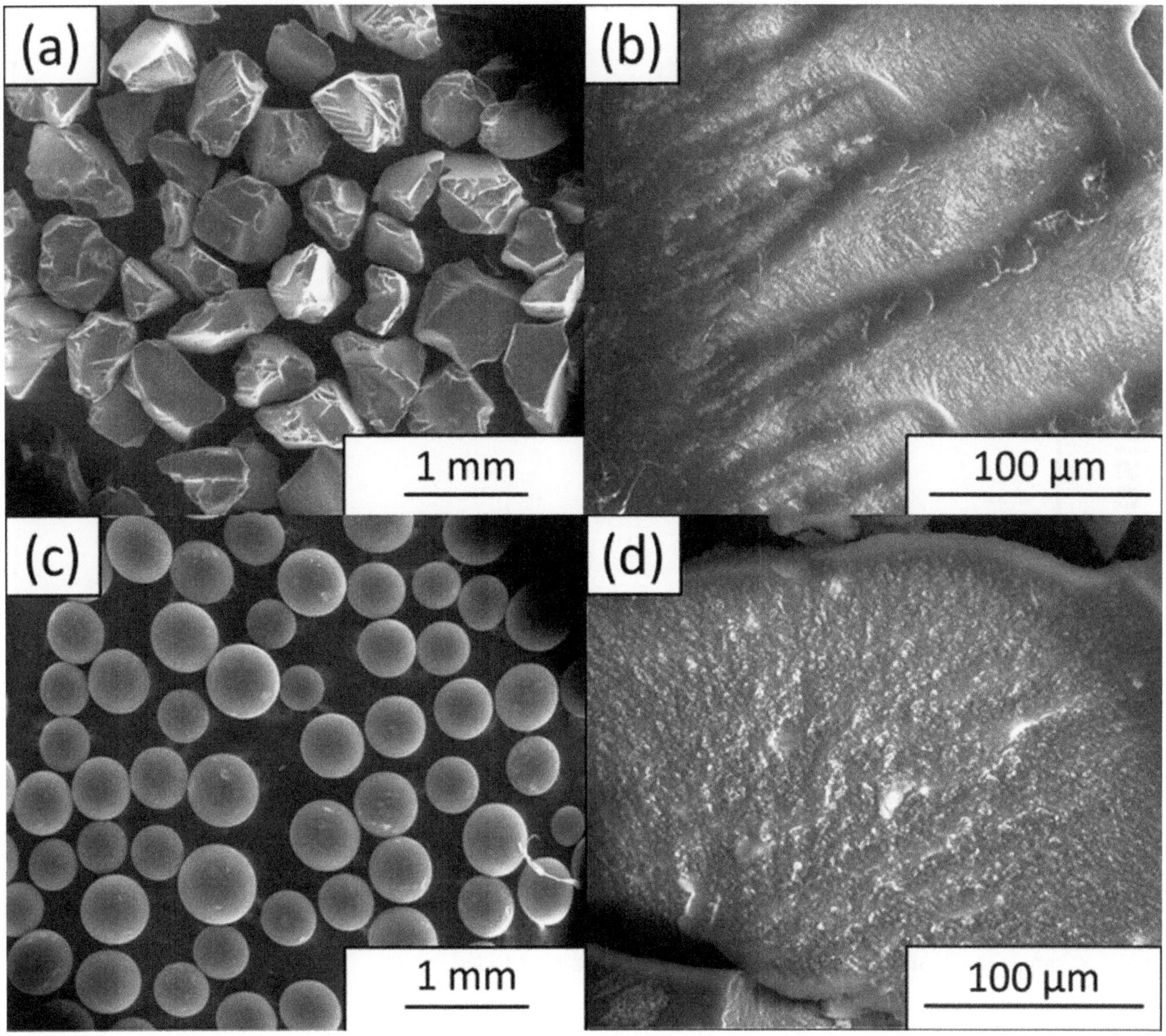

Figure 3. FESEM images of commercial (a and b) lipozyme and (c and d) novozyme.

commercially available lipozyme and novozyme, respectively, under similar experimental conditions.

Experimental Methods

1.1 Materials and characterization techniques

High purity sodium hydroxide (NaOH), ammonium hydroxide (NH_4OH, 25%), and acetic acid were purchased from Merck. Chitosan with a 75%–85% degree of deacetylation was purchased from Sigma Aldrich. Graphite flakes were purchased from Ashbury Graphite Mill Inc. The characterizations of the biocatalysts were performed using an FEI Nova NanoSEM 400 field emission scanning electron microscope (FESEM), Philips X-Ray diffractometer (XRD), Renishaw inVia Raman microscope, and TA-Q500 thermo-gravimetric analyzer.

1.2 Synthesis of chitosan/graphene oxide beads

The chitosan/graphene oxide beads were prepared using the following procedure. Initially, a chitosan solution (2% (w/v) of chitosan in a 2.5% acetic acid solution) was added drop-wise to a coagulant bath containing a 1 M NaOH solution, and then the beads were obtained after 4 h of coagulation. The chitosan beads were washed 5 times with deionized water in order to remove the excess acetic acid and NaOH. Then, the chitosan beads (100 μg/mL) were mixed with GO. The GO was prepared using a simplified Hummer's method [26] and was reduced in size by sonication using a 600 W horn sonicator for 1 h (represented as CS/GO-S beads). For comparison, the same protocol was used to prepare CS/GO beads without sonication. The beads were washed and freeze-dried for 24 h.

1.3 Functionalization of chitosan/graphene oxide beads

A 50 mg quantity of the chitosan/graphene oxide beads was dispersed into 50 mL of phosphate buffer saline (PBS) solution (pH 7), and then N-Hydroxysulfosuccinimide sodium (NHS) was added. The mixture was subjected to sonication for 15 min, and then 1-ethyl-(3-dimethylaminopropyl) carbodiimide (EDC) (20 mmol/L) was added. The resulting mixture was shaken at

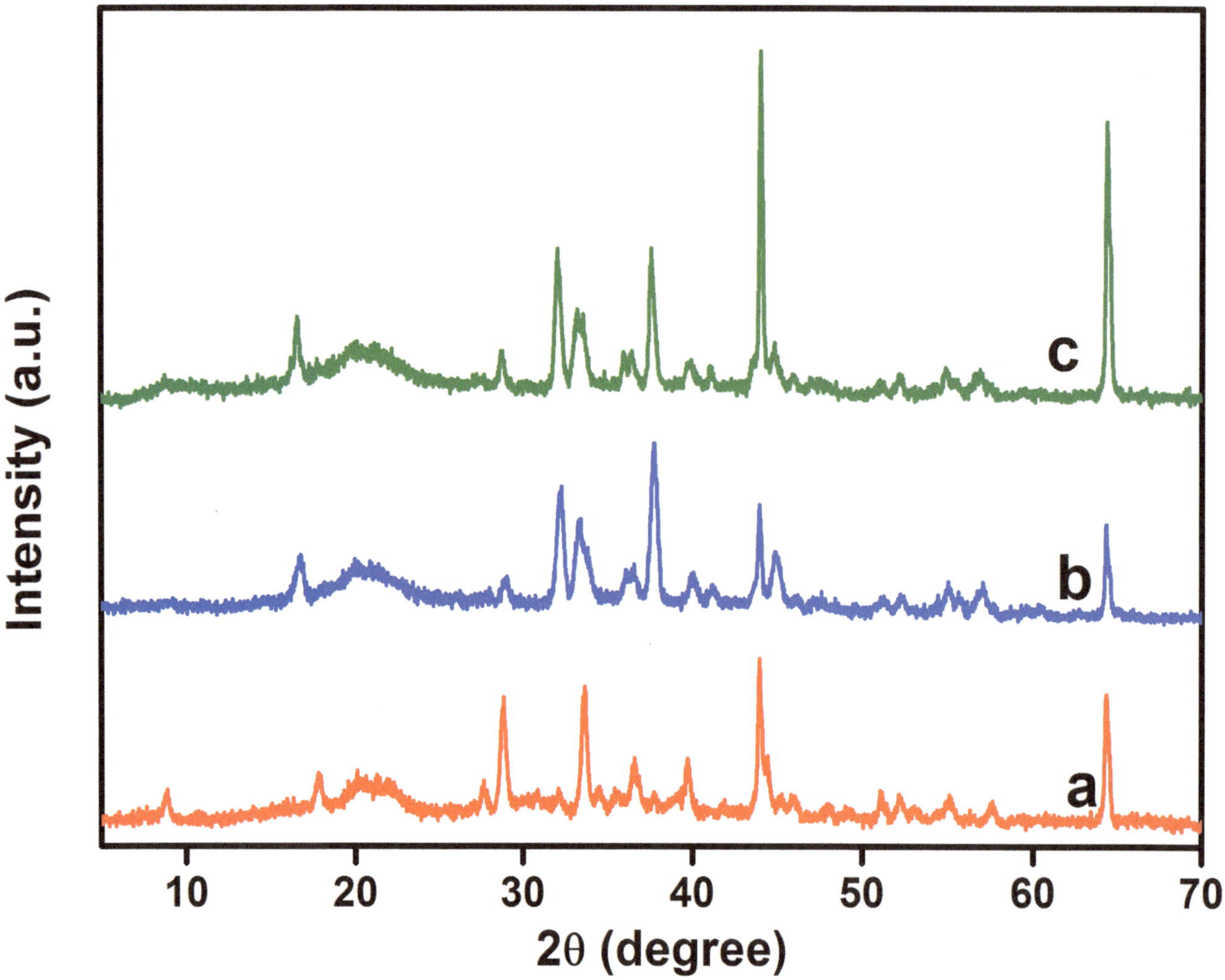

Figure 4. XRD patterns obtained for (a) CS, (b) CS/GO, and (c) CS/GO-S beads.

200 rpm for 60 min. The activated solution was then filtered through filter paper and rinsed thoroughly with PBS buffer to remove the excess EDC and NHS.

1.4 Immobilization of lipase with functionalized beads

The functionalized chitosan/graphene oxide beads were transferred to a solution of lipase (5 mg/mL). The immobilization of the lipase was performed at 25°C with 150 rpm stirring for 15 h. After conjugation, the mixture was filtered using filter paper and 6 washes were performed typically, with PBS solution added each time to remove the unbound lipase. The lipase immobilized with CS/GO with the aid of coupling agents is denoted as CS/GO-CL, and the lipase immobilized with CS/GO-S with the aid of coupling agents is denoted as CS/GO-SCL. As a positive control, CS/GO not treated with a coupling agent was prepared and used to immobilize lipase, and is denoted as CS/GO-L. Similarly, CS/GO-S not treated with a coupling agent was prepared and used to immobilize lipase, and is denoted as CS/GO-SL. Likewise, CS is denoted as CS-L and CS-CL upon the immobilization of lipase in the absence and presence of coupling agents.

1.5 Esterification of lauric acid and oleyl alcohol

1.5.1 Water bath method. The reaction mixture consisted of lauric acid (1 mmol), oleyl alcohol (1 mmol), and a biocatalyst (0.5 w/w). Hexane was added to make a total volume of 10 mL. The reaction mixture was incubated in a horizontal shaker water bath at a speed of 150 rpm at 40°C for 24 h [27]. All of the experiments were performed in triplicate, and control experiments were performed without biocatalysts.

1.5.2 Oil bath method. The reaction mixture consisted of lauric acid (1 mmol), oleyl alcohol (1 mmol), and a biocatalyst (0.5 w/w). Hexane was added to make a total volume of 10 mL. The reaction mixture was stirred in an oil bath at 60°C for 4 h [28]. All of the experiments were conducted in triplicate, and control experiments were performed without biocatalysts.

1.6 Analysis

The reaction was terminated by dilution with 3 mL of ethanol:acetone (50:50, v/v), and the biocatalyst was filtered through filter paper. The unreacted lautic acid in the reaction mixture was determined by titration with 0.1 M NaOH using a pH meter to an end point of pH 9.9±0.1. The activity of the

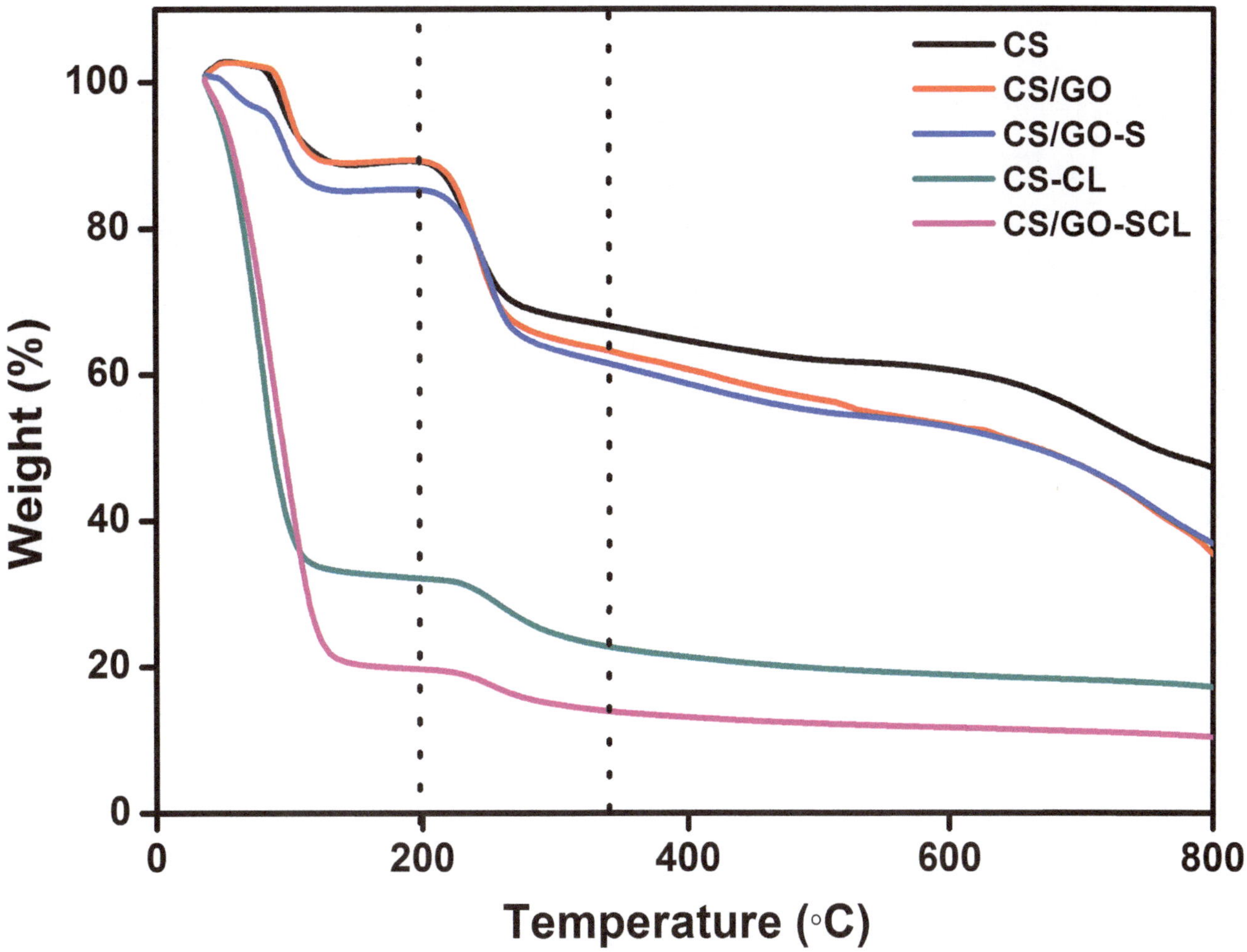

Figure 5. Thermogravimetric analysis results for prepared samples.

biocatalyst was expressed as a conversion percentage, as shown in Eq. (1).

$$\text{Percentage of Conversion} = \frac{V_{control} - V_{sample}}{V_{control}} \times 100\% \qquad (\text{Eq.}1)$$

Where $V_{control}$ = volume of 0.1 M NaOH needed to titrate the control and V_{sample} = volume of 0.1 M NaOH needed to titrate the sample.

The biocatalyst activity was determined by using the Kwon and Rhee method [29]. First, a substrate was prepared using olive oil and phosphate buffer (1:1 ratio, pH 7, 50 mM). A control was prepared by mixing 0.02 mL of $CaCl_2$, 2.5 mL of the olive oil–PBS mixture, and 1 mL of the PBS solution. A sample contained 0.02 mL of $CaCl_2$, 2.5 mL of the olive oil–PBS mixture, 1 mL of the PBS solution, and 5 mg of the biocatalyst. The mixture was then incubated at 20°C and 200 rpm for 30 min in a horizontal shaker water bath. Then, 1 mL of HCl and 5 mL of isooctane were added to each of the reaction mixtures and vortexed for 30 s. After the mixture had settled down and two layers formed, 4 mL of the upper layer was transferred into a new test tube, and 1 mL of copper pyridine was added. It was then vortexed for 30 s, and the mixture was allowed to settle for 30 min. A 1 mL quantity of the mixture was then placed into a cuvette, and absorbance reading at 715 nm was recorded in triplicate. A blank test was carried out using isooctane. The required formula for calculating the enzyme activity is shown in Eq. (2).

$$Y = \frac{Abs_{715}(\text{sample} - \text{control})}{0.020} / (\text{incubation time} \times \text{enzyme volume}) \qquad (\text{Eq.}2)$$

Results and Discussion

The morphologies of the chitosan beads with coagulation periods of 2, 4, and 20 h were studied using the FESEM and are shown in **Fig. 1**. The beads prepared after 2 h of coagulation were fragile and lacked a definite structure (**Fig. 1a**). At a microscopic level, in which 2000× magnification was used, the beads were found to lack porosity and had a featureless appearance (**Fig. 1b**). When the coagulation time was doubled, the beads reflected a much improved structure, which was firm and spherical, indicating an optimum aging time for the formation of well-defined spherical beads (**Fig. 1c**). A higher degree of porosity at the micron scale resulting from scaffolds made up of paper-like thin walls of concave pockets could be observed

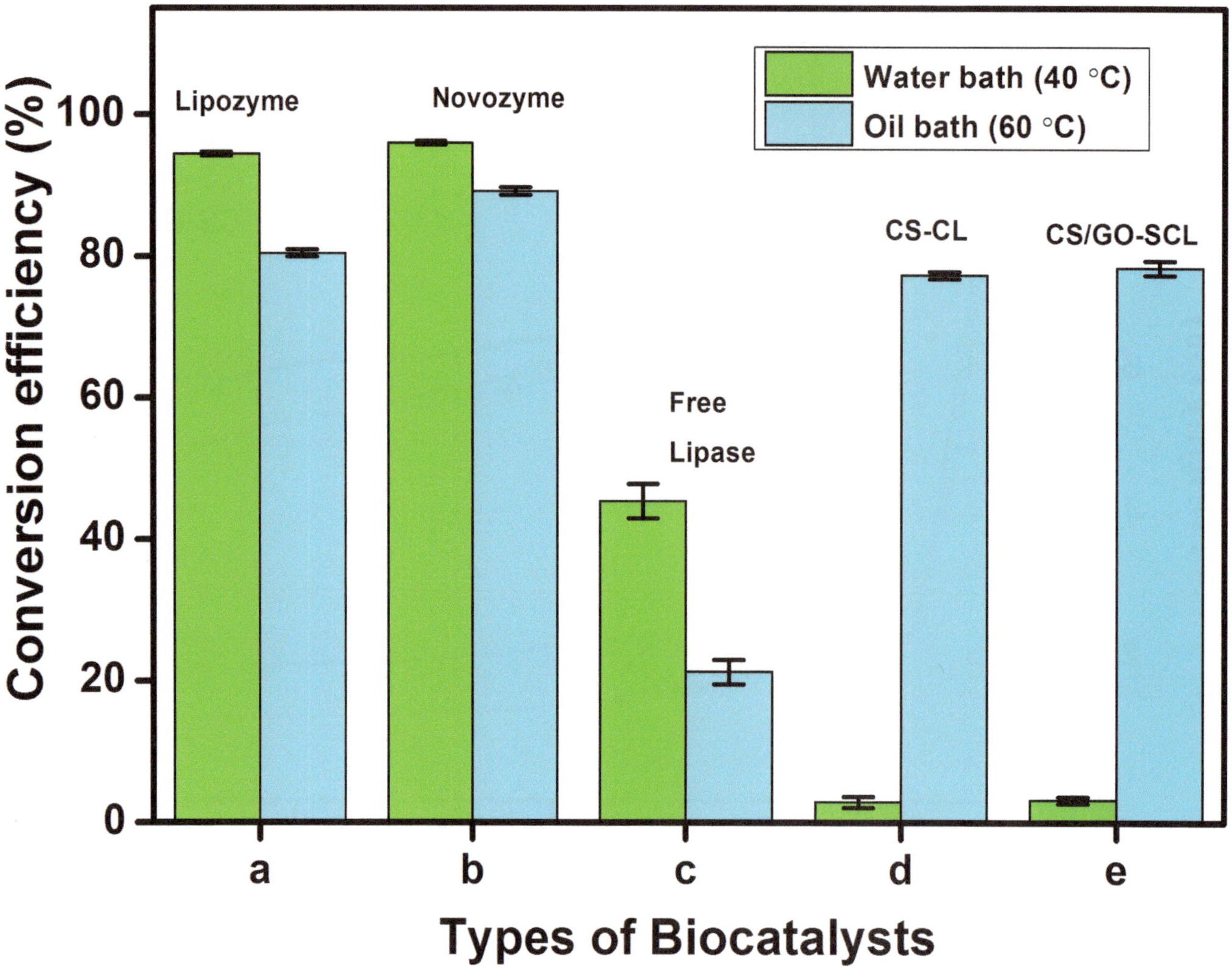

Figure 6. Lipase activities for lipozyme, novozyme, free lipase, CS-L, CS/GO-SL, CS-CL, and CS/GO-SCL.

(**Fig. 1d**). At a prolonged coagulation time of 20 h, the beads shrunk and lost their spherical integrity (**Fig. 1e**). Upon scrutiny, the morphology of the beads appeared to be in layers, suggesting that they once had thin walls with a continuous interconnected structure that had collapsed, causing the pores to be indistinguishable (**Fig. 1f**).

Even though the CS beads prepared at the optimized coagulation time of 4 h showed better formed spherical beads, they failed to maintain their mechanical strength because they were brittle when touched. Therefore, the CS beads were incorporated with GO to enhance their mechanical structure. The inclusion of GO changed the morphology of the beads, as shown in **Fig. 2**. The CS/GO-S bead had a more rigid and spherical structure (**Fig. 2a**), and its FESEM image (**Fig. 2b**) shows a spiky structure that resulted from the inclusion of sonicated GO nanoparticles, which could further enhance its mechanical strength. The CS/GO bead had a less rigid spherical structure (**Fig. 2c**), and its FESEM image (**Fig. 2d**) shows concave pockets with smaller pores compared to those of **Fig. 1d**. It also shows some nanostructures, which resulted from the presence of a small portion of GO nanosheets. For comparison purposes, the FESEM images for commercially available lipozyme and novozyme were recorded and are shown in **Fig. 3**. It can be seen from **Fig. 3(a and b)** that the lipozyme has an irregular polygonal shape, and it has an uneven surface with a compact structure, which is very different from the highly porous structure of the prepared beads. Novozyme has a definite round shape (**Fig. 3(c and d)**) with a relatively smooth surface and is as compact as the lipozyme. These two commercial biocatalysts are dense and completely absent of pores, which are in contrast with the prepared biocatalysts.

In **Fig. 4**, the XRD profile of CS presents two intense peaks at $2\theta = 28^\circ$ and 34°, which are assigned to JCPDS card No: 039–1894. The XRD profiles of both CS/GO and CS/GO-S differ only slightly from that of CS, which indicates that the GO nanosheets were exfoliated and uniformly dispersed within the polymer matrix. The typical peak of GO at around 9°–11° is missing [30], evidencing the complete mixing of GO within the matrix of CS. The XRD results indicate that a significant amount of the amino groups of CS were inserted between the GO layers. After an amino nucleophilic substitution reaction with the epoxy groups of GO, the exfoliation of GO occurred. Hence, the destruction of its layered structure resulted in the disappearance of the typical peak of GO [31]. Furthermore, the incorporation of the fillers did little to affect the crystalline structure of CS because of the low GO content. However, the peak at $2\theta = 28^\circ$ was greatly reduced in intensity for the nanocomposites, implying that the crytalline plane was altered by the presence of GO. Moreover, the

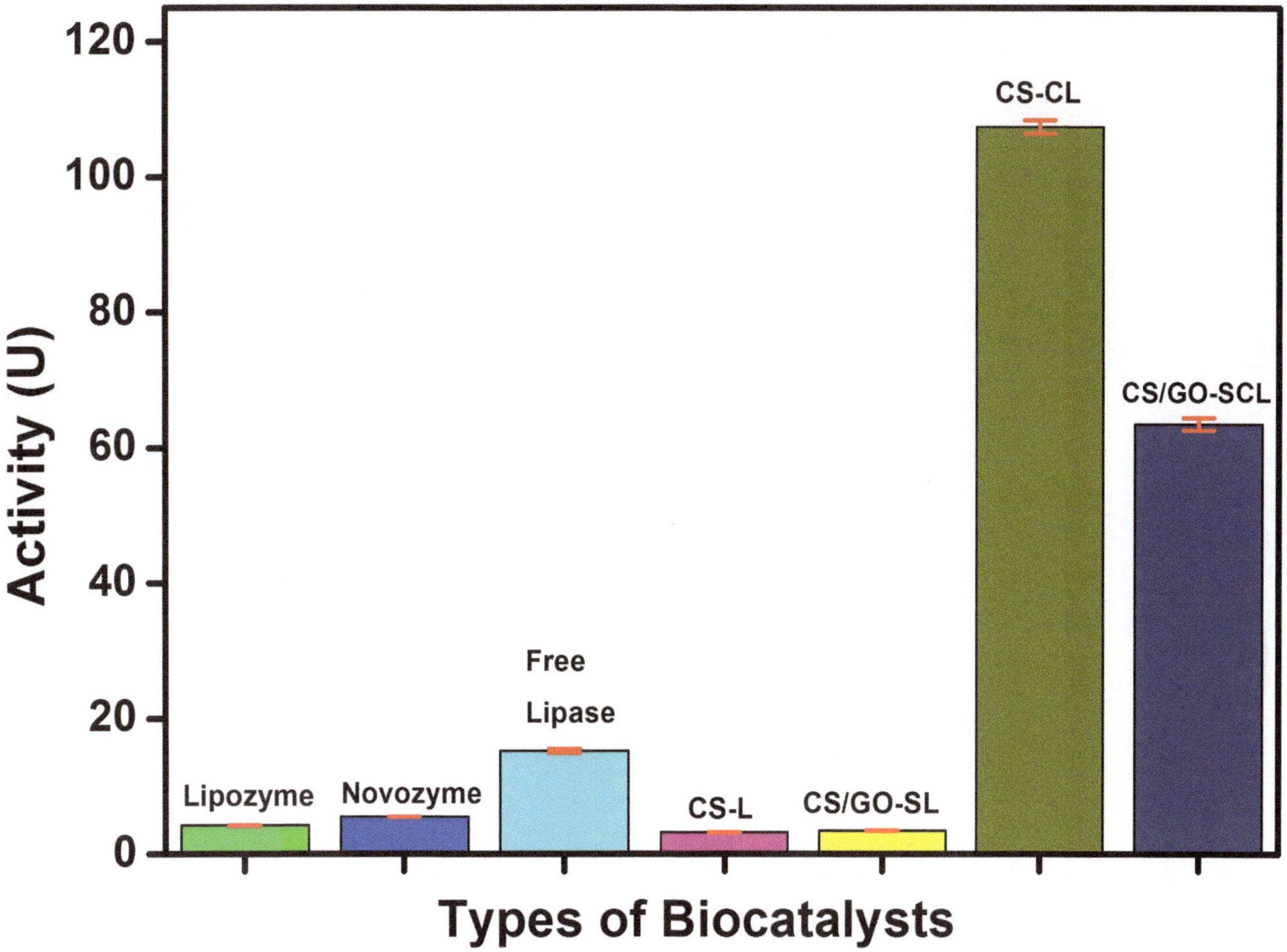

Figure 7. Percentage of conversion into wax ester after esterification process between lauric acid and oleyl alcohol catalyzed by (a) lipozyme, (b) novozyme, (c) free lipase, (d) CS-CL, and (e) CS/GO-SCL using water bath (on right) and oil bath (on left) methods.

sharp peak at $2\theta=37.5^\circ$ indicated that the regular stack of GO was broken, in addition to the absence of a graphite peak at $2\theta=26.58^\circ$ [32].

The thermal analysis of all the samples is clearly divided into 3 regions (**Fig. 5**). The first region from 40°C to 200°C [33,34] indicates the loss of water molecules, whereas the second region from 200°C to 350°C [34,35] is mainly because of the decomposition of the polysaccharide units of CS. The third region after 350°C indicates the degradation of 2surface –COOH groups [36], which was contributed by GO. For CS-CL and CS/GO-SCL, the water left behind with the matrix upon immobilization is reflected in the massive loss of water in the first region. The loss of lipase also contributed to this large decrease [33]. The weight difference trends for the second region were similar for all 5 curves regardless of the lipase immobilization, which implied that lipase was immobilized through physisorption [33].

Based on **Fig. 6**, the CS-L and CS/GO-SL possessed nearly the same lipase activities, which were 3.23 and 3.48 unit, respectively. This shows that the involvement of GO caused little increase in the lipase activity but improved the mechanical strength of the support [23]. During the lipase immobilization steps, the CS/GO-SL beads were able to maintain their shapes after 24 h of shaking, but some CS-L beads had their structures ruptured. The dismally low lipase activity was tremendously enhanced after using the coupling agents because the use of NHS and EDC can enhanced the lipase immobilization onto supports [37]. The CS-CL beads possessed a higher absorbance, with a calculated activity of 107.40 unit, compared to CS/GO-SCL beads, which had a lower enzyme activity of 63.53 unit. The significantly high activity of the coupled CS-CL and CS/GO-SCL could be explained in two ways. Initially, the biomimetic surface layer of the coupled beads was able to create a biocompatible environment for the immobilized lipase and thus retain the high lipase activity. Then, the hydrophilic biomimetic layer could expose the active sites of the lipase and reduce the random coupling of lipase and the beads [37]. The importance of hydrophilicity is clearly portrayed by the minimal enzyme activity of the commercial lipozyme and novozyme. During the enzyme assay, the hydrophobic lipozyme and novozyme adhered to the walls of the vials, which caused a lipase loss and data inaccuracy. The free lipase possessed a lipase activity of 15.22 unit. This proved that the beads had successfully increased the lipase activities by exposing the active sites upon immobilization onto the polymer matrix and thus acted as a good biocatalyst support [37].

Fig. 7 shows the percentage of esterification catalyzed at two different temperatures of 40°C and 60°C, as controlled by water bath [27] and oil bath [28] methods. The novozyme and lipozyme showed the highest conversions of ~96% and ~95% using the water bath method [38,39]. The conversion efficiencies of the novozyme and lipozyme were significantly higher than those of the other biocatalysts. Free lipase had a 45% conversion, whereas CS/GO-SCL and CS-CL possessed 3.16% and 2.88%, respectively. The poor conversion of the beads could plausibly be due to the poor mixing among the reactants, solvent, and biocatalyst because the biocatalysts floated on the reaction mixture in the simple agitation state. In contrast, the commercial biocatalysts mixed thoroughly with the reactants and solvent because they were fully dispersed in the reaction medium. The immersion of the biocatalyst in the reaction medium was influenced by the density of the biocatalyst. The commercial biocatalysts consisted of compact beads with a diameter of 450 μm, whereas the prepared porous biocatalysts had an average diameter of 430 μm.

The conversion efficiencies of CS-CL and CS/GO-SCL using the oil bath method were 77% and 78%, respectively, because of the improved lipase activity in the biocatalysts. However, the physical integrity of CS-CL was compromised at 60°C because the beads crumbled significantly at the end of the reaction compared to that of CS/GO-SCL. In addition to the agitation state during the reaction, the heat energy at the higher temperature of 60°C influenced the rate of reaction by increasing the collision frequency between the enzyme and support [28]. Moreover, a higher temperature reduced the mixture viscosity and improved the support diffusion process. However, a further increase in the incubation temperature, to more than 65°C, would lead to enzyme denaturation [28]. Even though the lipozyme and novozyme maintained the highest conversion efficiencies of 80% and 89%, respectively, the water bath method was better than the oil bath method for the commercial biocatalysts. The commercial biocatalysts decomposed to nothing at the oil bath temperature of 60°C. This resulted in the release of lipase, which then recoiled to the state of neat lipase, causing the active sites of the enzyme to be hidden once again.

Conclusion

Chitosan beads incorporated with sonicated GO provided a much improved platform for the immobilization of lipase compared to the non-sonicated GO. This was because the smaller size GO was well-dispersed within the matrix of the chitosan polymer, creating uniformity in the active sites throughout the beads. In the presence of coupling agents, NHS and EDC, the lipase activity improved as much as 5 fold and 3 fold for chitosan and CS/GO, respectively. The CS beads containing GO had a much better mechanical strength compared to those of the neat CS and commercially available lipozyme and novozyme because the CS/GO beads almost maintained their initial physical appearance after the vigorous esterification between lauric acid and oleyl alcohol. Even though the thermal analysis showed that lipase was physically adsorbed onto the prepared beads, they possessed the high catalytic activity of esterification. The conversion efficiency of the prepared CS/GO biocatalyst was more than 85% with the commercially available lipozyme and novozyme. Considering that the cost of 10 g of lipozyme is 400 USD, whereas that of novozyme exceeds 900 USD, the prepared lipase immobilized beads could provide a cheaper alternative for lipase usage.

Author Contributions

Conceived and designed the experiments: HNL MB HRFM. Performed the experiments: SCL HNL. Analyzed the data: SCL HNL MB HRFM AAT NMH AP CHC YA. Contributed reagents/materials/analysis tools: SCL HNL MB HRFM AAT NMH AP CHC YA. Contributed to the writing of the manuscript: SCL HNL NMH AP.

References

1. Behzad A, Nasim D (2014) Modified bentonite as catalyst for esterification of oleic acid and ethanol. Journal of the Taiwan Institute of Chemical Engineers 45: 1468–1473.
2. Orrego CE, Salgado N, Valencia JS, Giraldo GI, et al. (2010) Novel chitosan membranes as support for lipases immobilization: characterization aspects. Carbohydrate Polymers 79: 9–16.
3. Panagiota YS, Athanasios F, Michalis F, Maria K, Maria P, et al. (2013) Advances in lipase-catalyzed esterification reactions. Biotechnology Advances 31: 1846–1859.
4. Cabrera Z, Lorente GF, Lafuente RF, Palomo JM, Guisan JM (2009) Novozym 435 displays very different selectivity compared to lipase from *Candida antarctica B* adsorbed on other hydrophobic supports. Journal of Molecular Catalysis B: Enzymatic 57: 171–176.
5. Wang J, Meng G, Tao K, Feng M, Zhao X, et al. (2012) Immobilization of Lipases on Alkyl Silane Modified Magnetic Nanoparticles: Effect of Alkyl Chain Length on Enzyme Activity. PLoS ONE 7(8): e43478. doi:10.1371/journal.pone.0043478
6. Seema SB, Steven HN (2002) Immobilization of lipase using hydrophilic polymers in the form of hydrogel beads. Biomaterials 23: 3627–3636.
7. Xie WL, Wang JL (2012) Immobilized lipase on magnetic chitosan microspheres for transesterification of soybean oil. Biomass and Bioenergy 36: 373–380.
8. Mizuki T, Sawai M, Nagaoka Y, Morimoto H, Maekawa T (2013) Activity of Lipase and Chitinase Immobilized on Superparamagnetic Particles in a Rotational Magnetic Field. PLoS ONE 8(6): e66528. doi:10.1371/journal.pone.0066528
9. Won KH, Kim SB, Kim KJ, Park HW, Moon SJ (2005) Optimization of lipase entrapment in Ca-alginate gel beads. Process Biochemistry 40: 2149–2154.
10. Ines Belhaj BR, Zamen BR, Ali G, Hafedh B (2011) Esterification activity and stability of *Talaromyces thermophilus* lipase immobilized onto chitosan. Journal of Molecular Catalysis B: Enzymatic 68: 230–239.
11. Foresti ML, Ferreira ML (2007) Chitosan-immobilized lipases for the catalysis of fatty acid esterifications. Enzyme and Microbial Technology 40: 769–777.
12. Cabral PP, Fonseca MMR, Dias SF (2010) Esterification activity and operational stability of *Candida rugosa* lipase immobilized in polyurethane foams in the production of ethyl butyrate. Biochemical Engineering Journal 48: 246–252.
13. Jian X, Wang YJ, Yu H, Luo GS, Dai YY (2006) *Candida rugosa* lipase immobilized by a specially designed microstructure in the PVA/PTFE composite membrane. Journal of Membrane Science 281: 410–416.
14. Chiou SH, Wu WT (2004) Immobilization of *Candida rugosa* lipase on chitosan with activation of the hydroxyl groups. Biomaterials 25: 197–204.
15. Kawamura Y, Tanibe H, Imamura S, Harada J (1996) Lipase immobilized on a chitosan carrier. United States Patent 5508185.
16. Leuba, Jean L (1978) Enzymes immobilized on chitosan. United States Patent 4094743.
17. Roy SK, Todd JG, Glasser WG (1998) Hydrogel beads from chitosan. WIPO Patent Application WO/1998/041546.
18. Arena BJ (1981) Chitin- and chitosan-based immobilized metal catalysts. United States Patent 4274980.
19. Yi SS, Noh JM, Lee YS (2009) Amino acid modified chitosan beads: improved polymer supports for immobilization of lipase from *Candida rugosa*. Journal of Molecular Catalysis B: Enzymatic 57: 123–129.
20. Huang XJ, Dan G, Xu ZK (2007) Preparation and characterization of stable chitosan nanofibrous membrane for lipase immobilization. European Polymer Journal. 43: 3710–3718.
21. Kuo CH, Liu YC, Chang CM, Chen JH, Chang C, et al. (2012) Optimum conditions for lipase immobilization on chitosan-coated Fe_3O_4 nanoparticles. Carbohydrate Polymers 87: 2538–2545.
22. Hung TC, Giridhar R, Chiou SH, Wu WT (2003) Binary immobilization of *Candida rugosa* lipase on chitosan. Journal of Molecular Catalysis B: Enzymatic 26: 69–78.
23. Wang J, Ellsworth MW (2010) Graphene and graphene oxide aerogels. WIPO Patent Application WO/2010/065121.
24. Juan C, Cristo'bal V, Diego L, Enrique DS, Carlos L, et al. (2014) Selective ethanolysis of sunflower oil with Lipozyme RM IM, an immobilized *Rhizomucor miehei* lipase, to obtain a biodiesel-like biofuel, which avoids glycerol production through the monoglyceride formation. New Biotechnology 00: 00.
25. Kathleen FH, Goran TV, James CR, Zoltan N, Basudeb S (2014) Kinetics of the pre-treatment of used cooking oil using Novozyme 435 for biodiesel production. Chemical Engineering Research and Design 92: 713–719.
26. Huang NM, Lim HN, Chia CH, Yarmo MA, Muhamad MR (2011) Simple room-temperature preparation of high-yield large-area graphene oxide. International Journal of Nanomedicine 6: 3443–3448.

27. Gunawan ER, Basri M, Rahman MBA, Salleh AB, Rahman RNZA (2004) Lipase-catalyzed synthesis of palm-based wax esters. Journal of Oleo Science 53: 471–477.
28. Basri M, Kassim MA, Mohamad R, Ariff AB (2013) Optimization and kinetic study on the synthesis of palm oil ester using Lipozyme TL IM. Journal of Molecular Catalysis B: Enzymatic 85–86: 214–219.
29. Kwon DY, Rhee JS (1986) A simple and rapid colorimetric method for determination of free fatty acids for lipase assay. JAOCS 63(1): 89–92.
30. Lim HN, Huang NM, Loo CH (2012) Facile preparation of graphene-based chitosan films: enhanced thermal, mechanical and antibacterial properties. Journal of Non-Crystalline Solids 358(3): 525–530.
31. George ZK, Dimitrios NB, Eleni AD (2014) Advanced low-swelling chitosan/graphite oxide-based biosorbents. Materials Letters 128: 46–49.
32. Rifaqat AKR, Singh S, Singh BR, Khan W, Naqvi AH (2014) Synthesis and characterization of surface modified graphene–zirconium oxide nanocomposite and its possible use for the removal of chlorophenol from aqueous solution. Journal of Environmental Chemical Engineering 2: 199–210.
33. Lee DG, Kanagasabai MP, Kim M, Hwang S, Ahn IS, et al. (2009) Immobilization of lipase on hydrophobic nano-sized magnetite particles. Journal of Molecular Catalysis B: Enzymatic 57: 62–66.
34. Zang LM, Qiu JH, Wu XL, Zhang WJ, Sakai E, et al. (2014) Preparation of magnetic chitosan nanoparticles as support for cellulase immobilization. Industrial & Engineering Chemistry Research 53: 3448–3454.
35. Lu XX, Luo ZG, Yu SJ, Fu X (2012) Lipase-catalyzed synthesis of starch palmitate in mixed ionic liquids. Journal of Agricultural and Food Chemistry 60: 9273–9279.
36. Tripti R, Arpana B, Lalit MM, Amita RS, Datta M (2013) Robust nanobioconjugates of *Candida antarctica lipase B*–Multiwalled carbon nanotubes: characterization and application for multiple usages in non-aqueous biocatalysis. Bioresource Technology 140: 103–110.
37. Peng Y, Xu ZK, Jian W, Christophe I, Patrick S (2006) Nanofibrous poly(acrylonitrile-co-maleic acid) membranes functionalized with gelatin and chitosan for lipase immobilization. Biomaterials 27: 4169–4176.
38. Nikolaus W, Petra W, Kumar DM (2001) Steryl and stanyl esters of fatty acids by solvent-free esterification and transesterification in vacuo using lipases from *Rhizomucor miehei, Candida antarctica*, and *Carica papaya*. Journal of Agricultural and Food Chemistry 49: 5210–5216.
39. Estela HM, Christina O (2009) Enzymatic re-esterification of lower glycerides from soybean oil with conjugated linoleic acid (CLA). Journal of Agricultural and Food Chemistry 57: 701–708.

6

The Evolution of Enzyme Specificity in the Metabolic Replicator Model of Prebiotic Evolution

Balázs Könnyű[1], Tamás Czárán[2]*

1 Department of Plant Taxonomy and Ecology, Institute of Biology, Eötvös Loránd University, Budapest, Hungary, 2 Theoretical Biology and Ecology Research Group, Institute of Biology, Eötvös Loránd University, Budapest, Hungary

Abstract

The chemical machinery of life must have been catalytic from the outset. Models of the chemical origins have attempted to explain the ecological mechanisms maintaining a minimum necessary diversity of prebiotic replicator enzymes, but little attention has been paid so far to the evolutionary initiation of that diversity. We propose a possible first step in this direction: based on our previous model of a surface-bound metabolic replicator system we try to explain how the adaptive specialization of enzymatic replicator populations might have led to more diverse and more efficient communities of cooperating replicators with two different enzyme activities. The key assumptions of the model are that mutations in the replicator population can lead towards a) both of the two different enzyme specificities in separate replicators: efficient "specialists" or b) a "generalist" replicator type with both enzyme specificities working at less efficiency, or c) a fast-replicating, non-enzymatic "parasite". We show that under realistic trade-off constraints on the phenotypic effects of these mutations the evolved replicator community will be usually composed of *both types of specialists* and of a limited abundance of *parasites*, provided that the replicators can slowly migrate on the mineral surface. It is only at very weak trade-offs that generalists take over in a phase-transition-like manner. The parasites do not seriously harm the system but can freely mutate, therefore they can be considered as pre-adaptations to later, useful functions that the metabolic system can adopt to increase its own fitness.

Editor: David Liberles, University of Wyoming, United States of America

Funding: The project was supported by the Hungarian Research Foundation (OTKA) Grant No. K67907. The funders had no role in study design, data collection and analysis, decision to publish, or preparation of the manuscript.

Competing Interests: The authors have declared that no competing interests exist.

* E-mail: czaran@ludens.elte.hu

Introduction

The complex, specific and efficient biocatalysts (enzymes) present in all recent forms of life are obviously the products of long Darwinian adaptation, therefore they could not be present on prebiotic Earth. The strongest evolutionary pressure must have affected the spatial structure and the amino acids closest to the active centre of the enzymes, i.e., the details of the peptide structure most responsible for the catalytic function. Maintaining the specific amino acid sequence (primary structure) of an enzyme has evolutionary relevance only as long as it affects its catalytic function. It is mostly the evolutionary "wobbling" of the less important structural elements in distant taxa that explains the divergence of enzymes which are similar in terms of the spatial structure of their active centres and in their biological function (i.e., they catalyse the same reaction), but still have very different amino acid sequences.

Studies of early evolution have long acknowledged the essential role catalysts must have played in the origin of life. It was in Wächtershäuser's hypothesis of the *"prebiotic pizza"* where inorganic compounds were first assumed to carry the function of early biocatalysts [1–3]. According to the hypothesis the first chemical reactions ultimately leading to life would have taken place on mineral (e.g., pyrite) surfaces. The prebiotic pizza offers a solution to the thermodynamic problem of condensation reactions yielding water molecules among their products in an environment saturated with water (i.e., in the *"prebiotic soup"*; [4]). The surface of pyrite is also capable of adsorbing organic substrates, thereby increasing their concentrations and catalysing their reactions. Even though this inorganic catalytic effect might have been neither very specific nor very efficient, it might have been sufficient to aid the emergence of the first relatively complex, enzyme-like macromolecule catalysts, which could have worked bound to the surface themselves thereafter [5,6]. Whether these first catalytic macromolecules were peptides (like most biocatalysts today) or some very different chemical species is still debated [7,8]. The reason why the first evolvable biocatalysts could not be peptides is that the transmission of the information content of peptides to the next generation would have required a very complicated and very specific cellular apparatus - as a peptide-based genetic machinery - for which there is no evidence in extant organisms [4,9]. It seems more plausible to assume that the first enzymes were macromolecules produced on mineral surfaces, capable of playing the dual role of the phenotypic function of biocatalysis and that of genetic information transfer, thus evading the difficult problem of the lack of translation at the wake of life. The most likely candidate for this role is the RNA molecule, because it can bind to mineral surfaces due to its electrostatic charges, and its production on mineral surfaces has been proven experimentally [10–13]. The nucleobase sequence of RNA stores biologically meaningful information that, in principle, is not difficult to copy, and within-strand Watson-Crick type base-pairing may produce various complicated and reproducible spatial structures which may show very different catalytic activities depending on the base sequence of the molecule (*RNA World*, [14–19]). Recent empirical research has provided ample support to this hypothesis by demonstrating the wide spectrum of catalytic activities of RNA, which led to the hypothesis of a surface-bound RNA World [10–13,15,20,21].

Of course there remain many – so far unresolved – questions with regard to the chemical mechanisms implied by the RNA World hypothesis. To mention just a few: the synthesis of ribose, phosphates and the four nucleobases under plausible prebiotic environmental conditions is far from straightforward, but in the light of recent empirical work they seem not hopeless either. For example the *formose reaction* (a possible route for ribose synthesis) and the *Stercker-synthesis* (amino acids, nucleobases and many more biologically important compounds) require very different chemical milieu [22–25] suggesting that they should have been separated in space if these were to produce ribose and nucleobases for RNA replication. However, recently it was shown that there exists a different chemical route to nucleotides which works at neutral pH and provides an excellent yield [26,27]. Another notorious problem is that of the concatenation of nucleotides: without activator agents (e.g. imidazole) the proper 3′-5′ direction is not exclusive in the concatenation reaction. Schwartz and his co-workers have shown that phosphites are able to conduct the concatenation of nucleotides in the right direction in water without any activator agent present, and the phosphite would oxidise to phosphate during the reaction [28]. Mineral surfaces also promote the proper 3′-5′ direction of concatenation into long oligonucleotides (40–60-mers) [13,25] and the homochiral segregation of sugar molecules [29,30] without which the polymerization process would be blocked by enantiomeric cross-inhibition [31]. von Kiedrowski's suggests that RNA constituents must have been polymerized on mineral surfaces, but many other reactions essential on the way to life could have taken place in the water body of the prebiotic ocean [32].

Yet another problematic issue of the RNA World hypothesis is the "Holy Grail" of prebiotic evolution: the self-replication of RNA molecules, or, more precisely, the lack of an efficient RNA-replicase ribozyme. The problem is connected to *Eigen's paradox* [33] which states in the present context that the precise replication of long ribozymes requires long ribozymes to catalyse their replication. This remains true even if neutral-mutations are considered [34]. There are two possible solutions for the information integration problem: either many short RNA molecules need to cooperate in a compulsory fashion, or an RNA-replicase ribozyme has to evolve somehow "from scratch", and maintain its own sequence just like that of many other, cooperating ribozymes. Recent experimental results show that both ways are chemically feasible to some extent already. For example, four RNA molecules are shown to cooperate in catalysing their own ligation into a single strand ribozyme [35–37]. RNA replicase ribozymes have been seeked for for decades, but it was only very recently that a substantial step forward was taken in that respect: in an *in vitro* evolution project [38] a 189-nucleotide RNA molecule has been discovered that was able to elongate its own 95-nucleotide primer in a template-directed manner. The fidelity of the replication process was 99.4%.

In spite of such empirical "missing links" in connection to the RNA World hypothesis it seems quite probable that the earliest self-reproducing macromolecules with catalytic activities would have been RNA (or very similar) molecules. Theoretical studies suggest that simple (RNA-like) molecules of very weak specificity and efficiency could have evolved to much more specific and efficient enzymatic replicators, through gradual adaptation [39–41]. However, due to strict constraints on their spatial structure, the evolution of enzymes is not as simple as assumed in these models. The longer the replicator the more complex its spatial structure will be. This has two important consequences: first, it is not easy to find the "native" conformation of a long macromolecule, and second, a complicated spatial structure makes template-copying more difficult, because long molecules have to be unfolded before being replicated. Such long and complex molecules have a significant fitness handicap at replication compared to shorter and simpler, therefore easy-to-copy ones. However, there is an indirect way to compensate for the direct fitness loss due to structural complexity: complex replicators may have much better catalytic properties. Provided that – due to its catalytic effect – the complex replicator can significantly contribute to the production of its own monomers, its local monomer supply will be higher than that of its simpler, non-catalytic competitors. This indirect fitness advantage may be sufficient to overcompensate the fitness loss caused by decreased replicability, and the catalytic replicator may spread in the population. Of course the above argument applies only to single-step monomer production, i.e., we have to assume that the successful replicator enzyme **A** catalyzes the terminal step of the production of its own monomers – and that all the other resources of replication are present in the environment in sufficient concentrations (i.e., the replicator population is heterotrophic in all other respects). Such an ecological situation is very unlikely to last long: due to increased consumption, some of the substrates of the catalysed reaction will become in short supply as the increasing population of replicator enzyme **A** depletes them, creating selective pressure towards a second enzymatic activity **B** to produce the substrate of replicator **A**, and so on. This Liebig type argument constitutes part of Szathmáry's "progressive sequestration" scenario [42].

How can this second enzymatic activity **B** be obtained by a population of replicators showing enzyme activity **A**? There are essentially two plausible ways to get there: *"enzyme promiscuity"* and *"enzyme cooperation"*. "Enzyme promiscuity" means that a single enzyme is capable of catalyse two (or more) different reactions. This was considered an almost impossible option until very recently, because the dominant view on peptide folding allowed only for a single "native" conformation, and therefore just a single catalytic activity for a peptide enzyme. The conventional view seems to be considerably weakened by now, because quite a few peptide and RNA enzymes have been revealed to admit either more than one native conformations, each of which have to be present to accomplish a single function [43], or more than one native conformations, each with a different function [44–47]. Such "multi-functional" enzymes are pre-adapted for later diversification and specialization, to yield different single-functional types, basically by each type gradually losing all but a single activity, and the activity retained becoming more and more specific and efficient. This leads to the second option: the cooperation of two or more different enzymes within the same *"molecular community"*. Coexistence and (co)evolution within such molecular communities have been studied for more than 30 years now, mainly by means of mathematical and simulation models, all of which seek a solution to the basic ecological problem: how could a number of different molecular replicators – using common resources for their replication, therefore being strong competitors of each other – form a stable assembly? All the answers given so far assume some kind of cooperation within the replicator community, to offset the effect of competition which would certainly ruin molecular diversity otherwise. The hypercycle model assumes that the cooperation may be direct, each replicator specifically catalysing the replication of just one other member of the molecular community in a circular topology [40,33]. Kaufmann's autocatalytic network model [7] also assumes specific and direct catalytic help among different replicators, but not in the rigid circular topology of the hypercycle. According to our metabolic replicator model, the molecular species of the community may be coupled through indirect mutualism, in which case cooperation and coexistence are mediated by a common metabolism which is driven by the replicators as enzymes producing monomers for the community [48,49]. Whichever approach we take, we need to answer two questions: 1) what makes and keeps the molecular cooperators of the community distinct and specialized, and 2) what saves the community from being invaded and destroyed by parasitic replicators?

We attempt to answer these questions using a simple model of enzymatic speciation in a surface-bound, metabolically coupled replicator population. Each replicator might have two enzyme activities (E_1 and E_2, catalysing two different keystone reactions of the metabolism), but this versatility is paid in catalytic efficiency: both enzyme activities of the "generalists" are weak. E_1 can be increased by mutation, but only at the expense of E_2, and *vice versa*, i.e., the two enzyme activities are in trade-off. Schultes & Bartel [46] have provided solid empirical proof of the E_1/E_2 trade-off for two different folds of the same RNA sequence, each of which had a different catalytic activity: one fold was a ligase, the other catalysed a cleavage reaction. The two enzyme activities have shown strict trade-off: the ligase fold was very weak at cleavage, the cleavage fold was bad at ligation, and the artificial intermediate folds were weak in both activities.

In accordance with the steric constraints on replication mentioned above, we assume that efficient enzyme replicators have a complicated secondary structure which makes them difficult to copy. We assume that replication is a simple, template-directed, non-enzymatic process weakly catalysed by the surface itself. This amounts to postulating that enzyme activities and replication rates (fitnesses) are also traded off: good enzymes make poor templates for replication, and *vice versa*. With these restraints we ask under what circumstances the emergence and the persistence of a community of specialized metabolic replicators can be expected, and to what extent parasitic sequences (fast replicating, short molecules of weak catalytic activity) undermine the efficiency and stability of the metabolic system.

Methods

The model is a two-dimensional cellular automaton of toroidal lattice topology to avoid edge-effects. Each of the 300×300 square lattice sites may be empty or occupied by a single replicator molecule at any point of time. The basic assumptions with respect to replication and metabolism are similar to those of our earlier Metabolic Replicator Model [48]:

i. replicators bind to a mineral surface (represented by the lattice) and they are template-replicated there;
ii. each replicator has a "basic fitness" parameter, which is its replication rate (k) under ideal environmental conditions; this parameter specifies the quality of a replicator - as a template - in its own replication process.
iii. each replicator is able to catalyze one or two essential reactions of a hypothetical metabolic network;
iv. each of the two catalytic activities (E_1 and E_2) are necessary for metabolism to produce monomers for replication;
v. metabolism supplies monomers locally only to sites which have a non-zero sum of enzyme activity in *both* E_1 and E_2 within their metabolic neighbourhood (i.e., if any of the two activities is missing from the metabolic neighbourhood, the replicator in the focal site cannot replicate for lack of monomers).
vi. the random walk (*diffusion*) of replicators is constrained by their adherence to the mineral surface;

To study molecular speciation within the metabolic model, we allow for mutational changes in three traits of the replicators: the two enzyme activities (E_1 and E_2) and k, the replication rate under ideal conditions (i.e., for a local excess of monomers). The changes in the mutable traits are constrained by a three-way trade-off relation as described later (in 2.3). Based on their mutable traits (E_1, E_2 and k) replicators can be classified into four different phenotype categories:

1. *specialists:* with a single, strong enzyme activity (either E_1 or E_2) and a small replication rate k;
2. *generalists:* with two, roughly equal enzyme activities ($E_1 = E_2 \neq 0.0$), replication rate depending on the actual values of E_1 and E_2, and a small replication rate k;
3. *parasites:* replicators without metabolic activity ($E_1 = E_2 = 0.0$) but of high replication rate;
4. *all the rest:* intermediate phenotypes with different enzymatic activities and replication rates.

With these assumptions we ask which of these phenotype categories evolve in the model, and what determines the actual outcome of the speciation process?

At $t=0$, the simulations are initiated with 80% of the sites occupied by replicator molecules, the phenotypes of which are attributed at random (of course with the three-way trade-off constraints considered). We use an asynchronous random updating algorithm to determine the phenotype distribution of the replicator population at the next, $(t+1)^{th}$ generation. A schematic representation of the algorithm is given on Figure 1, and the details of the interaction modules are explained below.

1. "Death"

Replicators may disappear from the mineral surface for different reasons. They may simply detach and move away from the surface, or they may disintegrate due to chemical "corrosion" (e.g., hydrolysis). Hydrolysis is more frequent in longer molecules, but longer replicators remain adhered to the surface for a longer time. These two effects counteract each other, so we assume that the net dependence of death rates on replicator length is negligible. From the viewpoint of the metabolic system, these events are all "deaths", and we treat them as such. We assume that deaths occur at a constant rate p_d independent of other traits of the replicator itself and of the other ones in its neighbourhood. If the site being updated contains a molecule, then it disappears with probability p_d, leaving the site empty. We used $p_d = 0.1$ throughout our simulations.

2. Metabolism and competition

If the updated site is empty, then the replicators in its *replication neighbourhood* (i.e., in its von Neumann neighbourhood) compete for the site to put a copy of themselves there. The chance of replicator i to win the site is proportional to its actual fitness W_i, which is the product of its replication rate k_i and its local metabolic supply M_i:

$$W_i = k_i \cdot M_i \qquad (1)$$

The metabolic supply M_i depends on the two enzyme activities

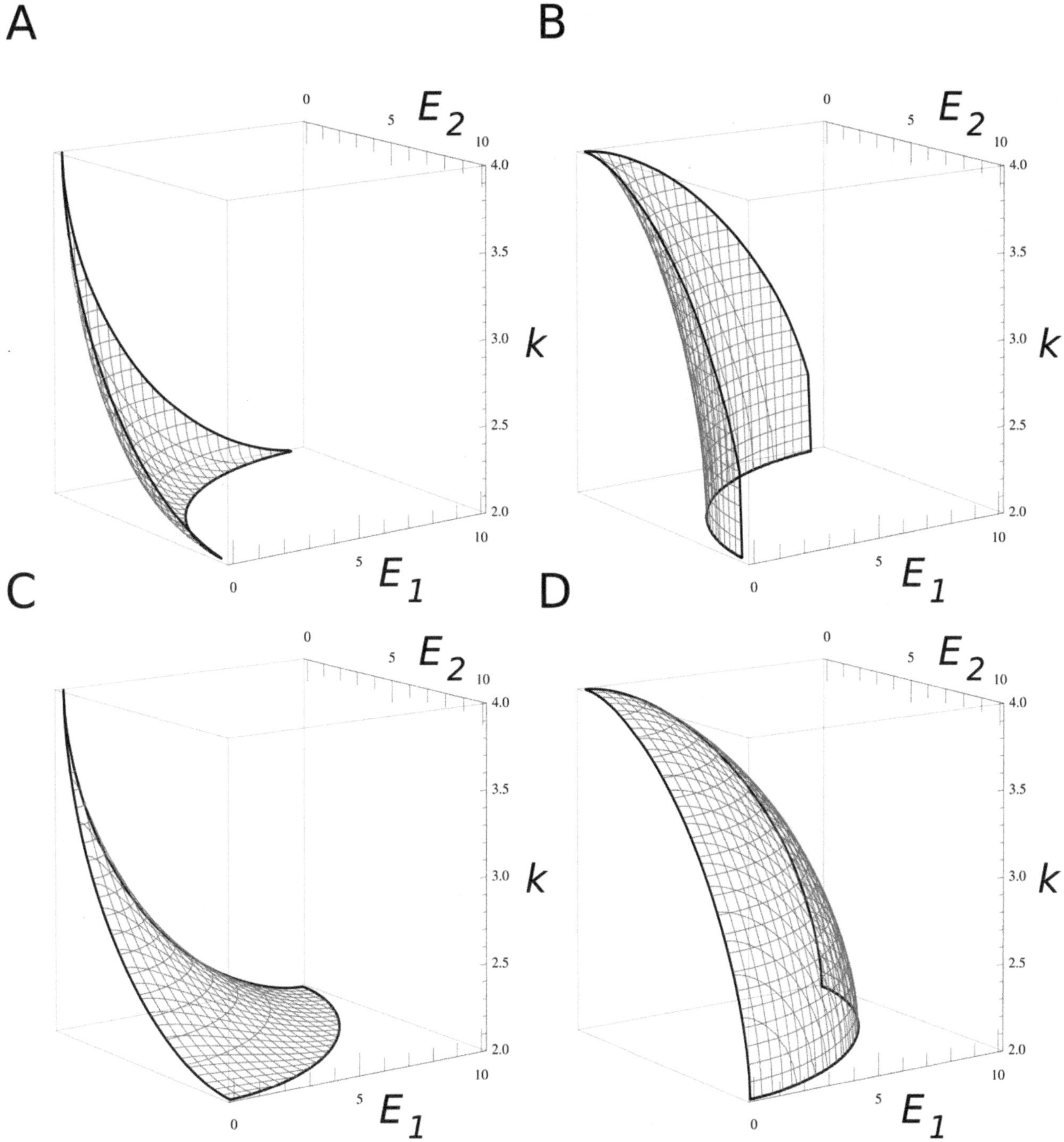

Figure 1. The $E_1/E_2/k$ trade-off relation. The E_1-E_2-k trade-off surface as defined by Eq. 6. The trade-off function constrains the phenotypes of newly emerging mutant replicators below the surface. A: convex function representing strong trade-off both between the two enzyme activities E_1/E_2 and between enzyme activities and replication rate, E/k. ($b=0.6$ and $g=0.6$). B: a function with convex (strong) E_1/E_2 trade-off and concave (weak) E/k trade-off ($b=0.6$ and $g=1.67$). C: concave (weak) E_1/E_2 and convex (strong) E/k trade-off ($b=1.67$ and $g=0.6$). D: both the E_1/E_2 and the E/k trade-offs are concave (weak) ($b=1.67$ and $g=1.67$). Other parameters: $E_{max}=10$, $k_{min}=2$, $k_{max}=4$.

within the *metabolic (Moore) neighbourhood* of replicator i in a multiplicative manner:

$$M_i=\sqrt{\sum_{k=1}^{n} E_{1k}\cdot\sum_{k=1}^{n} E_{2k}} \quad (2)$$

where n is the size of the metabolic neighbourhood. M_i is the geometric mean of the local enzyme activities, therefore with any one of the enzyme types missing from the metabolic neighbourhood of i the local metabolic supply is zero and replicator i cannot be copied.

Even those replicators of which both enzyme activities are non-zero (phenotypes 2. and 4. above) can catalyse only one of the corresponding reactions at a time, since the two different enzyme activities are attributed to two different secondary structures of the molecule. We assume that within the time span of a single generation the replicators do not change conformation, but they can do so between two generations with transition probability

$$p_{tr} = 1.0 - s = 1.0 - \frac{(E_1 - E_2)}{Max(E_1, E_2)}. \quad (3)$$

s measures (on the 0.0–1.0 scale) to what extent the replicator in question can be considered as a "specialist". Eq. 3 assumes that specialist replicators ($s = 1.0$) never change conformation from one generation to the next, whereas "generalists" (of roughly equal enzyme activities for both keystone reactions) almost always do. From the viewpoint of molecular speciation this is not necessarily a realistic assumption; we use it as a worst-case scenario of conformation change from the viewpoint of enzyme specialization: in pure populations of generalist replicators the regular swaps of individual enzyme activities (i.e., of conformation) keeps their metabolism running locally everywhere, whereas purely specialist populations might get stuck in one of the two conformations for a long time, and their local metabolism may break down for lack of the complementary enzyme activity.

The chance of replicator i to copy itself onto a neighbouring empty site is

$$p_i = \frac{W_i}{W_e + \sum_{j=1}^{l} W_j}, \quad (4)$$

where j runs through the replication neighbourhood of the empty site, and W_e quantifies the tendency of a site to remain empty ($W_e = 20.0$). W_e sets the effect of absolute enzymatic activities in the neighbourhood: weak claims for replication within the replication neighbourhood result in the "no replication" event with a higher probability. Thus the probability that an empty site remains empty is

$$p_e = \frac{W_e}{W_e + \sum_{j=1}^{l} W_j} \quad (5)$$

The next state of the empty site is determined by a random draw using the probabilities of Eq's. 4 and 5.

Note that the qualitative behaviour of the model is not affected if we use the *Moore*-type replication neighbourhood (with $l = 8$) instead of the *von Neumann*-type ($l = 4$).

3. Mutation and enzyme activity trade-offs

The winner of the local competition for space may be a copy of replicator i from within the replication neighbourhood of the empty site. The copy will be "mutated" with probability p_m. The mutant copy will be different from the template with respect to its primary (sequence) and secondary (folding) structures alike, which in turn might have an effect on the phenotype of the offspring, in terms of both enzymatic functionality and replicability. We allow for phenotype changes in three parameters of the replicators: the two enzyme activities (E_1 and E_2), and the basic replication rate k. As explained in the Introduction, these traits are in a three-way trade-off: efficient enzymes cannot be very good at template replication, and any one of the two enzyme activities can increase only at the expense of the other. These trade-off relations constrain the feasible part of the phenotype space (Figure 1) to under the trade-off constraint surface $C(E_1, E_2)$ given by Eq. 6:

$$C(E_1, E_2) = \left[E_{\max}^g - \left[E_1^b + E_2^b\right]^{\frac{1}{b}g}\right]^{\frac{1}{g}} \cdot \frac{k_{\max} - k_{\min}}{E_{\max}} + k_{min}, \quad (6)$$

where E_{max} ($E_{max} = 10.0$) is the absolute maximum of the enzyme activities, k_{max} ($k_{max} = 2.0$ or 2.5 or 4.0 in all simulations) is the highest, k_{min} ($k_{min} = 2.0$) is the lowest possible replication rate of any replicator in the metabolic system, and b and g determine the strength of the trade-offs which affects the shape of the trade-off surface. This formulation allows us to independently manipulate – through the parameters b and g – the convexity of the $E_1 - E_2$ and $E - k$ trade-off dependencies, respectively. For the enzyme activities, the strict trade-off constraint applies if $E_1 + E_2 < E_{max}$, which in turn means $0.0 < b < 1.0$. Then the trade-off function is convex (strong) on the $E_1 - E_2$ plane, and the smaller b is, the more restrictive the trade-off. $b = 1.0$ represents the limiting case of $E_1 + E_2 = E_{max}$, so that the two enzyme activities together are equivalent to a single specialist's activity. The trade-off is less restrictive for $E_1 + E_2 < 2E_{max}$, which translates to $1.0 < b < \infty$ and a concave (weak) trade-off function. The upper limit $b = \infty$ represents no trade-off at all: whichever role the enzyme plays, it works as if it were a perfect specialist. The same applies to the trade-off between enzyme activities and replication rates: g describes the strength and shape of the trade-off between $E_1 + E_2$ and k, and the smaller it is, the more convex (stronger) the trade-off relation.

Mutant phenotypes are assigned to the copies of a parent replicator template by a random draw of a point (E_1, E_2, k) from below the trade-off surface, so that $k(E_1, E_2) \le C(E_1, E_2)$, $E_2 \le \left(E_{\max}^b - E_1^b\right)^{1/b}$, $E_1 \le E_{\max}$. Mutants can be similar or quite different from their parents, which is reasonable to assume for replicator molecules of complex phenotypes: the effects of a mutation on the enzyme activities and the replication rate of an enzymatic replicator are difficult to predict, and they can be of any magnitude, depending on which part of the macromolecular structure is affected. Note that the average mutation is deleterious: selection shifts the phenotype distribution of the replicators towards the trade-off surface (i.e., to higher enzymatic efficiencies and replication rates), but mutants are drawn at random from below the surface.

4. Diffusion

Since the replicators are assumed to be bound to the mineral surface by reversible secondary (non-covalent) chemical bonds, their movement on the mineral surface is possible. We have modelled this movement by a simple site-swap algorithm: we choose two neighbouring replicators at random, and swap their positions. The intensity of the resulting diffusion process depends on how many site swaps are allowed per replication (generation) on average, which is defined in the diffusion parameter D of the model. Note that even at $D = 0.0$ a minimum of mixing occurs due to the fact that the copy of each template is placed into one of the neighbouring sites during replication, i.e., the offspring moves away from the parent. Again, we have tried two different neighbourhoods for site swapping, and found that choosing neighbours from the von Neumann or the Moore neighbourhood makes no detectable difference in the outcomes of simulation.

Results

The model predicts the coexistence of multiple replicator species and also enzyme specialization for a large part of its parameter space, even at relatively weak trade-off constraints and limited replicator mobility. Parasites are usually present in the evolved metabolic system at small densities

We have screened the effects of changing the crucial model parameters on the outcome of replicator evolution in terms of enzyme specialization and efficiency (i.e., specificity and parasitism). Having run the simulations for sufficiently long times we determined the phenotype distribution of the persistent replicators within the three-dimensional (E_1, E_2, k_{max}) trait space. Since the trait space under the trade-off constraint surface $C(E_1,E_2)$ (Eq. 6) is continuous, it is a matter of definition which of its parts we consider to contain "specialist", "generalist" and "parasitic" replicator phenotypes. We used the E_1/E_2 projection of the three-dimensional trait space to define these phenotype classes. The "absolute" specialists are those replicators which sit on the E_1 or the E_2 axis of the projected plane, whereas absolute generalists are found on the 45° straight line between these axes. The absolute parasite occupies the origin. Any replicator sufficiently close to these parts of the E_1/E_2 plane can be classified as belonging to the corresponding phenotype. Accordingly, we have defined the phenotype categories as shown on Figure 2.

Two of the five relevant model parameters – the shape of the trade-off function of the enzymatic activities (b) and the diffusion rate (D) – had a strong effect in the sense that changing them across a sufficiently wide range of their possible values results in qualitative changes in the outcomes. The remaining three parameters (the mutation rate p_m, the trade-off shape parameter g for replicator enzyme activity and replicability, and the highest possible replication rate k_{max}) are of much weaker effect. We explain these results below in more detail. Since changing the neighbourhood definition for the replication or the diffusion algorithm did not make substantial differences in the results of simulation, we used the von Neumann neighbourhood for replication and the Moore neighbourhood for diffusion throughout the simulations presented.

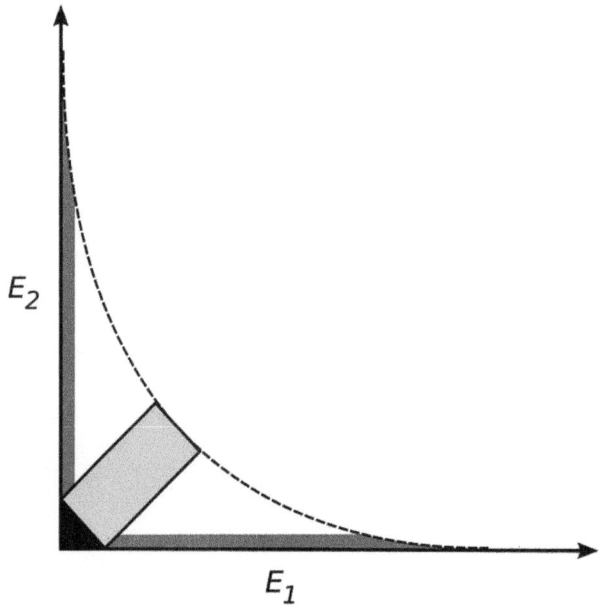

Figure 2. Types of replicators on the E_1–E_2 phase plane. *Specialists* (dark grey lines), *generalists* (light grey box), *parasites* (black triangle) and *all the rest* (white). The dashed line represents the E_1/E_2 trade-off relation.

1. The effects of the trade-off parameters (b and g) and the replication rate (k_{max})

The actual shape of the E_1/E_2 trade-off surface (i.e., the parameter b, Figure 1) has an almost all-or-none type effect on the enzyme specialization process: trade-off relations that are convex beyond a certain threshold (for $D=0$, $k_{max}=2.5$ and $g=1.0$ the threshold is at about $b{\sim}0.6$) yield a high frequency (>80%) of specialists; higher values of b result in an overwhelming dominance of generalists, but with their frequency almost completely detached from the actual value of b (Figure 3a). This phase-transition-like behavior is preserved at higher replicator mobility as well, but then the threshold of generalist dominance shifts to a very concave shape of the trade-off function: for $D=5$ it is at $b=1.67$ (Figure 3b). This means that at somewhat higher replicator mobility specialization pays even if the trade-off between the two enzyme specificities is rather loose (note that as b approaches infinity, the trade-off approaches zero).

The evolved enzyme activities and the corresponding (evolved) replication rate of the specialized replicators depend on the k_{max} parameter: the larger it is, the smaller the actual enzyme activities at equilibrium, because the direct evolutionary advantage of achieving a high replication rate compensates for the loss in the indirect advantage of having a better monomer supply within the metabolic neighbourhood. This leads to the evolutionary shift of the replicators towards a tendency of becoming parasites – the stronger the temptation (i.e., the higher k_{max}) the closer the replicator community creeps to parasitism. Yet, even at a high advantage of parasitism ($k_{max}=4.0$) the replicators remain enzymatically active and specialized, at least for convex trade-off relations. It is only at very concave E_1/E_2 trade-off functions (i.e., at $b>1.67$ for moderate $D=5$ replicator mobility) where generalists dominate the replicator community (expect for $g<1.0$ when parasites also appear), but this represents a case of very weak trade-off between the two enzyme activities, therefore it is biochemically unlikely to occur.

We have run a similar series of simulations changing g, the E/k trade-off parameter as well, with the general conclusion that g has a much weaker effect on catalytic specialization than b. Its straightforward but moderate and quantitative effect is that very convex E/k trade-off functions (very small g values) decrease the frequency of generalists and at high mobility it helps the parasites in spreading.

2. The effects of diffusion (D)

The mobility of replicators (expressed by the diffusion parameter D) has a twofold effect on the efficiency of metabolism, the first aspect of which is positive, and the second one negative (cf. [47]). First, it mixes the different types of enzymatic replicators on the surface, i.e., it dissolves the large, homogeneous patches of replicator clones (consisting of identical ribozymes) which would inevitably arise due to template and copy remaining neighbours in the absence of diffusion. Since the inside of a homogeneous patch of specialized enzymatic replicators lacks the complementary replicators needed to run the metabolism locally, the spatial mixing of specialized replicators is necessary to some extent for the metabolic system to work – metabolism would stop producing monomers and thus replication would be impossible almost everywhere on the surface (except for the borderlines between different patches) otherwise. However, generalist replicators (i.e., those which can switch between different enzyme activities) do not need to be spatially mixed to run local metabolisms, because even clone-mates can complement each other metabolically. This is why generalists almost always exclude specialists at very slow diffusion ($D=0.0$, Figure 4). (Note that the survival of specialists at

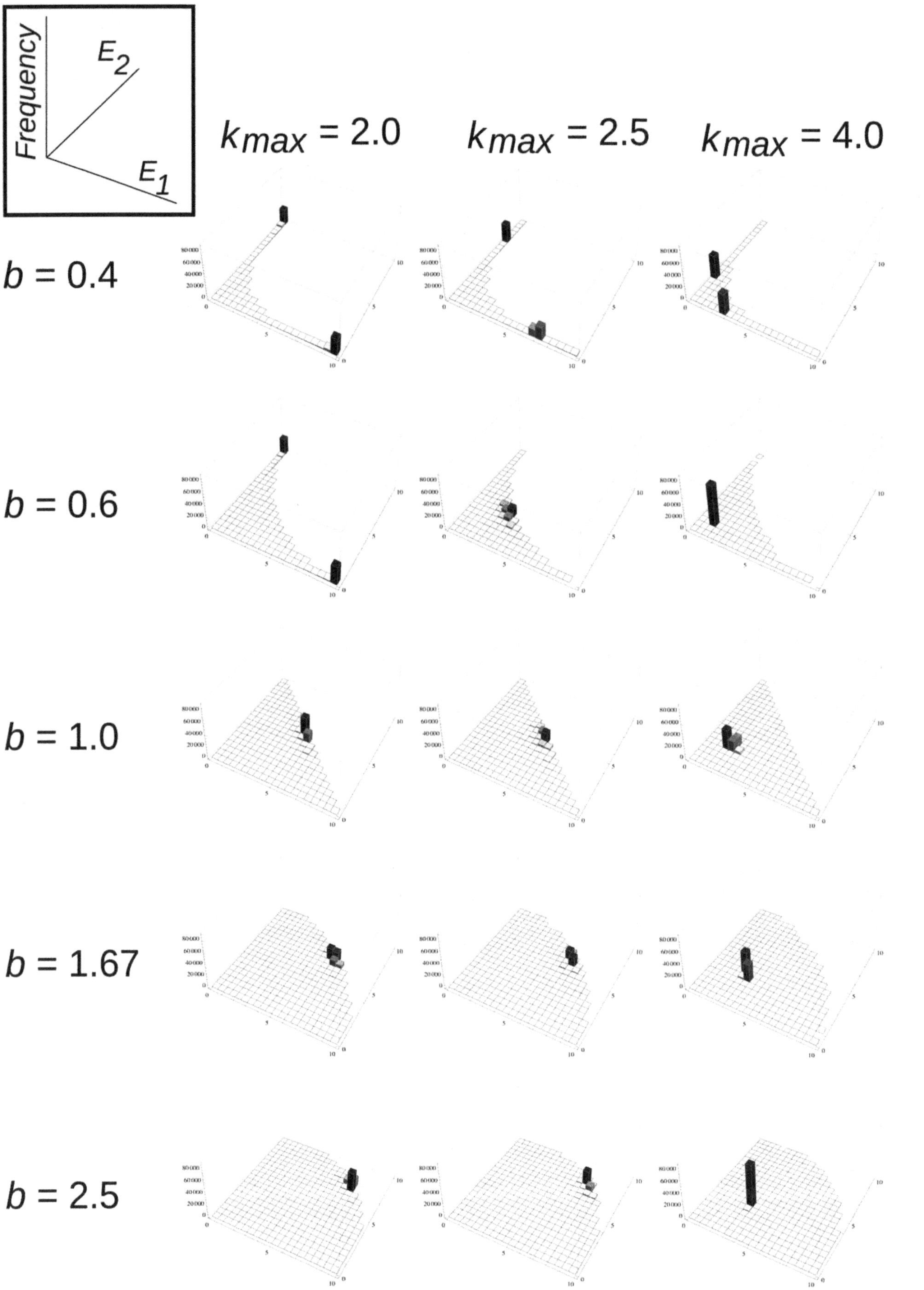
Frequency
E_2
E_1
$k_{max} = 2.0$
$k_{max} = 2.5$
$k_{max} = 4.0$
$b = 0.4$
$b = 0.6$
$b = 1.0$
$b = 1.67$
$b = 2.5$

Figure 3. Frequencies of replicator types. A: The frequencies of specialist and generalist replicators as a function of the b parameter, at $D=0$; B: the same, at $D=5$. Other parameters: $p_m=0.01$, $g=1.0$ and $k_{max}=2.5$ at the 150.000th generation. Note that the frequency of parasitic replicators is less then 1% everywhere in this parameter setting, so we have not plotted it here.

$b \leq 0.4$ is due to the minimal diffusion inherent in the replication algorithm). As the diffusion of replicators becomes faster, specialists become viable and –depending weakly on the shape parameter b of the trade-off function between the two enzyme activities – they may win the competition against generalists (Figure 5).

The second effect of increasing replicator mobility is negative for the metabolic system as a whole: parasitic replicators can invade. In a well-mixed population it is the parasitic replicator that has the highest fitness, because it has a high chance of finding itself in a metabolically complete assembly of cooperating replicators which it can exploit then for its own benefit. Since parasites, just as cooperators, can replicate only in a complete metabolic neighbourhood (to which they do not contribute at all), the spatial aggregation of kin replicators, i.e. slow diffusion, prevents the uncontrolled spread of parasites. In fact parasites need to be relatively mobile for taking a sizeable share of abundance within the lattice habitat, because they need to disperse far apart from their clone-mates to be efficient in exploiting the cooperators. Parasite aggregates are doomed to even faster extinction than enzymatic replicator aggregates. As a consequence, at small to moderate rates of diffusion we observe the coexistence of cooperators and parasites.

The balance of cooperators (enzymatic replicators of relatively small rate of population growth k) and parasites (replicators of high growth rate with no or very weak enzyme activities E_1 and/or E_2) within the replicator community depends on the contribution of two fitness components to population growth. One is the trivial *direct* contribution through the growth parameter k, the other is the *indirect* contribution to one's own fitness through the local production of monomers for replication that in turn depends on higher values of E_1 and/or E_2. With the possibility of mutational changes in these parameters (which is constrained by the E/k trade-off relation), the relative weights of the direct and indirect fitness components can be regulated through selection. Obviously, the actual weights of *direct* and *indirect selection* are determined by the diffusion parameter: the higher D is, the larger the effect of direct selection relative to that of the indirect one, which means more parasites and less efficient metabolism. This effect is clearly visible on Figs. 3 and 4: the replicator populations approach the origin of the E_1/E_2 trade-off plane and achieve high values of k as D increases. The "temptation" to become a parasite is the highest if losing some enzyme activity may result in a substantial increase of replication rate, i.e., at high values of k_{max}.

Discussion

In vitro selection experiments [20,21] aimed at producing RNA molecules of different phenotypes have suggested that the functional diversity of RNA molecule populations consisting even of rather short digital (nucleotide) sequences might have been sufficiently high for booting up life on prebiotic Earth [5,6]. It is very likely that, among the many possible functionalities that RNA molecules can possess, some may have evolved to catalyse the copying of the RNA molecule itself and of other RNA molecules, but the template-replicase ribozyme is still to be discovered [17,50]. Theoretical models [41,49,51] have demonstrated that, once some basic functionalities ensuring the self-reproduction of the inhabitants of the RNA world are in place, the way to obtaining more efficient functions (i.e., higher fitness) through Darwinian RNA evolution is open. The resulting communities of early replicators must have evolved towards higher efficiency through cooperation in the long run, but short-term competition among different RNA sequences was obviously inevitable, because the different RNA species must have used the same resources (monomers) for replication. The basically ecological problems of the competitive exclusion of slower replicating RNA sequences and the possible invasion of parasitic ones have been tackled in quite a few theoretical studies [33,48,51–55]. It is difficult, in many cases impossible, to separate evolutionary and ecological aspects of replicator evolution [56,57]. This statement applies to the present model as well, in which we have established ecological conditions for the evolution of specialized and efficient enzymatic replicators under different trade-off constraints among two enzyme specificities and the replication rates of metabolically coupled replicators. The trade-off function represents an inseparable relation between the genuine chemical-biochemical constraints on the metabolic roles that replicators can play in the system and the ecological trait of their potential population growth rate k_{max}.

Increasing enzymatic activity provides better monomer supply for the replicator in its own immediate neighbourhood, therefore it helps its own reproduction by supplying more resources. However, the price of this ecological advantage due to more efficient local catalysis is to be paid by the replicator in terms of its decreased replication rate k_{max}. This lower replication potential is a direct consequence of the stable and compact secondary structure which is a necessary trait for an efficient catalyst, but obviously makes the replicator more difficult to copy. Parasitic replicators which do not contribute to monomer production but are able to use up the monomer supply faster due to their loose, easily unfoldable and thus easy-to-copy secondary structure are at an ecological advantage compared to the metabolically active enzymatic replicators. Therefore we might expect the parasites to destroy the metabolic system altogether, finally leaving the surface devoid of replicators altogether. Yet, this is not what actually happens. We have found that increasing k_{max} (i.e., the existence of the E/k_{max} trade-off) results in the evolved replicator population becoming less efficient catalytically but still remaining active and specialized: the majority of the replicators edge closer to the origin of the E_1/E_2 trait plane (i.e., they shift towards parasitism) but they still stay on the specialist axes (cf. Figure 5). This shows that the indirect positive effect of metabolic activity becomes decisive for the dynamics of the system with the frequency of parasites increasing in the replicator population. In all versions of the metabolic model [48,49] – just as in the present one – it is the indirect metabolic cooperation of the replicators and their limited mobility that makes the system persistent in spite of the differences in replication rates. For the enzymatic replicators to be able to replicate they need their metabolic neighbourhood to be complete. This requires the presence of all *replicator species* within the metabolic neighbourhood in the previous models, and the presence of all *enzyme activities* in this one. In the absence of diffusion there is only one way to maintain all the necessary enzyme activities within a metabolic neighbourhood consisting of identical replicators: all of them must be "generalists" with both enzyme functions present in the same macromolecule – of course at a low activity on both accounts, because this is what the E_1/E_2 trade-off constraint permits. Even very modest replicator diffusion allows for enzymatic replicator

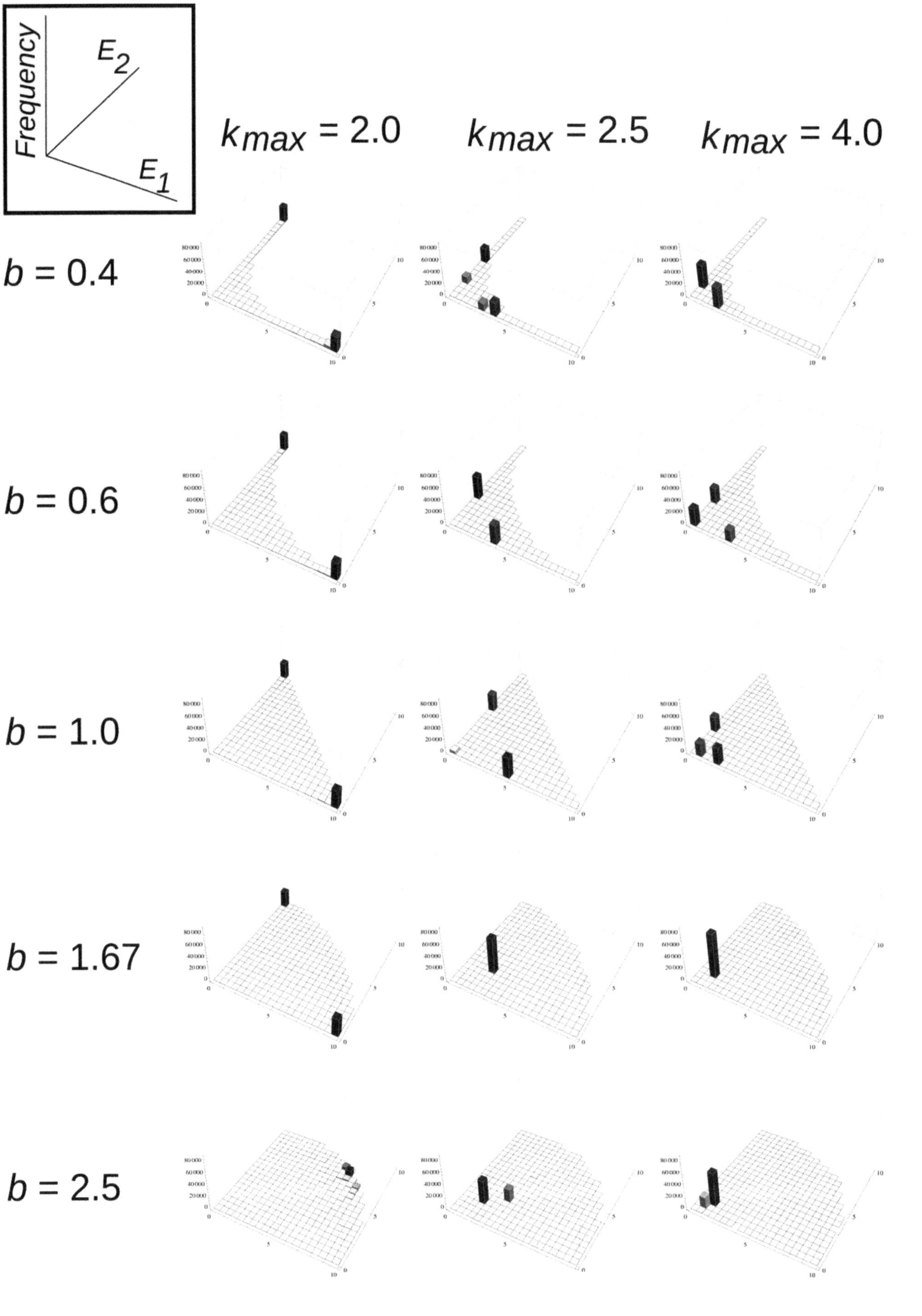

Frequency
E_2
E_1
k_{max} = 2.0
k_{max} = 2.5
k_{max} = 4.0
b = 0.4
b = 0.6
b = 1.0
b = 1.67
b = 2.5

Figure 4. The distribution of enzyme activities without diffusion at different b and k_{max} values. Horizontal axes: enzyme activities E_1 and E_2; vertical axis: the frequency of replicators in the lattice, with the corresponding combinations of E_1 and E_2.. Other parameters: $p_m = 0.01$, $D = 0$ and $g = 1.0$, at generation 150.000.

specialization, because it mixes different specialized replicator species, so that they can be present together in most metabolic neighbourhoods, allowing for a more efficient local metabolism and, consequently, for the exclusion of less efficient generalists (Figure 5). Our results also suggest that – between rather wide limits – the actual shape of the E/k_{max} trade-off function (the parameter *g*) does not have a decisive effect on enzymatic replicator specialization and the survival of the system. It is only for biochemically infeasible, very concave E/k_{max} trade-off shapes that specialized enzymatic replicators are displaced by generalists.

As we have briefly mentioned in Sect. 3.1., there are *direct* and *indirect* selection effects which influence the evolution of enzymatic functions. Indirect selection favours those functions of the replicators which have a positive influence on their cooperation through which they gain indirect benefits. Under indirect selection the replicators evolve towards becoming more efficient enzymes, the communities of which have more chances to spread due to their better resource supply. Efficient specialists are expected to show up only at a minimum of spatial mixing: replicator mobility helps to harvest indirect selective advantages. Direct selection attempts to maximize the growth rate of the replicators, which pushes the system towards the prevalence of parasites according to the E/k_{max} trade-off constraint. Direct selection is responsible for decreasing the enzyme activities of replicators, and, at the extreme, the dominance of parasites and the collapse of the metabolic system. Real parasites can spread only within metabolically active neighbourhoods, because pure parasite patches produce no resources for replication, and they die out. Therefore, parasites cannot evolve in the absence of diffusion, i.e., at zero replicator mobility. Increased diffusion helps the parasites in finding exploitable neighbourhoods and to avoid the pure-parasite dead end of the dynamics of the system. Under such circumstances parasites coexist with the enzymatic replicators [48,49]. That is, direct and indirect selection are both enhanced by diffusion [49]; indirect selection helps coexistence, whereas direct selection favours parasitism. We have shown that at a large interval of the D parameter axis these two effects are balanced, and the result is a community of more or less specialized and efficient enzymatic replicators, infected by a population of parasites.

Just like in the model of Könnyű and Czárán [49], the controlled presence of the parasite may be even advantageous for the metabolic system as a whole: lacking selective pressures sufficiently effective to completely eliminate it, and also lacking any positive selection to maintain its sequence information, the parasite may freely mutate. Thus it can wander about in the sequence space, and it has a chance to find a function there which is useful for the system as a whole. That is, the parasite is persistent and pre-adapted to many possible beneficial functions, including new metabolic enzyme activities, replicase activity, or membrane production. All these functions must have evolved at some stage of the origin of life, leading to the first membrane-covered macromolecule communities (proto-cells), in which a new and more efficient level of selection described in the stochastic corrector model [53,57] could have acted to maintain the optimal molecular composition of the earliest living creatures.

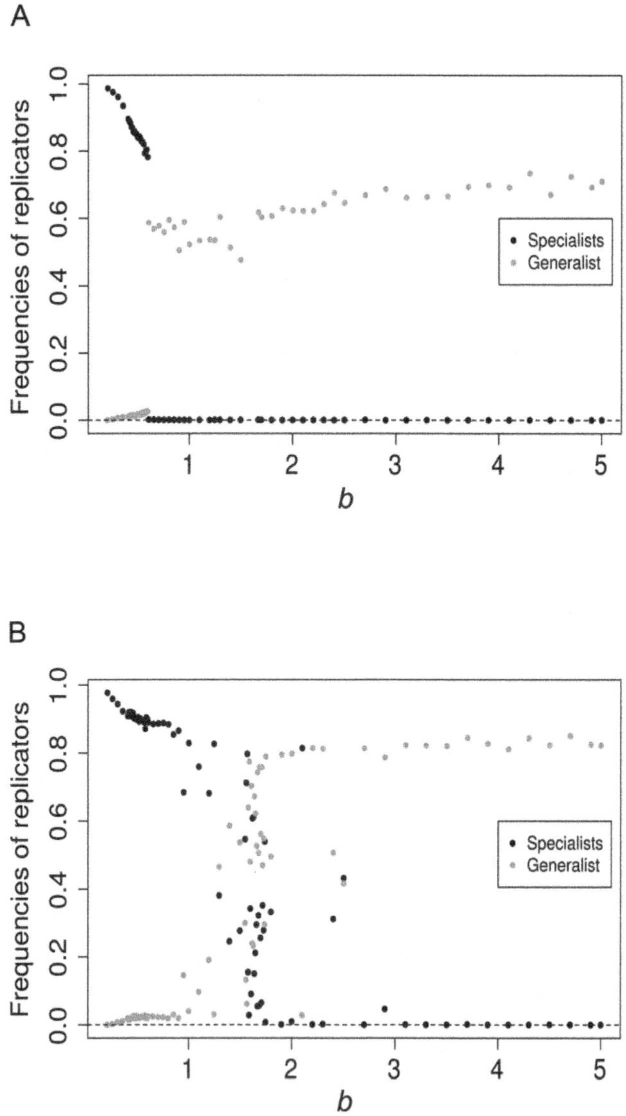

Figure 5. The distribution of enzyme activities with diffusion at different b and k_{max} values. Axes labeled as in Figure 4. Other parameters: $p_m = 0.01$, $D = 5$ and $g = 1.0$, at generation 150.000.

Acknowledgments

We wish to thank three anonymous referees for their valuable comments.

Author Contributions

Conceived and designed the experiments: BK TC. Performed the experiments: BK TC. Analyzed the data: BK TC. Wrote the paper: BK TC.

References

1. Wächtershäuser G (1990) Evolution of the first metabolic cycles. Proceedings of the National Academy of Sciences of the USA 87: 200–204.
2. Wächtershäuser G (2000) Life as we don't know it. Science 289: 1307–1308.
3. Wächtershäuser G (2007) On the chemistry and evolution of the pioneer organism. Chemistry and Biodiversity 4: 584–602.
4. Maynard-Smith J, Szathmáry E (1995) The major transitions of evolution. Oxford: W.H. Freeman/Spectrum.

5. Ma W, Yu C, Zhang W, Hu J (2007) Nucleotide synthetase ribozymes may have emerged first in the RNA world. RNA 13: 2012–2019.
6. Garay J (2010) Active centrum hypothesis: the origin of chiral homogenity and the RNA-World. Bio Systems, In press.
7. Kauffman S (1986) Autocatlytic sets of proteins. Journal of Theoretical Biology 119: 1–24.
8. Nelson KE, Levy M, Miller SL (2000) Peptide nucleic acids rather than RNA may have been the firs genetic molecule. Proceedings of the National Academy of Sciences of the USA 97: 3868–3871.
9. Orgel LE (2003) Some consequences of the RNA world hypothesis. Origins of Life and Evolution of Biospheres 33: 211–18.
10. Ferris JP, Hill RA, Liu R, Orgel LE (1996) Synthesis of long prebiotic oligomers on mineral surfaces. Nature 381: 59–61.
11. Franchi M, Ferris JP, Gallori E (2003) Cations as mediators of the adsorption of nuclec acids on clay surfaces in prebiotic environments. Origins of Life and Evolution of Biospheres 33: 1–16.
12. Franchi M, Gallori E (2005) A surface-mediated origin of RNA World: biogenic activities of clay-absorbed RNA molecules. Gene 346: 205–214.
13. Ferris JP (2006) Montmorillonite-catalysed formation of RNA oligomers: the possible role of catalysis in the origins of life. Philosohical Transactions of the Royal Society B 361: 1777–1786.
14. Gilbert W (1986) Origin of life: the RNA world. Nature 319: 618.
15. Joyce GF (2002) The antiquity of RNA-based evolution. Nature 418: 214–221.
16. Lilley DMJ (2003) The origins of RNA catalysis in ribozyme. Trends in Biochemical Sciences 28: 495–501.
17. Chen X, Li N, Ellington AD (2007) Ribozyme catalysis of metabolism in the RNA World. Chemistry & Biodiversity 4: 633–655.
18. Cech TR (2009) Crawling out of the RNA World. Cell 136: 599–602.
19. McConnell TS, Hercchlag D, Cech TR (1997) Effects of divalent metal ions on individual steps of the Tetrahymena ribozyme reaction. Biochemistry 36: 8293–8303.
20. Landweber LF, Simon PJ, Wagner TA (1998) Ribozyme engineering and early evolution. Bio Science 48: 94–103.
21. Bartel DP, Unrau PS (1999) Constructing an RNA world. Trends in Genetics 15: M9–M13.
22. Ferris JP, Joshi PC, Edelson EH, Lawless JG (1978) HCN: A plausible source of purines, pyrimidines and amino acids on the primitive Earth. Journal of Molecular Evolution 11: 293–311.
23. Ferris JP, Hagan JR (1984) HCN and chemical evolution: the possible role of cyano compounds in prebiotic synthesis. Tetrahedron 40: 1093–1120.
24. Weber AL (2008) Sugar-driven prebiotic sythesis of 3,5(6)-dimethylpyrazin-2-one: a possible nucleobase of a primitve replication process. Origins of Life and Evolution of Biospheres 38: 279–292.
25. Robertson MP, Joyce GF (2010) The origins of the RNA world. Cold Spring Harbor Perspectives in Biology. pp 1–22.
26. Powner MW, Gerland B, Sutherland JD (2009) Synthesis of activated pyrimidine ribonucleotides in prebiotic plausible conditions. Nature 459: 239–42.
27. Powner MW, Sutherland JD, Szostak JW (2010) Chemoselective multicomponent one-pot assembly of purine precursors in water. Journal of American Chemical Society 132: 16677–16688.
28. Schwartz AW (2006) Phosphorus in prebiotic chemistry. Philosohical Transactions of the Royal Society B 361: 1743–1749.
29. Bielski R, Tencer M (2007) A possible path to the RNA World: enantioselective and diastereoselective purification of ribose. Origins of Life and Evolution of Biospheres 37: 167–175.
30. Hazen RM, Filley TR, Goodfriend GA (2001) Selective adsorption of L- and D-amino acids on calcite: implications for biochemical homochirality. Proceedings of the National Academy of Sciences of the USA 98: 5487–5490.
31. Joyce GF, Visser GM, van Boeckel CA, van Boom JH, Orgel LE, et al. (1984) Chiral selection in poly(C)-directed synthesis of oligo(G). Nature 310: 602–604.
32. von Kiedrowski G (1996) Priomrdial soup or creps. Nature 381: 20–21.
33. Eigen M, Schuster P (1979) The Hypercycle. Berlin: Springer-Verlag.
34. Kun Á, Santos M, Szathmáry E (2005) Real ribozymes suggest a relaxed error threshold. Nature Genetics 37: 1008–1011.
35. Riley CA, Lehman N (2003) Generalized RNA-directed recombination of RNA. Chamistry & Biology 10: 1233–1243.
36. Hayden EJ, Lehman N (2006) Self-assemblyof a group I intron from inactive oligonucleotide fragments. Chamistry & Biology 13: 909–918.
37. Burton AS, Lehman N (2010) Enhancing the prebiotic relevance of a set of covalently self-assembling autorecombining RNAs through in vitro selection. Journal of Molecular Evolution 70: 233–241.
38. Wochner A, Attwater J, Coulson A, Holliger P (2011) Ribozyme-catalyzed transcription of an active ribozyme. Science 332: 209–212.
39. Kacser H, Beeby R (1984) Evolution of catalytic proteins or On the origin of enzyme species by means of natural selection. Journal of Molecular Evolution 20: 38–51.
40. Scheuring I (2000) Avoiding catch-22 of early evolution by stepwise increase in copy fidelity. Selection 1–3: 13–23.
41. Szabó P, Scheuring I, Czárán T, Szathmáry E (2002) In silico simulations reveal that replicators with limited dispersal evolve towards higher efficiency and fidelity. Nature 420: 340–343.
42. Szathmáry E (2007) Coevolution of metabolic networks and membranes: the scenario of progressive sequestration. Philosohical Transactions of the Royal Society B 362: 1781–1787.
43. Huang Z, Pei W, Han Y, Jayaseelan S, Shekhtman A, et al. (2009) One RNA aptamer sequence, two structures a collaborating pair that inhibits AMPA receptors. Nucleid Acids Research 37: 4022–4032.
44. O'Brien PJ, Herschlag D (1999) Catalytic promiscuity and the evolution of new enzymatic activities. Chemistry and Biology 6: R91–R105.
45. Ancel WL, Fontana W (2000) Plasticity, evolvability, and modularity in RNA. Journal of Experimental Zoology (Mol. Dev. Evol.) 288: 242–283.
46. Schultes EA, Bartel DP (2000) One sequence, two ribozymes: implication for the emergence of new ribozyme fold. Science 289: 448–452.
47. Khersonsky O, Roodveldt C, Tawfik DS (2006) Enzyme promiscuity, evolutionary and mechanistic aspects. Current Opinion in Chemical Biology 10: 498–508.
48. Czárán T, Szathmáry E (2000) Coexistence of replicators in prebiotic evolution. In: U D, R L, J MJA, eds. The Geometry of Ecological Interactions, IIASA and Cambridge University Press. pp 116–134.
49. Könnyű B, Czárán T, Szathmáry E (2008) Prebiotic replicase evolution in a surface-bound metabolic system: parasites as a source of adaptive evolution. BMC Evol Biol 8: 267.
50. McGinness K, Joyce GF (2003) In search of an RNA replicase rybozyme. Chemistry and Biology 10: 5–14.
51. Scheuring I, Czárán T, Szabó P, Károlyi Gy, Toroczkai Z (2003) Spatial models of prebiotic evolution: soup before pizza? Origins of Life and Evolution of Biospheres 33: 319–355.
52. Maynard-Smith J (1979) Hypercycles and the origin of life. Nature 280: 445–446.
53. Szathmáry E, Demeter L (1987) Group selection of early replicators and the origin of life. Journal of Theoretical Biology 128: 463–486.
54. Boerljis C, Hogeweg P (1991) Spiral wave structure in pre-biotic evolution: hypercicles stable against parasite. Physica D 48: 17–28.
55. Károlyi Gy, Péntek I, Scheuring I, Tél T, Toroczkai Z (2000) The antiquity of RNA-based evolution. Proceedings of the National Academy of Sciences of the USA 97: 13661–13665.
56. Meszena G, Szathmáry E (2001) Adaptive dinamics of parabolic replicators. Selection 1–2: 147–159.
57. Zintzaras E, Santos M, Szathmáry E (2002) Living under the challenge of information decay: the stochastic corrector model vs. hypercycles. Journal of Theoretical Biology 217: 167–181.

7

Structure-Guided Modification of *Rhizomucor miehei* Lipase for Production of Structured Lipids

Jun-Hui Zhang, Yu-Yan Jiang, Ying Lin, Yu-Fei Sun, Sui-Ping Zheng, Shuang-Yan Han*

Guangdong Key Laboratory of Fermentation and Enzyme Engineering, School of Bioscience and Bioengineering, South China University of Technology, Guangzhou, P. R. China

Abstract

To improve the performance of yeast surface-displayed *Rhizomucor miehei* lipase (RML) in the production of human milk fat substitute (HMFS), we mutated amino acids in the lipase substrate-binding pocket based on protein hydrophobicity, to improve esterification activity. Five mutants: Asn87Ile, Asn87Ile/Asp91Val, His108Leu/Lys109Ile, Asp256Ile/His257Leu, and His108Leu/Lys109Ile/Asp256Ile/His257Leu were obtained and their hydrolytic and esterification activities were assayed. Using Discovery Studio 3.1 to build models and calculate the binding energy between lipase and substrates, compared to wild-type, the mutant Asp256Ile/His257Leu was found to have significantly lower energy when oleic acid (3.97 KJ/mol decrease) and tripalmitin (7.55 KJ/mol decrease) were substrates. This result was in accordance with the esterification activity of Asp256Ile/His257Leu (2.37-fold of wild-type). The four mutants were also evaluated for the production of HMFS in organic solvent and in a solvent-free system. Asp256Ile/His257Leu had an oleic acid incorporation of 28.27% for catalyzing tripalmitin and oleic acid, and 53.18% for the reaction of palm oil with oleic acid. The efficiency of Asp256Ile/His257Leu was 1.82-fold and 1.65-fold that of the wild-type enzyme for the two reactions. The oleic acid incorporation of Asp256Ile/His257Leu was similar to commercial Lipozyme RM IM for palm oil acidolysis with oleic acid. Yeast surface-displayed RML mutant Asp256Ile/His257Leu is a potential, economically feasible catalyst for the production of structured lipids.

Editor: Shannon Wing-Ngor Au, The Chinese University of Hong Kong, China

Funding: This work was funded by National High-tech R&D Program of China (2012AA022205), Guangdong Key Technology R&D Program (No. 2012B010300008), the Fundamental Research Funds for the Central Universities (No. 2012ZZ0092).The funders had no role in study design, data collection and analysis, decision to publish, or preparation of the manuscript.

Competing Interests: The authors have declared that no competing interests exist.

* E-mail: syhan@scut.edu.cn

Introduction

Protein engineering or protein design is an important and effective strategy for developing biocatalysts. This strategy evolves or tailors enzymes to have desired properties such as higher enzyme activity, or better stability or selectivity [1,2]. Structure-guided design, which takes advantage of known protein structure, is combined with *in silico* methods to select target amino acids for mutation. This avoids the time and labor of high-throughput screening from a library of thousands of mutants [3,4].

Lipase esterification activity correlates with enzyme hydrophobicity to some extent. Increasing the hydrophobicity of a lipase can promote access of the substrate to the enzyme and can stabilize the substrate-enzyme complex by decreasing the binding energy in a nonaqueous phase [5]. In many cases, selection or modification of a lipase-immobilizing matrix or altering lipase regions to have increased hydrophobicity greatly enhances the lipase esterification activity [6–11]. In addition, directly modifying the lipase surface or lipase lid domain based on amino acid hydrophobicity increases esterification activity and can even change chain length selectivity [12–15]. Therefore, modifying the lipase-binding pocket using a structure-guided method to improve pocket hydrophobicity might increase lipase esterification activity.

Human milk fat (HMF) is mostly triacylglycerols, which are the main energy source in breast milk and infant milk formula. In typical HMF, the total fatty acid composition is about 21.8% palmitic acid and 33.9% oleic acid [16]. About 70% of the palmitic acid of HMF is in the sn-2 position; alternatively stated, 57.2% of the sn-2 position is occupied by palmitic acid [16]. The sn-1,3 position is mostly occupied by unsaturated fatty acids, mainly oleic acid (44%) followed by palmitic acid (18.7%) and stearic acid (14.2%) [17]. In fact, 1,3-dioleoyl-2-palmitoyl-glycerol (OPO, 19%) is the second most abundant triacylglycerol species in HMF [18]. OPO reduces the formation of calcium soaps, which causes stool hardness and constipation [19,20]. *Rhizomucor miehei* lipase (RML) is a typical lipase that catalyzes the esterification or hydrolysis of lipids along with other oil modification reactions such as acidolysis, alcoholysis, glycerolysis and interesterification [21]. The crystal structure of RML has been solved [22]. With its good regioselectivity and relatively high incorporation rate, RML is an excellent lipase for the production of HMFS [23]. We developed a whole cell catalyst, a *Pichia pastoris* that displays RML on the cell surface, and used this catalyst for esterification of flavor esters and biodiesel production [24,25]. Although yeast surface-displayed RML is inexpensive and recyclable, its esterification activity is still being developed.

In this study, we increased the hydrophobicity of the RML binding pocket based on a designed enzyme structure. We used computational software to model the mutated structure and evaluate the activity of mutant enzymes. These were tested for the production of HMFS in a solvent system with tripalmitin and oleic

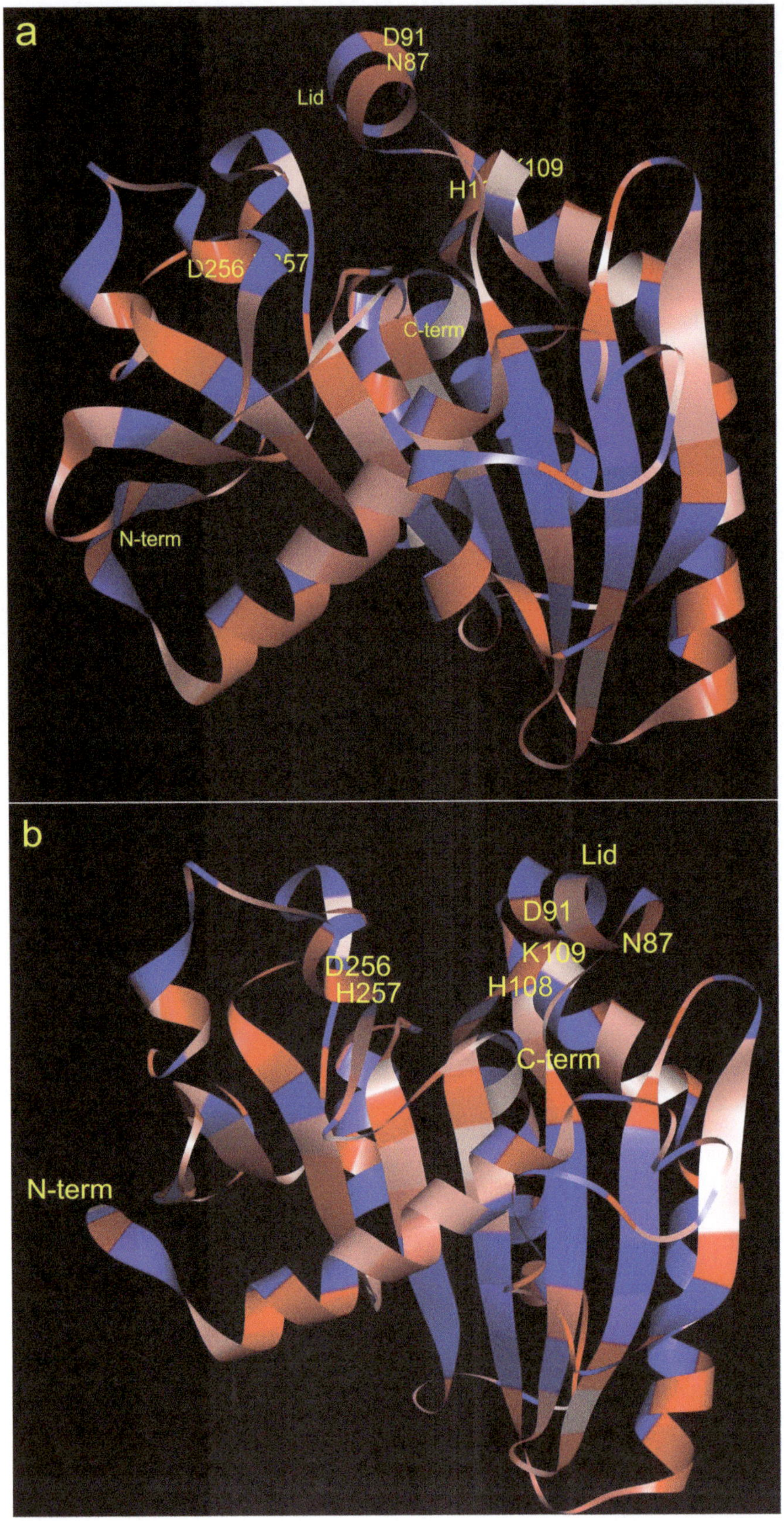

Figure 1. Amino acid hydrophobicity profile of RML. a: 3TGL, lid-closed form, b: 4TGL, lid-open form. Red, hydrophilic residues; blue, hydrophobic residues.

acid and in a solvent-free system with palm oil and oleic acid as substrates.

Materials and Methods

Chemicals

Yeast extract and tryptone were from OXOID (Basingstoke, UK). Peptone was from BD (Sparks, MD). Restriction enzymes, ligase and polymerase were from TAKARA (Dalian, China). *p*-nitrophenyl butyrate (pNPB) was from Sigma (St. Louis, MO). Organic acid, ethanol and other chemicals were analytical or higher grade. Tripalmitin was from Aladin (Shanghai, China) and palm oil was a gift from Shenzhen Jinyi Lipid Co. (Shenzhen, China). Commercial RML-lipozyme RM IM (*R. miehei* lipase immobilized on a macroporous anion-exchange resin) was from Novozyme (Tianjin, China).

Strains and Media

Escherichia coli TOP10 and *Pichia pastoris* GS115 were from Invitrogen (USA). Strain GS115/pKFSR was *R. miehei* lipase displayed on the *P. pastoris* GS115 cell surface by flocculation using the Flo1p-short (FS) anchor protein [24].

Luria-Bertani medium (1% tryptone, 0.5% yeast extract, 1% NaCl), with ampicillin (100 µg/ml) added when necessary, was used to incubate and select *E. coli* transformants. YPD (1% yeast extract, 2% peptone, 2% dextrose) was used to incubate *P. pastoris* GS115 and maintain strains on agar plates. Yeast transformants were selected in MD medium (1.34% yeast nitrogen base [YNB], 400 µg/L biotin, 2% dextrose, 1.5% agar). *P. pastoris* was cultivated in BMGY (1% yeast extract, 2% peptone, 0.1 M potassium phosphate, pH 6.0, 1.34% YNB, 400 µg/L biotin, 1% glycerol) and BMMY (1% yeast extract, 2% peptone, 0.1 M potassium phosphate pH 6.0, 1.34% YNB, 400 µg/L biotin, 1% methanol).

Mutant Construction

The overlap extension PCR method was used for site mutagenesis of *RML* with plasmid pKFSR [24] as template. RML was expressed with the surface-display anchor protein on the N-terminus and a Flag tag on the C-terminus.

Primer sequences were:

RML-F: 5′- CGC GGC ACG CGT GTT CCA ATT AAG AGA CAA TCT AACT -3′, RML-R: 5′-GCC AGC GAA TTC TTA CTT ATC GTC GTC ATC CTT GTA ATC AGT ACA CAA ACC AGT GTT AAT ACC -3′, with underlining showing restriction sites *Mlu*I and *Eco*RI respectively. N87I-F: 5′-ATT AGA *ATC* TGG ATT GCT GAT TTG ACT-3′, N87I-R: 5′-AGT CAA ATC AGC AAT CCA *GAT* TCT AAT -3′, N87I/D91V-F: 5′- ATT AGA *ATC* TGG ATT GCT *GTT* TTG ACT-3′, N87I/D91V-R: 5′- AGT CAA *AAC* AGC AAT CCA *GAT* TCT AAT -3′, H108L/K109I-F: 5′- AAG GTT *CTT ATC* GGT TTT TTGG -3′, H108L/K109I-R: 5′- CCA AAA AAC C*GA TAA* GAA CCTT -3′, D256I/H257L-F: 5′- CTG TTT TG*A TCC TT*T TGT CTT AC -3′, D256I/H257L-R: 5′- GTA AGA CAA *AAG GAT* CAA AAC AG -3′, with italic letters showing mutation sites.

The *RML* gene was amplified with *Mlu*I and *Eco*RI restriction sites on both ends, and inserted into digested vector pKFS. The mutated pKFSR plasmids were used to transform *E. coli* TOP10 for amplification and then linearized and used to transform *P. pastoris* GS115.

The *RML* gene in *P. pastoris* GS115 was amplified with the KOD FX PCR system (Toyobo, Shanghai, China) directly from colonies and sequenced by BGI (Guangzhou, China) to verify mutations.

Computational Analysis of Mutants

Discovery Studio V3.1 (Accelrys, San Diego, CA) was used for computational analysis of mutants. Structures were created by the protocol Macromolecules-Protein Design-Build Mutants. Binding energy was calculated by the protocol Receptor-Ligand Interaction-CDocker.

RML Expression

P. pastoris strains were inoculated into 50 ml of BMGY medium in a 500 ml flask. After 24 h of cultivation, the culture was centrifuged at 5000 rpm for 6 min and the pellet was resuspended in BMMY medium containing 1% methanol at an OD_{600} of 1. Methanol was added every 24 h to the culture to a final

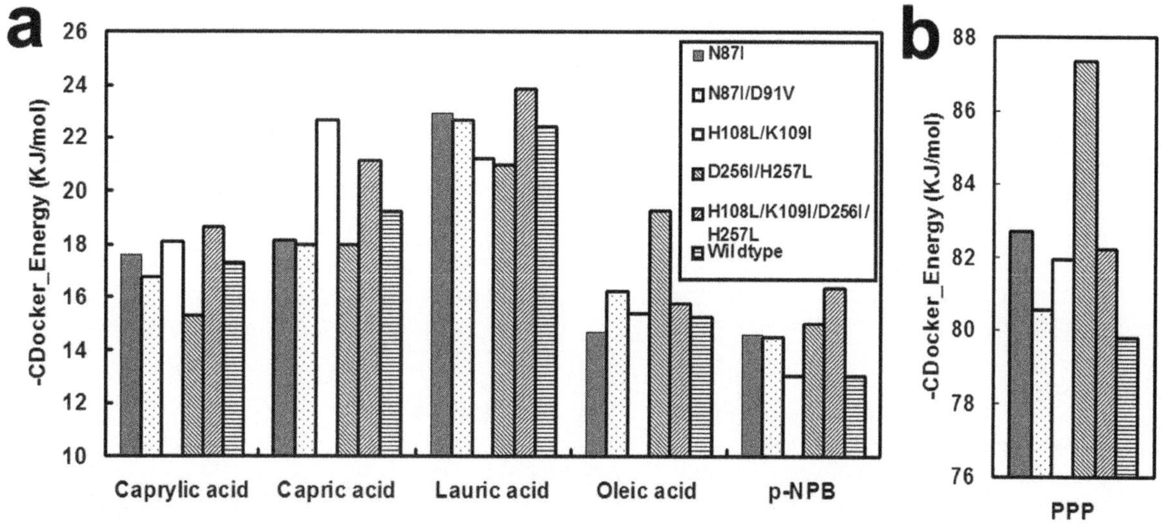

Figure 2. CDocker energy of RML mutants and wild-type with different substrates. a and b are separate because of discrepancies in the range of their Y-axes. a and b have the same legend. For comparison with the enzyme activity data, the energy value was converted to positive.

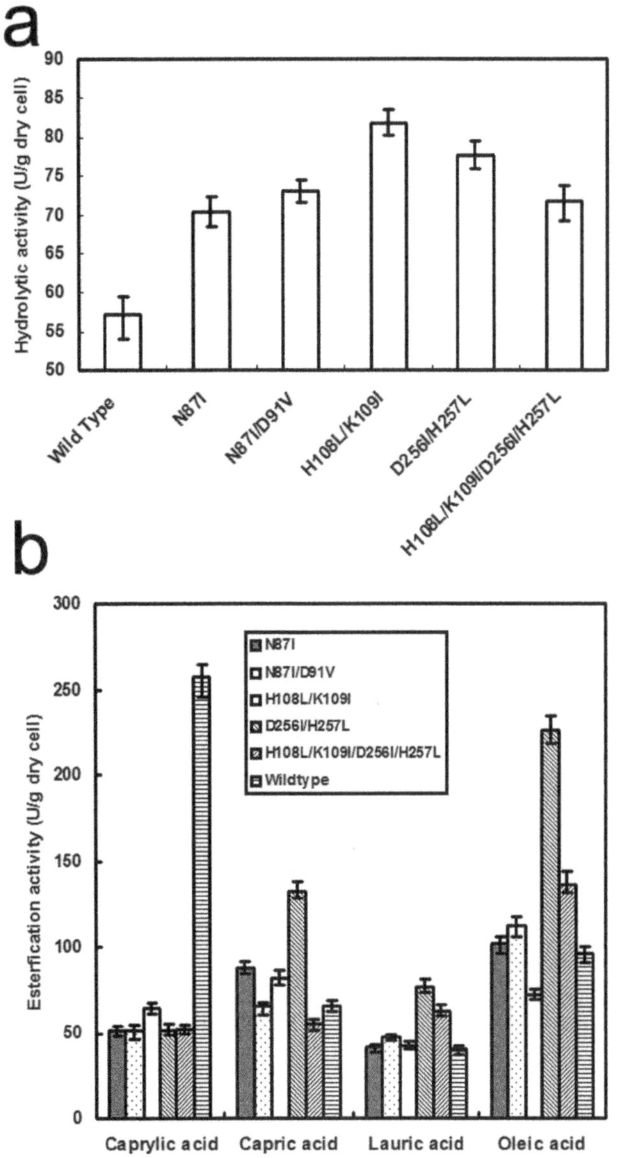

Figure 3. Flow cytometry of whole cell RML catalysts. a: Asn87Ile/Asp91Val, b: His108Leu/Lys109Ile, c: Asp256Ile/His257Leu, d: His108Leu/Lys109Ile/Asp256Ile/His257Leu, e: wild-type, f: negative control.

concentration of 1%. Cells were incubated for 6 days at 30°C with shaking at 250 rpm.

Cells displaying RML were centrifuged at 5000 rpm for 6 min, washed with water, and centrifuged again. Pellets were resuspended in 4 ml 50 mM Tris-Cl buffer (pH 8.0) and 0.5 ml 10% (*m/v*) trehalose and lyophilized (Christ Beta 1–8, Germany).

Enzyme Activity Assay

Hydrolytic activity was determined colorimetrically using pNPB as substrate. Cell-displayed lipases were dissolved in lipase assay buffer (50 mM Tris-HCl, pH 8.0), mixed with substrate solution (0.0625 mM pNPB, 0.1% Triton X-100 in lipase assay buffer), and reacted at 45°C for 5 min before measuring OD_{405} with a kinetic microplate reader. Reactions were terminated by 20 µl 20% trichloroacetic acid. After centrifugation at 10,000×*g* for 1 min, a 200 µl aliquot of supernatant was placed in a 96-well plate and measured. Product yields were determined from a standard curve using *p*-nitrophenol as a standard. One unit of hydrolytic activity was defined as the amount of enzyme that released 1 µmol *p*-nitrophenol from substrate per minute.

Synthetic activity was determined by the synthetic reaction of ethanol with organic acid such as caprylic acid, capric acid, lauric acid, or oleic acid. Substrates composition were 0.25 M ethanol and 0.2 M acid with heptane as solvent. Yeast displaying RML in lyophilized powder form (100 mg) were suspended in 5 ml substrate mixture in a 25 ml conical flask, stoppered to prevent evaporation of solvents and reactants. Reactions were carried out at 55°C at 200 rpm in an orbital shaker. After reactions, solutions were centrifuged at 12,000 rpm for 1 min. Supernatant was titrated with 0.02 M sodium hydroxide to determine the amount of residual acid. One unit of synthetic activity was defined as the amount of enzyme that synthesized 1 µmol of ethyl ester per minute.

Flow Cytometry Analysis

Yeast cells were harvested by centrifugation after methanol induction for 120 h. Immunofluorescence analysis was performed using a modified version of the Kobori et al. protocol [26]. Yeast cells were resuspended in phosphate-buffered saline (PBS, pH 7.4) supplemented with 1% (*m/v*) bovine serum albumin (BSA) to block the cell surface. A monoclonal antibody (Agilent, Santa Clara, CA), against FLAG tag (DYKDDDDK) was used as the primary antibody. The cell suspension (200 µl, $OD_{600} = 1$) was incubated with 1 µl of antibody (2 µg/µl) at room temperature for 2 h. Cells were washed twice with PBS and resuspended in 200 µl PBS (with 1% BSA), before being exposed to the secondary antibody (1 µl, 2 µg/µl) Alexa Fluor 488 conjugated goat-anti-mouse IgG (Invitrogen, Grand Island, NY), for 1 h at room temperature. Cells were washed three times with PBS. Part of the cell suspension was used for flow cytometry analysis (Quanta SC, Beckman Coulter). Each sample was counted for 10,000 cells and analyzed with Exp032 software (Beckman Coulter) to obtain the total and average fluorescence value.

Lipase Catalyzed Acidolysis

Tripalmitin and oleic acid in organic solvent: Acidolysis reactions were 120 mg tripalmitin and 240 mg oleic acid (minimum purity 85%), dissolved in 6 ml heptane. Either whole cell RML catalyst or commercial RML-Lipozyme RM IM was used at 54 mg. The mixture was placed in 25 ml Erlenmeyer shaking flasks for incubation in a water bath shaker at 200 rpm, 55°C. Samples were withdrawn from the reaction mixture at predetermined time intervals for analysis.

Palm oil and oleic acid in solvent-free system: Acidolysis reactions used 3 g palm oil, 6 g oleic acid, 0.9 g lipase without solvent. The mixture was placed in 50 ml Erlenmeyer shaking flasks for incubation in a water bath with magnetic agitation at 200 rpm, 55°C. Samples were taken from the reaction mixture at predetermined time intervals for analysis.

All reactions were performed at least in triplicate, and means were used to evaluate results.

Fatty Acid Assay

Triglyceride isolation: Samples were isolated by thin-layer chromatography (TLC, silicic acid 60G, thickness 1 mm, 20 cm×20 cm) with petroleum ether, diethyl ether, and acetic acid (90:10:0.5, v/v/v) as developing solvent [27]. Bands containing triacylglycerols (TAGs) were removed by scraping and extracted with isooctane. Isolated TAGs were used to analyze the fatty acid compositions.

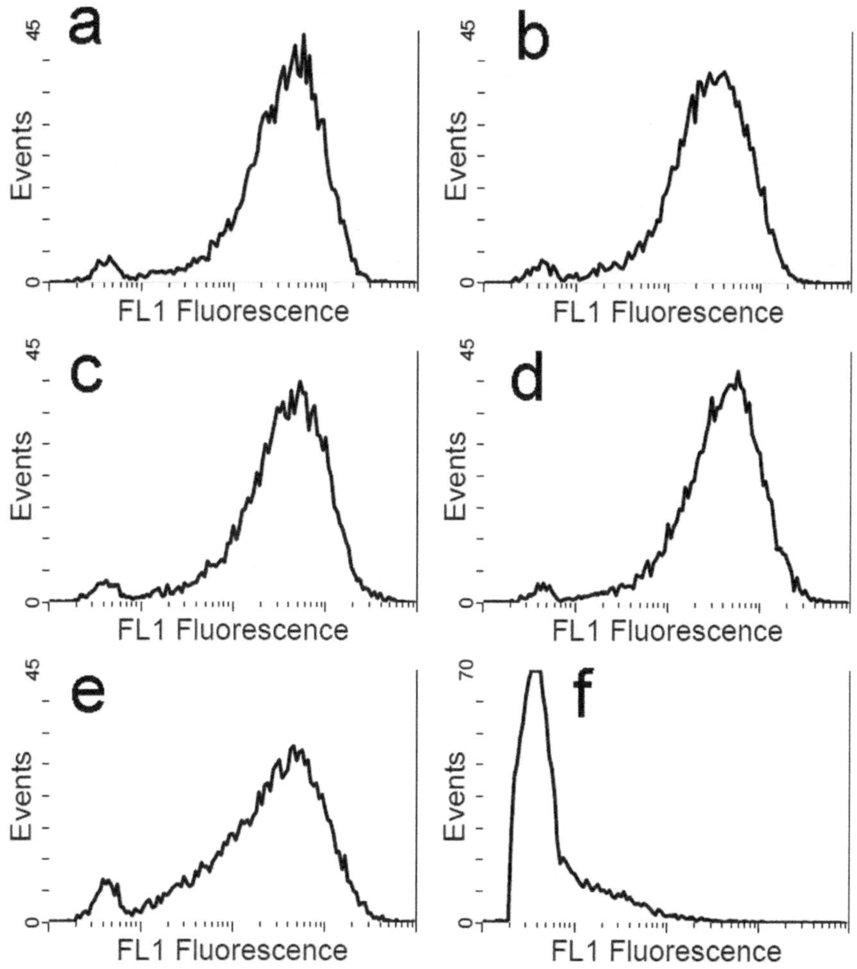

Figure 4. Enzyme activity of RML wild-type and mutants. a: hydrolytic activity, b: esterification activity.

Methylation and fatty acid (FA) composition analysis by gas chromatography (GC): Bands corresponding to TAGs were removed by scraping and methylated to fatty acid methyl esters (FAME). Samples were saponified with 0.5 M NaOH in methanol (4 ml) at 80°C for 5 min, and methylated with 14% BF_3 in methanol (5 ml) for 10 min. After cooling to room temperature, isooctane (3 ml) and saturated NaCl were added and mixed, and the upper isooctane layer was collected. An Agilent 7890A gas chromatograph equipped with a hydrogen flame-ionization detector and a DB-FFAP silica capillary column (0.25 mm×30 m, Agilent, Santa Clara, CA, USA) was used to analyze fatty acid composition. The injector and detector were set at 250°C. The carrier gas was nitrogen, and the total gas flow rate was 23 ml/min. Injector temperatures was 250 °C and detector temperature was 260 °C. Column temperature was 150 °C for 3 min increasing to 215 °C for 10 min at the rate of 10 °C/min. A sample of 1 µl was injected, and the FAME content was calculated using C17:0 as an internal standard.

Results and Discussion

Selection of Amino Acids for Mutation

The catalytic triad of RML is formed by Ser144, His257 and Asp203. The RML lid is a 15-amino acid helix from residues 82 to 96, in which residues 83–84 and 91–95 are hinges. Opening the lid exposes the binding pocket, which has the shape of a shallow bowl [28].

From Protein Data Bank (PDB), we obtained the open structure (PDB ID: 4TGL) and the closed structure (PDB ID: 3TGL) of RML. In the binding pocket and the lid (Fig. 1), Asn87 (−3.5), Asp91 (−3.5), His108 (−3.2), Lys109 (−3.9), Asp256 (−3.5) and His257 (−3.2) are exposed on the inner side of the binding pocket and have low hydrophobic values (listed in the bracket after the residue). To increase the hydrophobicity of the binding pocket, we mutated the above residues to Val (4.2), Leu (3.8) and Ile (4.5), which have higher hydrophobic values [29].

We organized the six residues into three groups according to their positions [30]. Considering that Asn87 was on the lid helix and Asp91 was in the hinge area, we mutated Asn87Ile/Asp91Val and Asn87Ile alone to avoid influencing lid movement. Since His108-Lys109 and Asp256-His257 were adjacent residues, we mutated both residues simultaneously to His108Leu/Lys109Ile and Asp256Ile/His257Leu.

Computational Analysis of Mutants

Discovery Studio 3.1 was used to simulate the protein structures of five RML mutants: Asn87Ile, Asn87Ile/Asp91Val, His108Leu/Lys109Ile, Asp256Ile/His257Leu, His108Leu/Lys109Ile/Asp256Ile/His257Leu. From the structures, we selected the one with the lowest energy to dock with hydrolytic and esterification

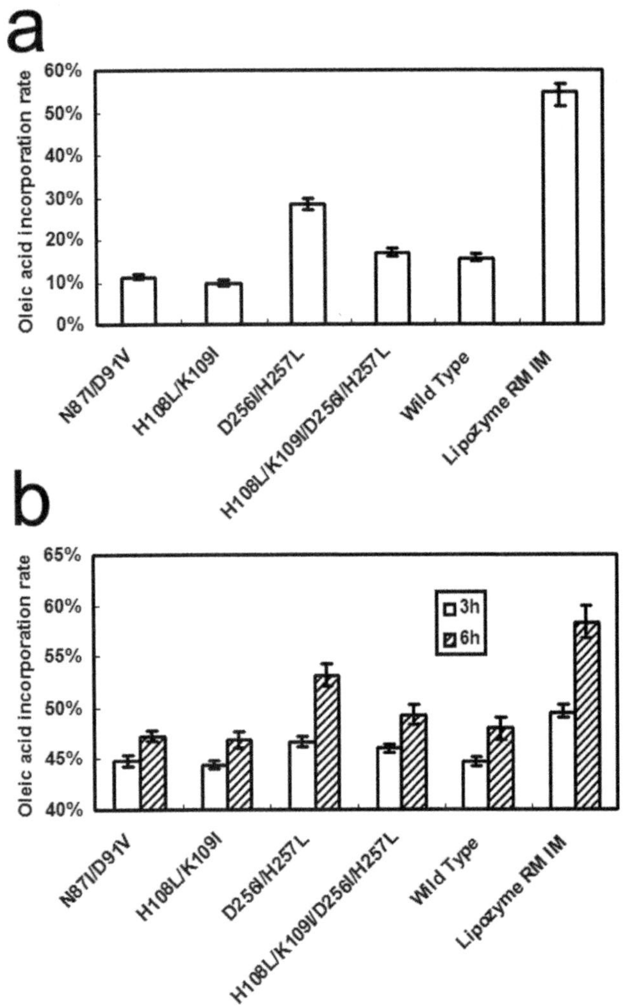

Figure 5. OA incorporation of whole cell catalyst and commercial lipase. a: OA incorporation in a reaction of PPP with OA for 24 h; b: OA incorporation in a reaction of palm oil with OA for 3 h and 6 h.

substrates (Fig. 2). CDocker is a semi-flexible docking program in which the ligand is flexible. CDocker searched ligand configurations randomly using a dynamic method, and configurations in the active site of the enzyme were optimized by simulated annealing.

The results showed that the CD_docker energy of the mutants with most substrates was not always lower than wild-type. However, for oleic acid, pNPB and tripalmitin (PPP), several mutants had significantly lower CD_docker energy than wild-type. We assayed enzyme activity in addition to the computation analysis.

Enzyme Activity

To determine whether the variation in enzyme activity was caused by the mutant structure or a change in the level of expression on the cell surface, we used flow cytometry to evaluate cell-surface fluorescence. The average fluorescence in relative fluorescence units was 475.32 for cells displaying Asn87Ile/Asp91Val, 461.09 for His108Leu/Lys109Ile, 483.88 for Asp256Ile/His257Leu, 476.10 for His108Leu/Lys109Ile/Asp256Ile/His257Leu, and 466.74 for wild-type (Fig. 3). Liang et al. demonstrated a linear relationship between recombinant yeast cell fluorescence and the number of the enzyme proteins on the cell surface [31]. Taking account of the systematic error in our method, the fluorescence of the mutants and wild-type showed no obvious difference. This result indicated the same offset value as the negative control. Therefore, we concluded that expression of the mutants and wild-type RML was within the same order of magnitude.

The hydrolytic activity of the five mutants and wild-type RML were assayed (Fig. 4a). The hydrolytic activity was 70.2 U/g for yeast displaying Asn87Ile, 73.1 U/g for Asn87Ile/Asp91Val, 81.8 U/g for His108Leu/Lys109Ile, 77.6 U/g for Asp256Ile/His257Leu, 71.7 U/g for His108Leu/Lys109Ile/Asp256Ile/His257Leu and 57.1 U/g for wild-type RML. Activity for all mutants was higher than for wild-type RML, and among them the mutants His108Leu/Lys109Ile and Asp256Ile/His257Leu showed the greatest enhancements of 43.2% and 36.0% compared to wild-type.

Esterification activity was also determined according to a synthesis reaction with ethanol and several organic acids with different carbon chain lengths from C8 to C18:1 (Fig. 4b). For caprylic acid, the wild-type activity was greater than for the mutants; mutant activity was about 34% of wild-type activity. For the other acids however, most mutant activities were higher than the wild-type. The activity of Asp256Ile/His257Leu was significantly higher than wild-type when reacting with three acids, but not caprylic acid.

Among the five mutants, His108Leu/Lys109Ile had the highest hydrolytic activity and Asp256Ile/His257Leu had the highest esterification activity. Mutants Asn87Ile and Asn87Ile/Asp91Val showed higher hydrolytic and esterification activity (except caprylic acid as substrate) than wild-type, but less than Asp256Ile/His257Leu. His108, Lys109, Asp256 and His257 are all located in the binding pocket but not the lid. This suggested that mutation of the binding pocket changed the enzyme characteristics. However, the His108Leu/Lys109Ile/Asp256Ile/His257Leu mutation did not show a synergetic effect (Fig. 4). In fact, the hydrolytic and esterification activities of the quadruple mutant were higher than wild-type, but lower than the double mutant Asp256Ile/His257Leu.

Comparison with Computational Analysis

Although the enzyme activity did not completely match the values from CDocker_Energy, we drew some conclusions from the results.

For most substrates, the binding energy of mutants was decreased compared to the wild-type enzyme, indicating that the enzyme activity might be increased (Fig. 2). However, for caprylic acid, enzyme activity of all the mutants was reduced (Fig. 4b). For other substrates, the activity of the quadruple mutant might be higher than the double mutants by predicted docking energy, but the actual esterification activity of the quadruple mutant was lower than the double mutant Asp256Ile/His257Leu (Fig. 4b). The reason for this result might be that the protein structure in organic media was different from the crystal structure, or that the program accuracy was decreased for predicting the results of additional mutations [32]. The prediction also showed that mutation of additional residue of His108Leu/Lys109Ile did not add to the effect of Asp256Ile/His257Leu on the esterification activity. The docking energy of Asn87Ile was predicted to be lower than Asn87Ile/Asp91Val, which predicted that the mutation Asp91Val in the hinge might hinder binding of the substrate in the pocket (Fig. 2).

Mutation in the binding pocket affected acyl carbon chain-length specificity in esterification by RML. In our case, the change

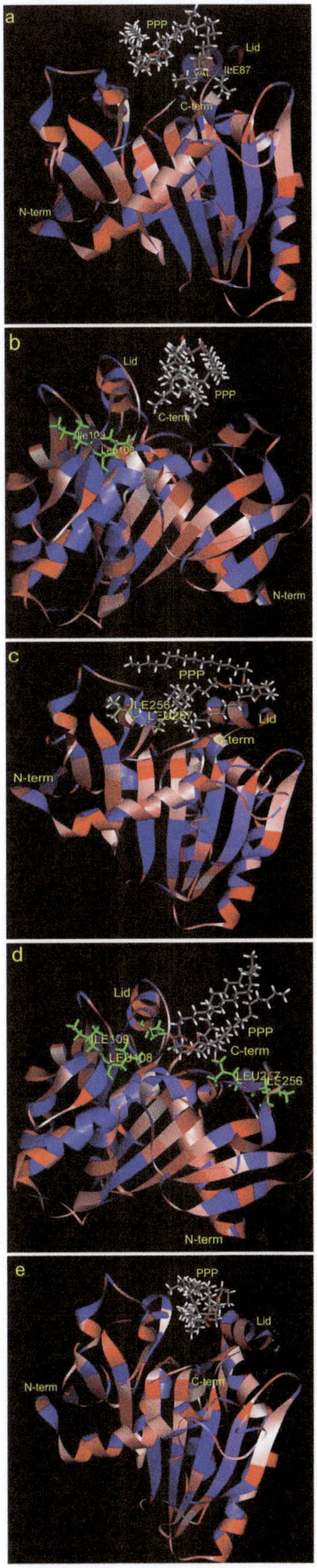

Figure 6. Profiles of RML mutants and wild-type docking with PPP. a: Asn87Ile/Asp91Val, b: His108Leu/Lys109Ile, c: Asp256Ile/His257Leu, d: His108Leu/Lys109Ile/Asp256Ile/His257Leu, e: wild-type.

in specificity resulting from hydrophobicity variance maybe relate to three factors. First, hydrophobic interactions between the mutants and substrates are different. A change in hydrophobicity in binding pocket influences the orientation and degree of binding of the substrate. Second, the size of the alcohol and acyl groups affects the approachability of the binding pocket; this might result in changes in selectivity for the optimal acid. Next, mutation of the lid or hinge influences the opening and closing of the lid. Santarossa et al. found that mutations in the *Pseudomonas fragi* lipase lid region influence chain-length specificity of the substrate [14]. The relative activity of C8/C4 increased with mutation of one, two, or three residues, but the ratio of C12/C4 was not related to the number of mutations. Discovery Studio and its predecessor Insight II are powerful tools in drug design and enzyme engineering. However, the software accuracy is not perfect, resulting in discrepancies between the activity of some mutants and the docking results [32].

For oleic acid and tripalmitin (PPP), which are the two substrates for HMFS modification, the mutant Asp256Ile/His257Leu showed a distinct decrease in docking energy compared to the wild-type and the other mutants, predicting a significant increase in enzyme activity (Fig. 2). The prediction was verified by the esterification of ethanol and oleic acid (Fig. 4b). Based on the measured enzyme activity and the predicted docking energy, the four mutants including Asn87Ile/Asp91Val, His108Leu/Lys109Ile, Asp256Ile/His257Leu, and His108Leu/Lys109Ile/Asp256Ile/His257Leu were used for OPO synthesis and oil modification for HMFS production.

Whole Cell Catalysts in HMFS Modification

For the production of structured lipids, TAG acidolysis with fatty acids catalyzed by 1,3-specific lipase was a useful strategy. The process was TAG hydrolysis, then DAG or MAG esterification with acid with possible acyl immigration from the 1,3-position to the 2-position of glycerol [17]. The hydrolytic and esterification activity of the lipase were both important for product yield. When lipase esterification activity was increased, the incorporation of acid and the structured lipid yield increased.

We first evaluated the performance of the mutants for synthesizing structured lipid-OPO in organic solvent with tripalmitin and oleic acid as substrates, with cell-displayed wild-type RML and commercial RML-Lipozyme RM IM as controls. After 24 h of reaction, oleic acid incorporation was 15.57% for wild-type, 11.93% for Asn87Ile/Asp91Val, 9.65% for His108Leu/Lys109Ile, 28.27% for Asp256Ile/His257Leu, and 16.94% for His108Leu/Lys109Ile/Asp256Ile/His257Leu (Fig. 5a). Incorporation did not increase for all mutants compared to the wild-type. Of four of the mutants, the incorporation of Asp256Ile/His257Leu was the highest, at 1.82-fold the wild-type level. This incorporation was close to the result of Tecelão [33], which was 32.1% at 24 h. The incorporation of Asn87Ile/Asp91Val and His108Leu/Lys109Ile were decreased 23.4% and 38.0% relative to the wild-type. The quadruple mutant showed an increase of 8.8% of the wild-type rate. This result coincided with the Discovery Studio prediction for Asp256Ile/His257Leu and the quadruple mutant (Fig. 2).

Increasing the hydrophobicity of the lipase or the immobilizing support was a feasible way to improve lipase esterification activity. Our results also indicated a difference in oleic acid (OA) incorporation by the mutant with the highest activity, Asp256Ile/His257Leu, and Lipozyme RM IM (Fig. 5). Immobilization was an effective way to increasing RML stability and activity [11,34,35]. Lipozyme RM IM carrier, an anion-exchange resin, has hydrophobic properties and disperses well in solvent, contributing to esterification activity [36]. Hydrophobicity might be a reason for differences in activity between yeast-displayed RML and Lipozyme RM IM.

Production of HMFS in a Solvent-free System

Oil acidolysis with OA was a viable and useful way to simplify HMFS production. We used cell-displayed RML wild-type and mutants for acidolysis palm oil with oleic acid in a solvent-free system. OA incorporation was 47.11% for Asn87Ile/Asp91Val, 46.80% for His108Leu/Lys109Ile, 49.22% for His108Leu/Lys109Ile/Asp256Ile/His257Leu, and 47.99% for wild-type (Fig. 5b). Asp256Ile/His257Leu showed excellent performance with an oleic acid incorporation of 53.18% after 6 h. This incorporation was similar to reports from Esteban and Telecão [16,33]. Esteban had an oleic acid incorporation of about 48% after 6 h in a reaction of a palmitic acid-rich TAG with an oleic acid-rich free fatty acid (FFA) catalyzed by Lipozyme RM IM [16]. Telecão had an OA incorporation of about 41.5% after 48 h in a reaction of tripalmitin with oleic acid catalyzed by *Carica papaya* lipase extracted from petiole leaves [33].

Oleic acid is about 40% in the fatty acids of initial palm oil. Compared to palm oil substrate, the increase in incorporation of oleic acid was 7.11% by Asn87Ile/Asp91Val, 6.80% by His108Leu/Lys109Ile, 13.18% by Asp256Ile/His257Leu, 9.22% by His108Leu/Lys109Ile/Asp256Ile/His257Leu and 7.99% by wild-type. In this reaction system, the catalysis efficiency of Asp256Ile/His257Leu was the highest at 1.65-fold of wild-type.

Of note, Asp256Ile/His257Leu showed an oleic acid incorporation of 53.18%, which was close to the 58.32% of Lipozyme RM IM when palm oil and oleic acid were the substrates.

Hydrophobicity and Enzyme Activity

In our study, we increased the hydrophobicity of the lipase substrate-binding pocket; this was a small area compared to the entire cell-catalyst surface. In fact, other approaches to changing the hydrophobicity of whole cells such as displaying hydrophobin on the yeast surface together with the lipase, might also increase esterification activity, since yeast cells act as the matrix to immobilize the enzyme. A combination of traditional immobilization technology with protein structure design, specifically immobilizing the enzyme after site-directed mutagenesis on the hydrophobic support, could be a powerful way to improve enzyme activity [37].

However, an increase in hydrophobicity does not always result in enhancement of esterification activity [38]. In addition to the hydrophobicity, open or closed enzyme form; lipase type, specifically active center conformation; solvent polarity or ion strength; and substrate molecular size or chain length are all important factors that affect lipase esterification activity. For immobilized enzymes, the kind of immobilizing support; the immobilization pattern; the presence of detergents; and the immobilization strategy, such determining as the ion strength or pH of solutions, can all influence lipase form, structure, and rigidity, affecting activity or selectivity [35].

Even in cases in which hydrophobicity enhances enzyme activity, the mechanisms are complex and different. One of the most important mechanisms is activation of lipases in the presence of a hydrophobic support interface to form a lid-open state [7,9,35]. In addition, using an immobilized carrier could lower the retention of water or acids, which also contributes to improved activity. The hydrophobicity of the carrier helps the disperse the enzyme in the substrate [36]. Modification of lid hydrophobicity balanced the hydrophobic/hydrophilic structure of the lid in the environment [12]. Mutations on lipase lid increased hydrophobic and electrostatic interactions between the lid and substrate [15].

Compared with previous reports, we found that modification of the binding pocket improved substrate-enzyme interaction. Docking results showed that docking energy decreased after mutation. However, simply promoting binding was not the only factor in the activity increase, because binding with the active center or recognition center, or reversible or irreversible binding had different consequences. Lid movement is the essential factor affecting enzyme activity of lipases with lids. Our results suggested that lid movement might be changed in mutations, since activity of Asn87Ile and Asn87Ile/Asp91Val were lower than Asp256Ile/His257Leu (Fig. 4).

The pattern and orientation of RML mutants docking with PPP were different (Fig. 6). For Asn87Ile/Asp91Val, His108Leu/Lys109Ile and wild-type, PPP reacted with lipase in a horizontal orientation and with a shallower entry of the substrate into the pocket than Asp256Ile/His257Leu and His108Leu/Lys109Ile/Asp256Ile/His257Leu. His108Leu is located at the bottom of the binding pocket. In His108Leu/Lys109Ile/Asp256Ile/His257Leu, an alcohol moiety entered the pocket and the acyl part protruded. Asp256Ile/His257Leu is located on the opposite side of the lid, which caused the acyl moiety to be closer to the binding pocket. The three acyl chains separated farther than the other mutants. Since Leu255 and Leu258 are both hydrophobic amino acids, when connected to Asp256Ile/His257Leu, the binding pocket was most suitable for long-chain substrates. These factors meant that Asp256Ile/His257Leu had higher activity than the other mutants in this study.

Conclusion

We found that, with the same enzyme load, displayed RML had an OA incorporation comparable to Lipozyme RM IM for HMFS production. The whole-cell catalyst has a simpler productive technology and a higher expression level. As an enzyme modified based on hydrophobicity, the yeast surface-displayed RML mutant Asp256Ile/His257Leu is a promising catalyst for HMFS production, and could be a potential alternative catalyst for other oil modifications.

Author Contributions

Conceived and designed the experiments: JZ YL SH. Performed the experiments: JZ YJ. Analyzed the data: JZ YJ. Contributed reagents/materials/analysis tools: YL YS SZ. Wrote the paper: JZ YJ.

References

1. Lutz S (2010) Beyond directed evolution – semi-rational protein engineering and design. Curr Opin Biotechnol 21: 734–743.
2. Kourist R, Brundiek H, Bornscheuer UT (2010) Protein engineering and discovery of lipases. Eur J Lipid Sci Technol 112: 64–74.
3. Turner NJ (2009) Directed evolution drives the next generation of biocatalysts. Nat Chem Biol 5: 567–573.
4. Reetz MT, Jaeger KE (1999) Superior biocatalysts by directed evolution. In Biocatalysis – From Discovery to Application, Springer: Birlin/Heideberg, Germany. 200: 31–57.

5. Basri M, Ampon K, Wan Yunus WMZ, Razak CNA, Salleh AB (1995) Enzymic synthesis of fatty esters by hydrophobic lipase derivatives immobilized on organic polymer beads. J Am Oil Chem Soc 72: 407–411.
6. Cho S, Rhee JS (1993) Immobilization of lipase for effective interesterification of fats and oils in organic solvent. Biotechnol Bioeng 41: 204–210.
7. Reetz MT (1997) Entrapment of biocatalysts in hydrophobic sol-gel materials for use in organic chemistry. Adv Mater 9: 943–954.
8. Yokozeki K, Yamanaka S, Takinami K, Hirose Y, Tanaka A, et al. (1982) Application of immobilized lipase to region-specific interesterification of triglycerides in organic solvent. Eur J Appl Microbiol Biotechnol 14: 1–5.
9. Palomo JM, Muñoz G, Fernández-Lorente G, Mateo C, Fernández-Lafuente R, et al. (2003) Interfacial adsorption of lipases on very hydrophobic support (octadecyl- Sepabeads): immobilization, hyperactivation and stabilization of the open form lipases. J Mol Catal B: Enzym 19–20: 279–286.
10. Ruckenstein E, Wang X (1993) Lipase immobilized on hydrophobic porous polymer supports prepared by concentrated emulsion polymerization and their activity in the hydrolysis of triacylglycerides. Biotechnol Bioeng 42: 821–828.
11. Garcia-Galan C, Berenguer-Murcia A, Fernández-Lafuente R, Rodrigues RC (2011) Potential of different enzyme immobilization strategies to improve enzyme performance. Adv Synth Catal 353: 2885–2904.
12. Thomas A, Allouche M, Basyn F, Brasseur R, Kerfelec B (2005) Role of the lid hydrophobicity pattern in pancreatic lipase activity. J Biol Chem 280: 40074–40083.
13. Otero C, Rúa ML, Robledo L (1995) Influence of the hydrophobicity of lipase isoenzymes from *Candida rugosa* on its hydrolytic activity in reverse micelles. FEBS Lett 360: 202–206.
14. Santarossa G, Lafranconi PG, Alquati C, DeGioia L, Alberghina L, et al. (2005) Mutations in the "lid" region affect chain length specificity and thermostability of a *Pseudomonas fragi* lipase. FEBS Lett 579: 2383–2386.
15. Gao B, Xu T, Lin J, Zhang L, Su E, et al. (2011) Improving the catalytic activity of lipase LipK107 from *Proteus* sp. by site-directed mutagenesis in the lid domain based on computer simulation. J Mol Catal B: Enzym 68: 286–291.
16. Esteban L, Jiménez MJ, Hita E, González PA, Martín L, et al. (2011) Production of structured triacylglycerols rich in palmitic acid at *sn*-2 position and oleic acid at *sn*-1,3 positions as human milk fat substitutes by enzymatic acidolysis. Biochem Eng J 54: 62–69.
17. Sahin N, Akoh CC, Karaali A (2006) Human milk fat substitutes containing omega-3 fatty acids. J Agric Food Chem 54: 3717–3722.
18. Zou XQ, Huang JH, Jin QZ, Guo Z, Liu YF, et al. (2013) Model for human milk fat substitute evaluation based on triacylglycerol composition profile. J Agric Food Chem 61: 167–75.
19. Chen ML, Vali SR, Lin JY, Ju YH (2004) Synthesis of the structured lipid 1,3-dioleoyl-2-palmitoylglycerol from palm oil. J Am Oil Chem Soc 81: 525–532.
20. Sørensen ADM, Xu XB, Zhang L, Kristensen JB, Jacobsen C (2010) Human milk fat substitute from butterfat: production by enzymatic interesterification and evaluation of oxidative stability. J Am Oil Chem Soc 87: 185–194.
21. Rodrigues RC, Fernández-Lafuente R (2010) Lipase from *Rhizomucor miehei* as an industrial biocatalyst in chemical process. J Mol Catal B: Enzym 64: 1–22.
22. Derewenda U, Brzozowski AM, Lawson DM, Derewenda ZS (1992) Catalysis at the interface: the anatomy of a conformational change in a triglyceride lipase. Biochem 31: 1532–1541.
23. Srivastava A, Akoh CC, Chang SW, Lee GC, Shaw JF (2006) *Candida rugosa* lipase LIP1-catalyzed transesterification to produce human milk fat substitute. J Agric Food Chem 54: 5175–5181.
24. Han ZL, Han SY, Zheng SP, Lin Y (2009) Enhancing thermostability of a *Rhizomucor miehei* lipase by engineering a disulfide bond and displaying on the yeast cell surface. Appl Microbiol Biotechnol 85: 117–126.
25. Huang DF, Han SY, Han ZL, Lin Y (2012) Biodiesel production catalyzed by *Rhizomucor miehei* lipase-displaying *Pichia pastoris* whole cells in an isooctane system. Biochem Eng J 63: 10–14.
26. Kobori H, Sato M, Osumi M (1992) Relationship of actin organization to growth in the two forms of the dimorphic yeast *Candida tropicalis*. Protoplasma 167: 193–204.
27. Teichert SA, Akoh CC (2011) Stearidonic acid soybean oil enriched with palmitic acid at the sn-2 position by enzymatic interesterification for use as human milk fat analogues. J Agric Food Chem 59: 5692–5701.
28. Pleiss J, Fischer M, Schimid RD (1998) Anatomy of lipase binding sites: the scissile fatty acid binding site. Chem Phys Lipids 93: 67–80.
29. Kyte J, Doolittle RF (1982) A simple method for displaying the hydropathic character of a protein. J Mol Biol 157: 105–132.
30. Reetz MT, Prasad S, Carballeira JD, Gumulya Y, Bocola M (2010) Iterative saturation mutagenesis accelerates laboratory evolution of enzyme stereoselectivity: rigorous comparison with traditional methods. J Am Chem Soc 132: 9144–9152.
31. Liang XX, Wang BB, Sun YF, Lin Y, Han SY, et al (2013) Quantitative evaluation of *Candida antarctica* lipase B displayed on the cell surface of a *Pichia pastoris* based on an FS anchor system. Biotechnol Lett 35: 367–374.
32. Xu T, Zhang L, Wang X, Wei D, Li T (2009) Structure-based substrate screening for an enzyme. BMC Bioinformatics 10: 257–263.
33. Tecelão C, Rivera I, Sandoval G, Ferreira-Dias S (2012) *Carica papaya* latex: A low-cost biocatalyst for human milk fat substitutes production. Eur J Lipid Sci Technol 114: 266–276.
34. Huge-Jensen B, Galluzzo DR, Jensen RG (1988) Studies on free and immobilized lipases from *Mucor miehei*. J Am Oil Chem Soc 65: 905–910.
35. Rodrigues RC, Ortiz Z, Berenguer-Murcia Á, Torres R, Fernández-Lafuente R (2013) Modifying enzyme activity and selectivity by immobilization. Chem Soc Rev DOI: 10.1039/c2cs35231a.
36. Ibrahim NA, Guo Z, Xu X (2008) Enzymatic interesterification of palm stearin and coconut oil by a dual lipase system. J Am Oil Chem Soc 85: 37–45.
37. Hernandez K, Fernández-Lafuente R (2011) Control of protein immobilization: Coupling immobilization and site-directed mutagenesis to improve biocatalyst or biosensor performance. Enzyme Microb Technol 48: 107–122.
38. Lie E, Molin G (1991) Hydrolysis and esterification with immobilized lipase on hydrophobic and hydrophilic zeolites. J Chem Technol Biotechnol 50: 549–553.

8

Efficient Production of (*R*)-2-Hydroxy-4-Phenylbutyric Acid by Using a Coupled Reconstructed D-Lactate Dehydrogenase and Formate Dehydrogenase System

Binbin Sheng[1], Zhaojuan Zheng[2], Min Lv[1], Haiwei Zhang[1], Tong Qin[1], Chao Gao[1]*, Cuiqing Ma[1], Ping Xu[1¤]

1 State Key Laboratory of Microbial Technology, Shandong University, Jinan, People's Republic of China, **2** College of Chemical Engineering, Nanjing Forestry University, Nanjing, People's Republic of China

Abstract

Background: (*R*)-2-Hydroxy-4-phenylbutyric acid [(*R*)-HPBA] is a key precursor for the production of angiotensin-converting enzyme inhibitors. However, the product yield and concentration of reported (*R*)-HPBA synthetic processes remain unsatisfactory.

Methodology/Principal Findings: The Y52L/F299Y mutant of NAD-dependent D-lactate dehydrogenase (D-nLDH) in *Lactobacillus bulgaricus* ATCC 11842 was found to have high bio-reduction activity toward 2-oxo-4-phenylbutyric acid (OPBA). The mutant D-nLDH$^{Y52L/F299Y}$ was then coexpressed with formate dehydrogenase in *Escherichia coli* BL21 (DE3) to construct a novel biocatalyst *E. coli* DF. Thus, a novel bio-reduction process utilizing whole cells of *E. coli* DF as the biocatalyst and formate as the co-substrate for cofactor regeneration was developed for the production of (*R*)-HPBA from OPBA. The biocatalysis conditions were then optimized.

Conclusions/Significance: Under the optimum conditions, 73.4 mM OPBA was reduced to 71.8 mM (*R*)-HPBA in 90 min. Given its high product enantiomeric excess (>99%) and productivity (47.9 mM h^{-1}), the constructed coupling biocatalysis system is a promising alternative for (*R*)-HPBA production.

Editor: John R. Battista, Louisiana State University and A & M College, United States of America

Funding: This work was supported by the National Natural Science Foundation of China (31270856, 31170052, 31270090 and J1103515), the Independent Innovation Foundation of Shandong University (2012GN017) and the Excellent Middle-Aged and Youth Scientist Award Foundation of Shandong Province (BS2013SW025). The funders had no role in study design, data collection and analysis, decision to publish, or preparation of the manuscript.

Competing Interests: The authors have declared that no competing interests exist.

* Email: jieerbu@sdu.edu.cn

¤ Current address: State Key Laboratory of Microbial Metabolism, School of Life Sciences and Biotechnology, Shanghai Jiao Tong University, Shanghai, People's Republic of China

Introduction

(*R*)-2-hydroxy-4-phenylbutyric acid [(*R*)-HPBA] and ethyl (*R*)-2-hydroxy-4-phenylbutyrate [(*R*)-HPBE] can be used as the key precursors for the production of angiotensin-converting enzyme (ACE) inhibitors [1–5]. ACE inhibitors such as benazepril, enalapril, lisinopril, ramipril, and quinapril are widely used in the first-line therapy of hypertension and congestive heart failure [6–9]. Owing to the substantial demand for these drugs, various chemical or biological processes have been developed to produce (*R*)-HPBA or (*R*)-HPBE. In recent years, great success has been achieved in asymmetric synthesis of (*R*)-HPBE catalyzed by recombinant reductases [10,11]. For example, whole cells of a recombinant *Escherichia coli* strain harboring CgKR2 and glucose dehydrogenase (GDH) were applied in preparing (*R*)-HPBE with high concentration, desirable enantiomeric excess (ee) (>99%) and yield [10]. Compared with that of (*R*)-HPBE, the product yield and concentration of the reported (*R*)-HPBA synthesis processes remained unsatisfactory [1,12].

In previous studies, enzymatic resolution and asymmetric reduction were used in the biological production of (*R*)-HPBA. Compared with enzymatic resolution catalyzed by hydrolases, especially lipases [4,9,13], asymmetric bio-reduction of 2-oxo-4-phenylbutyric acid (OPBA) by dehydrogenases is more desirable because of its excellent stereoselectivity and high theoretical yield up to 100% [1,14]. For practical production of (*R*)-HPBA from OPBA through bio-reduction, highly efficient reductases and cofactor regeneration systems are needed.

In contrast to the (*R*)-HPBE preparation processes, which often utilize a specific carbonyl reductase, the production of (*R*)-HPBA from OPBA is catalyzed by 2-ketoacid reductases, especially NAD-dependent D-lactate dehydrogenase (D-nLDH) [12,15]. However, as an unnatural substrate of D-nLDH, OPBA could not be efficiently catalyzed by the biocatalyst because of its large aromatic

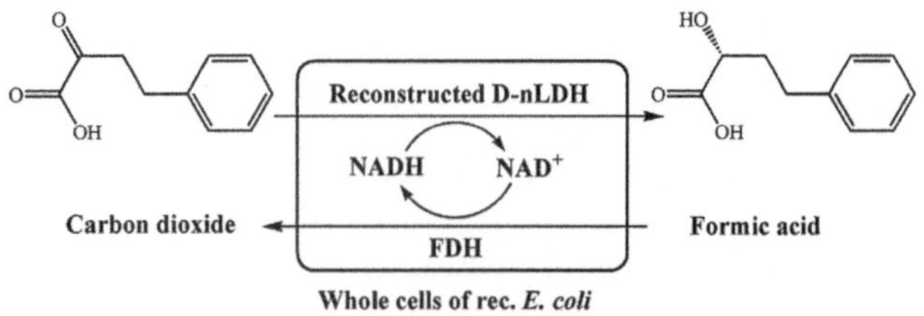

Figure 1. Scheme for (*R*)-HPBA production from OPBA by using a coupled system of reconstructed D-nLDH and FDH.

group at C-4.On the other hand, cofactor regeneration systems that utilize glucose as a co-substrate in (*R*)-HPBE production may not be the proper choice in the (*R*)-HPBA production. The addition of glucose to the reaction system may result in the production of organic acids (such as gluconic acid and lactic acid) as byproducts and increase the complexity of the (*R*)-HPBA separation process [16,17]. In a previous study, a partially purified D-nLDH was used to transform OPBA to (*R*)-HPBA. The cofactor NADH was regenerated by formate dehydrogenase (FDH) present in whole cells of *Candida boidinii* ATCC 32195. Although this NADH regeneration system produced CO_2 as the only byproduct, which facilitated the isolation of (*R*)-HPBA, the whole cells of *C. boidinii* should be pre-treated with toluene to make them permeable [12].

In our previous studies, the D-nLDH in *Lactobacillus bulgaricus* ATCC 11842 was rationally re-designed and then used for the bio-reduction of substrates with large aliphatic or aromatic groups at C-3 [14]. In this study, the activities of different D-nLDH mutants toward OPBA (2-oxo carboxylic acids with an aromatic group at C-4) were assayed. The most active reconstructed D-nLDH was co-expressed with FDH from *C. boidinii* NCYC 1513 in *E. coli* BL21 (DE3). Then, a novel process utilizing whole cells of recombinant *E. coli* was developed for efficient production of (*R*)-HPBA from OPBA (Fig. 1).

Materials and Methods

Materials

OPBA was purchased from Gracia Chemical Technology Co., Ltd. Chengdu (China). Isopropyl-β-D-1-thiogalactopyranoside (IPTG), phenylmethanesulfonyl fluoride (PMSF), and (*R*)-HPBA were purchased from Sigma-Aldrich. (*S*)-HPBA was purchased from J&K Chemical. All other chemicals in this study were of reagent grade.

Microorganisms and growth conditions

The bacterial strains, plasmids, and oligonucleotide primers used in this study are listed in Table 1. *E. coli* DH5α and BL21 (DE3) were used for general cloning and expression procedures, respectively. *E. coli* WD, *E. coli* D1, and *E. coli* D2 were used to express wild D-nLDH, D-nLDH^{F299Y}, and D-nLDH$^{Y52L/F299Y}$, respectively [14]. *E. coli* PD containing the vector pETDuet-1 was used as a control. The plasmid pETDuet-*ldhD*$^{Y52L/F299Y}$-*fdh* was constructed as follows: the *ldhD*$^{Y52L/F299Y}$ gene was amplified using primers D.f and D.r with plasmid pETDuet-*ldhD*$^{Y52L/F299Y}$ as a template. The *fdh* gene was amplified using primers F.f and F.r with genomic DNA of *C. boidinii* NCYC 1513 as a template. The resulting PCR products *ldhD*$^{Y52L/F299Y}$ and *fdh* were digested with *Nco*I-*Bam*HI and *Nde*I-*Xho*I, respectively, and cloned into MCS1 and MCS2 of pETDuet-1 successively to construct pETDuet-*ldhD*$^{Y52L/F299Y}$-*fdh*. The plasmid pETDuet-*ldhD*$^{Y52L/F299Y}$-*fdh* was then transformed into *E. coli* BL21 (DE3) to construct *E. coli* DF. All of the *E. coli* strains were grown in Luria-Bertani (LB) medium, and ampicillin was added at a concentration of 100 μg ml^{-1} if necessary.

Table 1. Strains, plasmids, and oligonucleotide primers used in this study.

Strain, plasmid, or primer	Relevant characteristics	Source or reference
Strain		
E. coli DH5α	φ80 *lacZ*ΔM15 Δ(*lacZYA-argF*) U169 *recA*1 *endA*1 *hsdR*17 *supE*44λ-*thi*-1	Invitrogen
E. coli BL21(DE3)	F^- *ompT gal dcm lon hsdS*$_B$(r_B^- m_B^-) λ(*DE3*)	Novagen
C. boidinii NCYC 1513	Wild-type, source of *fdh* gene	NCYC[a]
E. coli PD	*E. coli* BL21(DE3) containing vector pETDuet-1	This study
E. coli WD	*E. coli* BL21(DE3) expressing wild type D-nLDH	[14]
E. coli D1	*E. coli* BL21(DE3) expressing D-nLDH^{F299Y}	[14]
E. coli D2	*E. coli* BL21(DE3) expressing D-nLDH$^{Y52L/F299Y}$	[14]
E. coli DF	*E. coli* BL21(DE3) expressing D-nLDH$^{Y52L/F299Y}$ and FDH	This study
Plasmid		
pETDuet-1	Expression vector, Amp^r	Novagen
pETDuet-*ldhD*$^{Y52L/F299Y}$	N-terminal His-tagged *ldhD*$^{Y52L/F299Y}$ gene in pETDuet-1	[14]
pETDuet-*ldhD*$^{Y52L/F299Y}$-*fdh*	Both *ldhD*$^{Y52L/F299Y}$ and *fdh* without His-tag in pETDuet-1	This study
Oligonucleotide primer	Sequence (5′→3′)	
D.f	CCATGGTGACTAAAATTTTTGCTTACGCA (*Nco*I)	
D.r	GGATCCTTAGCCAACCTTAACTGGAGTTT (*Bam*HI)	
F.f	CATATGAAGATCGTTTTAGTCTTATATGATGCTGGTA (*Nde*I)	
F.r	CTCGAGTTATTTCTTATCGTGTTTACCGTAAGCTTTG (*Xho*I)	

[a]NCYC, National Collection of Yeast Cultures.

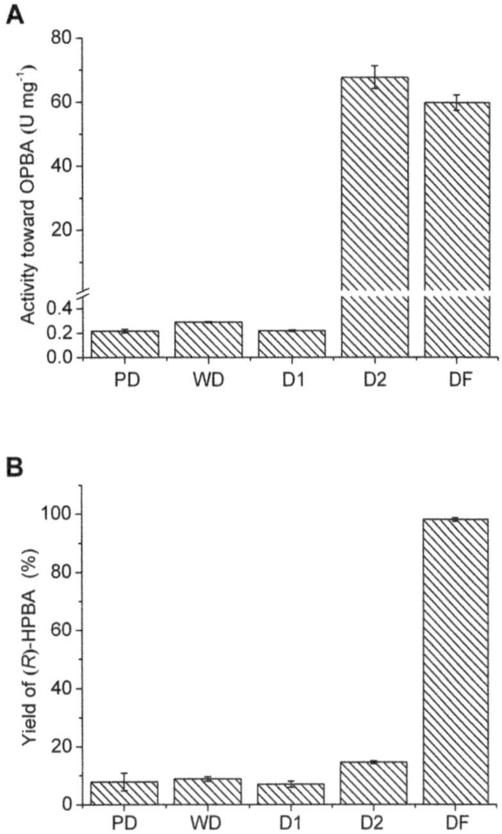

D-nLDH		+	+	+	+
F299Y			+	+	+
Y52L				+	+
FDH					+

Figure 2. Feasibility of (*R*)-HPBA production through cofactor regeneration by reconstructed D-nLDH and FDH. (A) OPBA reduction activities in the crude extract of different *E. coli* strains. (B) Asymmetric reduction of OPBA by whole cells of different *E. coli* strains. For *E. coli* PD, *E. coli* WD, *E. coli* D1, and *E. coli* D2, NADH regeneration was conducted by the direct addition of 50 mM glucose. For *E. coli* DF, formate of 50 mM was added in the reaction broth for NADH regeneration.

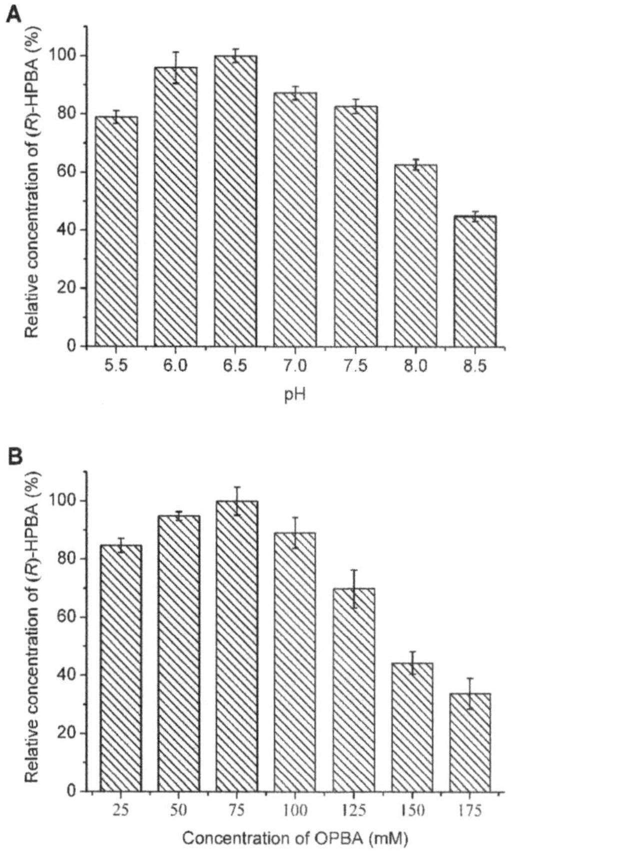

Figure 3. Optimization of the biocatalysis conditions. (A) pH. (B) Concentration of OPBA.

Biocatalyst preparation

The recombinant strains of *E. coli* PD, *E. coli* WD, *E. coli* D1, *E. coli* D2, and *E. coli* DF were all cultured in LB medium (100 μg ml^{-1} ampicillin) at 37°C to an optical density of 0.6 at 600 nm. IPTG (1 mM) was then added to induce protein expression, and cultures were grown at 16°C for a further 12 h. Cells were harvested by centrifugation at 6,000 rpm for 10 min, washed twice with 67 mM phosphate buffer solution (pH 7.4), and then subjected to successive biotransformation.

Optimization of biocatalysis conditions

To optimize the biotransformation conditions, 5-ml reaction mixtures were incubated at 37°C and 120 rpm in a 25-ml flask. The pH was adjusted from 5.5 to 8.5. The concentrations of OPBA and formate were 25–175 mM. The concentration of the whole cells was 1–8 g dry cell weight (DCW) l^{-1}. Samples (0.2 ml) were collected periodically and centrifuged at 12,000 rpm. The concentrations of OPBA and (*R*)-HPBA in the supernatant were analyzed by a high-performance liquid chromatography (HPLC) system (Agilent 1100 series, Hewlett-Packard, USA).

Analytical procedures

Cells of *E. coli* PD, *E. coli* WD, *E. coli* D1, *E. coli* D2, and *E. coli* DF were harvested, suspended in 67 mM phosphate buffer solution (pH 7.4) containing 1 mM PMSF, and then disrupted by sonication (Sonics 500 W; 20 KHz) for 5 min in an ice bath. Thereafter, intact cells and cell debris were removed by centrifugation, and the resultant crude extracts were subjected to successive D-nLDH activity assays. The reduction activities of D-nLDH wild-type and mutants toward OPBA were assayed at 37°C in 1 ml of 50 mM Tris-HCl buffer (pH 7.5) containing 0.2 mM NADH, 10 mM OPBA, and the crude extracts of *E. coli* PD, *E. coli* WD, *E. coli* D1, *E. coli* D2, and *E. coli* DF. The rate of NADH decrease was determined by measuring the absorbance change at 340 nm [14,18,19]. One unit of D-nLDH activity was defined as the amount that catalyzed the oxidation of 1 μmol of NADH per minute. The protein concentration was determined by the Lowry procedure by using bovine serum albumin as the standard [20].

OPBA and (*R*)-HPBA were measured by HPLC (Agilent 1100 series) equipped with an Agilent Zorbax SB-C18 column (150×4.6 mm, 5 μm) and a variable-wavelength detector at 210 nm. The mobile phase consisted of 1 mM H_2SO_4 and acetonitrile with a ratio of 85:15 (v/v) at a flow rate of 0.7 ml min^{-1} at 30°C. Stereoselective assays for (*R*)-HPBA and

Table 2. Effects of concentration of whole cells on biotransformation[a].

Cell concentration (g DCW l^{-1})	1	3	5	6	7	8
Reaction time (min)	285	140	75	55	50	45
(*R*)-HPBA concentration (mM)	33.0	61.8	59.9	59.4	60.3	61.1
Productivity[b] (mM min^{-1} g^{-1} DCW)	0.116	0.147	0.160	0.180	0.172	0.170

[a]Value is the average value of three separate assays.
[b]Productivity was calculated when the reaction was conducted to approximately 80% of the theoretical yield except for the reaction at 1 g DCW l^{-1} whole cells.

(*S*)-HPBA were performed by HPLC analysis by using a chiral column (MCI GEL CRS10W, Japan) and a tunable UV detector at 254 nm. The mobile phase was 2 mM $CuSO_4$ and acetonitrile with a ratio of 85:15 (v/v) at a flow rate of 0.5 ml min^{-1} and a temperature of 25°C. The ee of (*R*)-HPBA was defined as [((*R*)-HPBA−(*S*)-HPBA)/((*R*)-HPBA+(*S*)-HPBA)] ×100%.

Results and Discussion

Activity of D-nLDH wild-type and mutants toward OPBA

To evaluate the possibility of transforming OPBA into (*R*)-HPBA by D-nLDH, the wild type D-nLDH from *L. bulgaricus* ATCC 11842 and its mutants were overexpressed in *E. coli* BL21 (DE3). Crude extracts of *E. coli* PD, *E. coli* WD, and *E. coli* D1 exhibited rather low OPBA reduction activity (Fig. 2A). The Y52L/F299Y mutant of D-nLDH caused the specific activity of the crude extract of *E. coli* D2 to be 233.2–312.3 fold higher than that in extracts of *E. coli* PD, *E. coli* WD, and *E. coli* D1. These results suggest that the mutant D-nLDH$^{Y52L/F299Y}$ is rather active toward OPBA and may have the potential to efficiently produce (*R*)-HPBA from OPBA.

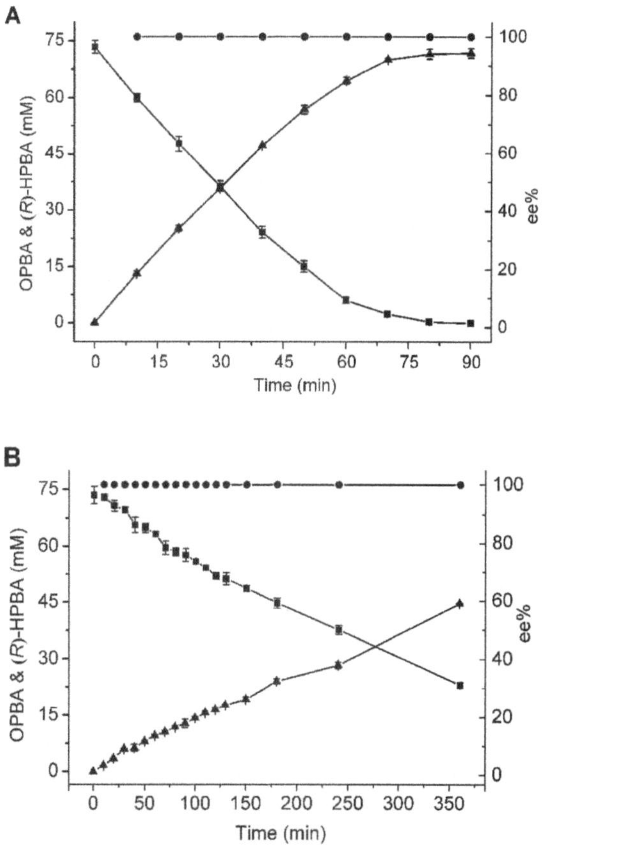

Figure 4. Time course of highly optically pure (*R*)-HPBA production from OPBA under optimal conditions. (A) Biotransformation using whole cells of *E. coli* DF as a biocatalyst and formate for cofactor regeneration. (B) Biotransformation using whole cells of *E. coli* D2 as a biocatalyst and glucose for cofactor regeneration. (■), OPBA; (▲), (*R*)-HPBA; (●), ee.

Feasibility of (*R*)-HPBA production through the cofactor regeneration system

Asymmetric reduction of OPBA by whole cells of *E. coli* PD, *E. coli* WD, *E. coli* D1, *E. coli* D2, and *E. coli* DF was investigated to further explore the potential by using D-nLDH in the synthesis of (*R*)-HPBA. OPBA at 50 mM was used as the substrate. Whole cells of *E. coli* PD, *E. coli* WD, *E. coli* D1, and *E. coli* D2 at a concentration of 8 g DCW l^{-1} were added to the reaction broth. The reaction was conducted at 37°C for 2 h. Here, NADH was regenerated through the direct addition of 50 mM glucose in the reaction system. Whole cells of *E. coli* D2 exhibited higher (*R*)-HPBA producing capability than did cells of *E. coli* PD, *E. coli* WD, and *E. coli* D1(Fig. 2B). However, the (*R*)-HPBA productivity (3.7 mM h^{-1}) was still rather low because of the low efficiency of the NADH regeneration system. Additionally, organic acids, including pyruvic acid, lactic acid, and acetic acid, accumulated in the reaction broth (Fig. S1).

FDH is a good choice for NADH regeneration in a biocatalysis system because its substrate, formate, has a low cost and its product, carbon dioxide, is easily separated [21–25]. In this work, FDH was coexpressed with D-nLDH$^{Y52L/F299Y}$ in *E. coli* DF and the (*R*)-HPBA production capability of the novel biocatalyst was investigated. Formate (50 mM) was added to the reaction broth for the regeneration of NADH. Although the activity of D-nLDH$^{Y52L/F299Y}$ in the crude extract of *E. coli* DF was lower than in the extract of *E. coli* D2, whole cells of *E. coli* DF exhibited much higher (*R*)-HPBA producing capability than other biocatalysts (Fig. 2A and Fig. 2B). (*R*)-HPBA at 49.0 mM was obtained from 50 mM OPBA. The productivity of (*R*)-HPBA was 24.5 mM h^{-1}. Thus, whole cells of *E. coli* DF were selected as biocatalysts for (*R*)-HPBA production in the subsequent experiments.

Optimization of biocatalysis conditions

To achieve a higher product concentration, the biocatalytic conditions for (*R*)-HPBA production from OPBA by using whole cells of *E. coli* DF were optimized. The influence of the reaction pH was determined in reaction mixtures containing 13 g DCW l^{-1} whole cells of *E. coli* DF, 50 mM OPBA, 50 mM sodium formate, and 200 mM phosphate buffer (pH ranging from 5.5 to 8.5). After bioconversion at 37°C for 15 min, the highest (*R*)-HPBA production was detected at pH 6.5 (Fig. 3A).

Table 3. Comparison of recently reported processes for (*R*)-HPBA or (*R*)-HPBE production through bio-reduction.

Biocatalyst	Product (mM)	Productivity (mM h^{-1})	ee (%)	Co-substrate	References
Whole cells of *Candida boidinii* CIOC21	20	1.7	99	5% glucose	[27]
Whole cells of *Bacillus pumilus* Phe-C3	29.6	1.1	97.1	2% glucose	[28]
Whole cells of *Candida krusei* SW2026	79.5	5.0	97.4	5% glucose	[26]
Saccharomyces cerevisiae pretreated with α-phenacyl chloride	4.8	0.2	92	-	[29]
Saccharomyces cerevisiae pretreated with α-phenacyl chloride	167.7	3.5	87.5	1.5% ethanol (v/v)	[30]
Lyophilized cells of *E. coli* BL21/ pCgKR 2 and lyophilized GDH powders	1000	140.1	>99	27% glucose	[10]
Whole cells of *E. coli* BL21 coexpressing IolS and GDH	50	3.8	99.5	3.6% glucose	[31]
Whole cells of *E. coli* BL21 coexpressing IolS and GDH	1600	132.1	99.5	20% glucose	[11]
D-LDH from *Staphylococcus epidermidis* and FDH from *Candida boidinii**	182	38.2	>99.8	2.2% ammonium formate	[15]
Partially purified D-LDH (EC 1.1.1.28) and whole cells of *Candida boidinii* ATCC 32591 containing FDH*	56.7	49.9	ND	0.8% sodium formate	[12]
Whole cells of *E. coli* BL21 coexpressing YiaE and GDH*	100	4.2	98	3.6% glucose	[1]
Whole cells of *E. coli* DF*	71.8	47.9	>99	0.5% sodium formate	This study

*Substrates were OPBA. Substrates of the other processes were OPBE. ND represents no data.

To determine the effect of the OPBA concentration, reactions with eight different OPBA and sodium formate concentrations (25, 50, 75, 100, 125, 150, and 175 mM) were conducted at pH 6.5 and 37°C for 30 min. The highest (*R*)-HPBA production was detected when 75 mM OPBA was used (Fig. 3B). The effect of the biocatalyst concentration was also investigated to determine the optimal range. The biotransformation was conducted with 75 mM OPBA, 75 mM sodium formate, 200 mM phosphate buffer (pH 6.5), and whole cells of *E. coli* DF at six different concentrations (1, 3, 5, 6, 7, and 8 g DCW l^{-1}). When the reactions were conducted to approximate 80% theoretical yield, the highest specific productivity was observed at a biocatalyst concentration of 6 g DCW l^{-1} (Table 2).

Production of (*R*)-HPBA under optimal conditions

On the basis of the results presented above, an optimal bioconversion system for production of optically pure (*R*)-HPBA from OPBA was developed. Biotransformation was conducted at 37°C in 200 mM phosphate buffer (pH 6.5) with 6 g DCW l^{-1} whole cells of *E. coli* DF as the biocatalyst. As shown in Fig. 4A, 71.8 mM (*R*)-HPBA with a high enantiomeric purity (ee >99%, Fig. S2) was obtained from 73.4 mM OPBA in 90 min. When whole cells of *E. coli* D2 only expressing D-nLDH$^{Y52L/F299Y}$ were used as the biocatalyst, and glucose was added for NADH regeneration, only 44.7 mM (*R*)-HPBA was produced with a yield of 60.9% after 360 min (Fig. 4B).

Many biocatalysts have been used in the enantioselective production of (*R*)-HPBE and (*R*)-HPBA through bio-reduction [1,12,26–28]. Compared with (*R*)-HPBE production processes, the product concentrations of the reported (*R*)-HPBA synthesis processes were rather low (Table 3) [1,10–12,15,29,30]. In the previous study, purified D-LDH from *Staphylococcus epidermidis* and FDH from *Candida boidinii* were applied for (*R*)-HPBA production. (*R*)-HPBA at a concentration of 182 mM was produced, which is the highest reported yield of (*R*)-HPBA to date [15]. However, problems concerning the application of the process, such as the complicated enzyme purification and costly cofactor addition, remain. In the present work, mutant D-nLDH and FDH were co-expressed in *E. coli* DF and used for (*R*)-HPBA production from OPBA. The productivity (47.9 mM h^{-1}) and ee (>99%) of the product were rather high for (*R*)-HPBA production. Additionally, given the simple composition of the biocatalytic system, separation of (*R*)-HPBA from the biocatalytic system would be relatively inexpensive. Therefore, the novel process established in this study could also be used as a promising route for the production of highly optically pure (*R*)-HPBA.

Conclusions

In summary, whole cells of *E. coli* DF coexpressing D-nLDH$^{Y52L/F299Y}$ from *L. bulgaricus* ATCC 11842 and FDH from *C. boidinii* NCYC 1513 exhibited catalytic capability for (*R*)-HPBA production from OPBA. After optimization of the biotransformation conditions, 73.4 mM OPBA was reduced to 71.8 mM (*R*)-HPBA with a high productivity of 47.9 mM h^{-1} and an excellent ee (>99%). The constructed coupled biocatalysis system developed in this work may be a promising alternative for the production of the key medical intermediate (*R*)-HPBA.

Author Contributions

Conceived and designed the experiments: BS CG CM PX. Performed the experiments: BS ML HZ TQ. Analyzed the data: BS ZZ CG CM. Contributed reagents/materials/analysis tools: CG CM PX. Contributed to the writing of the manuscript: BS CG CM PX.

References

1. Yun H, Choi HL, Fadnavis NW, Kim BG (2005) Stereospecific synthesis of (*R*)-2-hydroxy carboxylic acids using recombinant *E. coli* BL21 overexpressing YiaE from *Escherichia coli* K12 and glucose dehydrogenase from *Bacillus subtilis*. Biotechnol Prog 21: 366–371.
2. de Lacerda PSB, Ribeiro JB, Leite SGF, Coelho RB, Lima ELD, et al. (2006) Microbial enantioselective reduction of ethyl-2-oxo-4-phenyl-butanoate. Biochem Eng J 28: 299–302.
3. de Lacerda PSB, Ribeiro JB, Leite SGF, Ferrara MA, Coelho RB, et al. (2006) Microbial reduction of ethyl 2-oxo-4-phenylbutyrate. Searching for *R*-enantioselectivity. New access to the enalapril like ACE inhibitors. Tetrahedron: Asymmetry 17: 1186–1188.
4. Chen B, Yin HF, Wang ZS, Liu JY, Xu JH (2010) A new chemo-enzymatic route to chiral 2-hydroxy-4-phenylbutyrates by combining lactonase-mediated resolution with hydrogenation over Pd/C. Chem Commun 46: 2754–2756.
5. Huang Y, Liu N, Wu X, Chen Y (2010) Dehydrogenases/reductases for the synthesis of chiral pharmaceutical intermediates. Curr Org Chem 14: 1447–1460.
6. Iwasaki G, Kimura R, Numao N, Kondo K (1989) A practical and diastereoselective synthesis of angiotensin converting enzyme inhibitors. Chem Pharm Bull 37: 280–283.
7. Lin WQ, He Z, Jing Y, Cui X, Liu H, et al. (2001) A practical synthesis of ethyl (*R*)- and (*S*)-2-hydroxy-4-phenylbutanoate and D-homophenylalanine ethyl ester hydrochloride from L-malic acid. Tetrahedron: Asymmetry 12: 1583–1587.
8. Nakamura K, Yamanaka R, Matsuda T, Harada T (2003) Recent developments in asymmetric reduction of ketones with biocatalysts. Tetrahedron: Asymmetry 14: 2659–2681.
9. Larissegger-Schnell B, Glueck SM, Kroutil W, Faber K (2006) Enantio-complementary deracemization of (+/−)-2-hydroxy-4-phenylbutanoic acid and (+/−)-3-phenyllactic acid using lipase-catalyzed kinetic resolution combined with biocatalytic racemization. Tetrahedron 62: 2912–2916.
10. Shen ND, Ni Y, Ma HM, Wang LJ, Li CX, et al. (2012) Efficient synthesis of a chiral precursor for angiotensin-converting enzyme (ACE) inhibitors in high space-time yield by a new reductase without external cofactors. Org Lett 14: 1982–1985.
11. Ni Y, Su Y, Li H, Zhou J, Sun Z (2013) Scalable biocatalytic synthesis of optically pure ethyl (*R*)-2-hydroxy-4-phenylbutyrate using a recombinant *E. coli* with high catalyst yield. J Biotechnol 168: 493–498.
12. Bai Y, Yang ST (2005) Biotransformation of *R*-2-hydroxy-4-phenylbutyric acid by D-lactate dehydrogenase and *Candida boidinii* cells containing formate dehydrogenase coimmobilized in a fibrous bed bioreactor. Biotechnol Bioeng 92: 137–146.
13. Kalaritis P, Regenye RW, Partridge JJ, Coffen DL (1990) Kinetic resolution of 2-substituted esters catalyzed by a lipase ex. *Pseudomonas fluorescens*. J Org Chem 55: 812–815.
14. Zheng Z, Sheng B, Gao C, Zhang H, Qin T, et al. (2013) Highly stereoselective biosynthesis of (*R*)-α-hydroxy carboxylic acids through rationally re-designed mutation of D-lactate dehydrogenase. Sci Rep 3: 3401.
15. Schmidt E, Ghisalba O, Gygax D, Sedelmeier G (1992) Optimization of a process for the production of (*R*)-2-hydroxy-4-phenylbutyric acid–an intermediate for inhibitors of angiotensin converting enzyme. J Biotechnol 24: 315–327.
16. Zheng Z, Ma C, Gao C, Li F, Qin J, et al. (2011) Efficient conversion of phenylpyruvic acid to phenyllactic acid by using whole cells of *Bacillus coagulans* SDM. PLoS One 6: e19030.
17. Gao C, Zhang L, Xie Y, Hu C, Zhang Y, et al. (2013) Production of (3*S*)-acetoin from diacetyl by using stereoselective NADPH-dependent carbonyl reductase and glucose dehydrogenase. Bioresour Technol 137: 111–115.
18. Singhvi M, Jadhav A, Gokhale D (2013) Supplementation of medium with diammonium hydrogen phosphate enhanced the D-lactate dehydrogenase levels leading to increased D-lactic acid productivity. Bioresour Technol 146: 736–739.
19. Zheng Z, Sheng B, Ma C, Zhang H, Gao C, et al. (2012) Relative catalytic efficiency of *ldhL*- and *ldhD*-encoded products is crucial for optical purity of lactic acid produced by *Lactobacillus* strains. Appl Environ Microbiol 78: 3480–3483.
20. Markwell MAK, Haas SM, Bieber LL, Tolbert NE (1978) A modification of the Lowry procedure to simplify protein determination in membrane and lipoprotein samples. Anal Biochem 87: 206–210.
21. Bai Y, Yang ST (2007) Production and separation of formate dehydrogenase from *Candida boidinii*. Enzyme Microb Technol 40: 940–946.
22. Schütte H, Flossdorf J, Sahm H, Kula M-R (1976) Purification and properties of formaldehyde dehydrogenase and formate dehydrogenase from *Candida boidinii*. Eur J Biochem 62: 151–160.
23. Wang Y, Li L, Ma C, Gao C, Tao F, et al. (2013) Engineering of cofactor regeneration enhances (2*S*,3*S*)-2,3-butanediol production from diacetyl. Sci Rep 3: 2643.
24. Fröhlich P, Albert K, Bertau M (2011) Formate dehydrogenase – a biocatalyst with novel applications in organic chemistry. Org Biomol Chem 9: 7941–7950.
25. Yu S, Zhu L, Zhou C, An T, Jiang B, et al. (2013) Enzymatic production of D-3-phenyllactic acid by *Pediococcus pentosaceus* D-lactate dehydrogenase with NADH regeneration by *Ogataea parapolymorpha* formate dehydrogenase. Biotechnol Lett 36:627–631.
26. Zhang W, Ni Y, Sun Z, Zheng P, Lin W, et al. (2009) Biocatalytic synthesis of ethyl (*R*)-2-hydroxy-4-phenylbutyrate with *Candida krusei* SW2026: A practical process for high enantiopurity and product titer. Process Biochem 44: 1270–1275.
27. Chen Y, Lin H, Xu X, Xia S, Wang L (2008) Preparation the key intermediate of angiotensin-converting enzyme (ACE) inhibitors: High enantioselective production of ethyl (*R*)-2-hydroxy-4-phenylbutyrate with *Candida boidinii* CIOC21. Adv Synth Catal 350: 426–430.
28. He CM, Chang DL, Zhang J (2008) Asymmetric reduction of substituted α- and β-ketoesters by *Bacillus pumilus* Phe-C3. Tetrahedron: Asymmetry 19: 1347–1351.
29. Shi Y, Fang Y, Ren Y, Guan H, Zhang JY (2009) Applying alpha-phenacyl chloride to the enantioselective reduction of ethyl 2-oxo-4-phenylbutyrate with baker's yeast. J Chem Technol Biotechnol 84: 681–688.
30. Shi YG, Fang Y, Wu HP, Li F, Zuo XQ (2009) Improved production of ethyl-(*R*)-2-hydroxy-4-phenylbutyrate with pretreated *Saccharomyces cerevisiae* in water/organic solvent two-liquid phase systems. Biocatal Biotransfor 27: 211–218.
31. Su Y, Ni Y, Wang J, Xu Z, Sun Z (2012) Two-enzyme coexpressed recombinant strain for asymmetric synthesis of ethyl (*R*)-2-hydroxy-4-phenylbutyrate. Chin J Catal 33: 1650–1660.

9

pKa Modulation of the Acid/Base Catalyst within GH32 and GH68: A Role in Substrate/Inhibitor Specificity?

Shuguang Yuan[1,2], Katrien Le Roy[1], Tom Venken[2], Willem Lammens[1], Wim Van den Ende[1]*, Marc De Maeyer[2]*

1 Laboratory of Molecular Plant Physiology, Institute of Botany and Microbiology, KU Leuven, Heverlee, Belgium, **2** Laboratory for Biomolecular Modelling, Department of Chemistry, Division of Biochemistry, Molecular and Structural Biology, KU Leuven, Heverlee, Belgium

Abstract

Glycoside hydrolases of families 32 (GH32) and 68 (GH68) belong to clan GH-J, containing hydrolytic enzymes (sucrose/fructans as donor substrates) and fructosyltransferases (sucrose/fructans as donor and acceptor substrates). In GH32 members, some of the sugar substrates can also function as inhibitors, this regulatory aspect further adding to the complexity in enzyme functionalities within this family. Although 3D structural information becomes increasingly available within this clan and huge progress has been made on structure-function relationships, it is not clear why some sugars bind as inhibitors without being catalyzed. Conserved aspartate and glutamate residues are well known to act as nucleophile and acid/bases within this clan. Based on the available 3D structures of enzymes and enzyme-ligand complexes as well as docking simulations, we calculated the pKa of the acid-base before and after substrate binding. The obtained results strongly suggest that most GH-J members show an acid-base catalyst that is not sufficiently protonated before ligand entrance, while the acid-base can be fully protonated when a substrate, but not an inhibitor, enters the catalytic pocket. This provides a new mechanistic insight aiming at understanding the complex substrate and inhibitor specificities observed within the GH-J clan. Moreover, besides the effect of substrate entrance on its own, we strongly suggest that a highly conserved arginine residue (in the RDP motif) rather than the previously proposed Tyr motif (not conserved) provides the proton to increase the pKa of the acid-base catalyst.

Editor: Beata G. Vertessy, Institute of Enzymology of the Hungarian Academy of Science, Hungary

Funding: SY was supported by DBOF/08/31. WVdE and MDM are supported by G.0379.09 (FWO Vlaanderen). KLR is supported by PDO-08-00246 (FWO Vlaanderen). WL was supported by PDMK/09/104. The funders had no role in study design, data collection and analysis, decision to publish, or preparation of the manuscript.

Competing Interests: The authors have declared that no competing interests exist.

* E-mail: wim.vandenende@bio.kuleuven.be (WV); marc.demaeyer@fys.kuleuven.be (MDM)

Introduction

Carbohydrates play an important role in a diverse array of biological processes. Their functional and structural variety implies a large amount of enzymes involved in their modification, synthesis and breakdown. The classification of 'carbohydrate-active enzymes' (CAZy) (http://www.cazy.org/) [1,2] is based on their biological properties, including glycoside polysaccharide lyases, hydrolases, carbohydrate esterases, carbohydrate-binding modules and glycosyltransferases. Glycoside hydrolases (GH) split the glycosidic bond between two carbohydrates or between a carbohydrate and an aglycon moiety. They play important roles both in plants and micro-organisms. In plants for instance, they are involved in carbohydrate partitioning affecting overall growth and development, pollen development and fertilization [3,4]. GHs are further classified in families. Because there is a direct relationship between sequence and folding similarities [5], such a classification helps to reveal the evolutionary relationships between such enzymes. Families GH32 and GH68 belong to clan GH-J, one of the 14 clans defined in CAZy. Both families contain of 5-fold β-propeller in their tertiary structure, harboring the catalytic site. GH32 members contain an additional C-terminal β-sheet domain absent in GH68. Naumoff proposed that clan GH-J should be combined with α-arabinases and β-xylosidases [5] (families GH43 and GH62) into one "β-fructosidase superfamily".

Within GHs, inverting (Figure 1A) and retaining (Figure 1B) mechanisms can be discriminated, depending on the outcome of the reaction [6]. In all cases, the reaction involves two acidic residues. In the inverting mechanism, the configuration of the anomeric carbon is inverted in a single step, using a nucleophile that activates a water molecule (Figure 1A). In contrast, the GHs that retain the configuration of the anomeric carbon operate via a double displacement mechanism in which a covalent glycosyl-enzyme intermediate is formed and hydrolyzed via an oxocarbenium ion-like transition state, which is believed to be stabilized by a transition state stabilizer (Figure 1B). GH32 and GH68 are retaining enzymes using Asp (e.g. WMN**D**PNG motif in GH32) as nucleophile and Glu (e.g. **E**C motif in GH32) as proton donor (acid/base catalyst) as first established for yeast invertase [7–10]. As first established for yeast invertase, the process involves the protonation of the glycosidic oxygen by an acid/base catalyst and the attack on the anomeric carbon of the substrate by the nucleophile [10]. The transition state is believed to be stabilized by a transition state stabilizer (an Asp in the **R**D**P** motif) [11].

GH-J members show extended structural similarities, especially in the vicinity of the active site. However, an enormous variation

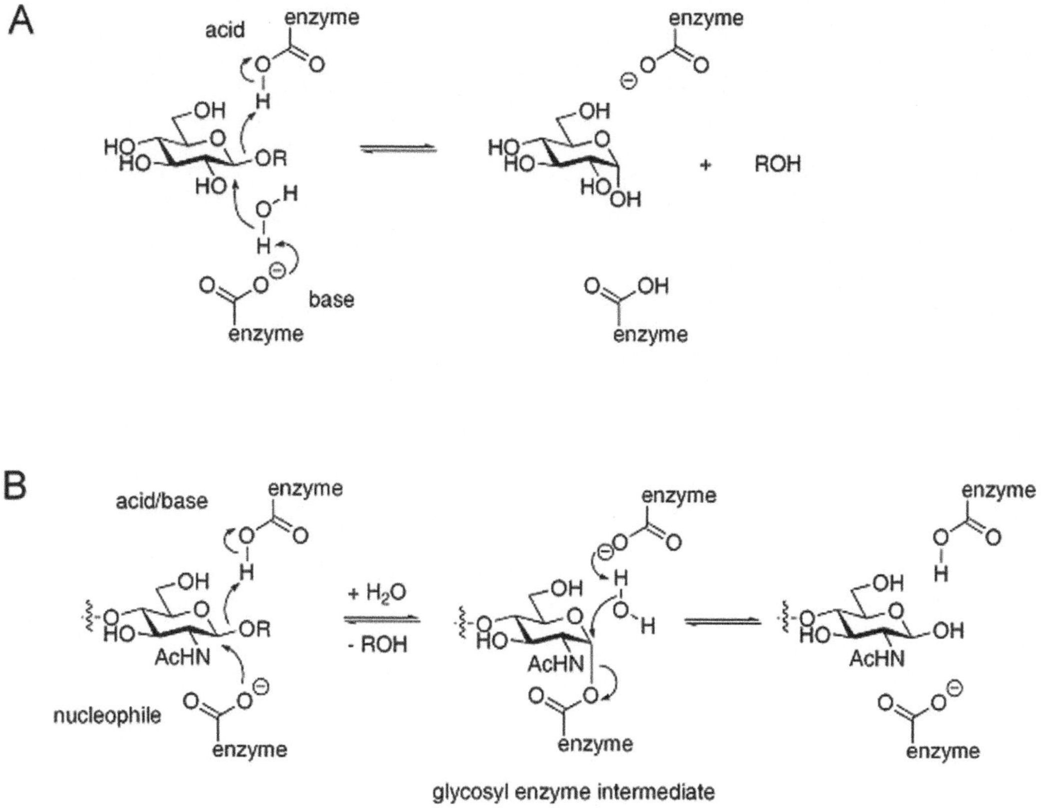

Figure 1. Inverting and retaining catalytic mechanisms in GHs. (A) Inverting mechanism: the conformation of the glycoside residue inverts after reaction (B) Retaining mechanism as occurring in GH32 and GH68: the glycoside residue keeps the same conformation after reaction.

in substrate specificities is observed in clan GH-J including levansucrases, inulosucrases (GH68), plant and microbial invertases, microbial endo- and exo-type inulinases and levanases and an array of plant fructosyltransferases (1-SSTs, 1-FFTs, 6-SFTs, 6^G-FFTs) and plant fructan exohydrolases (1-FEHs, 6-FEHs, 6&1-FEHs) [12]. Essentially, these enzymes use sucrose or fructans as donor substrates and sucrose, fructans or water as acceptor substrates. An array of 3-D structures became available within clan GH-J [13–25] (see further Table 1 and 2), boosting structure-function research in this area. Substrate specificity is often influenced by one or a few amino-acid substitutions [26]. However, when it comes to substrate specificity, many questions remain, including a full understanding why some sugars act as inhibitors rather as substrates for some enzymes. This suggests that "binding without catalysis" frequently occurs. So far, only one hypothesis has been formulated to explain the absence of catalysis despite binding (e.g. sucrose as inhibitor in 1-FEH IIa, [26]). Deeper insights would greatly contribute to a further rational enzyme design within clan GH-J.

Proton donors (acids) and acceptors (bases) provide and accept protons during catalytic processes. This involves pKa changes of residues around the catalytic pocket [7,9]. Here, it is described that the protonation state of the acid/base catalyst, as resolved by pKa calculation, is not favorable for catalysis in many GH-J members. However, this can be changed upon substrate entrance. A Tyr next to the acid/base was proposed to modulate the pKa values of the acid/base [27], but pKa calculations and Molecular Dynamics simulations in this work indicate that an Arg next to the acid/base is a more suitable candidate. This Arg (**R**DP motif) is completely conserved in clan GH-J while the Tyr (**Y**ASK motif) is not [28]. Similar findings have been reported before in another enzyme clan [29]. Based on all these findings, an improved catalytic mechanism is proposed within the GH-J clan.

Results

AtcwINV-1: existing enzyme-sucrose complexes and sucrose docking

The GH32 member AtcwINV-1, the most important cell wall invertase (sucrose as preferential donor) in the model plant *Arabidopsis thaliana*, became one of the best structurally studied GH-J enzymes, since various mutated and inactive AtcwINV-1/sucrose complexes with similar sucrose binding positions were obtained next to the apo-enzyme (2AC1) [30]. We checked the predictive power of sucrose docking simulations in this system and indeed confirmed that Glide [31,32] in Schrödinger 2011 could position sucrose in a correct way in the active site of AtcwINV1, as observed in the mutated crystal structures 2QQU, 2QQV and 2QQW (Figures 2 & 3). Moreover, a simulation of the covalent enzyme-fructosyl intermediate (Figure 4) was obtained by using Schrödinger 2011.

The substrate binding mode in AtcwINV1

In the docked AtcwINV1 (2AC1)/sucrose complex, the sucrose molecule is stabilized by a series of H-bonds including a 2.8 Å H-bond between the glycosidic oxygen O1 and acid/base E203, a distance allowing proton transfer from E203 to the glycosidic oxygen. A 2.7 Å H-bond is observed between the O1′ of the fructose part of nucleophile D23. Two 2.6 Å H-bonds between O3′/O4′ and the transition state stabilizer D149 are observed.

Table 1. pKa value of crystallized GH32 family members.

AtcwINV1 of *Arabidopsis thaliana* pH 5 [23]							
	Description	**E203**	**R148**	**Y279**	**Ligand**	**Sub/Inhi[1]**	**$[A^-]/[HA]$**
2AC1	APO	4.7	14.0	18.6	na[2]	na	2.0
2QQU	D239A	3.6	11.7	16.3	Sucrose	Sub	25.1
2QQV	E203A	na	11.4	15.1	Sucrose	Sub	na
2QQW	D23A	7.5	13.2	17.5	Sucrose	Sub	3×10^{-3}
TRANS	TS	5.7	14.1	18.6	Fructose	na	0.2
Dock	Dock to 2AC1	6.0	13.3	18.6	Sucrose[3]	Sub	0.1
1-FEH IIa of *Cichorium intybus* pH 5 [15]							
	Description	**E201**	**R146**	**Y274**	**Ligand**	**Sub/Inhi**	**$[A^-]/[HA]$**
1ST8	APO	3.0	13.6	15.2	na	na	100
2ADD	Inhibitor	3.3	12.9	15.1	Sucrose	Inhi	50.1
Dock	Dock to 1ST8	5.4	12.9	14.1	1-Kestose[3]	Sub	0.3
Dock	Dock to 1ST8	5.3	13.0	14.6	Inulobiose[3]	Sub	0.2
Exo-inulinase of *Aspergillus awamori* pH 5 [17,19]							
	Description	**E241**	**R188**	**Y313**	**Ligand**	**Sub/Inhi**	**$[A^-]/[HA]$**
1Y4W	APO	3.7	12.8	16.9	na	na	20.0
Dock	Dock to 1Y4W	4.7	12.1	16.9	Sucrose[3]	Sub	0.5
Dock	Dock to 1Y4W	8.0	11.8	17.3	1-Kestose[3]	Sub	1×10^{-3}
β-fructosidase of *Thermotoga maritima* pH 6 [14,18]							
	Description	**E190**	**R137**	**Y240**	**Ligand**	**Sub/Inhi**	**$[A^-]/[HA]$**
1UYP	APO	5.0	17.5	16.9	na	na	10.0
Dock	Dock to 1UYP	6.5	16.7	16.4	Sucrose[3]	Sub	0.3
Fructosyltransferase of *Aspergillus japonicus* pH 6 [22]							
	Description	**E292**	**R190**	**Y369**	**Ligand**	**Sub/Inhi**	**$[A^-]/[HA]$**
3LF7	APO	7.9	15.7	18.1	na	na	0.01
Dock	Dock to 3LF7	11.1	14.7	18.2	Sucrose[3]	Sub	7.9×10^{-6}
Dock	Dock to 3LF7	10.8	14.6	18.0	1-Kestose[3]	Sub	1.6×10^{-5}
Dock	Dock to 3LF7	11.4	14.6	18.2	Nystose[3]	Sub	4.0×10^{-6}
Fructofuranosidase of *Schwanniomyces occidentalis* pH 5 [21]							
	Description	**E230**	**R178**	**Y293**	**Ligand**	**Sub/Inhi**	**$[A^-]/[HA]$**
3KF3	APO	3.8	12.9	17.4	na	na	15.9
Dock	Dock to 3KF3	6.6	12.4	17.4	Sucrose[3]	Sub	0.03
β-fructofuranosidase of *Bifidobacterium longum* pH 6 [24]							
	Description	**E235**	**R180**	**Y302**	**Ligand**	**Sub/Inhi**	**$[A^-]/[HA]$**
3PIG	APO	4.1	13.7	15.1	na	na	79.4
Dock	Dock to 3PIG	7.1	13.0	15.4	Sucrose[3]	Sub	0.08
Dock	Dock to 3PIG	7.6	12.8	15.5	1-Kestose[3]	Sub	0.03

[1]Sub/Inhi: substrate or inhibitor.
[2]na: not applicable for this case.
[3]ligand docked into related APO crystal structure.

Further, two additional H-bonds are found between the O6′ of fructose and the nitrogen in the W47 side chain (2.9 Å) and between the O2 of Glc and E203 (2.8 Å) (Figure 2). Moreover, D239 forms a weak H-bond with O3 of Glc (3.2 Å). Superimposing the docking result with the mutated AtcwINV-1/sucrose complexes [30] shows that the fructose moiety of sucrose takes an identical position while some slight variation is observed among the complexes for the position of glucose (Figure 3). Molecular dynamics (MD) studies confirmed the stability of the conserved H-bond distances of this binding mode (Figure S1).

Table 2. pKa value of crystallized GH68 family members.

Levansucrase of *Bacillus subtilis* pH 6 [13,20]							
	Description	**E342**	**R262**	**Y411**	**Ligand**	**Sub/Inhi[1]**	**[A−]/[HA]**
1OYG	APO	5.9	9.6	14.4	na[2]	na	1.26
Dock	Dock to 1OYG	8.5	9.8	13.6	Sucrose[3]	Sub	0.003
Dock	Dock to 1OYG	8.0	12.0	14.3	Raffinose[3]	Sub	0.01
Levansucrase of *Bacillus megaterium* pH 7 [25]							
	Description	**E352**	**R256**	**Y421**	**Ligand**	**Sub/Inhi**	**[A−]/[HA]**
3OM6	APO	4.8	15.9	15.0	na	na	158.5
Dock	Dock to 3OM6	7.0	15.5	15.9	Sucrose[3]	Sub	1.0
Levansucrase of *Gluconacetobacter diazotrophicus* pH 5 [16]							
	Description	**E401**	**R308**	**Y484**	**Ligand**	**Sub/Inhi**	**[A−]/[HA]**
1W18	APO	1.7	14.3	20.9	na	na	1995.3
Dock	Dock to 1W18	4.9	13.2	20.8	Sucrose[3]	Sub	1.3

[1]Sub/Inhi: substrate or inhibitor.
[2]na: not applicable for this case.
[3]ligand docked into related APO crystal structure.

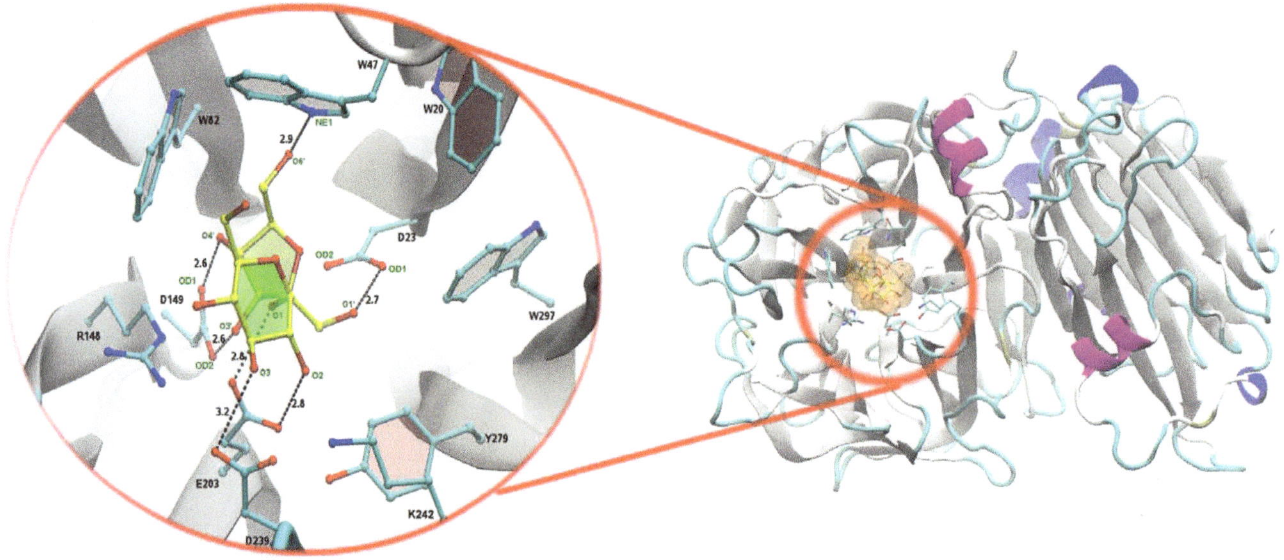

Figure 2. Structure of AtcwINV1 docked with sucrose. Right panel: global view of the structural organization of AtcwINV1. Left panel: a detailed view on the active site showing the position of docked sucrose (yellow) and its interactions with neighbouring amino-acids. Distance units are in angstrom (Å).

Simulating the enzyme-fructosyl transition state

The covalent transition state is stabilized by several strong H-bonds (Figure 4): 2.6 Å and 2.7 Å H-bonds are found between O3′ and O4′ in the covalently bound fructose and the transition stabilizer D149; O6′ and O1′ show H-bonds with W47 (2.7 Å) and D23 (2.7 Å) respectively; a weak H-bond is found between O3′ and the acid/base E203. The position of the covalent fructose in the transition state is rather similar to the one in the active AtcwINV1/sucrose complex (Figure 5A). However, the position of C2 in the transition state shifted 1.5 Å to the inner part of the active pocket (Figure 5B). Compared to the active AtcwINV1/sucrose complex, the conformation of C2′-OD2 flipped 180° which is in agreement with the double displacement mechanism [6].

Relation between pH and pKa. pKa calculation of amino acids in proteins

An acid dissociation constant, Ka, is a quantitative measure of the strength of an acid in solution. It is the equilibrium constant for the acid-base/dissociation reaction. Three methods are available to calculate the pKa of amino-acid residues in proteins: methods based on the Poisson-Boltzmann equation (PBE), empirical methods and MD-based methods [33–35]. We applied the most widely used pKa calculation tool [36], namely a VMD [37] plug-in, PROPKA2 [38,39]. It is widely used for protein or protein/ligand pKa calculations and it is also embedded in various tools such as: CCG MOE (Molecular Operating Environment), PKD, PDB2PQR, WEBPDB and VEGA-ZZ. This fast and efficient software is based on empirical methods. With this method, good correlations were found with experimental data [38–40]. The

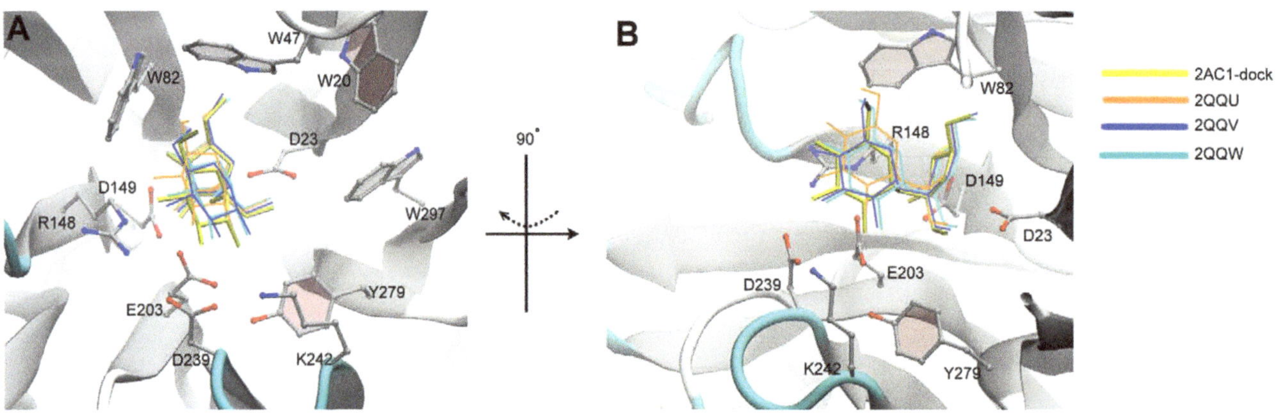

Figure 3. Superimposed AtcwINV1-sucrose complexes with docked sucrose. AtcwINV1-sucrose complexes were derived from Lammens et al. (2008) [30] and compared to the position of docked sucrose (Figure 2). Two views (A,B) are presented. B is obtained after 90° rotation around the vertical axis. Yellow: native AtcwINV-1 docked with sucrose; Orange: crystal structure of mutated AtcwINV1 (D239A, PDB: 2QQU) with sucrose; Blue: crystal structure of mutated AtcwINV1 (E203A, PDB: 2QQV) with sucrose; Cyan: crystal structure of mutated AtcwINV1 (D23A, PDB: 2QQW) with sucrose. While the position of the fructose moiety is conserved, the position of the glucose moiety showed more variation.

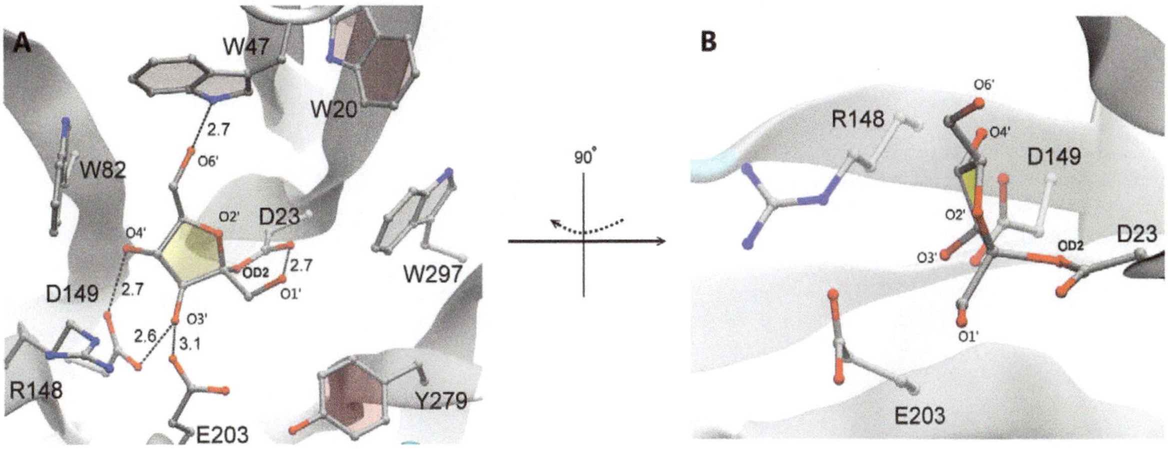

Figure 4. Estimation of the transition state of AtcwINV1. (A) Simulation of the covalent enzyme-fructosyl transition state in AtcwINV1. A covalent bond is formed between fructose and the nucleophile D23. Distance units are in angstrom (Å). (B) This view is generated by 90° rotation around the vertical axis.

method was found superior compared to other pKa calculation tools [40]. PROPKA2 calculates the pKa of a group through the combination of an environmental perturbation, ΔpKa, and the unperturbed pKa value of the group, $pK_{Model.}$

$$pKa = pKa_{Model} + \Delta pKa \quad (1)$$

where pK_{Model} and ΔpKa are determined empirically. For each ionizable group, ΔpKa consists of five terms including global (GlobalDes) and local desolvation (LocalDes), hydrogen bonds with side-chain groups (SDC-HB), hydrogen bonds with amide backbone (BKB- HB), and interactions with charged groups (ChgChg).

$$\Delta pKa = \Delta pK_{GlobalDes} + \Delta pK_{LocalDes} + \Delta pK_{SDC\text{-}HB} + \Delta pK_{BKB\text{-}HB} + \Delta pK_{ChgChg} \quad (2)$$

Simple distance functions and, for backbone hydrogen bonds, distance/angle functions with a constant empirical pKa shift are introduced to compute the above ΔpKa terms. This scheme neglects the effect of nonsigmoidal titration curves for final pKa values. However, this effect is generally smaller than 1.0 pH unit errors observed for most other pKa prediction methods [41]. All available GH32 and GH68 structures were used for these calculations.

pKa calculations of the acid/base E203 in AtcwINV1 (complexes) and the transition state

pKa values for the acid-base catalyst E203 and its neighboring key residues R148 (**R**DP motif) and Y279 (**Y**ASK motif) were calculated for the AtcwINV1 apo-enzyme, the mutants in complex with sucrose, the apo-enzyme with docked sucrose and the transition state (Table 1). The protonation state ($[A^-]/[HA]$) of the acid/base E203 is also indicated (Table 1), calculated at pH 5.0, the pH optimum of AtcwINV1 [23]. It can be concluded that E203 is only partially protonated in the apo-enzyme. Indeed, the ratio between deprotonated and protonated state $[A^-]/[HA]$ is 2.0. Sucrose entrance (docked sucrose in 2AC1) leads to a more complete protonation ($[A^-]/[HA] = 0.1$), facilitating catalysis. A similar pKa increase of E203 was observed in the D23A mutant in

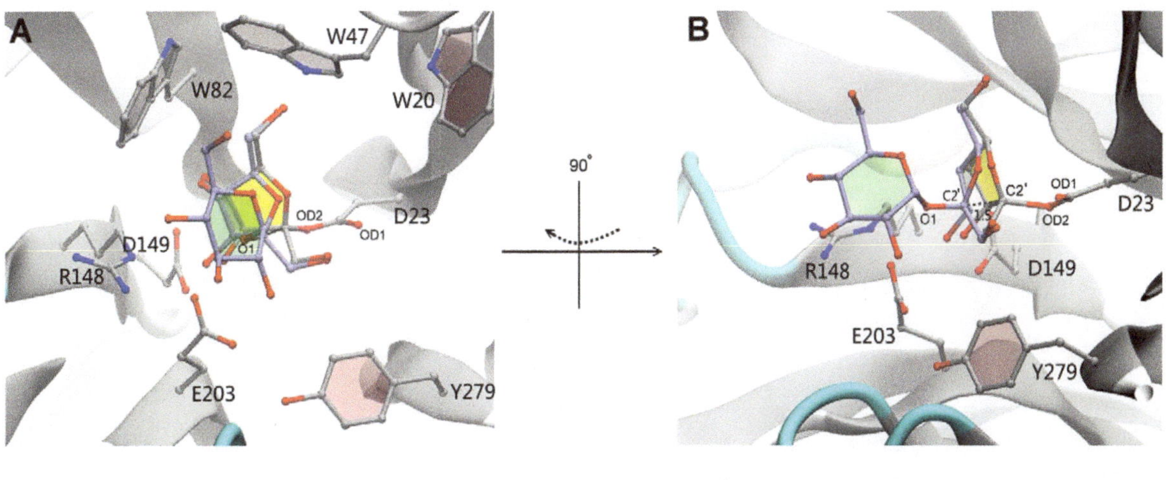

Figure 5. Superimposed AtcwINV1/sucrose and transition state. (A) Superimposal of AtcwINV1/docked sucrose (as derived from Figure 2) and the covalent enzyme-fructosyl transition state (derived from Figure 4). Gray: transition state; ice-blue: sucrose. (B) is generated by 90° rotation around the vertical axis. Distance units are in angstrom (Å).

complex with sucrose and for the transition state (Table 1). By contrast, the D239A mutant in complex with sucrose shows a dramatic decrease in the pKa value for E203 (i.e. $[A^-]/[HA] = 25.1$), counteracting efficient catalysis. This result is in line with the poor catalytic properties of the D239 mutant [26]. D239 and K242 (or a homologue R residue) form a D/K or D/R couple (Figure 2) in GH32 enzymes using sucrose as a preferential donor substrate [12,26]. So far it was assumed that mutating the amino acids, making up this couple, would drastically interfere with the sucrose binding as a donor substrate. The possibility to generate a D239A sucrose complex [30] suggests that at least partial sucrose binding is still possible. The results presented here suggest that the pKa value for E203 in D239A is so low that most of the E203 would be deprotonated (Table 1), leading to a very inefficient catalysis on (that part of) sucrose molecules that would still be able to bind. It should be noted that the D239A mutation greatly changed the active pocket electrostatic properties, making it more hydrophobic and less accessible for a water molecule acting as a final fructosyl acceptor (Figure 6). It can be concluded that the poor catalytic properties of the D239A mutant can be explained by three different reasons: (1) a reduced sucrose binding efficiency [26] (2) a decreased pKa of E203 and (3) a more hydrophobic active site affecting water entrance (Figure 6).

pKa modulation and Molecular Dynamics on AtcwINV1

The pKa of R148 is generally 3 to 5 units lower than the one of Y279, which means that R148 is almost 10^3–10^5 times more acidic (Table 1) and as such able to release a proton more efficiently as compared to Y279. This strongly suggests that R148 functions as a pKa modulator, rather than the previously proposed Y279 residue (the homologue in Ci1-FEH IIa is Y274 [27]). Furthermore, 2AC1 structural refinement with Brugel [42] and visualization with PYMOL [43] showed that the C = O double bond in E203 is pointing to Y279 while the C-O$^-$ has the ability to accept a proton from R148 (Figure 7).

A 5ns MD simulation was pursued by program Desmond [44] for the docked AtcwINV1/sucrose complex. Intriguingly, a water molecule is frequently observed between R148 and D239 (Figure 8). At pH 5.0, the pH optimum of AtcwINV1, this water can be protonated (H_3O^+) and attracted to an area with negative electrostatic surface near the catalytic pocket (Figure 6). It can be speculated that this particular H_3O^+ could assist in stabilizing the position of D239. Additionally, this H_3O^+ could also act as a proton donor to R148, after R148 donated its own proton to E203, upon entrance of the substrate (see also below).

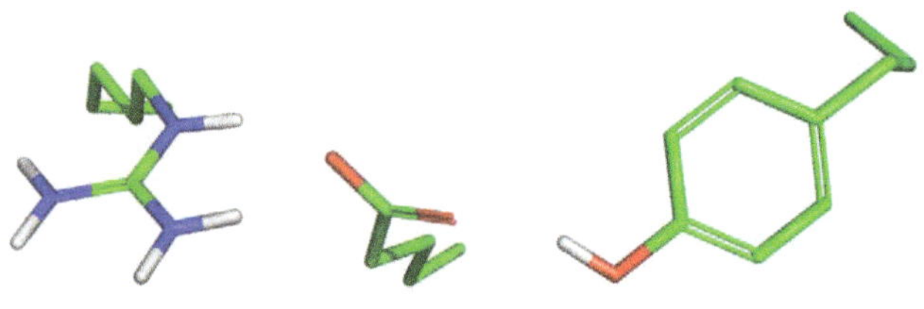

Figure 7. E203 interactions in AtcwINV1. E203 in AtcwINV1 (PDB: 2AC1) interacts with Y279 via its keto (C = O) function and with R148 through its C-O$^-$ bond.

pKa calculations on the acid/base E201 in Ci1-FEH IIa and its complex with sucrose as inhibitor. Comparison with the Aspargillus awamori exo-inulinase

In contrast to invertases, plant fructan exohydrolases (FEHs) use fructans as preferential donor substrates. For instance 1-kestose (a fructan trisaccharide Glc-Fru-Fru) and inulobiose (a fructan disaccharide Fru-Fru) are suitable substrates. Sucrose (Glc-Fru) is not a donor substrate. In many cases, FEHs are strongly inhibited by sucrose. This is also the case for 1-FEH IIa of *Cichorium intybus* (Ci1-FEH IIa) [23], the first plant GH32 enzyme of which the 3D structure was resolved [27]. The pH optimum of Ci1-FEH IIa is 5.0, similar to the one of AtcwINV1. The protonation state of E201 in the apo-enzyme (PDB: 1ST8) is low (Table 1), while entrance of the substrates 1-kestose and inulobiose results in a more complete protonation, allowing enzyme catalysis at pH 5.0. The docked complexes with 1-kestose and inulobiose are shown in Figure S2, indicating the involved H-bonds. Similar to the AtcwINV1- D239A mutant in complex with sucrose, the Ci1-FEH IIa-sucrose complex 2ADD shows a pKa value well below the pH optimum, preventing catalysis. Hitherto, the inhibitory effect of sucrose in 1-FEH IIa and the absence of catalytic activity on sucrose were explained by the rotated position of the Glc moiety, bringing O2 in a favorable position to make a hydrogen bond with E201, counteracting catalytic activity [27]. The results provided in this manuscript now offer an alternative point of view

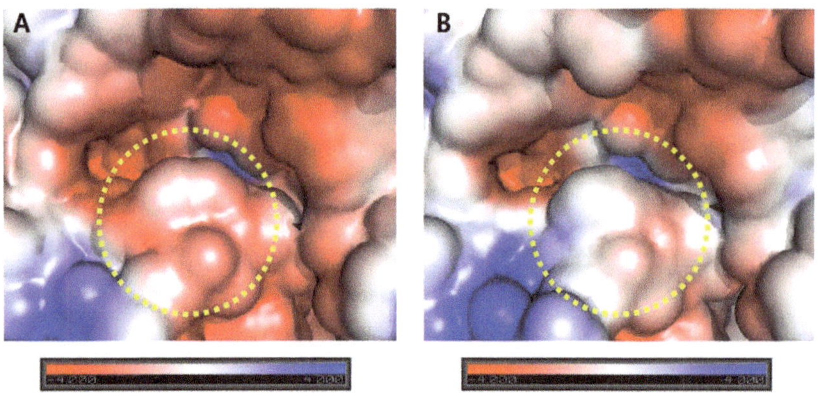

Figure 6. Electrostatic maps for AtcwINV1 and its D239A mutant. Electrostatic maps were generated by the APBS [55] tool in PYMOL. AtcwINV1 (Panel A, PDB: 2AC1) and its D239A mutant (Panel B, PDB: 2QQU) are presented. The region occupied by the D239A mutant (dotted circle) is much more hydrophobic than in the wild type.

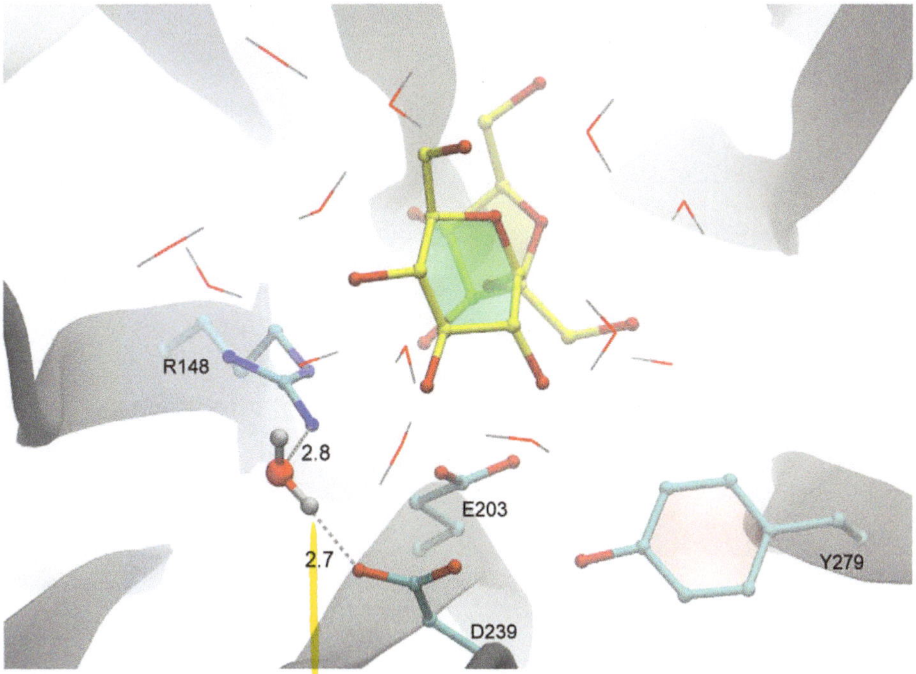

Figure 8. A water molecule between R148 and D239. A water molecule is frequently observed between R148 and D239 during MD simulations. Distance units are in angstrom (Å).

to explain the absence of catalysis when sucrose is provided to Ci1-FEH IIa.

To further corroborate this alternative hypothesis, pKa values were also evaluated on a conformational ensemble of two 5 ns MD simulates: 1-FEH IIa/sucrose and 1-FEH IIa/1-kestose. Over 100 snapshots throughout the whole simulation were extracted for pKa analysis (Figure 9). The mean for 1-FEH IIa/sucrose is 2.9 with standard deviation 0.30; while the mean for 1-FEH IIa/1-kestose is 5.0 with comparatively higher standard deviation 0.88 owing to the flexibility of 1-kestose and residues nearby (Figures S3C and S4C). Statistical analysis showed that PROPKA derived pKa values are stable throughout the MD simulations (Figure 9) and consistent with pKa values 3.3 for 1-FEH IIa/sucrose and 5.4 for 1-FEH IIa/1-kestose presented in Table 1. Moreover, RMSD calculations on both sucrose and 1-kestose complexes confirmed overall stability with minor fluctuations (Figures S3 and S4): the RMSF for E201 in both cases are below 0.4 Å. Additionally, plotting pKa values and experimental k_{cat} showed good correlation (Figure S5).

Both the plant enzymes AtcwINV1 and Ci1-FEH IIa are examples of GH-J members with well-defined preferential donor substrate specificity, towards sucrose and fructans, respectively. Clearly in these enzymes the evolutionary process is almost reaching the point of "unifunctionality" with only minor side activities present next to a clear, main activity. This is different in microbial GH-J members often showing less strict substrate specificities. For instance, it is well known that the exo-inulinase of *Aspergillus awamori* is not very specific to fructans, since it also accepts sucrose as a donor substrate, at least to some extent [17]. This substrate specificity fits nicely with the observed pKa values for the acid-base catalyst E241 of this enzyme and its complexes (Figure S2). Indeed, the apo-enzyme shows a pKa below the pH optimum, while the docking with sucrose shows a pKa close to the pH optimum and a docking with 1-kestose shows a pKa far above the optimal pH, promoting a more efficient catalysis (Table 1).

Studies on other GH-J members

We extended our calculations to all available structures within the GH-J clan, listed in Table 1 (GH32) and Table 2 (GH68), respectively. First, it can be concluded that the acid/base is not fully protonated before the substrate enters the active site. The fructosyltransferase of *A. japonicus* (Table 1) is the only exception since its acid/base E292 is already protonated in the apo-enzyme. Maybe this is related to the presence of an unusual His residue (Figure S2) in its active site. Second, in all cases the pKa values of the acid/base increases upon substrate entry, this also holds true for the fructosyltransferase from *A. japonicus*. Third, the conserved Arg (RDP motif) next to the acid/base is usually much lower in pKa than the Tyr (YASK motif).

Discussion

Model for an extended catalytic mechanism within GH-J. A role for substrate binding and for a conserved Arg as pKa modulator?

The classic catalytic mechanism, based on original yeast invertase (GH32) research, assumes that the acid/base catalyst is sufficiently protonated to allow catalysis. Here, all (except one) pKa calculations predicted an insufficient protonation of the apo-enzymes. Therefore, we present here an extension on the original reaction mechanism (Figure 10). For simplicity, we focus on the AtcwINV1 and Ci1-FEH IIa cases, but comparable mechanisms might hold true for other GH-J members. In AtcwINV1, the acid/base E203 is only partially protonated before sucrose enters the active pocket. The pKa value of E203 increases when the substrate comes in and E203 will get protonated by R148, which on itself can extract a proton from a nearby H_3O^+ (Figure 10A). Next, the glycosidic oxygen in sucrose is protonated by acid/base E203, the nucleophile D23 attacks the anomeric carbon, forming a covalent enzyme-fructosyl complex and releases glucose from the active

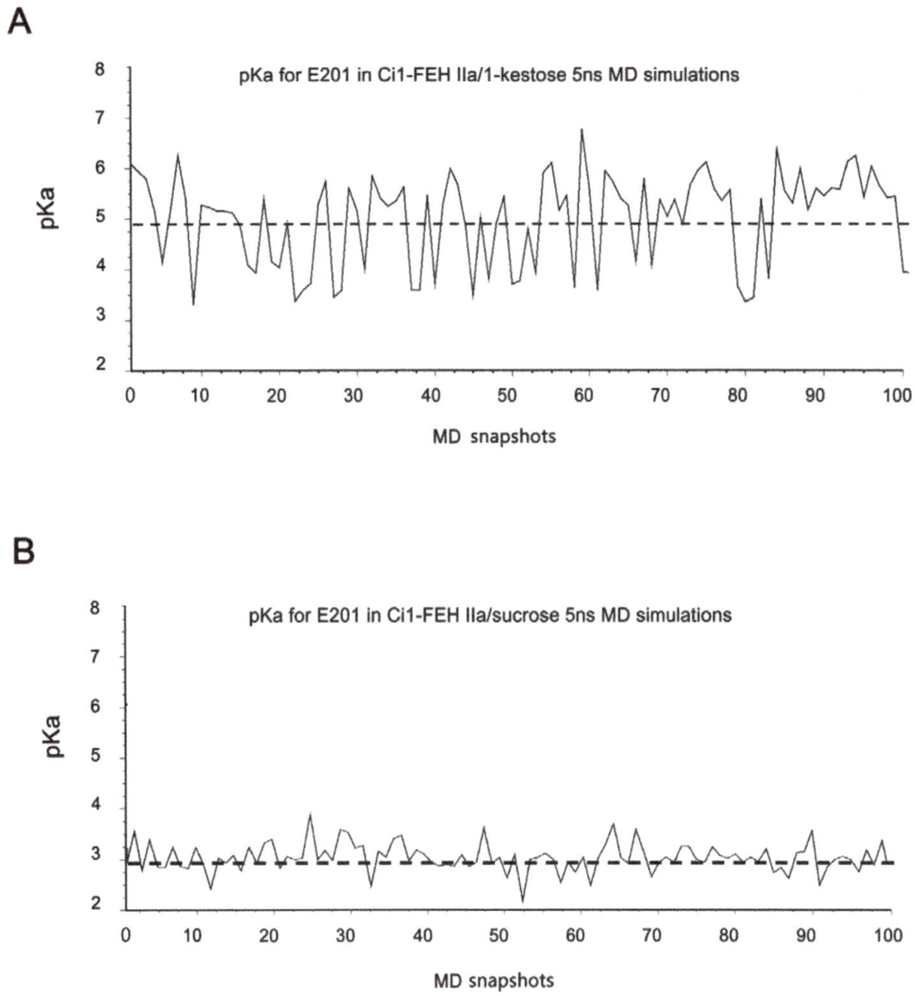

Figure 9. pKa fluctuation for 100 snapshots extracted from the trajectory throughout the 5 ns MD simulations. (A) pKa for E201 in 1-FEH IIa/1-kestose. Its mean (dash line) is 5.0 with standard deviation 0.88; (B) pKa for E201 in 1-FEH IIa/1-sucrose. The mean for this case is 2.9 with standard deviation 0.30.

pocket (Figure 10B and Figure 10C). Finally, a water molecule is split, reprotonating the acid/base and leaving the hydroxyl group on fructose. During the leaving of the fructose moiety out of the active site the proton on E203 is mainly transferred to bulk water. (Figure 10D and Figure 10E). Upon the entrance of a new substrate the cycle will start again with the transfer of a bulk water proton to R148 and further to E203 (Figure 10A). In severe contrast, in case of Ci1-FEH IIa, the binding of sucrose does not increase the pKa, possibly explaining why sucrose acts as an inhibitor and not as a substrate for this enzyme.

These findings indicate that the pKa values of the active-site residues in an enzyme can be critical for the catalytic mechanism and actual substrate specificity. This is not merely limited within clan GH-J, but extends to many pH dependent enzymes including xylanases [45], amylases [46], lysozymes [47], proteinases [48], protein kinases [49] and others. Therefore, understanding pKa values of key role amino acids in the active site of such enzymes aims at understanding an array of biological processes and assists in the process of rational enzyme design. Various tools and methods focusing on pKa determinations have been developed for such purposes [50–52].

In summary, within the clan GH-J, most apo acid/bases are in a (partially) deprotonated state before substrate binding. The acid/base pKa value increases when a suitable substrate comes into the catalytic pocket. An acid/base neighboring and strictly conserved Arg, which can accept protons from the solvent environment, might transfer a proton to the acid/base. Thereafter, the protonated acid/base can further transfer the proton to the substrate for catalysis. When an inhibitor binds to the active pocket, the pKa value of acid/base is not increased, preventing catalysis. Favorite substrates seem to correlate well with a favorable increase of the acid/base pKa values. This strongly suggests that the exact binding modus of a sugar determines whether it will act as a substrate or as an inhibitor, by virtue of its effect on the pKa of the acid/base catalyst.

Materials and Methods

Protein Preparation

Structures were imported into the Maestro9.2 [53] program. All crystallographic solvent molecules and glycosyl chains were deleted. Hydrogens were added to the structure according to the pH environment as reported in literature. The protein preparation utility in Maestro9.2 was used to run a restrained minimization

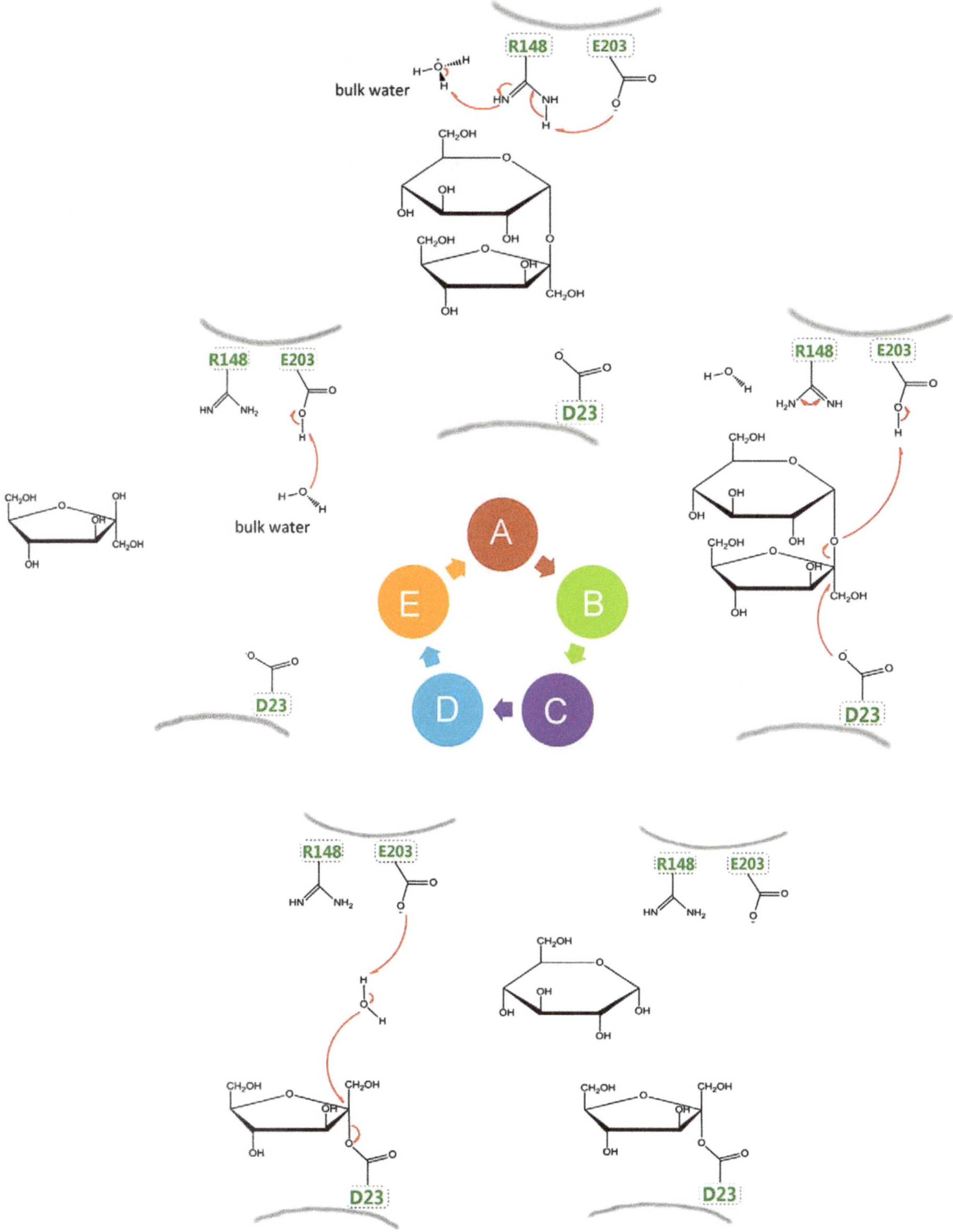

Figure 10. Proposed extensions on the reaction mechanism for the catalytic process in clan GH-J, as derived from AtcwINV1. (A) When sucrose binds in the active site of AtcwINV1, the pKa of E203 increases and it is protonated by a neighboring R148 which can extract a proton from a nearby water molecule (Figure 8) ; (B&C) the glycosidic oxygen in sucrose is protonated by acid/base E203, the nucleophile D23 attacks the anomeric carbon, forming a covalent enzyme-fructosyl complex and releasing glucose from the active pocket; (D&E) a water molecule is split, reprotonating the acid/base and leaving the hydroxyl group on fructose. The enzyme is ready to accept another sucrose molecule after fructose has left the active pocket.

which removed unfavorable steric contacts and improved the quality of the protein hydrogen bonding network without large rearrangements of the protein heavy atoms.

Docking

It is observed that the position of the fructose moiety is highly conserved within all the available crystal structures. Therefore, all oligosaccharides were initially placed such that the fructose moiety in the catalytic pocket is oriented in this conserved position. Cubic boxes centered on the ligand mass center with a size of 10 Å, 12 Å and 14 Å for the disaccharides, trisaccharides and tetrasaccharides, respectively, defined the docking binding regions. Extra-precision (XP) docking and scoring was executed in all cases. Twenty poses per ligand were included in post-docking minimization for bond length and angle as well as torsional angle optimization. Poses were rescored using a scaled Coulomb-van der Waals term and the GlideScore system. The best-scored poses were chosen as the optimal solution.

Transition complex construction

The covalent transition state complex was built manually in Maestro9.2 [53] based on the previous docking result. The handmade transition state was submitted to Prime [54] in Maestro9.2 for energy minimization refinement, using the OPLS_2005 force field and Surface Generalized Born (SGB) continuum solvation model.

pKa calculation

The prepared structures were submitted to PROPKA2 [38,39] in VMD1.87 for pKa calculation. Default parameters were used throughout the calculations. Glycosyl chains were stripped off in all calculations to present the input data in the same way to the program. However, pKa values for structures with glycosyl chains were also performed but their values were not significantly different.

Molecular Dynamics

All MD simulations were performed using the Desmond package, version 3.0.3. The OPLS_2005 force field was applied and topologies for the sucrose molecule were recognized automatically in Desmond. Each protein was placed in a trigonal box and solvated with TIP3P water molecules solvated with 0.15 M NaCl with a distance of 10 Å between any protein atom and the box edge, resulting in ≈55,000 atoms. Chloride counter-ions were added to compensate for the net positive charge of the protein. Unfavorable contacts in each system were relieved using 5000 steps of conjugate gradient energy minimization. A 100 ps length equilibration step at constant pressure (NPT ensemble) was executed with position restraints on all heavy protein atoms to allow relaxation of the solvent molecules. Next, a 5 ns production MD on each system was executed. During all simulations, pressure and temperature were coupled separately for protein and solvent atoms (including ions) to an external bath using the Berendsen coupling method at 298 K and 1 bar. The temperature and pressure constants were fixed to 0.1 ps and 1 ps, respectively. Periodic boundary conditions were imposed in all three directions. Short-range non-bonded interactions were cut off at 10 Å, while long-range electrostatic interactions were calculated using the particle-mesh Ewald (PME) summation schemes. The M-SHAKE algorithm was used to constrain all bonds in each system with a 2 fs integration step. Root Mean Square Deviation (RMSD) analyses were performed using the tools available in the Desmond suite.

Supporting Information

Figure S1 Conserved H-bond distances for AtcwINV1/ sucrose MD simulations. (A) Distance between D23(OD1) and sucrose(O1′); (B) Distance between D149(OD1) and sucrose(O4′); (C) Distance between W47(NE1) and sucrose(O6′); (D) Distance between D149(OD2) and sucrose(O3′). Distance units are in angstrom (Å).

Figure S2 Docked poses for clan GH-J members. (A) Sucrose docked in exo-inulinase of *Aspergillus awamori*; (B) 1-kestose docked in exo-inulinase of *Aspergillus awamori*; (C) Crystal structure of sucrose in 1-FEH IIa of *Cichorium intybus*; (D) 1-kestose docked in 1-FEH IIa of *Cichorium intybus*; (E) Sucrose docked in β-fructosidase of *Thermotoga maritima;* (F) Sucrose docked in fructosyltransferase of *Aspergillus japonicus;* (G) 1-kestose docked in fructosyltransferase of *Aspergillus japonicus;* (H) Nystose docked in fructosyltransferase of *Aspergillus japonicus;* (I) Sucrose docked in fructofuranosidase of *Schwanniomyces occidentalis;* (J) Sucrose docked in β-fructofuranosidase of *Bifidobacterium longum;* (K) 1-kestose docked in β-fructofuranosidase of *Bifidobacterium longum;* (L) Sucrose docked in levansucrase of *Bacillus subtilis;* (M) Raffinose docked in levansucrase of *Bacillus subtilis;* (N) Sucrose docked in levansucrase of *Bacillus megaterium;* (O) Sucrose docked in levansucrase of *Gluconacetobacter diazotrophicus.*

Figure S3 RMSD (Å) and RMSF (Å) for 1-FEH IIa/ sucrose MD simulations. (A) RMSD of protein (1-FEH IIa) backbone; (B) RMSD of E201 heavy atoms with mean 0.24 and standard deviation 0.08; (C) RMSD of sucrose heavy atoms with mean 0.60 and standard deviation 0.18; (D) RMSF for each residue of 1-FEH IIa.

Figure S4 RMSD (Å) and RMSF (Å) for 1-FEH IIa/1-kestose MD simulations. (A) RMSD of protein (1-FEH IIa) backbone; (B) RMSD of E201 heavy atoms with mean 0.36 and standard deviation 0.10; (C) RMSD of 1-kestose heavy atoms with mean 0.84 and standard deviation 0.21; (D) RMSF for each residue of 1-FEH IIa.

Figure S5 Correlation between pKa calculations and $\mathbf{k_{cat}}$. X-axis, pKa calculation for AtcwINV1; Y-axis, $\mathbf{k_{cat}}$ value for AtcwINV1 from experimental data according to Le Roy K, *et.al* 2007 [26]. The error bars for the pKa are estimations based on the calculated standard deviation average for the 2 MD runs shown in figure 9.

Author Contributions

Conceived and designed the experiments: SY WV MDM. Performed the experiments: SY KLR WL TV MDM. Analyzed the data: SY TV WV MDM. Contributed reagents/materials/analysis tools: SY KLR WL TV MDM. Wrote the paper: SY WV MDM.

References

1. Henrissat B (1991) A classification of glycosyl hydrolases based on amino acid sequence similarities. Biochem J 280(Pt 2): 309–316.
2. Coutinho PM, Henrissat B, eds. Carbohydrate-active enzymes: an integrated database approach. pp 3–12.

3. Hoj PB, Fincher GB (1995) Molecular evolution of plant beta-glucan endohydrolases. Plant J 7: 367–379.
4. Minic Z, Jouanin L (2006) Plant glycoside hydrolases involved in cell wall polysaccharide degradation. Plant Physiol Biochem 44: 435–449.
5. Grigoriev IV, Kim SH (1999) Detection of protein fold similarity based on correlation of amino acid properties. Proc Natl Acad Sci U S A 96: 14318–14323.
6. Davies G, Henrissat B (1995) Structures and mechanisms of glycosyl hydrolases. Structure 3: 853–859.
7. Lawson SL, Wakarchuk WW, Withers SG (1997) Positioning the acid/base catalyst in a glycosidase: studies with *Bacillus circulans* xylanase. Biochemistry 36: 2257–2265.
8. Vocadlo DJ, Davies GJ, Laine R, Withers SG (2001) Catalysis by hen egg-white lysozyme proceeds via a covalent intermediate. Nature 412: 835–838.
9. Zechel DL, Withers SG (2001) Dissection of nucleophilic and acid-base catalysis in glycosidases. Curr Opin Chem Biol 5: 643–649.
10. Koshland DE, Jr., Stein SS (1954) Correlation of bond breaking with enzyme specificity; cleavage point of invertase. J Biol Chem 208: 139–148.
11. Altenbach D, Nuesch E, Ritsema T, Boller T, Wiemken A (2005) Mutational analysis of the active center of plant fructosyltransferases: festuca 1-SST and barley 6-SFT. FEBS Lett 579: 4647–4653.
12. Lasseur B, Schroeven L, Lammens W, Le Roy K, Spangenberg G, et al. (2009) Transforming a fructan:fructan 6G-fructosyltransferase from perennial ryegrass into a sucrose:sucrose 1-fructosyltransferase. Plant Physiol 149: 327–339.
13. Chambert RHE, Petit-Glatron M (1995) Kinetics of the unfolding-folding transition of *Bacillus subtilis* levansucrase precursor. FEBS Lett 360: 307–309.
14. Liebl W, Brem D, Gotschlich A (1998) Analysis of the gene for beta-fructosidase (invertase, inulinase) of the hyperthermophilic bacterium *Thermotoga maritima*, and characterisation of the enzyme expressed in *Escherichia coli*. Appl Microbiol Biotechnol 50: 55–64.
15. De Roover J, Van Laere A, De Winter M, Timmermans JW, Van den Ende W (1999) Purification and properties of a second fructan exohydrolase from the roots of *Cichorium intybus*. Physiol Plant 106: 28–34.
16. Trujillo LE, Arrieta JG, Dafhnis F, Garcia J, Valdes J, et al. (2001) Fructo-oligosaccharides production by the *Gluconacetobacter diazotrophicus* levansucrase expressed in the methylotrophic yeast *Pichia pastoris*. Enzyme Microb Technol 28: 139–144.
17. Kulminskaya AA, Arand M, Eneyskaya EV, Ivanen DR, Shabalin KA, et al. (2003) Biochemical characterization of *Aspergillus awamori* exoinulinase: substrate binding characteristics and regioselectivity of hydrolysis. Biochim Biophys Acta 1650: 22–29.
18. Alberto F, Bignon C, Sulzenbacher G, Henrissat B, Czjzek M (2004) The three-dimensional structure of invertase (beta-fructosidase) from *Thermotoga maritima* reveals a bimodular arrangement and an evolutionary relationship between retaining and inverting glycosidases. J Biol Chem 279: 18903–18910.
19. Nagem RA, Rojas AL, Golubev AM, Korneeva OS, Eneyskaya EV, et al. (2004) Crystal structure of exo-inulinase from *Aspergillus awamori*: the enzyme fold and structural determinants of substrate recognition. J Mol Biol 344: 471–480.
20. Esawy MA, Mahmoud DAR, Fattah AFA (2008) Immobilisation of *Bacillus subtilis* NPC33a levansucrase and some studies on its properties. Braz J Chem Eng 25: 237–246.
21. Alvaro-Benito M, Polo A, Gonzalez B, Fernandez-Lobato M, Sanz-Aparicio J (2010) Structural and kinetic analysis of *Schwanniomyces occidentalis* invertase reveals a new oligomerization pattern and the role of its supplementary domain in substrate binding. J Biol Chem 285: 13930–13941.
22. Chuankhayan P, Hsieh CY, Huang YC, Hsieh YY, Guan HH, et al. (2010) Crystal structures of *Aspergillus japonicus* fructosyltransferase complex with donor/acceptor substrates reveal complete subsites in the active site for catalysis. J Biol Chem 285: 23251–23264.
23. Hothorn M, Van den Ende W, Lammens W, Rybin V, Scheffzek K (2010) Structural insights into the pH-controlled targeting of plant cell-wall invertase by a specific inhibitor protein. Proc Natl Acad Sci U S A 107: 17427–17432.
24. Bujacz A, Jedrzejczak-Krzepkowska M, Bielecki S, Redzynia I, Bujacz G (2011) Crystal structures of the apo form of beta-fructofuranosidase from *Bifidobacterium longum* and its complex with fructose. FEBS J 278: 1728–1744.
25. Strube CP, Homann A, Gamer M, Jahn D, Seibel J, et al. (2011) Polysaccharide synthesis of the Levansucrase SacB from *Bacillus megaterium* is controlled by distinct surface motifs. J Biol Chem 286: 17593–17600.
26. Le Roy K, Lammens W, Verhaest M, De Coninck B, Rabijns A, et al. (2007) Unraveling the difference between invertases and fructan exohydrolases: a single amino acid (Asp-239) substitution transforms *Arabidopsis* cell wall invertase1 into a fructan 1-exohydrolase. Plant Physiol 145: 616–625.
27. Verhaest M, Van den Ende W, Le Roy K, De Ranter CJ, Van Laere A, et al. (2005) X-ray diffraction structure of a plant glycosyl hydrolase family 32 protein: fructan 1-exohydrolase IIa of *Cichorium intybus*. Plant J 41: 400–411.
28. Naumoff DG (2001) beta-fructosidase superfamily: homology with some alpha-L-arabinases and beta-D-xylosidases. Proteins 42: 66–76.
29. Fukamizo T, Juffer AH, Vogel HJ, Honda Y, Tremblay H, et al. (2000) Theoretical calculation of pKa reveals an important role of Arg205 in the activity and stability of Streptomyces sp. N174 chitosanase. J Biol Chem 275: 25633–25640.
30. Lammens W, Le Roy K, Van Laere A, Rabijns A, Van den Ende W (2008) Crystal structures of *Arabidopsis thaliana* cell-wall invertase mutants in complex with sucrose. J Mol Biol 377: 378–385.
31. Friesner RA, Banks JL, Murphy RB, Halgren TA, Klicic JJ, et al. (2004) Glide: a new approach for rapid, accurate docking and scoring. 1. Method and assessment of docking accuracy. J Med Chem 47: 1739–1749.
32. Friesner RA, Murphy RB, Repasky MP, Frye LL, Greenwood JR, et al. (2006) Extra precision glide: docking and scoring incorporating a model of hydrophobic enclosure for protein-ligand complexes. J Med Chem 49: 6177–6196.
33. Onufriev A, Case DA, Ullmann GM (2001) A novel view of pH titration in biomolecules. Biochemistry 40: 3413–3419.
34. Gunner MR, Mao J, Song Y, Kim J (2006) Factors influencing the energetics of electron and proton transfers in proteins. What can be learned from calculations. Biochim Biophys Acta 1757: 942–968.
35. Khandogin J, Brooks CL, 3rd (2006) Toward the accurate first-principles prediction of ionization equilibria in proteins. Biochemistry 45: 9363–9373.
36. Rostkowski M, Olsson MH, Sondergaard CR, Jensen JH (2011) Graphical analysis of pH-dependent properties of proteins predicted using PROPKA. BMC Struct Biol 11: 6.
37. Humphrey W, Dalke A, Schulten K (1996) VMD: visual molecular dynamics. J Mol Graph 14: 33–38, 27–38.
38. Li H, Robertson AD, Jensen JH (2005) Very fast empirical prediction and rationalization of protein pKa values. Proteins 61: 704–721.
39. Bas DC, Rogers DM, Jensen JH (2008) Very fast prediction and rationalization of pKa values for protein-ligand complexes. Proteins 73: 765–783.
40. Davies MN, Toseland CP, Moss DS, Flower DR (2006) Benchmarking pK(a) prediction. BMC Biochem 7: 18.
41. Mason AC, Jensen JH (2008) Protein-protein binding is often associated with changes in protonation state. Proteins 71: 81–91.
42. Delhaise P, Bardiaux M, Wodak SJ (1984) Interactive computer animation of macromolecules. J Mol Graphics 2: 103–106.
43. The PyMOL Molecular Graphics System, Version 1.3, Schrödinger, LLC.
44. Jensen MO, Dror RO, Xu H, Borhani DW, Arkin IT, et al. (2008) Dynamic control of slow water transport by aquaporin 0: implications for hydration and junction stability in the eye lens. Proc Natl Acad Sci U S A 105: 14430–14435.
45. Kongsted J, Ryde U, Wydra J, Jensen JH (2007) Prediction and rationalization of the pH dependence of the activity and stability of family 11 xylanases. Biochemistry 46: 13581–13592.
46. Nielsen JE (2007) Analysing the pH-dependent properties of proteins using pKa calculations. J Mol Graph Model 25: 691–699.
47. Alexov E (2003) Role of the protein side-chain fluctuations on the strength of pair-wise electrostatic interactions: comparing experimental with computed pK(a)s. Proteins 50: 94–103.
48. Hajjar E, Dejaegere A, Reuter N (2009) Challenges in pKa predictions for proteins: the case of Asp213 in human proteinase 3. J Phys Chem A 113: 11783–11792.
49. Bjarnadottir U, Nielsen JE (2010) Calculating pKa values in the cAMP-dependent protein kinase: the effect of conformational change and ligand binding. Protein Sci 19: 2485–2497.
50. Polgar T, Magyar C, Simon I, Keseru GM (2007) Impact of ligand protonation on virtual screening against beta-secretase (BACE1). J Chem Inf Model 47: 2366–2373.
51. Tynan-Connolly BM, Nielsen JE (2007) Redesigning protein pKa values. Protein Sci 16: 239–249.
52. Kukic P, Farrell D, Sondergaard CR, Bjarnadottir U, Bradley J, et al. (2010) Improving the analysis of NMR spectra tracking pH-induced conformational changes: removing artefacts of the electric field on the NMR chemical shift. Proteins 78: 971–984.
53. Banks JL, Beard HS, Cao Y, Cho AE, Damm W, et al. (2005) Integrated modeling program, Applied Chemical Theory (IMPACT). J Comput Chem 26: 1752–1780.
54. Jacobson MP, Pincus DL, Rapp CS, Day TJ, Honig B, et al. (2004) A hierarchical approach to all-atom protein loop prediction. Proteins 55: 351–367.
55. Baker NA, Sept D, Joseph S, Holst MJ, McCammon JA (2001) Electrostatics of nanosystems: application to microtubules and the ribosome. Proc Natl Acad Sci U S A 98: 10037–10041.

10

Cysteine Methylation Controls Radical Generation in the Cfr Radical AdoMet rRNA Methyltransferase

Martin R. Challand[1], Enrico Salvadori[2,3], Rebecca C. Driesener[4], Christopher W. M. Kay[2,3]*, Peter L. Roach[4,5]*, James Spencer[1]*

1 School of Cellular and Molecular Medicine, University of Bristol Medical Sciences Building, Bristol, United Kingdom, **2** Institute of Structural and Molecular Biology, University College London, London, United Kingdom, **3** London Centre for Nanotechnology, University College London, London, United Kingdom, **4** Chemistry, University of Southampton, Highfield, Southampton, United Kingdom, **5** Institute for Life Sciences, University of Southampton, Highfield, Southampton, United Kingdom

Abstract

The 'radical *S*-adenosyl-L-methionine (AdoMet)' enzyme Cfr methylates adenosine 2503 of the 23S rRNA in the peptidyltransferase centre (P-site) of the bacterial ribosome. This modification protects host bacteria, notably methicillin-resistant *Staphylococcus aureus* (MRSA), from numerous antibiotics, including agents (e.g. linezolid, retapamulin) that were developed to treat such organisms. Cfr contains a single [4Fe-4S] cluster that binds two separate molecules of AdoMet during the reaction cycle. These are used sequentially to first methylate a cysteine residue, Cys338; and subsequently generate an oxidative radical intermediate that facilitates methyl transfer to the unreactive C8 (and/or C2) carbon centres of adenosine 2503. How the Cfr active site, with its single [4Fe-4S] cluster, catalyses these two distinct activities that each utilise AdoMet as a substrate remains to be established. Here, we use absorbance and electron paramagnetic resonance (EPR) spectroscopy to investigate the interactions of AdoMet with the [4Fe-4S] clusters of wild-type Cfr and a Cys338 Ala mutant, which is unable to accept a methyl group. Cfr binds AdoMet with high (~ 10 μM) affinity notwithstanding the absence of the RNA cosubstrate. In wild-type Cfr, where Cys338 is methylated, AdoMet binding leads to rapid oxidation of the [4Fe-4S] cluster and production of 5'-deoxyadenosine (DOA). In contrast, while Cys338 Ala Cfr binds AdoMet with equivalent affinity, oxidation of the [4Fe-4S] cluster is not observed. Our results indicate that the presence of a methyl group on Cfr Cys338 is a key determinant of the activity of the enzyme towards AdoMet, thus enabling a single active site to support two distinct modes of AdoMet cleavage.

Editor: Katrin Karbstein, The Scripps Research Institute, United States of America

Funding: This work was supported by the Leverhulme Trust (www.leverhulme.ac.uk; award ref. F/00 182/CO to JS, PLR and RC) and the United Kingdom Biotechnology and Biological Sciences Research Council (www.bbsrc.ac.uk; award ref. BB/J017906/1 to JS, PLR and MRC). The funders had no role in study design, data collection and analysis, decision to publish, or preparation of the manuscript.

Competing Interests: The authors have declared that no competing interests exist.

* E-mail: c.kay@ucl.ac.uk (CWMK); p.l.roach@soton.ac.uk (PLR); Jim.Spencer@bristol.ac.uk (JS)

Introduction

The radical *S*-adenosyl-L-methionine (AdoMet) enzyme Cfr is one of a growing number of enzymes that have been discovered to catalyse methyl transfer reactions to unactivated C–H bonds [1]. Cfr methylates the unreactive C2 and C8 carbon atoms of adenosine 2503 (A2503; *Escherichia coli* numbering) of the 23S RNA component of the large subunit of the bacterial ribosome (rRNA) [2,3]. Methylation at C8 protects Cfr-producing bacteria from the action of a variety of antibiotics, including the oxazolidinone linezolid, pleuromutilins and chloramphenicol, that interact with the peptidyltransferase site (P-site) of the bacterial ribosome [4]. The *cfr* gene is carried on mobile genetic elements that have facilitated the rapid dissemination of this antibiotic resistance mechanism throughout the bacterial community including, since 2007, strains of methicillin resistant *Staphylococcus aureus* (MRSA) [5], and other Gram-positive and Gram-negative bacteria including Enterococci [6,7] and *E. coli* [8]. The advent of transferable linezolid resistance is particularly significant as it highlights the availability of a pool of pre-existing resistance determinants to a fully synthetic agent with no natural product origin [9]. In most bacteria, including *S. aureus*, A2503 is methylated at the C2 position by a related enzyme, RlmN [10], from which Cfr is thought to have evolved. Recent investigations [11–14] have established that these enzymes achieve this challenging reaction by distinctive catalytic mechanisms.

Members of the radical AdoMet enzyme superfamily have attracted much recent interest as awareness grows of their wide distribution and the extensive range of biochemical reactions that they are able to catalyse ([15] and references therein). These enzymes contain a [4Fe-4S] cluster in the active site, usually ligated via three of the iron atoms to a conserved CXXXCXXC sequence motif. Structural analyses of several radical AdoMet enzymes [16–23] have demonstrated a common AdoMet binding mode where AdoMet binds via its amino acid moiety to the fourth iron atom of the [4Fe-4S] cluster (termed the 'unique iron'; ([24] and references therein)) in a bidentate manner. This brings the sulfonium ion of AdoMet into close proximity with the cluster (typically between 3–4 Å). AdoMet binding in this manner raises the mid-point potential for a one electron reduction of the [4Fe-4S] cluster from an overall +2 to +1 oxidation state into the range where this can be facilitated by a physiological reducing agent (such as flavodoxin in *E. coli*) [25,26]. Subsequent electron transfer from the reduced [4Fe-4S]$^{1+}$ cluster to the juxtaposed sulfonium ion of AdoMet triggers homolysis of the adjacent 5′C-S bond to yield methionine and a highly reactive primary 5′-deoxyadenosyl radical (5′dA•). This potent oxidant can initiate a wide range of chemically challenging biological processes by abstracting a hydrogen atom from unreactive centres to generate a substrate radical. Substrates for radical generation range from small molecule precursors, or biosynthetic intermediates, for diverse primary and secondary metabolites [27], to proteins [28] and nucleic acids [29].

Recent reports [11–13,18] have provided substantial mechanistic details for the methyl transfer reaction(s) catalysed by Cfr and RlmN. It is now established that each turnover requires two molecules of AdoMet, one serving as the source of the methyl group and the other of 5′dA• (Figure 1). Experimental evidence indicates that the methyl group is transferred from AdoMet to the 23S rRNA via a conserved cysteine residue (Cys355 in RlmN or Cys338 in Cfr), located within a flexible loop at the C-terminus of the polypeptide chain [18]. Recent spectroscopic investigations [14] demonstrate that both Cfr and RlmN contain a single [4Fe-4S] cluster. AdoMet bound to this cluster can both undergo reductive cleavage to 5′dA•, and react with the thiol moiety of (Cfr) Cys338 in an S_N2 substitution reaction. The resulting methyl cysteine thioether (mCys) is then the substrate for reaction with 5′dA•, making Cfr and RlmN unusual radical AdoMet enzymes in that they catalyse hydrogen atom abstraction from a post-translationally modified residue of their own polypeptide chain.

Establishing how the single [4Fe-4S] cluster of these enzymes controls the turnover of AdoMet to permit two distinct reactivities (Figure 1, methyl transfer (step 1) and 5′dA• formation (step 2)) at specific points of the overall reaction cycle is then a central question in understanding the reactions of these mechanistically intriguing, and clinically important, enzymes. In pursuit of this goal we here use continuous-wave electron paramagnetic resonance (cw-EPR) and absorbance spectroscopy to study the [4Fe-4S] centres of wild-type Cfr and of a Cys338 Ala mutant (which is incapable of accepting a methyl group from AdoMet) in the presence and absence of AdoMet, and to determine dissociation constants for AdoMet binding. Our results show that AdoMet binding leads to rapid oxidation of the [4Fe-4S] cluster of wild-type Cfr and results in uncoupled cleavage of AdoMet to 5′-deoxyadenosine. However, in the Cys338 Ala Cfr mutant, which still possesses an intact [4Fe-4S] centre with spectroscopic properties similar to wild-type Cfr, AdoMet binding leads neither to oxidation of the [4Fe-4S] cluster nor to significant 5′-deoxyadenosine production. Our results indicate that methylation of Cfr Cys338 may be a prerequisite for generation of the 5′dA• and thus a means by which alternative reactivities towards AdoMet are controlled and uncoupled cleavage (in part) avoided.

Materials and Methods

Protein Expression

Recombinant Cfr was expressed in *Escherichia coli* by a modified version of our previously reported protocol [30]. The *cfr* open reading frame, fused to an N-terminal hexhistidine tag (His_6-Cfr), was sub-cloned from plasmid pACYC:Cfr [30] into the NdeI and XhoI sites of the expression vector pCDFDuet-1 (Novagen) to create plasmid pMC115. pMC115 and plasmid pDB1282 (a pAra13 derivative encoding the *isc* operon of Fe-S cluster assembly proteins from *Azotobacter vinelandii*; [31]) were co-transformed into *E. coli* Rosetta2(DE3)pLysS (Merck Millipore) and plated on L-agar (Maitland) containing 34 µg/mL chloramphenicol, 50 µg/mL streptomycin and 100 µg/mL ampicillin. A single colony was used to inoculate a 50 mL starter culture in L-broth (Maitland) plus antibiotics as above, which was cultured overnight and used as a 1% inoculum for fresh L-broth (4×1 L). These bacterial cultures were incubated in an orbital shaker (150 rpm) at 37°C until absorbance at 600 nm reached approximately 0.3. At this point 10 mM arabinose was added to induce expression of the isc proteins and $FeSO_4.7H_2O$ and L-cysteine added to final concentrations of 2 mM and 200 µM, respectively. Culture at 37°C was continued until absorbance at 600 nm reached approximately 0.8, at which point the flasks were cooled to 18°C, Cfr expression induced by the addition of IPTG (1 mM) and culture continued at 18°C for a further 12 h. Bacteria were then harvested by centrifugation (5,000 g, 30 min, 4°C), the supernatant decanted and the wet cell paste stored at −80°C until purification.

Site Directed Mutagenesis

The Cfr Cys338 Ala mutant was prepared using the Quikchange Lightning site directed mutagenesis kit (Agilent) following the manufacturer's protocol, using primers Cfr C338A Fwd (5′-caatttgggattgatattgacgctgctgctggtcaattatatggtaattatc-3′) and Cfr C338A Rev (5′-gataattaccatataattgaccagcagcagcgtcaatatcaatcccaaattg-3′). The presence of the mutation, and the absence of any secondary mutations, was confirmed by DNA sequencing

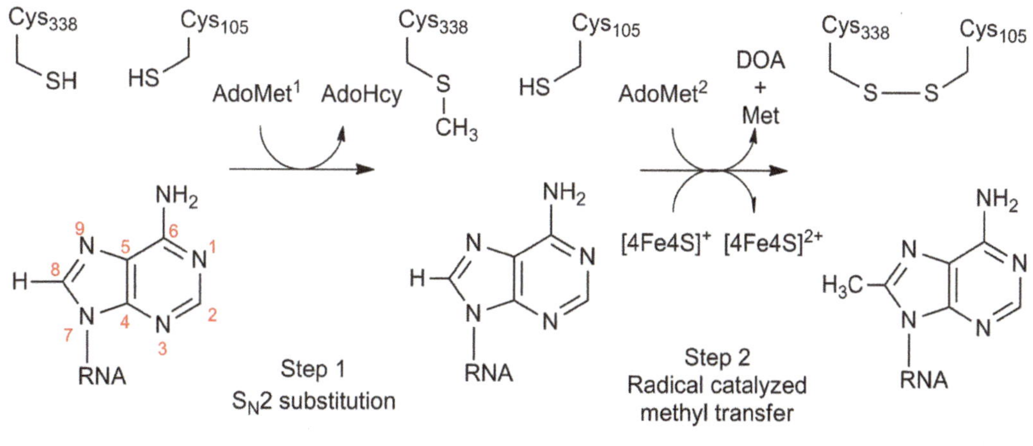

Figure 1. Utilisation of AdoMet by Cfr. Cfr consumes two AdoMet equivalents per reaction cycle to support both methyl transfer to Cfr Cys338 (AdoMet[1]; step 1) and subsequently generation of the 5′dA· radical (AdoMet[2]; step2).

(The Sequencing Service, University of Dundee, U.K). Expression and purification of Cfr Cys338 Ala followed identical procedures to those described for the wild-type enzyme.

Anaerobic Protein Purification and Reconstitution

Anaerobic procedures were carried out in a glove box (Belle Technology, Weymouth, U.K.) in a nitrogen atmosphere within a laboratory maintained at a constant 20°C. Chromatography steps utilised an AKTAPrime Plus workstation (GE Life Sciences). An anaerobic cell lysate was prepared by suspending wet cell paste (0.3 g/L) in anaerobic buffer A (50 mM bis-Tris-propane pH 8.4, 20% (v/v) glycerol, 0.25 M NaCl, 20 mM imidazole). An EDTA free protease inhibitor tablet (Roche), Benzonase (New England Biolabs; ~ 2,500 units)) and lysozyme (0.5 mg/mL) were added and the suspension stirred at r.t. for 30 min before cooling to ~10°C in a chilled water bath. Cells were disrupted by sonication, transferred to a gas tight centrifuge bottle and the insoluble debris removed by centrifugation (15,000 g, 30 min, 4°C). The cleared supernatant was applied to a 20 mL chelating Sepharose column (GE Life Sciences), previously charged with $NiSO_4$ and equilibrated with anaerobic buffer A. The column was washed with anaerobic buffer A until absorbance at 280 nm approached the baseline. Bound His_6-Cfr was eluted on a 40 mL 20 mM – 250 mM imidazole gradient followed by isocratic elution at 250 mM imidazole collecting 3 mL fractions. Cfr containing fractions (coloured brown) were pooled and immediately exchanged into anaerobic buffer C (50 mM bis-Tris-propane pH 8.4, 10% (v/v) glycerol, 0.5 M KCl, 5 mM DTT) by size exclusion chromatography (50 mL Sepharose G-50 column (GE Healthcare)). Progress of the purification procedure was monitored by SDS-PAGE.

Dithiothreitol (DTT; 5 mM) was added to purified Cfr (20–25 mL in buffer C, 1.5–2.5 mg/mL). After 15 min gentle stirring 4 molar equivalents of iron(III) chloride were added dropwise from a freshly prepared 10 mM stock solution. The resulting dark red solution was stirred for a further 15 min followed by dropwise addition of 4 molar equivalents of $Na_2S.9H_2O$, also from a freshly prepared 10 mM stock solution in anaerobic water. After 60–90 min the dark brown solution was centrifuged (15,000 g, 5 min, 4°C) to remove precipitated protein and excess FeS. Reconstituted Cfr was stored at −80°C and thawed anaerobically as required. Iron content (of samples desalted on 2.5 mL Sephadex G-25 columns (PD-10; GE Life Sciences)) was determined by the method of Fish [32] using $FeSO_4$.7H2O as a standard. Protein concentrations were determined using the Bradford assay [33] using bovine serum albumin (BSA) as a standard.

Mass Spectrometry

Samples were desalted on 50 µm POROS R2 resin (Applied Biosystems) prior to mass spectrometry. Mass spectrometry analysis was by flow injection on a MicroTOF-Q mass spectrometer (Bruker) using an electrospray source operated in positive-ion mode. Summed spectra were deconvoluted using the maximum-entropy module of the Bruker data analysis software suite.

AdoMet Production

Expression of *E. coli* AdoMet synthetase was carried out as described previously [34] with slight modifications. Briefly, an overnight starter culture of *E. coli* DM22 (pk8) (generous gift of J. Broderick, Montana State University) in 2YT media (Oxoid) with oxytetracycline (10 µg/mL, LB/Oxytet) was used as a 1% inoculum into fresh LB/Oxytet media (5 L). After incubation at 37°C in an orbital shaker (180 rpm) for 10 h the cells were isolated by centrifugation (7,500 rpm, 30 min, 4°C) typically yielding 2 g/L of wet cell paste. A crude cell lysate was prepared by resuspending the cell paste (0.3 g/L) in buffer D (100 mM Tris pH 8.0, 1 mM EDTA, 50 mg/mL lysozyme), the mixture was then stirred at room temperature for 30 min and lysed by 10×1 min sonication cycles. The lysate was cleared by centrifugation (7,500 rpm, 20 min, 4°C) and stored as 1 mL aliquots at −80°C.

Enzymatic AdoMet synthesis was achieved using a previously described method [34] with minor modifications. Reactions (20 mL) were prepared by dissolving ATP (13 mM) (Melford Laboratories Ltd) in buffer E (100 mM Tris pH 8.0, 26 mM $MgCl_2$, 50 mM KCl, 1 mM EDTA) before addition of acetonitrile (20% (v/v)) and L-methionine (10 mM). AdoMet synthetase crude cell lysate (1 mL, ~ 30 mg crude protein, as estimated by Bradford assay) was then added and the reaction gently stirred at room temperature for 4–5 h. Reaction progress was monitored by reverse-phase thin-layer chromatography (3:1 water:acetonitrile, 0.1% acetic acid; R_f values: AdoMet –0.25, 5′-methylthioadenosine (MTA) –0.40, adenine –0.55, ATP –0.93). The pH was adjusted to ~ 7 by addition of 1% TFA (trifluoroacetic acid, ~ 0.5 mL) and the mixture chilled on ice for 20 min before storage at −80°C.

The optimised purification protocol included acidification of the reaction mixture to ~pH 5 by addition of 1% TFA (1–1.5 mL) and removal of precipitated polypeptides by centrifugation (7,500 rpm, 20 min, 4°C). The resulting supernatant was filtered through a 0.22 µm filter (Millipore) and 10 mL applied, in 20 mM TFA, to an 8 mL Source 15S cation exchange column (GE Healthcare Life Sciences) previously equilibrated with 100 mM TFA. The column was further washed with 20 mM TFA until absorbance at 280 nm returned to baseline. Elution was effected by a 10 mL gradient to 1 M TFA. AdoMet typically eluted as a single peak at between 0.7–0.9 M TFA. AdoMet-containing fractions were identified and confirmed by TLC (AdoMet R_f value 0.75 under these acidic conditions), pooled and lyophilized prior to storage at −80°C. Overall yield was 20–30% and purity confirmed by HPLC (showing $<0.1\%$ adenine and $<1.5\%$ MTA) and ^1H NMR: (D_2O, 400 MHz) δppm 2.25–2.44 (m, 2 H, H_β), 2.92 (s, 0.1 H, (*R,S*) SMe), 2.96 (s, 3 H, (*S,S*) SMe), 3.40–3.50 (m, 1 H, H_γ), 3.60–3.70 (m, 1 H, H_γ), 3.85–3.99 (m, 3 H, H_α, $H_{5'}$), 4.48–4.54 (m, 1 H, $H_{4'}$), 4.54–4.59 (m, 1 H, $H_{3'}$), 4.78 (t, $^3J = 4.50$ Hz, 1 H, $H_{2'}$), 6.12 (d, $^3J = 4.00$ Hz, 1 H, $H_{1'}$), 8.40–8.41 (m, 2 H, $H_{2,8}$). Integration of the distinct signals of the AdoMet-methyl groups at 2.96 ppm and 2.92 ppm, respectively, in the ^1H NMR spectrum [35] estimated the ratio of (*S,S*) and (*R,S*) AdoMet as being typically ~ 95% of the active, (*S,S*) diastereoisomer. Stock solutions of purified AdoMet for use in spectroscopic experiments were prepared in appropriate anaerobic buffers and concentrations estimated from UV absorbance at 260 nm ($\varepsilon = 15{,}400\ M^{-1}\ cm^{-1}$ [36,37]).

Cw-EPR Spectroscopy

Reconstituted Cfr was concentrated in a stirred pressure cell (Amicon) using a 10 kDa molecular weight cut-off membrane (Amicon), exchanged into buffer F (50 mM HEPES pH 8.1, 10% (v/v) glycerol, 0.5 M KCl), diluted to 5 mg/mL (125 µM) and aliquots frozen at −80°C. Samples for spectroscopy were thawed anaerobically, mixed with sodium dithionite (1 mM) and/or purified AdoMet (400 µM) and incubated at room temperature for 30 min. 150 µL was transferred to an EPR tube (Wilmad, U.K.) which was sealed with a rubber septum, removed from the anaerobic chamber and immediately frozen in liquid nitrogen. Samples were maintained at liquid nitrogen temperatures at all times prior to obtaining EPR spectra. EPR measurements used a

Bruker EMXplus spectrometer operating at 9.4 GHz (X-band) equipped with a 4122SHQE resonator, with an Oxford Instruments ESR900 cryostat for measurements in the temperature interval 10–40 K. Spectra were acquired with a magnetic field sweep from 0 to 600 mT, a microwave power of 2 mW, a modulation amplitude of 0.5 mT and a modulation frequency of 100 kHz. Simulations of the cw-EPR spectra, to obtain the g-tensor principal components, were undertaken using the Easyspin toolbox [38] running in a Matlab (MathWorks, U.S.A.) environment.

HPLC Assays of DOA and AdoHcy Formation

Reconstituted Cfr was concentrated in a stirred pressure cell (Amicon) using a 10 kDa molecular weight cut-off membrane (Amicon). To ensure complete methylation of the wild-type enzyme AdoMet (Sigma; 1 mM) was added and the sample incubated for 30 min at room temperature. At this point a 50 µL aliquot was removed for quantification of *S*-adenosyl homocysteine (AdoHcy) production using HPLC as described below. The mixture was exchanged into buffer G (50 mM Bis-Tris-Propane pH 8.4, 10% (v/v) glycerol, 0.5 M KCl) by desalting on a disposable Sephadex G-25 column (Nap-10; GE Life Sciences) to remove any small molecules (including FeS, DTT, AdoMet and AdoHcy). Activity assays (final volume 100 µL in buffer G) were prepared by mixing either wild-type Cfr or Cfr Cys338 Ala (12.5 µM) and enzymatically synthesised and purified AdoMet (500 µM). Assays were equilibrated at 37°C for 5 min before initiation by addition of sodium dithionite (1 mM). Reactions were stopped over a range of selected time points by addition of 20% (v/v) perchloric acid (5 µL). Precipitated protein was removed by centrifugation (benchtop microcentrifuge, 13,000 rpm, 5 min) and the supernatant immediately analysed by HPLC injecting 10 µL onto a Gemini C18 (4.6×250 mm 5 µM, 110 Å) reverse phase column (Phenomenex). The mobile phase was an initial 8 min isocratic elution of 2.5% acetonitrile in 20 mM ammonium acetate pH 6.2 followed by a linear gradient to 25% acetonitrile over 12 min. DOA and AdoHcy concentrations were estimated from peak areas calibrated against synthetic standards (1–100 µM).

The data for wild-type Cfr were fitted to a first order process:

$$[DOA] = [DOA]_{\max}(1 - e^{-kt}) \quad (1)$$

where $[DOA]$ is the observed concentration of DOA; $[DOA]_{\max}$ is the maximum observed DOA concentration, t is time and k is the observed first order rate constant. Under the conditions of the experiment (500 µM AdoMet), at $t = 0$, the Cfr will be saturated such that [Cfr] = [Cfr.AdoMet]. Therefore, the initial turnover number k_{cat} can be estimated from the following equation:

$$k_{cat} = \frac{k * [DOA]_{\max}}{[E_0]} \quad (2)$$

where $[E_0]$ is the concentration of Cfr.

Absorbance Spectroscopy

Absorbance spectroscopy was carried out in 1 cm cuvettes using an Ocean Optics (Duiven, The Netherlands) USB2000 spectrophotometer and Mini-D2-GS light source connected by optical fibres to a cuvette holder contained within an anaerobic chamber [39]. Experiments were carried out in buffer G. Small aliquots (≤10 µL) of purified AdoMet.TFA were titrated into Cfr samples (1 mL; 60 µM). Samples were gently mixed and equilibrated for 2 min prior to recording absorbance spectra. Data were corrected for dilution and fitted to the tight binding quadratic equation [40]:

$$A = A_{init} - \Delta A \frac{[E_0] + [L] + K_d - \sqrt{([E_0] + [L] + K_d)^2 - 4 * [E_0] * [L]}}{2 * [E_0]} \quad (3)$$

Where A, A_{init} and ΔA denote absorbance, maximal absorbance and maximal change in absorbance, respectively; $[E_0]$ is the concentration of Cfr, $[L]$ is the concentration of ligand and K_d the dissociation constant.

Results and Discussion

The procedures described yield purified recombinant Cfr in typical yields of 5 mg/L after reconstitution, with an iron content of 4.3+/−0.2 molar equivalents for the wild-type enzyme. We ascribe our previous suggestion that Cfr may contain more than one [4Fe-4S] cluster [30] to increased adventitious iron binding at the elevated pH (9.5) at which our previous protein purification was carried out. Purification and reconstitution under the less alkaline conditions (pH 8.4) reported here yields material with an iron content consistent with other recent reports [14].

Recombinant Cfr was further characterised to determine the methylation state of the reconstituted wild-type enzyme. Cfr expressed using the protocol described yielded an electrospray mass spectrum centred on a mass of 41,084.7 Da, compared to calculated masses of 41,073.96 Da for the unmethylated, and 41 087.99 Da for the fully methylated, protein. Observation of a mass between those of unmethylated and methylated Cfr, but closer to the value expected for the methylated form, indicates that the protein is isolated as a mixture of the two species in which the methylated form predominates.

The transfer of methyl groups from AdoMet to the acceptor residue (Cys338) of Cfr also results in the formation of the product AdoHcy. By using HPLC analysis to quantify the AdoHcy formed, the extent of Cys338 methylation in Cfr can be estimated. When reconstituted Cfr was incubated with excess AdoMet, and the formation of AdoHcy quantified by HPLC, 0.12+/−0.03 mole equivalents of AdoHcy were observed to accumulate. Consistent with recent publications [14], these data indicate that our expression, purification and reconstitution procedure yields wild-type Cfr with Cys338 predominantly (≥85%) methylated. On the basis of these results we consider that investigations of AdoMet association with wild-type Cfr are likely to report binding to an active site configured for reductive AdoMet cleavage and formation of the 5′dA• intermediate, and ultimately DOA product (Figure 1, $AdoMet^2$, step 2), rather than to an active site configured for methylation of Cys338 (Figure 1, $AdoMet^1$, step 1).

We next used spectroscopic methods to investigate the [4Fe-4S] cluster of wild-type Cfr and its interactions with AdoMet. AdoMet purified from *in vitro* enzyme-catalysed syntheses (as described) was used in order to minimise the presence of contaminants arising from AdoMet racemisation and/or degradation [41]. Chemically reconstituted wild-type Cfr [30] was characterised by UV-Vis and cw-EPR spectroscopies. The visible absorbance spectrum of reconstituted Cfr has a broad peak around 400 nm, typical of intact [4Fe-4S] clusters ([42,43]; Figure 1A) as found in radical AdoMet enzymes [44–47] and other iron-sulfur proteins [48–50]. Titration of AdoMet into this material causes a reduction in absorbance, with the resulting difference spectrum (Figure 2A, inset)) containing a clear feature around 390 nm. Addition of AdoMet has a similar effect upon the absorbance spectrum of the

[4Fe-4S] cluster of biotin synthase [51] while crystal structures of AdoMet complexes of several radical AdoMet enzymes [24], including biotin synthase [17] and RlmN [18], show direct binding of AdoMet to the [4Fe-4S] cluster via bidentate coordination to the unique iron atom (a conformation favourable for reductive cleavage of the 5′C-S bond). We conclude that the observed absorbance changes are thus likely to result from a direct interaction between AdoMet and the [4Fe-4S] cluster of wild-type Cfr.

Observation of a measureable absorbance change upon AdoMet binding led us to investigate the concentration dependence of this spectroscopic signal. Monitoring absorbance at 390 nm during titration of wild-type Cfr with purified AdoMet (Figure 2B) produces a binding isotherm which could be fitted (equation 3) to yield a dissociation constant (K_d) of 8.4 μM. (Note that, although use of a quadratic equation (equation 3) is mandated under these (tight binding) conditions, the experimental data did not well define the stoichiometry of the interaction and the enzyme concentration ([E_0]; equation 3) was accordingly fixed at 60 μM as determined by Bradford assays). Interestingly, the measured value for K_d is consistent with that determined for AdoMet binding to biotin synthase in the presence ($K_d = 2.0$ μM), but not the absence ($K_d = 1$ mM) of the co-substrate dethiobiotin [51]). This discrepancy may reflect the differing catalytic mechanisms of the two enzymes. In biotin synthase co-operative substrate binding (i.e. prior dethiobiotin binding as a prerequisite for high-affinity association of AdoMet) has been proposed to aid in preventing inappropriate generation of the 5′dA• and uncoupled turnover of AdoMet. In contrast, current mechanisms for Cfr and RlmN employ the 5′dA• radical to activate the Cys338-bound methyl group and form a protein methyl radical [13]. As the wild-type enzyme utilised for these experiments is predominantly methylated (as shown by mass spectrometry and SAH quantification experiments), the cosubstrate for the radical generation step (methylated Cys338) is already present and radical formation may proceed even (as here) in the absence of RNA.

To provide a more detailed description of the interaction of AdoMet with Cfr, cw-EPR spectroscopy was used to investigate the electronic structure of the Cfr [4Fe-4S] cluster and the effects upon it of interaction with AdoMet. Wild-type Cfr (105 μM) that had been reduced with an excess (10 molar equivalents) of sodium dithionite for 15 minutes yielded a nearly axial EPR spectrum (Figure 3A) with g values (Table 1) of 2.03(5), 1.93(2) and 1.89(0) ($g_{av} = 1.95$) indicative of a $[4Fe\text{-}4S]^{1+}$ cluster. The temperature dependence of the signal [52] is consistent with the presence of a [4Fe-4S], rather than a [2Fe-2S], cluster: the spectrum reduces in intensity at temperatures higher than 30 K and is almost undetectable above 50 K (data not shown). Similar spectra have been reported for the [4Fe-4S] clusters of a number of other radical AdoMet enzymes [44,45,53,54]. However, in the presence of an excess (3 molar equivalents) of AdoMet, we observed almost complete disappearance of the reduced $[4Fe\text{-}4S]^{1+}$ signal (Figure 3B), implying that the cluster was most likely oxidised to $[4Fe\text{-}4S]^{2+}$ under these conditions. This behaviour contrasts with that observed for many other radical AdoMet enzymes [34,39,44,53,55,56], where, in the absence of cosubstrate, the EPR spectrum from the [4Fe-4S] cluster is maintained or even enhanced in the presence of AdoMet alone.

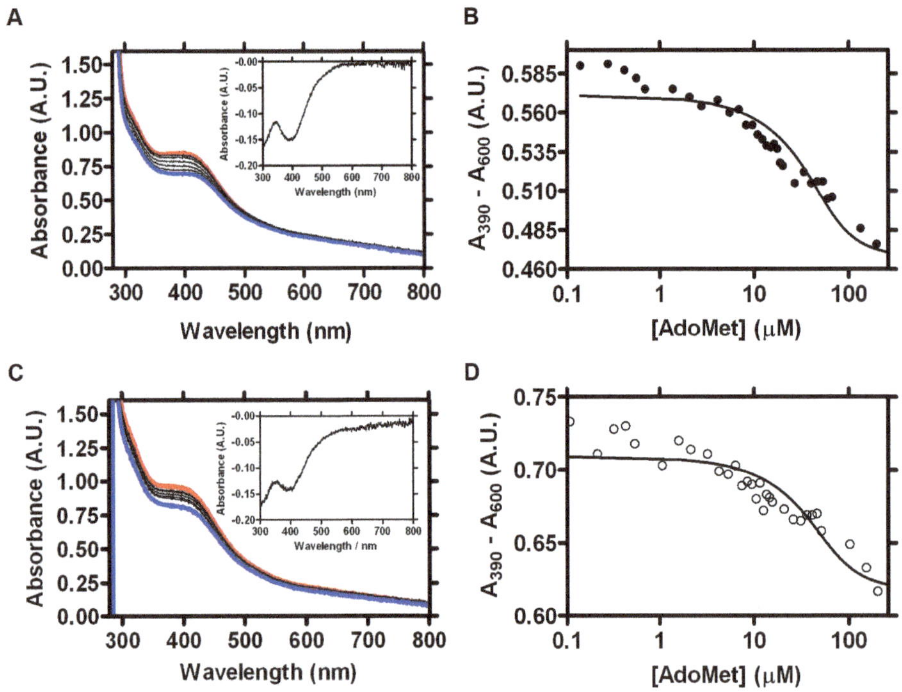

Figure 2. AdoMet Binding to the Cfr [4Fe-4S] Cluster. Interaction of AdoMet with the Cfr [4Fe-4S] cluster as monitored by absorbance spectroscopy. A. Absorbance spectra of reconstituted wild-type Cfr (60 μM; red trace) and on serial addition of AdoMet (black traces) to a final concentration of 260 μM (blue trace). Inset: difference spectrum obtained by subtracting last (blue; AdoMet) from first (red; no AdoMet) spectrum. B. Absorbance of reconstituted wild-type Cfr at 390 nm (corrected for baseline effects at 600 nm) plotted as a function of AdoMet concentration. Solid line is fit to equation 3 ($R^2 = 0.87$) giving a dissociation constant of 8.4 μM. C. Absorbance spectra of reconstituted Cys338 Ala Cfr (60 μM; red trace) and on serial AdoMet addition (black traces) to 260 μM (blue trace). Inset: difference spectrum obtained by subtracting last (blue; AdoMet) from first (red; no AdoMet) spectrum. D. Absorbance of reconstituted Cys338 Ala Cfr at 390 nm (corrected for baseline effects at 600 nm) plotted as a function of AdoMet concentration. Solid line is fit to equation 3 ($R^2 = 0.83$) giving dissociation constant of 11.7 μM.

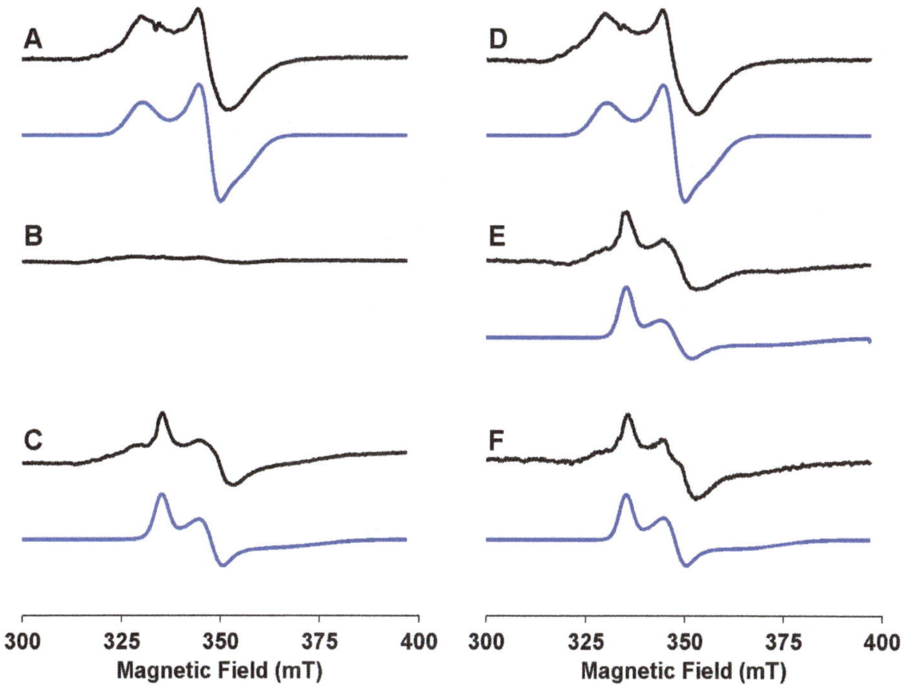

Figure 3. EPR Spectra of the Cfr [4Fe-4S] cluster. Cw-EPR spectra of wild-type Cfr (left hand column) and Cfr Cys338 Ala (right hand column). Experimental spectra are shown as black lines with corresponding simulations beneath (blue lines). A. Wild-type Cfr (105 μM) B. Wild-type Cfr (105 μM) plus AdoMet (300 μM). C. Wild-type Cfr plus AdoHcy (300 μM). D. Cfr Cys338 Ala (105 μM). E. Cfr Cys338 Ala plus AdoMet (300 μM). F. Cfr Cys338 Ala plus AdoHcy (300 μM). Note that all samples were reduced with sodium dithionite (1 mM).

These data indicate that wild-type Cfr is able to bind AdoMet with high affinity in the absence of the RNA cosubstrate, and that this probably leads to oxidation of the Cfr [4Fe-4S] cluster. The simplest explanation for this observed oxidation is that bound AdoMet undergoes reductive cleavage of the 5′C-S bond (Figure 1 step 2) consistent with previous observations [30] of uncoupled turnover of AdoMet to 5′-deoxyadenosine (DOA) by Cfr in the presence of sodium dithionite. Uncoupled turnover of AdoMet to DOA has been described for a number of other radical AdoMet enzyme systems [44,57–59]. To further investigate this hypothesis, we used RP-HPLC analysis at selected time points to quantify the uncoupled formation of DOA. After 1 h incubation at 37°C, almost 10 molar equivalents of DOA had formed with respect to Cfr concentration (Figure 4). This time course could be fitted to a first order process (Equation 1) and an initial turnover number (k_{cat}) derived (Equation 2) of 42×10^{-4} s^{-1}. The estimated value of k_{cat} is comparable to those determined for a number of other class 3 radical AdoMet enzymes [15]. (Class 3 radical AdoMet enzymes utilise the 5′-deoxyadenosine radical stoichiometrically with respect to the number of hydrogen atom abstraction steps required during each enzyme turnover, and therefore generate DOA as a byproduct [60]).

It has recently been proposed that the single [4Fe-4S] cluster of Cfr (and RlmN) is able to bind AdoMet for both methyl transfer to Cys338 (Figure 1, step 1) and subsequent formation of 5′dA• ([14]; Figure 1, step 2). Understanding how this is achieved requires description of the interaction(s) of AdoMet with Cfr in the presence and absence of a methyl group on Cys338. However, investigation of AdoMet binding to the [4Fe-4S] cluster of unmethylated Cfr is complicated by the difficulty of isolating the

Table 1. g-Values for EPR Spectra of Cfr and Ligand Complexes.

	g1	g2	g3	g_{av}
Cfr wild-type	2.03 (5)	1.93 (2)	1.89 (0)	1.95
Cfr wild-type+AdoMet	–	–	–	–
Cfr wild-type+AdoHcy	2.00 (0)	1.92 (6)	1.82 (3)	1.92
Cfr Cys338 Ala	2.03 (5)	1.93 (2)	1.89 (0)	1.95
Cfr Cys338 Ala+AdoMet	2.00 (0)	1.92 (5)	1.79 (0)	1.90
Cfr Cys338 Ala+AdoHcy	2.00 (0)	1.92 (7)	1.82 (3)	1.92

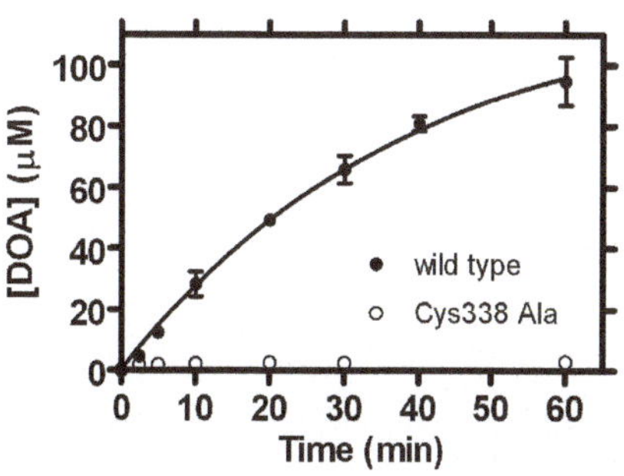

Figure 4. DOA production by Wild-type and Cys338 Ala Cfr. Production of 5′-deoxyadenosine (DOA) from uncoupled cleavage of AdoMet by wild-type (filled circles) or Cys338 Ala (open circles) Cfr assayed by reverse-phase HPLC. Solid line shows fit of wild-type data to equation 2 ($R^2=0.99$) yielding a k_{cat} value of 42×10^{-4} s^{-1}. Data points are the means of two independent measurements of separately prepared samples, shown with their standard error.

enzyme in a non-methylated state [14] and the ability of the enzyme to rapidly methylate on exposure to AdoMet. Accordingly, we used the methylation-incompetent Cys338 Ala mutant [14] as a system in which to study the interaction of AdoMet with the [4Fe-4S] cluster of the unmethylated enzyme (Figure 1 step 1). Cys338 Ala Cfr was expressed, purified and reconstituted using identical methodologies to the wild-type enzyme, and incorporates similar quantities of iron. To investigate the catalytic properties of Cys338 Ala Cfr, we used HPLC analysis to quantify production of AdoHcy and DOA when the AdoMet substrate was present in excess under reducing conditions. In contrast to the wild-type enzyme, Cfr Cys338 Ala was unable to catalyze production of AdoHcy or DOA, even after extended incubation in the presence of sodium dithionite (After 1 hour incubation, the amount of DOA produced by Cfr Cys338 Ala is 33-fold less than that obtained with the wild-type enzyme (Figure 4)). These data confirm that Cys338 Ala Cfr is unable to accept a methyl group from AdoMet, and that the formation of 5′dA• is significantly impaired.

The Cfr Cys338 Ala mutant was then subjected to spectroscopic investigation in order to assess the impact of the mutation upon the [4Fe-4S] centre and its interaction with AdoMet. The visual absorbance spectrum of the reconstituted mutant enzyme (Figure 2C) closely resembles that of wild-type Cfr, consistent with presence of an intact [4Fe-4S] cluster. Furthermore, the response of this mutant to titration with AdoMet (Figure 2C) is also very similar to that of the wild-type enzyme (above; Figure 2A) with a saturable reduction in visible absorbance generating a difference spectrum with a prominent feature around 390 nm. As for wild-type Cfr, these data were fitted to yield a dissociation constant for AdoMet binding of 11.7 μM (Figure 2D). On this basis we conclude that the Cys338 Ala mutation, and by implication the presence of a methyl group upon Cys338, has little effect on the affinity of Cfr for AdoMet.

While absorbance spectroscopy demonstrates that Cys338 Ala Cfr contains a [4Fe-4S] cluster that retains the ability to bind AdoMet, this does not reveal whether the mutation introduces changes to the structure or electronic properties of the cluster. To address this question cw-EPR spectroscopy was used to investigate the electronic structure of the [4Fe-4S] cluster of Cys338 Ala Cfr in the presence and absence of AdoMet. The EPR spectrum of chemically reconstituted and reduced Cys338 Ala Cfr (Figure 3D) closely resembles that of wild-type Cfr (g-values 2.03(5), 1.93(2) and 1.89(0); Table 1), indicating that the absence of the methyl group on Cys338) does not substantially alter the co-ordination environment of the [4Fe-4S] cluster. However, in contrast to the wild-type enzyme, reduction of Cfr Cys338 Ala in the presence of AdoMet (3 molar equivalents) gave an observable, although weaker, EPR spectrum (Figure 3E). Binding of AdoMet to the [4Fe-4S] cluster induces significant changes in the EPR spectrum: the signal symmetry becomes more rhombic with g-values 2.00(0), 1.92(5) and 1.79(0) ($g_{av} = 1.90$). The principal change is in the g3 value, which decreases from 1.89 to 1.79 on AdoMet binding, indicating a possible loss of symmetry in the cluster environment. These data indicate that direct interaction with AdoMet can perturb the electronic structure of the Cys338 Ala Cfr [4Fe-4S] cluster, but does not lead to its oxidation. The spectrum obtained in the presence of AdoMet to some extent resembled those observed for some other radical AdoMet enzymes (such as viperin, anaerobic ribonucleotide reductase and anaerobic sulfatase-maturating enzyme [44,53,56]) when AdoMet is co-ordinated to the [4Fe-4S] cluster. In these systems, presence of the second substrate is a pre-requisite for oxidation of the cluster in the presence of AdoMet and concomitant AdoMet cleavage.

As a further investigation of the effect of Cys338 methylation on the Cfr [4Fe-4S] cluster and its interactions with ligands, the EPR spectra of reduced wild-type and Cys338 Ala Cfr were measured in the presence of AdoHcy (Figure 3C;F). These experiments yielded near-identical EPR spectra, which most closely resembled those of Cys338 Ala Cfr in the presence of AdoMet (Figure 3E, for g-values see Table 1). This experiment confirms that, as well as exerting little effect upon the electronic properties of the Cfr [4Fe-4S] cluster, the Cys338 Ala mutation does not fundamentally disrupt the manner in which Cfr binds ligands. Thus the differences in observable properties of wild-type (methylated) and Cys338 Ala (non-methylated) Cfr described above reflect specific effects upon interactions with AdoMet rather than a general alteration of the ligand-binding properties of the enzyme.

Taken together, these data support the hypothesis that AdoMet binds to the [4Fe-4S] cluster of Cfr (and by implication the related RlmN enzyme) for two distinct activities: methyl transfer to Cys338 (or Cys355 in RlmN; Figure 1 step 1) and formation of 5′dA• (Figure 1 step 2; Figure 5 step 1). The EPR spectra presented here support a model in which the absence of a methyl group on Cys338 neither significantly perturbs the Cfr [4Fe-4S] cluster nor affects its affinity for AdoMet. However, only when mCys338 is present does AdoMet binding lead to re-oxidation of the [4Fe-4S] cluster, reductive cleavage of AdoMet and ultimately to DOA formation. Under the experimental conditions described (1 mM sodium dithionite), the absence of the mCys residue severely impairs reductive cleavage of AdoMet and the [4Fe-4S] cluster remains in the reduced, +1, oxidation state. One possible explanation for these observations is that the presence of the mCys residue may provide additional steric constraints on bound AdoMet, compared to its positioning in the unmethylated active site, that orient it more effectively for reductive cleavage to 5′dA•. Imposition of a conformation that forces the sulfonium of AdoMet into closer proximity to the [4Fe-4S] cluster may facilitate the necessary electron transfer due to increased orbital overlap with the 5′C-S σ* orbital, presumably resulting in a lower activation energy for AdoMet cleavage. As the rate of electron transfer may be very sensitive to the interatomic distance and the overlap of the electron donor and acceptor orbitals, even small adjustments in atomic positions, such as those necessary to accommodate the thioether methyl group of methylated Cys338, may exert significant effects. In the crystal structure of the RlmN:AdoMet complex (where Cys355 is methylated) the AdoMet sulfonium approaches to 3.2 Å from the unique Fe atom of the [4Fe-4S] cluster [18], in an orientation resembling those observed in ternary (enzyme:AdoMet:substrate) complexes of other radical AdoMet enzyme systems [16,17,21,23,24].

Hydrogen abstraction by 5′dA• from mCys338 has been characterised as thermodynamically favourable [61] by comparison of bond dissociation energies for a methyl thioether ($HS\text{-}CH_2\text{-}H = 397$ kJ/mol; as required to generate the radical, **4**, on Cfr mCys338, **1**, (Figure 5)) compared to a primary aliphatic group ($R\text{-}CH_2\text{-}H = 440$ kJ/mol; as involved in interconversion of 5′dA•, **2**, and DOA, **3**). In the light of these thermodynamic data, it is then tempting to speculate that reductive cleavage of AdoMet by Cfr leads to hydrogen atom abstraction from mCys338, **1**, (Figure 5 step 2). This hypothesis requires the observed uncoupled turnover to proceed via the methyl thioether radical, **4** (Figure 5 step 2, A; an intermediate in the catalytic mechanism proposed by Grove *et al* [13]) rather than by direct quenching of 5′dA• by a reaction with an alternative (non-specific) agent (Figure 5 step 2, B). Thus, mCys338 may be considered to fulfill a dual role, both providing a steric constraint that orients bound AdoMet for reductive cleavage and an appropriate target for the 5′dA• so generated. Using the

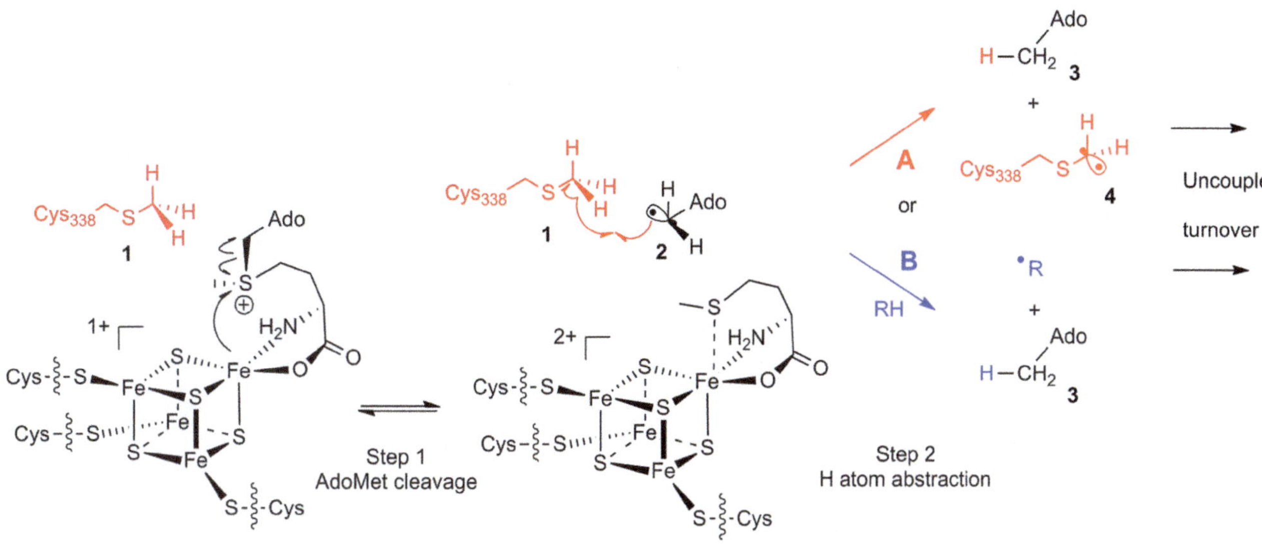

Figure 5. Mechanisms for Radical Generation and Uncoupled AdoMet Turnover by Cfr. Oxidative cleavage of AdoMet generates 5′dA·, **2**, which abstracts a hydrogen atom to yield the detected 5′-deoxyadenosine product, **3**. Uncoupled formation of 5′-deoxyadenosine could occur by quenching of 5′dA· through abstraction of a hydrogen atom from either i) mCys338, **1**, yielding the thermodynamically favourable radical intermediate **4** (pathway A, shown in red) or ii) a non-specific functional group (pathway B, shown in blue). Further details are given in the text.

methods employed here we could not however confirm the presence of the methyl thioether radical as, under these conditions, the EPR signal of the reduced [4Fe-4S] cluster of wild-type Cfr was completely quenched within 90 seconds of AdoMet addition (data not shown). The RlmN crystal structure [18] shows that this putative radical intermediate would lie in close proximity to the solvent accessible active site cleft and may be subject to rapid quenching by stepwise reduction and protonation. This may explain why it is not readily spectroscopically observable. Use of stoichiometric quantities of reductant and/or rapid quenching techniques may be required to confirm the presence of this radical intermediate.

The results reported here also highlight significant differences between Cfr and RlmN and many other radical AdoMet enzymes. In the majority of such enzymes studied to date, 5′dA• initiates catalysis by abstracting a hydrogen atom directly from the second substrate. In some systems, presence of the second substrate is necessary for high affinity AdoMet binding [51] and reductive cleavage [62–64], providing a mechanism by which these enzymes avoid uncoupled formation of the reactive 5′dA•. In contrast, here high-affinity AdoMet binding to Cfr is observed notwithstanding the absence of the RNA cosubstrate, and the rate of DOA formation (42×10^{-4} s^{-1}) is comparable to reported k_{cat} values for product formation by other radical AdoMet systems ([15] and references therein). Cfr and RlmN are unusual examples of radical AdoMet enzymes insofar as 5′dA• abstracts a hydrogen atom from a functional group covalently linked to its own polypeptide, rather than directly from a second substrate, in this case rRNA. In other systems (e.g. pyruvate formate lyase [23]) productive AdoMet binding requires the second substrate to be present, as shown by the substantial reorientation (a reduction in the AdoMet sulfonium:unique Fe distance from 6.1 Å to 3.2 Å) and ordering of AdoMet in ternary, as opposed to binary, complexes. Our spectroscopic investigations of Cfr instead indicate that, while mCys338 is required for reductive AdoMet cleavage, absence of the methyl group does not affect either the electronic structure of the [4Fe-4S] centre or its interactions with ligands (AdoHcy) as shown by the close resemblance of cw-EPR spectra of the wild-type and Cys338 Ala enzymes, both unliganded (Figure 3A and 3D) and in the presence of AdoHcy (Figures 3C and 3F). These data are consistent with the earlier conclusion that a single AdoMet binding site can support two distinct activities [14], and demonstrate that methylation has little effect upon active site conformation. The implication is that the (presumably subtle) conformational changes induced by the methylation of Cys338 in Cfr yield a similar change in reactivity towards AdoMet to that induced by secondary substrate binding to other radical AdoMet enzymes [23].

The observed behaviour of Cfr also differs from that reported for the related methyltransferase RlmN. In contrast to the results reported here for Cfr, Grove *et al.* show that mutation to Ala of the Cys338 equivalent, Cys355, in RlmN reduces, but does not abolish, the ability of the enzyme to catalyze uncoupled AdoMet cleavage in the presence of non-physiological reductant (sodium dithionite) [13]. Although no Cfr crystal structure is available, comparison with that recently determined for RlmN [18] indicates that Cys338 is located on a mobile loop close to the C-terminus of the protein. While the C-terminus of RlmN is formed by an extended α-helix, this helix is absent from Cfr, potentially increasing the mobility of the polypeptide chain in the vicinity of Cys338. Increased flexibility in this region may be required to enable alternative modes of RNA binding that permit Cfr to catalyse methyl transfer onto two positions, C2 and C8 of A2503, which lie on different sides of the adenosine ring system (Figure 1). The failure of the Cfr Cys338 Ala mutant to support AdoMet cleavage may then reflect local, but functionally significant, conformational differences between the two enzymes. More detailed structural and kinetic studies of both enzymes will be necessary to resolve this issue.

Concluding Remarks

Radical AdoMet enzymes are now recognised to employ a wide variety of mechanisms to modify a diverse range of (usually unreactive) substrates. This diversity extends to methyl transfer reactions, that are now believed to be achieved by a variety of mechanisms [65]. Our work here shows that use by Cfr and RlmN of a methylated cysteine residue not only permits methyl transfer to intrinsically unreactive positions on the RNA substrate but also enables them to control the timing of radical generation in the catalytic cycle, making possible the use of a single active site to achieve two distinct reactions.

Acknowledgments

We thank Steven Howell (Division of Protein Structure, National Institute of Medical Research, Mill Hill, London) for performing the mass spectrometry experiments and Richard Cammack (Pharmaceutical Science Research Division, King's College London) for assistance and advice related to cw-EPR experiments.

Author Contributions

Conceived and designed the experiments: MRC CWMK PLR JS. Performed the experiments: MRC ES. Analyzed the data: MRC ES CWMK JS. Contributed reagents/materials/analysis tools: RD. Wrote the paper: MRC JS.

References

1. Zhang Q, Li Y, Chen D, Yu Y, Duan L, et al. (2011) Radical-mediated enzymatic carbon chain fragmentation-recombination. Nat Chem Biol 7: 154–160.
2. Giessing AM, Jensen SS, Rasmussen A, Hansen LH, Gondela A, et al. (2009) Identification of 8-methyladenosine as the modification catalyzed by the radical SAM methyltransferase Cfr that confers antibiotic resistance in bacteria. RNA 15: 327–336.
3. Kehrenberg C, Schwarz S, Jacobsen L, Hansen LH, Vester B (2005) A new mechanism for chloramphenicol, florfenicol and clindamycin resistance: methylation of 23S ribosomal RNA at A2503. Mol Microbiol 57: 1064–1073.
4. Long KS, Poehlsgaard J, Kehrenberg C, Schwarz S, Vester B (2006) The Cfr rRNA methyltransferase confers resistance to Phenicols, Lincosamides, Oxazolidinones, Pleuromutilins, and Streptogramin A antibiotics. Antimicrob Agents Chemother 50: 2500–2505.
5. Toh SM, Xiong L, Arias CA, Villegas MV, Lolans K, et al. (2007) Acquisition of a natural resistance gene renders a clinical strain of methicillin-resistant Staphylococcus aureus resistant to the synthetic antibiotic linezolid. Mol Microbiol 64: 1506–1514.
6. Diaz L, Kiratisin P, Mendes RE, Panesso D, Singh KV, et al. (2012) Transferable plasmid-mediated resistance to linezolid due to cfr in a human clinical isolate of Enterococcus faecalis. Antimicrob Agents Chemother 56: 3917–3922.
7. Liu Y, Wang Y, Wu C, Shen Z, Schwarz S, et al. (2012) First report of the multidrug resistance gene cfr in Enterococcus faecalis of animal origin. Antimicrob Agents Chemother 56: 1650–1654.
8. Wang Y, He T, Schwarz S, Zhou D, Shen Z, et al. (2012) Detection of the staphylococcal multiresistance gene cfr in Escherichia coli of domestic-animal origin. J Antimicrob Chemother 67: 1094–1098.
9. Brickner SJ, Barbachyn MR, Hutchinson DK, Manninen PR (2008) Linezolid (ZYVOX), the first member of a completely new class of antibacterial agents for treatment of serious gram-positive infections. J Med Chem 51: 1981–1990.
10. Toh SM, Xiong L, Bae T, Mankin AS (2008) The methyltransferase YfgB/RlmN is responsible for modification of adenosine 2503 in 23S rRNA. RNA 14: 98–106.
11. Yan F, Fujimori DG (2011) RNA methylation by radical SAM enzymes RlmN and Cfr proceeds via methylene transfer and hydride shift. Proc Natl Acad Sci U S A 108: 3930–3934.
12. Yan F, Lamarre JM, Rohrich R, Wiesner J, Jomaa H, et al. (2010) RlmN and Cfr are Radical SAM Enzymes Involved in Methylation of Ribosomal RNA. J Am Chem Soc 132: 3953–3964.
13. Grove TL, Benner JS, Radle MI, Ahlum JH, Landgraf BJ, et al. (2011) A radically different mechanism for S-adenosylmethionine-dependent methyltransferases. Science 332: 604–607.
14. Grove TL, Radle MI, Krebs C, Booker SJ (2011) Cfr and RlmN contain a single [4Fe-4S] cluster, which directs two distinct reactivities for S-adenosylmethionine: methyl transfer by SN2 displacement and radical generation. J Am Chem Soc 133: 19586–19589.
15. Challand MR, Driesener RC, Roach PL (2011) Radical S-adenosylmethionine enzymes: Mechanism, control and function. Nat Prod Rep 28: 1696–1721.
16. Benjdia A, Heil K, Barends TR, Carell T, Schlichting I (2012) Structural insights into recognition and repair of UV-DNA damage by Spore Photoproduct Lyase, a radical SAM enzyme. Nucleic Acids Res 40: 9308–9318.
17. Berkovitch F, Nicolet Y, Wan JT, Jarrett JT, Drennan CL (2004) Crystal structure of biotin synthase, an S-adenosylmethionine-dependent radical enzyme. Science 303: 76–79.
18. Boal AK, Grove TL, McLaughlin MI, Yennawar NH, Booker SJ, et al. (2011) Structural basis for methyl transfer by a radical SAM enzyme. Science 332: 1089–1092.
19. Hanzelmann P, Schindelin H (2004) Crystal structure of the S-adenosylmethionine-dependent enzyme MoaA and its implications for molybdenum cofactor deficiency in humans. Proc Natl Acad Sci U S A 101: 12870–12875.
20. Layer G, Moser J, Heinz DW, Jahn D, Schubert WD (2003) Crystal structure of coproporphyrinogen III oxidase reveals cofactor geometry of Radical SAM enzymes. EMBO J 22: 6214–6224.
21. Lepore BW, Ruzicka FJ, Frey PA, Ringe D (2005) The x-ray crystal structure of lysine-2,3-aminomutase from Clostridium subterminale. Proc Natl Acad Sci U S A 102: 13819–13824.
22. Nicolet Y, Rubach JK, Posewitz MC, Amara P, Mathevon C, et al. (2008) X-ray structure of the [FeFe]-hydrogenase maturase HydE from Thermotoga maritima. J Biol Chem 283: 18861–18872.
23. Vey JL, Yang J, Li M, Broderick WE, Broderick JB, et al. (2008) Structural basis for glycyl radical formation by pyruvate formate-lyase activating enzyme. Proc Natl Acad Sci U S A 105: 16137–16141.
24. Vey JL, Drennan CL (2011) Structural insights into radical generation by the radical SAM superfamily. Chem Rev 111: 2487–2506.
25. Hinckley GT, Frey PA (2006) Cofactor dependence of reduction potentials for [4Fe-4S]2+/1+ in lysine 2,3-aminomutase. Biochemistry 45: 3219–3225.
26. Wang SC, Frey PA (2007) Binding energy in the one-electron reductive cleavage of S-adenosylmethionine in lysine 2,3-aminomutase, a radical SAM enzyme. Biochemistry 46: 12889–12895.
27. Roach PL (2011) Radicals from S-adenosylmethionine and their application to biosynthesis. Curr Opin Chem Biol 15: 267–275.
28. Selmer T, Pierik AJ, Heider J (2005) New glycyl radical enzymes catalysing key metabolic steps in anaerobic bacteria. Biol Chem 386: 981–988.
29. Atta M, Mulliez E, Arragain S, Forouhar F, Hunt JF, et al. (2010) S-Adenosylmethionine-dependent radical-based modification of biological macromolecules. Curr Opin Struct Biol 20: 684–692.
30. Booth MP, Challand MR, Emery DC, Roach PL, Spencer J (2010) High-level expression and reconstitution of active Cfr, a radical-SAM rRNA methyltransferase that confers resistance to ribosome-acting antibiotics. Protein Expr Purif 74: 204–210.
31. Dos Santos PC, Johnson DC, Ragle BE, Unciuleac MC, Dean DR (2007) Controlled expression of nif and isc iron-sulfur protein maturation components reveals target specificity and limited functional replacement between the two systems. J Bacteriol 189: 2854–2862.
32. Fish WW (1988) Rapid colorimetric micromethod for the quantitation of complexed iron in biological samples. Methods Enzymol 158: 357–364.
33. Bradford MM (1976) A rapid and sensitive method for the quantitation of microgram quantities of protein utilizing the principle of protein-dye binding. Anal Biochem 72: 248–254.
34. Walsby CJ, Hong W, Broderick WE, Cheek J, Ortillo D, et al. (2002) Electron-nuclear double resonance spectroscopic evidence that S-adenosylmethionine binds in contact with the catalytically active [4Fe-4S](+) cluster of pyruvate formate-lyase activating enzyme. J Am Chem Soc 124: 3143–3151.
35. Stolowitz ML, Minch MJ (1981) S-adenosyl-L-methionine and S-adenosyl-L-homocysteine, an NMR study. J Am Chem Soc 103: 6015–6019.
36. Schluckebier G, Kozak M, Bleimling N, Weinhold E, Saenger W (1997) Differential binding of S-adenosylmethionine S-adenosylhomocysteine and Sinefungin to the adenine-specific DNA methyltransferase M.TaqI. J Mol Biol 265: 56–67.
37. Shapiro SK, Ehninger DJ (1966) Methods for the analysis and preparation of adenosylmethionine and adenosylhomocysteine. Anal Biochem 15: 323–333.
38. Stoll S, Schweiger A (2006) EasySpin, a comprehensive software package for spectral simulation and analysis in EPR. J Magn Reson 178: 42–55.
39. Kriek M, Martins F, Leonardi R, Fairhurst SA, Lowe DJ, et al. (2007) Thiazole synthase from Escherichia coli: an investigation of the substrates and purified proteins required for activity in vitro. J Biol Chem 282: 17413–17423.
40. Muller B, Restle T, Reinstein J, Goody RS (1991) Interaction of fluorescently labeled dideoxynucleotides with HIV-1 reverse transcriptase. Biochemistry 30: 3709–3715.
41. Iwig DF, Booker SJ (2004) Insight into the polar reactivity of the onium chalcogen analogues of S-adenosyl-L-methionine. Biochemistry 43: 13496–13509.
42. Lippard SJ, Berg JM (1994) Principles of bioinorganic chemistry. Mill Valley, Calif.: University Science Books. xvii, 411 p. p.
43. Hagen KS, Watson AD, Holm RH (1983) Synthetic Routes to Fe2s2, Fe3s4, Fe4s4, and Fe6s9 Clusters from the Common Precursor [Fe(Sc2h5)4]2- - Structures and Properties of [Fe3s4(Sr)4]3- and [Fe6s9(Sc2h5)2]4-, Examples of the Newest Types of Fe-S-Sr Clusters. Journal of the American Chemical Society 105: 3905–3913.
44. Duschene KS, Broderick JB (2010) The antiviral protein viperin is a radical SAM enzyme. FEBS Lett 584: 1263–1267.
45. Kulzer R, Pils T, Kappl R, Huttermann J, Knappe J (1998) Reconstitution and characterization of the polynuclear iron-sulfur cluster in pyruvate formate-lyase-

activating enzyme. Molecular properties of the holoenzyme form. J Biol Chem 273: 4897–4903.
46. Ugulava NB, Gibney BR, Jarrett JT (2001) Biotin synthase contains two distinct iron-sulfur cluster binding sites: chemical and spectroelectrochemical analysis of iron-sulfur cluster interconversions. Biochemistry 40: 8343–8351.
47. Zhang Q, Chen D, Lin J, Liao R, Tong W, et al. (2011) Characterization of NocL involved in thiopeptide nocathiacin I biosynthesis: a [4Fe-4S] cluster and the catalysis of a radical S-adenosylmethionine enzyme. J Biol Chem 286: 21287–21294.
48. Bruschi M, Hatchikian C, Le Gall J, Moura JJ, Xavier AV (1976) Purification, characterization and biological activity of three forms of ferredoxin from the sulfate-reducing bacterium Desulfovibrio gigas. Biochim Biophys Acta 449: 275–284.
49. Chandramouli K, Unciuleac MC, Naik S, Dean DR, Huynh BH, et al. (2007) Formation and properties of [4Fe-4S] clusters on the IscU scaffold protein. Biochemistry 46: 6804–6811.
50. Yeeles JT, Cammack R, Dillingham MS (2009) An iron-sulfur cluster is essential for the binding of broken DNA by AddAB-type helicase-nucleases. J Biol Chem 284: 7746–7755.
51. Ugulava NB, Frederick KK, Jarrett JT (2003) Control of adenosylmethionine-dependent radical generation in biotin synthase: a kinetic and thermodynamic analysis of substrate binding to active and inactive forms of BioB. Biochemistry 42: 2708–2719.
52. Rupp H, Rao KK, Hall DO, Cammack R (1978) Electron Spin Relaxation of Iron-sulphur Proteins Studied by Microwave Power Saturation. Biochim Biophys Acta 537: 255–269.
53. Ollagnier S, Mulliez E, Schmidt PP, Eliasson R, Gaillard J, et al. (1997) Activation of the anaerobic ribonucleotide reductase from Escherichia coli. The essential role of the iron-sulfur center for S-adenosylmethionine reduction. J Biol Chem 272: 24216–24223.
54. Shepard EM, Duffus BR, George SJ, McGlynn SE, Challand MR, et al. (2010) [FeFe]-hydrogenase maturation: HydG-catalyzed synthesis of carbon monoxide. J Am Chem Soc 132: 9247–9249.
55. Lieder KW, Booker S, Ruzicka FJ, Beinert H, Reed GH, et al. (1998) S-Adenosylmethionine-dependent reduction of lysine 2,3-aminomutase and observation of the catalytically functional iron-sulfur centers by electron paramagnetic resonance. Biochemistry 37: 2578–2585.
56. Benjdia A, Subramanian S, Leprince J, Vaudry H, Johnson MK, et al. (2010) Anaerobic sulfatase-maturating enzyme–a mechanistic link with glycyl radical-activating enzymes? FEBS J 277: 1906–1920.
57. Farrar CE, Siu KK, Howell PL, Jarrett JT (2010) Biotin synthase exhibits burst kinetics and multiple turnovers in the absence of inhibition by products and product-related biomolecules. Biochemistry 49: 9985–9996.
58. Ollagnier-de Choudens S, Sanakis Y, Hewitson KS, Roach P, Munck E, et al. (2002) Reductive cleavage of S-adenosylmethionine by biotin synthase from Escherichia coli. J Biol Chem 277: 13449–13454.
59. Tse Sum Bui B, Lotierzo M, Escalettes F, Florentin D, Marquet A (2004) Further investigation on the turnover of Escherichia coli biotin synthase with dethiobiotin and 9-mercaptodethiobiotin as substrates. Biochemistry 43: 16432–16441.
60. Wang SC, Frey PA (2007) S-adenosylmethionine as an oxidant: the radical SAM superfamily. Trends Biochem Sci 32: 101–110.
61. Hioe J, Zipse H (2012) Hydrogen Transfer in SAM-Mediated Enzymatic Radical Reactions. Chemistry 18: 16463–16472.
62. Moss ML, Frey PA (1990) Activation of lysine 2,3-aminomutase by S-adenosylmethionine. J Biol Chem 265: 18112–18115.
63. Challand MR, Martins FT, Roach PL (2010) Catalytic activity of the anaerobic tyrosine lyase required for thiamine biosynthesis in Escherichia coli. J Biol Chem 285: 5240–5248.
64. Driesener RC, Challand MR, McGlynn SE, Shepard EM, Boyd ES, et al. (2010) [FeFe]-hydrogenase cyanide ligands derived from S-adenosylmethionine-dependent cleavage of tyrosine. Angew Chem Int Ed Engl 49: 1687–1690.
65. Zhang Q, van der Donk WA, Liu W (2012) Radical-mediated enzymatic methylation: a tale of two SAMS. Acc Chem Res 45: 555–564.

Screen for *IDH1, IDH2, IDH3, D2HGDH* and *L2HGDH* Mutations in Glioblastoma

Daniel Krell[1,2], Mawuelikem Assoku[1,2], Malcolm Galloway[3], Paul Mulholland[2], Ian Tomlinson[1], Chiara Bardella[1]*

1 Molecular and Population Genetics Laboratory, Wellcome Trust Centre for Human Genetics, University of Oxford, Oxford, United Kingdom, **2** Department of Medical Oncology, University College London Hospitals, London, United Kingdom, **3** Department of Cellular Pathology, The Royal Free Hospital, London, United Kingdom

Abstract

Isocitrate dehydrogenases (IDHs) catalyse oxidative decarboxylation of isocitrate to α-ketoglutarate (α-KG). IDH1 functions in the cytosol and peroxisomes, whereas IDH2 and IDH3 are both localized in the mitochondria. Heterozygous somatic mutations in *IDH1* occur at codon 132 in 70% of grade II–III gliomas and secondary glioblastomas (GBMs), and in 5% of primary GBMs. Mutations in *IDH2* at codon 172 are present in grade II–III gliomas at a low frequency. *IDH1* and *IDH2* mutations cause both loss of normal enzyme function and gain-of-function, causing reduction of α-KG to D-2-hydroxyglutarate (D-2HG) which accumulates. Excess hydroxyglutarate (2HG) can also be caused by germline mutations in D- and L-2-hydroxyglutarate dehydrogenases (*D2HGDH* and *L2HGDH*). If loss of IDH function is critical for tumourigenesis, we might expect some tumours to acquire somatic *IDH3* mutations. Alternatively, if 2HG accumulation is critical, some tumours might acquire somatic *D2HGDH* or *L2HGDH* mutations. We therefore screened 47 glioblastoma samples looking for changes in these genes. Although *IDH1* R132H was identified in 12% of samples, no mutations were identified in any of the other genes. This suggests that mutations in *IDH3, D2HGDH* and *L2HGDH* do not occur at an appreciable frequency in GBM. One explanation is simply that mono-allelic *IDH1* and *IDH2* mutations occur more frequently by chance than the bi-allelic mutations expected at *IDH3, D2HGDH* and *L2HGDH*. Alternatively, both loss of IDH function and 2HG accumulation might be required for tumourigenesis, and only *IDH1* and *IDH2* mutations have these dual effects.

Editor: Joseph Charles Glorioso, University of Pittsburgh School of Medicine, United States of America

Funding: This work was supported by Oxford Comprehensive Biomedical Core (BRC), the charity Mothers and Daughters, the University College London Hospitals, Experimental Cancer Medicine Centre, and a core centre grant from the Wellcome Trust (075491/Z/04). The funders had no role in study design, data collection and analysis, decision to publish or preparation of the manuscript.

Competing Interests: The authors have declared that no competing interests exist.

* E-mail: chiara.bardella@well.ox.ac.uk

These authors contributed equally to this work.

Introduction

Gliomas are the most common primary brain tumour, accounting for 70% of all primary central nervous system neoplasms. They show wide diversity with respect to location, morphology, genetic status and response to therapy.

Grade I gliomas occur more in children than in adults, are generally curable with complete surgical resection and rarely evolve into higher-grade lesions. WHO grade II or III gliomas are invasive, progress to higher-grade lesions and have a poor outcome. Glioblastoma (GBM), the most common and most malignant glioma, has a very poor prognosis and may develop rapidly without evidence of a less malignant precursor lesion (primary glioblastoma), or less commonly through progression from a lower grade tumour (secondary glioblastoma) [1,2].

Isocitrate dehydrogenases (IDHs) catalyse the oxidative decarboxylation of isocitrate to α-ketoglutarate (α-KG) and reduce NAD^+ or $NADP^+$ to NADH or NADPH. IDH1 and IDH2 are homodimeric, $NADP^+$-dependent enzymes that share considerable sequence similarity and an almost identical protein structure [3]. IDH3 is a heterotetramer composed of two α, one β and one γ subunit. IDH3 is a NAD^+-dependent enzyme. IDH1 is localised to the cytoplasm and peroxisomes, it is highly expressed in the liver and to a lesser level in other tissues [4]. As well as being involved in catalysing the conversion of isocitrate to α-KG, IDH1 is thought to play a role in cellular metabolic processes such as lipid and glucose metabolism [5,6], and has been shown to be involved in cellular defence against reactive oxygen species and radiation [7,8,9].

IDH2 is localised to the mitochondria, is highly expressed in heart, muscle, activated lymphocytes and moderately in other tissues [4]. IDH2 plays a key role in the regulation of the tricarboxylic acid cycle (TCA) and like IDH1, has been shown to have a protective role against insults such as oxidative stress [9,10,11].

IDH3 is also localised to the mitochondria and plays a central role in the TCA cycle.

In 2008, a genome-wide sequencing study identified somatic mutations in *IDH1* in 18 (12%) of 149 patients with GBM [12] and it was subsequently found that GBMs without *IDH1* mutations often have mutations affecting *IDH2* [13]. Both *IDH1* and *IDH2* mutations are more frequent in grade II-III gliomas and secondary glioblastoma (70–75%) than in primary glioblastoma (5%) [14], are present at higher frequencies in younger patients, and are associated with a relatively favourable prognosis [13,15]. *IDH1* and *IDH2* mutations are mono-allelic, somatic, missense changes. Mutations in *IDH1* almost always affect R132, which is the binding site for isocitrate [12]. Mutations in *IDH2* exclusively affect R172 and R140; the former of these arginines is analogous to *IDH1* R132 [13] [16].

Initial investigations demonstrated that mutated IDH1 had a reduced affinity for isocitrate [17]. Furthermore, expression of mutant *IDH1* in cultured cells has been shown to reduce the formation of the enzyme products, α-KG and NADPH [13,17,18]. These structural and biochemical findings suggested that *IDH* mutation results in loss of function, the most plausible mechanism being a dominant negative effect through the formation of catalytically inactive heterodimers of mutant and wild type proteins [17]. Deficiency of α-KG might lead to defective function of α-KG-dependent enzymes such as HIF prolyl hydroxylases.

Subsequent evidence has emerged suggesting that mutant IDH1 and IDH2 are not simply inactive enzymes, but instead possess novel enzymatic activity [16,19,20]. Dang *et al.* demonstrated that mutant, but not wild type, IDH1 catalyses the reduction of α-KG to D-2HG while converting NADPH to $NADP^+$. D-2HG accumulates in glioblastoma cells with *IDH1* R132H mutations [19]. The role of 2HG in tumourigenesis is unclear, although it may act as an oncometabolite, perhaps competitively inhibiting α-KG-dependent enzymes [21] [22]. One of the rationales for a role of 2HG in tumour formation is derived from the observation of patients suffering from hereditary 2-hydroxyglutaric aciduria. This is a rare inherited metabolic disorder, caused by homozygous inactivating germline 2-hydroxyglutarate dehydrogenase (*2HGDH*) mutations, that is weakly associated with brain tumours.

If IDH loss-of-function alone is critical for tumourigenesis, we might expect *IDH3* mutations to occur in brain tumours, including those sub-types with relatively low frequencies of *IDH1* and *IDH2* mutations. Alternatively, if accumulation of 2HG alone is critical, we might expect mutations in *D2HGDH* and/or *L2HGDH*. A further, more speculative line of reasoning is that, if patients with mutations in *IDH1* and *IDH2* have normal D2HGDH and L2HGDH function, excess 2HG should simply be converted back to α-KG by 2HGDH. The failure of this to occur might be due to saturation of the 2HGDH enzymes, but raises the possibility that patients with *IDH1* and *IDH2* mutations might also require inactivating mutations in *D2HGDH* and *L2HGDH* for 2HG to accumulate.

We therefore analysed a set of GBM for mutations in *IDH1*, *IDH2*, *IDH3*, *D2HGDH* and *L2HGDH*.

Results and Discussion

We studied 47 glioblastomas (WHO grade IV). Heterozygous mutations of *IDH1* were found in 6/47 tumours (12%). All 6 mutations were single base substitutions c.395G>A occurring at residue R132, resulting in an arginine to histidine (p.R132H) substitution (Figure 1). This frequency is consistent with previously described data. No mutations were found in *IDH2*, in keeping with the lower frequency of such changes than mutations in *IDH1*. We sequenced all the exons of *IDH3A*, *IDH3B* and *IDH3G*, which encode the α, β, and γ subunits of the IDH3 heterotetramer, and all the exons of *D2HGDH* and *L2HGDH*. In all cases examined we did not find any mutations. Known SNPs were found within each gene (details not shown).

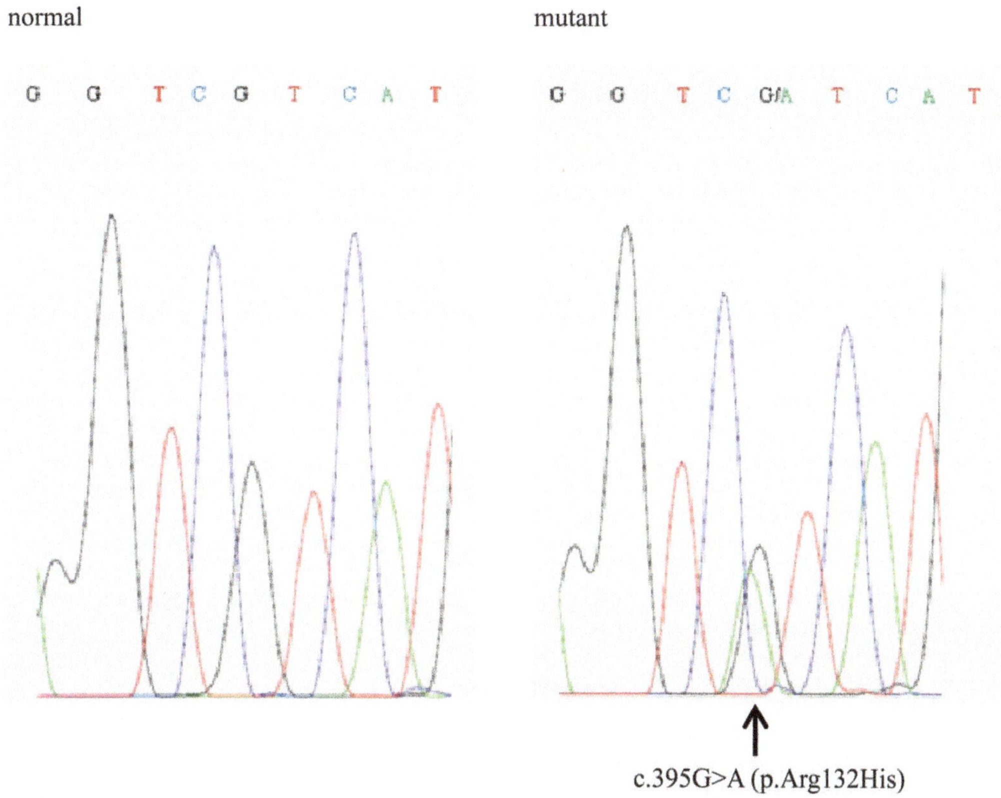

Figure 1. Identification of the G395A mutation at codon 132 by sequence analysis of the *IDH1* gene. The electropherogram shows a representative example of the heterozygous, single base G-to-A substitution at nucleotide position 395 of the *IDH1* gene (right panel) and the corresponding wild-type sequence (left panel). The mutation was detected in 6 out of 47 glioblastomas analyzed.

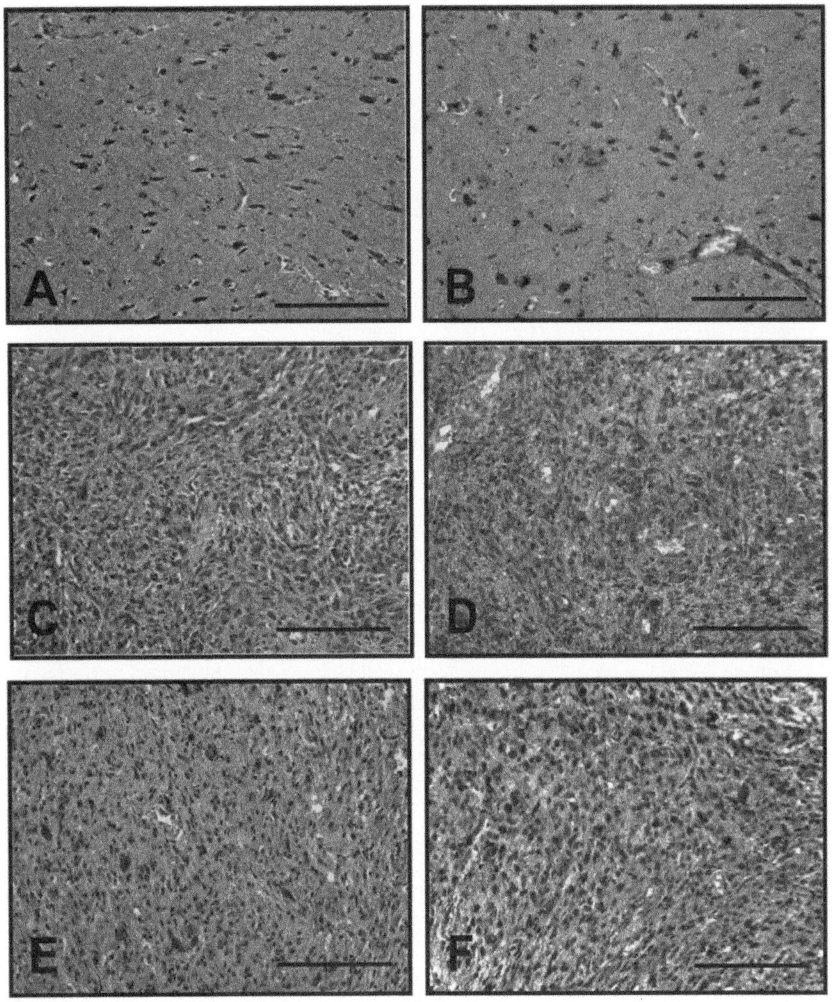

Figure 2. D2HGDH and L2HGDH are expressed in wildtype *IDH1/2* and *IDH1* R132H glioblastomas. Expression of D2HGDH and L2HGDH in normal brain is shown respectively in panels A and B. D2HGDH and L2HGDH were detected in brain tumour samples carrying either wild-type *IDH1/IDH2* (C and D) or mutated *IDH1* allele (E and F) by immunohistochemistry. Scale bar represents 50 μm in all panels.

We then wondered whether *D2HGDH* or *L2HGDH* might be inactivated by other mechanisms such as promoter methylation or miRNA over-expression. We therefore evaluated the expression of the corresponding proteins in human glioblastomas samples (Figure 2). D2HGDH and L2HGDH were detected in GBMs carrying wild type *IDH1* and *IDH2* genes, as well as in *IDH1*-mutants tumours. It is highly unlikely that the loss of D2HGDH or L2HGDH expression plays an important role in the pathogenesis of human GBMs.

Mutations in *IDH3, D2HGDH* and *L2HGDH* (or loss of expression of the 2HGDHs) do not therefore occur at an appreciable frequency in GBM. One explanation is simply that mono-allelic *IDH1* and *IDH2* mutations occur more frequently by chance than the bi-allelic mutations expected at *IDH3*, *D2HGDH* and *L2HGDH*. Alternatively, both loss of IDH function and 2HG accumulation are required for tumourigenesis, and only *IDH1* and *IDH2* mutations have these dual effects. It is also possible that 2HG accumulation is an epiphenomenon and that IDH3 loss-of-function does not lead to sufficient deficiency of α-KG to promote tumourigenesis. There might also be an unknown mechanism of tumourigenesis that is specific to defects in IDH1 or IDH2.

Moreover our data suggest that in patients with mutations of *IDH1* and *IDH2* and normal D2HGDH and L2HGDH function, the excess 2HG is not converted back to α-KG by 2HGDH, perhaps due to the saturation of the latter enzyme. A recent report identified heterozygous germline *IDH2* mutations in patients with idiopathic D-2-hydroxyglutaric aciduria, carrying no mutations in *D2HGDH* and consistently increased D-2HG levels in body fluids [23]; interestingly, the patient did not develop brain tumours.

The mechanisms by which *IDH1* and *IDH2* mutations cause tumourigenesis remain largely unclear, as do the putative alternative functional deficiencies in *IDH*-wildtype gliomas. However, even if 2HG has no direct pathogenic role it has potential as a specific marker of *IDH1* or *IDH2* mutations, to diagnose or monitor *IDH*-mutant glioma [24].

Materials and Methods

Sample collection

All brain tumours were obtained from the neuropathology department at The Royal Free Hospital, Hampstead, London. All 47 samples analysed were confirmed to be WHO grade IV glioblastoma. We have worked solely on anonymously samples. Study of these has been approved by Oxfordshire REC B 05/Q1605/66.

DNA extraction

DNA was extracted from paraffin embedded samples using DNeasy Blood and Tissue from Quiagen[R] (Alameda, CA), following the manufacturer's instructions.

Sequencing analysis

Mutation screening of each gene was performed by direct sequencing of genomic DNA in forward and reverse orientations using the Applied Biosystems BigDye terminator reaction kit and the AB 3730xl sequencing machine (Applied Biosystems, Foster City, CA). Primer sequences were designed to encompass the coding region and splice sites of exon 4 of *IDH1* and *IDH2*, and all exons of *IDH3A* (RefSeq: NM_005530.2), *IDH3B* (RefSeq: NM_174855.1), *IDH3G* (RefSeq: NM_004135.2 for the transcript variant 1 and RefSeq: NM_174869.1 for the transcript variant 2), *L2HGH* (RefSeq: NM_024884.2) and *D2HGDH* (RefSeq: NM_024884.2). Primer sequences and PCR conditions are available on request.

Immunohistochemistry

Formalin-fixed, paraffin-embedded tissue sections (4 μm) were de-waxed in xylene and rehydrated through graded alcohols to water. Endogenous peroxidase was blocked using 1.6% H_2O_2 for 20 minutes. For antigen retrieval, sections were pressure cooked in 10 mmol/L citrate buffer (pH6.0) for 5 minutes. Sections were blocked with 10% serum for 30 minutes. Slides were incubated with primary polyclonal antibodies anti human D2HGDH (ProteinTech Group, Chicago, IL 60612, 1:50), or anti human L2HGDH (ProteinTech Group, Chicago, IL 60612, 1:100) for 1 hour. Goat anti-rabbit secondary antibody was applied for one hour at room temperature. Sections were then incubated in ABC (Vector labs) for 30 minutes. DAB solution was applied for 2–5 minutes and development of the colour reaction was monitored microscopically. Slides were counterstained with haematoxylin, dehydrated, cleared and then mounted. Images were taken at 20× magnification.

Author Contributions

Conceived and designed the experiments: CB PM IT. Performed the experiments: DK MA. Analyzed the data: DK MA. Contributed reagents/materials/analysis tools: MG. Wrote the paper: DK MA CB.

References

1. Ohgaki H (2009) Epidemiology of brain tumours. Methods Mol Biol 472: 323–342.
2. Wen PY, Kesari S (2008) Malignant gliomas in adults. N Engl J Med 359: 492–507.
3. Xu X, Zhao J, Xu Z, Peng B, Huang Q, et al. (2004) Structures of human cytosolic NADP-dependent isocitrate dehydrogenase reveal a novel self-regulatory mechanism of activity. J Biol Chem 279: 33946–33957.
4. Reitman ZJ, Yan H (2010) Isocitrate dehydrogenase 1 and 2 mutations in cancer: alterations at a crossroads of cellular metabolism. J Natl Cancer Inst 102: 932–941.
5. Haselbeck RJ, McAlister-Henn L (1993) Function and expression of yeast mitochondrial NAD- and NADP-specific isocitrate dehydrogenases. J Biol Chem 268: 12116–12122.
6. Joseph JW, Jensen MV, Ilkayeva O, Palmieri F, Alarcon C, et al. (2006) The mitochondrial citrate/isocitrate carrier plays a regulatory role in glucose-stimulated insulin secretion. J Biol Chem 281: 35624–35632.
7. Minard KI, McAlister-Henn L (1999) Dependence of peroxisomal beta-oxidation on cytosolic sources of NADPH. J Biol Chem 274: 3402–3406.
8. Lee SM, Park SY, Shin SW, Kil IS, Yang ES, et al. (2009) Silencing of cytosolic NADP(+)-dependent isocitrate dehydrogenase by small interfering RNA enhances the sensitivity of HeLa cells toward staurosporine. Free Radic Res 43: 165–173.
9. Lee SH, Jo SH, Lee SM, Koh HJ, Song H, et al. (2004) Role of NADP+-dependent isocitrate dehydrogenase (NADP+-ICDH) on cellular defence against oxidative injury by gamma-rays. Int J Radiat Biol 80: 635–642.
10. Kil IS, Kim SY, Lee SJ, Park JW (2007) Small interfering RNA-mediated silencing of mitochondrial NADP+-dependent isocitrate dehydrogenase enhances the sensitivity of HeLa cells toward tumour necrosis factor-alpha and anticancer drugs. Free Radic Biol Med 43: 1197–1207.
11. Yang ES, Park JW (2009) Regulation of ethanol-induced toxicity by mitochondrial NADP(+)-dependent isocitrate dehydrogenase. Biochimie 91: 1020–1028.
12. Parsons DW, Jones S, Zhang X, Lin JC, Leary RJ, et al. (2008) An integrated genomic analysis of human glioblastoma multiforme. Science 321: 1807–1812.
13. Yan H, Parsons DW, Jin G, McLendon R, Rasheed BA, et al. (2009) IDH1 and IDH2 mutations in gliomas. N Engl J Med 360: 765–773.
14. Kloosterhof NK, Bralten LB, Dubbink HJ, French PJ, van den Bent MJ (2010) Isocitrate dehydrogenase-1 mutations: a fundamentally new understanding of diffuse glioma? Lancet Oncol.
15. Hartmann C, Meyer J, Balss J, Capper D, Mueller W, et al. (2009) Type and frequency of IDH1 and IDH2 mutations are related to astrocytic and oligodendroglial differentiation and age: a study of 1,010 diffuse gliomas. Acta Neuropathol 118: 469–474.
16. Ward PS, Patel J, Wise DR, Abdel-Wahab O, Bennett BD, et al. (2010) The common feature of leukemia-associated IDH1 and IDH2 mutations is a neomorphic enzyme activity converting alpha-ketoglutarate to 2-hydroxyglutarate. Cancer Cell 17: 225–234.
17. Zhao S, Lin Y, Xu W, Jiang W, Zha Z, et al. (2009) Glioma-derived mutations in IDH1 dominantly inhibit IDH1 catalytic activity and induce HIF-1alpha. Science 324: 261–265.
18. Ichimura K, Pearson DM, Kocialkowski S, Backlund LM, Chan R, et al. (2009) IDH1 mutations are present in the majority of common adult gliomas but rare in primary glioblastomas. Neuro Oncol 11: 341–347.
19. Dang L, White DW, Gross S, Bennett BD, Bittinger MA, et al. (2009) Cancer-associated IDH1 mutations produce 2-hydroxyglutarate. Nature 462: 739–744.
20. Gross S, Cairns RA, Minden MD, Driggers EM, Bittinger MA, et al. (2010) Cancer-associated metabolite 2-hydroxyglutarate accumulates in acute myelogenous leukemia with isocitrate dehydrogenase 1 and 2 mutations. J Exp Med 207: 339–344.
21. Figueroa ME, Abdel-Wahab O, Lu C, Ward PS, Patel J, et al. (2010) Leukemic IDH1 and IDH2 mutations result in a hypermethylation phenotype, disrupt TET2 function, and impair hematopoietic differentiation. Cancer Cell 18: 553–567.
22. Xu W, Yang H, Liu Y, Yang Y, Wang P, et al. (2011) Oncometabolite 2-Hydroxyglutarate Is a Competitive Inhibitor of alpha-Ketoglutarate-Dependent Dioxygenases. Cancer Cell 19: 17–30.
23. Kranendijk M, Struys EA, van Schaftingen E, Gibson KM, Kanhai WA, et al. (2010) IDH2 mutations in patients with D-2-hydroxyglutaric aciduria. Science 330: 336.
24. Dang L, Jin S, Su SM (2010) IDH mutations in glioma and acute myeloid leukemia. Trends Mol Med 16: 387–397.

12

QM/MM Molecular Dynamics Study of the Galactopyranose → Galactofuranose Reaction Catalysed by *Trypanosoma cruzi* UDP-Galactopyranose Mutase

Gustavo Pierdominici-Sottile[1]*, Rodrigo Cossio Pérez[1], Johan F. Galindo[2], Juliana Palma[1]

1 Departamento de Ciencia y Tecnología, Universidad Nacional de Quilmes, Bernal, Argentina, 2 Quantum and Computational Chemistry Group, Departamento de Química, Universidad Nacional de Colombia, Bogotá, Colombia

Abstract

The enzyme UDP-Galactopyranose Mutase (UGM) catalyses the conversion of galactopyranose into galactofuranose. It is known to be critical for the survival and proliferation of several pathogenic agents, both prokaryotic and eukaryotic. Among them is *Trypanosoma cruzi*, the parasite responsible for Chagas' disease. Since the enzyme is not present in mammals, it appears as a promising target for the design of drugs to treat this illness. A precise knowledge of the mechanism of the catalysed reaction would be crucial to assist in such design. In this article we present a detailed study of all the putative steps of the mechanism. The study is based on QM/MM free energy calculations along properly selected reaction coordinates, and on the analysis of the main structural changes and interactions taking place at every step. The results are discussed in connection with the experimental evidence and previous theoretical studies.

Editor: Emily Parker, University of Cantebury, New Zealand

Funding: This work was mainly supported by CONICET and Universidad de Quilmes. A part of this work was supported by the grant pict 2010-#2597 of the Agencia Nacional de Promocion Cientifica y Tecnologica for the University of Quilmes. The funders had no role in study design, data collection and analysis, decision to publish, or preparation of the manuscript.

Competing Interests: The authors have declared that no competing interests exist.

* Email: gsottile@unq.edu.ar

Introduction

Chagas' disease, also known as American trypanosomiasis, affects approximately 8 million people worldwide. It is endemic in Latin America but in the last decades it has also spread towards North America and Europe [1]. Its pathogenic agent is the flagellate protozoan *Trypanosoma cruzi* (*T. cruzi*), which is transmitted to humans by the faeces of triatomine insects. The disease was first described by Dr. Carlos Chagas in Brazil in 1909. Despite this early discovery there are still no drugs capable of curing it. Nifurtimox and Benznidazole are used in the acute phase of the disease. However none of them are efficient and both have strong side effects [2–4]. Most patients discontinue the treatment when the side effects become too severe. For these reasons new and more efficient drugs are needed.

Galactose is a common monosaccharide. In mammals it is exclusively found as galactopyranose (Gal*p*), the six-membered ring hemiacetal form. On the other hand, in *T. cruzi* and many other human pathogens such as *Mycobacterium tuberculosis*, *Escherichia coli*, *Leishmania major*, *Aspergillus fumigatus*, *Salmonella typhimurium* and *Klebsiella pneumoniae* [5–8], it is found as galactofuranose (Gal*f*), the five-membered ring hemiacetal form [5, 6, 8–17]. The sole source of Gal*f* in these species is the enzyme UDP-Galactopryranose Mutase (UGM), which catalyses the isomerization between UDP-Gal*p* and UDP-Gal*f*, the precursor of Gal*f* [18, 19]. It is known that Gal*f* is an essential component of the cell wall and the extracellular matrix of these pathogens [13, 19]. Suppression of the UGM gene in many of them caused attenuated virulence and increased sensitivity to drugs [20–22]. In *T. cruzi*, particularly, Gal*f* is attached to glycoinositolphospholipids and glycosylphosphatidylinositol anchor proteins [23, 24], which are highly expressed throughout the life cycle of the parasite and are essential for its survival and proliferation [25–27]. When *T. cruzi* is incubated with specific antibodies against Gal*f*, the binding of the parasite to the mammalian cells is blocked, leading to an 80% decrease in infectivity [13]. Since neither Gal*f* nor UGM have ever been found in mammals, UGM has gathered significant interest as a target for drugs design [28]. Due to this interest, it has been subjected to several structural and mechanistic studies [28–30].

In 2001 was presented the first known crystallographic structure of a UGM. It corresponded to *E. coli*, [31]. After that, other bacterial structures were also determined [31–36]. Eukaryotic UGMs received less attention. The first structure of that kind, corresponding to *Aspargillus fumigatus*, was published in 2012 [37]. Shortly after, the one of *T. cruzi* (*Tc*UGM) became also available [38]. The comparison between eukaryotic and prokaryotic UGMs revealed that they share a common folding and a GxGxxG motif, necessary to bind the cofactor, flavin adenine dinucleotide (FAD) [39]. Moreover, the cofactor conformation and its interaction with the enzyme environment is highly conserved in both groups. However, the interactions with the substrate differ significantly and the sequence identity is pretty low

(15%) [38]. In the active site, only 5 out of 13 residues are shared. Besides eukaryotic UGMs are approximately 100 residues longer than prokaryotic ones. This additional part of the chain forms extra secondary structures, modifying the active site flexibility and the oligomerization state of the enzyme [39].

Fig. 1 shows the main species of the catalysed reaction. The transformations between these species we will be denoted as "stages" of the mechanism. The first and last stages consist of just one reaction step while the second and third stages involve two. All the steps of the mechanism under analysis are presented in Fig. 2. According to different experimental studies the reaction initiates with the formation of a flavin-galactose adduct (conversion from I to II in Fig. 1) [28, 34, 40, 41]. This requires the rupture of the Gal*p*-UDP bond and the creation of a bond between Gal*p* and the nitrogen at position 5 of the reduced flavin adenine dinucleotide (FADH), $N5_{FADH}$ [40, 42–44].

It was experimentally found that no conversion between Gal*p* and Gal*f* occurred when the native cofactor was replaced by 5-deaza-FAD [45]. Since this modified cofactor can only participate in two-electron transfers, it was argued that the mechanism in UGM should involved a one electron transfer. In particular, it was suggested that an oxocarbenium ion was first formed, followed by a single electron transfer, and that the recombination of the radicals so formed would produce the flavin-galactose adduct. However, it was then argued that the evidence presented does not exclude the possibility of a nucleophilic attack of $N5_{FADH}$ onto the anomeric C of Gal*p*, $C1X_{GAL}$, with a S_N2 type mechanism [46]. Positional isotope effects experiments, together with studies that employed FAD analogues with different electron density on $N5_{FADH}$, uphold this hypothesis [46]. Besides, the analysis of the crystallographic structures, as well as recent investigations on *Tc*UGM, give further support to this mechanism [28, 34, 40, 43]. The next stage (conversion from II to III in Fig. 1), involves the opening of the ring to form an iminium ion [40, 43]. This intermediate species has been trapped using $NaCNBH_3$ in two independent studies [28, 40]. Naively, one would suggest that the iminium is formed by a direct proton transfer from $N5_{FADH}$ to the cyclic oxygen of galactose, $O5X_{GAL}$. However, as noted by Huang *et. al.*, such transference involves the passage through a four-membered ring structure which is rather high in energy. As an alternative, the same authors proposed that the proton is first passed from $N5_{FADH}$ to $O4_{FADH}$, and then transferred to Gal*p* to initiate the opening of the ring [41]. Once the iminium intermediate is formed, two stages are needed to complete the reaction. They can be considered as the reverse of the two previous stages, except for the fact that galactose is now in the furanose form. Thus, stage three involves the sugar ring closure to form Gal*f* (conversion from III to IV in Fig 1). Sobrado *et. al.* indicated that this is the stage that determines the rate of the whole process [28]. Stage four consists of the breaking of the flavin-substrate bond along with the binding of UDP to the sugar (conversion from IV to V in Fig. 1).

Huang *et. al.* performed a theoretical study on the mechanism of the reaction catalysed by UGM [41]. They carried out electronic structure computations on active site models built from the PDB structure of *Klebsiella pneumoniae* UGM (*Kp*UGM). The largest of their models contained 26 active site residues plus the substrate, the cofactor and several crystallographic water molecules. A quantum mechanics-molecular mechanics level theory (QM/MM) was employed to characterize the structures of reactants, products, intermediate species and transitions states appearing in the mechanism. More recently, the involvement of several active site residues on the catalytic activity of *Tc*UGM was evaluated through site directed mutageneis experiments [29].

In this article we present a QM/MM molecular dynamics study of the reaction catalysed by *Tc*UGM. We applied the umbrella sampling technique to obtain the free energy profiles along different reaction coordinates, conveniently defined to describe every step of the mechanism. QM/MM free energy computations have become a widely employed tool to gain information on the atomistic details of enzymatic reactions. One of their main assets is the ability to reveal both, energetic and dynamical contributions to catalysis. We also analysed the most significant conformational changes and interactions taking place at each step. This includes the monitoring of bond distances, dihedral angles, H-bonds, partial charges, bond orders as well as the Cremer-Pople angles that describe the conformations of the pyranose and furanose rings [47]. Finally, we implemented an energy decomposition method to evaluate the contribution of the active site residues to the lowering of the barriers at every step. The results of the simulations are discussed in connection with previous experimental findings, as well as with the theoretical analysis of Huang *et. al.*

Results and Discussion

In Fig. 3 we present a sketch of the free energy changes (ΔG_r^0) and free energy barriers ($\Delta G_r^\ddagger$) for the successive steps of the mechanism presented in Fig. 2. The profile shows that the barrier for the ring opening (step 3) is sensibly smaller than that of the ring closure (step 4). In fact, the barrier for step 4 is the highest. This is in agreement with the experimental findings of Sobrado *et. al.* [28]. The profile also indicates that products are more stable than reactants. The same result was found in the computations of Huang *et. al.* [41]. For the reverse reaction the largest barrier corresponds to the tautomerization of FADH. We also note that for both, forward and backward reactions, the appearance of the iminium ion species presents a small barrier.

In the following sections we describe in detail the outcome of the QM/MM computations for all the stages of the catalysed reaction. When pertinent, the results are compared with those recently reported for *Kp*UGM [41]. We note, however, that a meaningful

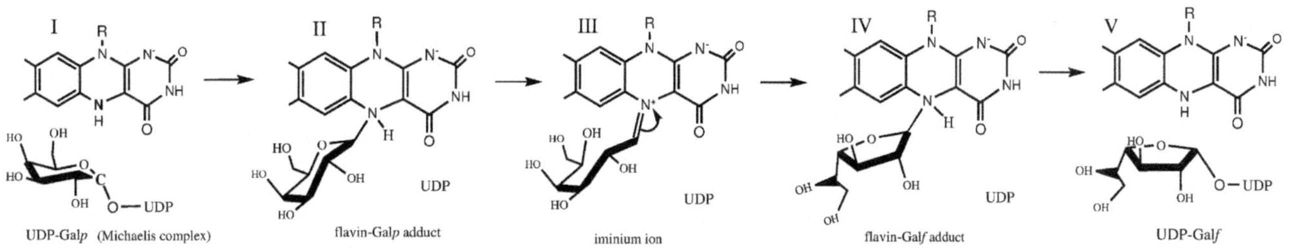

Figure 1. Main species of the reaction mechanism.

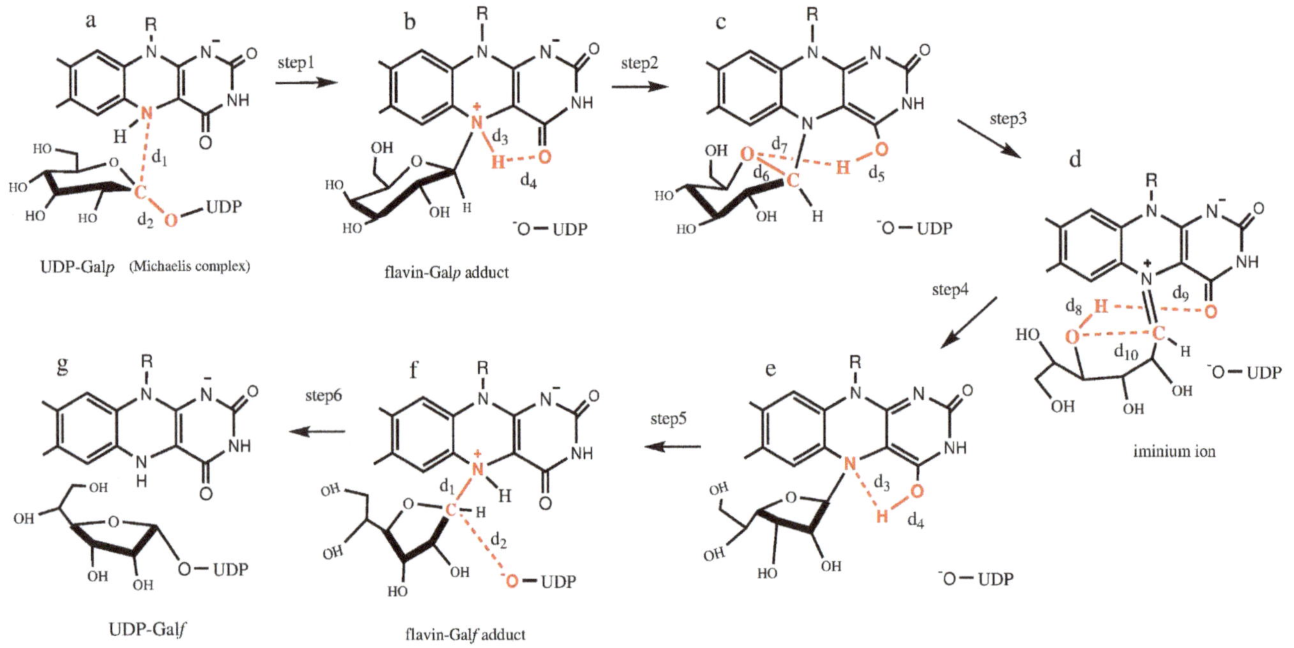

Figure 2. Detailed mechanism for the reaction catalysed by *Tc*UGM. The mechanism includes the intermediates detected by experiments as well as those whose existence was inferred from theoretical considerations. Red color is used to denote the bonds being broken (solid line) or formed (dashed line), as well as the atoms involved. The distances between these atoms are labelled because they are used to define the reaction coordinates.

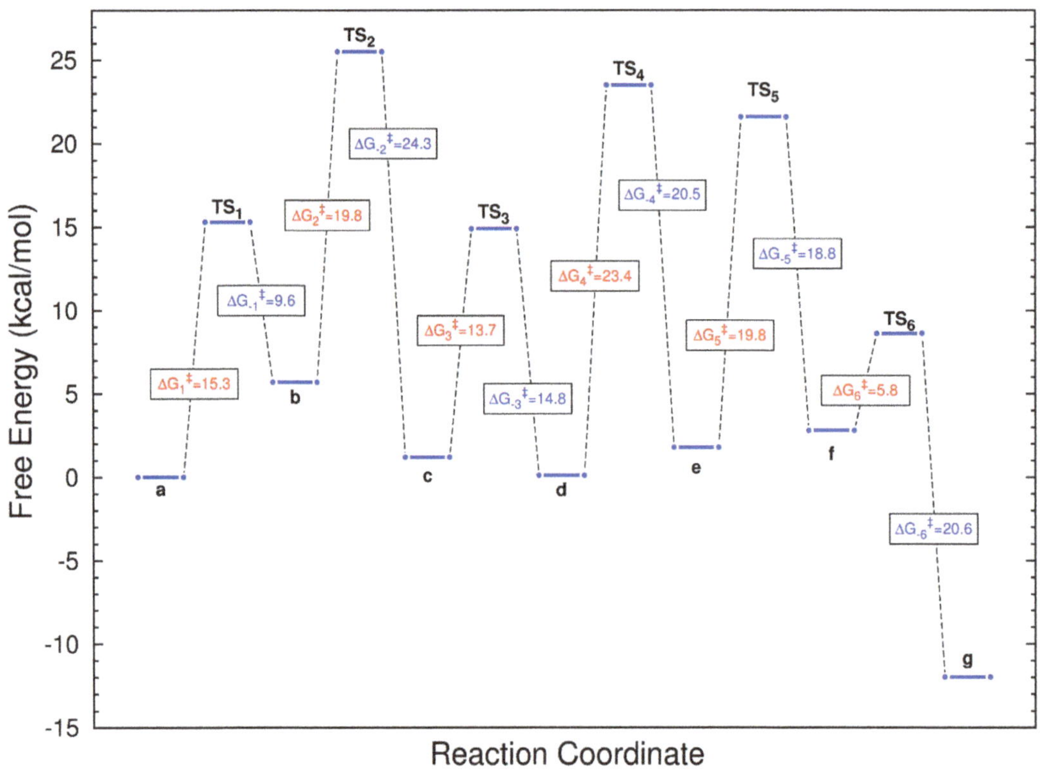

Figure 3. Sketch of the free energy profile for the whole mechanism. Activation free energies ($\Delta G_r^{\ddagger}$) for all the steps of the mechanism under analysis. Data for the direct reaction are shown in red while those of the reverse reaction are shown in blue. Letters a to g denote the species presented in Fig. 2. The highest point of the profile corresponds to the transition state for the tautomerization of FADH (conversion from b to c of Fig. 2). The largest barrier corresponds to the ring closure into the furanose form (conversion from d to e of Fig. 2).

comparison between these computations requires keeping in mind the aspects in which they differ. Among the differences we have: (1) that *Kp*UGM and *Tc*UGM bear low sequence homology: 18.3% for the whole protein and only 5 out of 13 residues for the active site [39]; (2) that free energy computations include dynamical effects that are not considered in electronic structure computations; (3) that residue His62 was protonated in the present work but was set as neutral in the work of Huang *et. al.*; (4) that we modelled the whole *Tc*UGM crystal structure in explicit solvent while Huang *et. al.* considered an active site model consisting of the cofactor, the substrate, 26 active site residues and 8 water molecules; (5) that the quantum subsystems and the DFT levels of theory employed in each computation were different.

In what follows the numbers of the residues and the names of the atoms correspond to the crystal structure of *Tc*UGM taken from the Protein Data Bank (PDB code 4DSH). Figure 2 should serve as a guide for the reading of the following sections. It describes every step of the mechanism, highlighting the distances involved in the definition of the corresponding reaction coordinates. On the other hand, at the supplementary information section, we provide pdb files with the average structures of reactants, products and all the intermediates of the reaction (Text S1 to Text S7). These structures can be used to obtain information that was not included in the main text in order to keep the article at a reasonable length. For the same reason, several pictures depicting the evolution of important distances and angles along the different steps of the reaction are given at the supplementary information section.

Stage 1: Formation of the flavin-Gal*p* adduct

This stage consists of just one concerted step in which $N5_{FADH}$ bonds covalently to the anomeric C of Gal*p* while the UDP moiety detaches from it. In panel a) of Figure S1 the evolution of the distances involved in the definition of the reaction coordinate ($z_1 = d_2 - d_1$) along this step, are shown. At the transition state, TS1, we obtained $d_1 = 2.52 \pm 0.05$ Å and $d_2 = 2.15 \pm 0.05$ Å. This corresponds to bond orders of 0.30 and 0.19 for the $C1X_{GAL}$-$O3B_{UDP}$ and $C1X_{GAL}$-$N5_{FADH}$ bonds, respectively. These orders support the hypothesis that the reaction proceeds via a dissociative S_N2 mechanism, as has been suggested by several independent experimental studies [28, 34, 46]. The calculated $\Delta G_r^{\ddagger}$ and ΔG_r^0 are 15.3 ± 0.3 kcal/mol and 5.7 ± 0.2 kcal/mol, respectively. These values qualitatively agree with those of Huang *et. al.* who found $\Delta E^{\ddagger} = 18.9$ kcal/mol and $\Delta E = 9.63$ kcal/mol for the same process. The partial charges of key atoms for reactants (Michaelis complex), TS1 and products (flavin-Gal*p* adduct) are presented in Table 1. It is observed that the cyclic oxygen, $O5X_{GAL}$, losses considerable electron density in going from reactants to TS1, but it partially recovers it when the adduct is finally reached. $C1X_{GAL}$, on the other hand, gains substantial electron density along the whole process. Finally, the partial charge of $N5_{FADH}$ increases from −0.18 to 0.14 while its configuration changes from planar to tetrahedral. We note that the substantial loss of electron density of the nucleophile nitrogen in this step was predicted by the experiments in which FAD analogues with different electron-withdrawing/donating groups where used to determine the S_N2 character of this step [46]. These changes weaken the $N5_{FADH}$-H bond facilitating the transference of the proton during the next step. The evolution of the Cremer-Pople angles is shown in Fig. S1 panel b). At the Michaelis complex, the pyranose ring shows a $^4C_1$ conformation and its θ angle oscillates around 0^0. However, after surmounting the transition state, this angle increases by $\sim 50^0$ while ϕ diminishes to -60^0. This correspond to a $^2H_3$ six membered ring conforma-

Table 1. Partial charges of key atoms for the first two steps of the mechanism.

Atom	a	TS1	b	TS2	c	TS3	d
$O3B_{UDP}$	−0.45(0.02)	−0.65(0.02)	−0.76(0.02)	−0.77(0.03)	−0.75(0.02)	−0.76(0.02)	−0.78(0.03)
$C1X_{GAL}$	0.40(0.01)	0.31(0.02)	0.18(0.04)	0.26(0.01)	0.25(0.01)	0.01(0.03)	−0.07(0.05)
$O5X_{GAL}$	−0.40(0.01)	−0.23(0.02)	−0.31(0.05)	−0.36(0.02)	−0.41(0.01)	−0.42(0.04)	−0.47(0.02)
$N5_{FADH}$	−0.18(0.02)	−0.21(0.01)	0.05(0.03)	−0.11(0.02)	−0.13(0.02)	0.12(0.04)	0.20(0.03)
H_{FADH}	0.25(0.01)	0.25(0.01)	0.24(0.01)	0.26(0.04)	0.37(0.01)	0.37(0.03)	0.33(0.02)
$O4_{FADH}$	−0.72(0.02)	−0.69(0.02)	−0.43(0.02)	−0.61(0.03)	−0.44(0.02)	−0.57(0.04)	−0.66(0.03)
$C4X_{FADH}$	−0.15(0.01)	−0.18(0.02)	−0.27(0.03)	−0.22(0.03)	−0.14(0.02)	−0.25(0.04)	−0.26(0.02)
$C4_{FADH}$	0.36(0.021)	0.40(0.01)	0.33(0.02)	0.43(0.02)	0.33(0.01)	0.46(0.03)	0.48(0.04)
$C5X_{FADH}$	0.10(0.01)	0.08(0.01)	0.00(0.02)	0.08(0.02)	0.09(0.01)	0.05(0.02)	0.02(0.03)

Charges are in atomic units. Labels a, b and c denote the species presented in Fig. 2. Numbers within parenthesis indicate the standard deviation of the charges.

tion. This conformation persists until the formation of the flavin-galactose adduct, when the pyranose ring returns to $^4C_1$.

The phosphate group of UDP bears strong H-bond interactions with Tyr395 and Tyr429 during the whole step. It also forms a H-bond with Arg327 but, at the Michaelis complex, this interaction is rather weak. However, once the covalent bond between UDP and the sugar is broken, the interaction gains strength because of the negative charge acquired by the reactive oxygen of the phosphate (see Table 1). Thus, while only 32.3% of the structures sampled before TS1 present a H-bond between Arg327 and the phosphate, the percentage raises to 69.7% for those sampled between TS1 and products. This indicates that Arg327 plays an important role in stabilizing TS1, as well as the products of the current step. Further support for this conclusion comes from Table 2 which shows that Arg327 has the most negative $\Delta E^i_{R \to TS}$ value. The location of this arginine within the active site can be seen in Figure 4. Arg327 is conserved in UGMs of both, prokaryotic and eukaryotic organisms [39]. For *Kp*UGM, it was found that its substitution by Ala completely abolishes the enzyme activity [48]. For *Tc*UGM the same substitution was found to reduce the catalytic activity, measured by k_{cat}/K_M, in 50% [29]. This effect is due to a reduction in k_{cat} since the mutated enzyme has a larger affinity by the substrate, as indicated by the decrease in K_M [29]. Our results agree with these experimental findings and suggest that the effect could be due to the role played by Arg327 in the catalysis of the flavin-galactose adduct formation. Finally, we note that the H-bonds between the phosphate and Tyr395, Tyr429 and Arg327 were found to be present in all of the remaining steps. Thus, they restrain the mobility of the phosphate group so that it is ready to re-bind the sugar moiety once the furanose ring is formed. This fact could explain the detrimental effect on k_{cat} observed when Tyr395 and Tyr429 are substituted by Ala [29], and could also contribute to the negative effect observed upon the substitution of Arg327 with Ala.

Another interaction that is worth mentioning is the H-bond between the H atom bonded to $N5_{FADH}$ and the carbonyl oxygen of Gly61. This interaction already exists at the reactants configuration but becomes stronger once TS1 is reached. The distance between the carbonyl oxygen of Gly61 and the H atom is 2.30 $\pm$ 0.39 Å before TS1, but decreases to 1.89 $\pm$ 0.11 Å after it (see Fig. S1 panel b). Because of this, the H-bond is present in

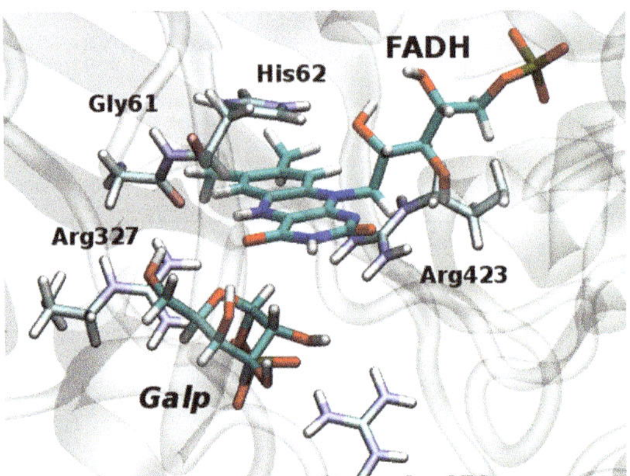

Figure 4. *Tc*UGM active site. Sketch of the active site of *Tc*UGM showing the relative positions of the FADH cofactor, the Gal*p* moiety of the substrate plus residues Gly61, His62, Arg176, Arg327 and Arg423.

26.4% of the configurations sampled before TS1 and 59.2% of those sampled afterwards. Since the interaction is stronger for TS1 than for reactants, it certainly helps to reduce the barrier to reaction. Unfortunately, the stabilizing effect of glycine residues cannot be evaluated with the energy decomposition method employed in this work. Because of this, Gly61 is not mentioned in Table 2. It has been found that the replacement of Gly61 with Ala or Pro has a profound detrimental effect on the activity of *Tc*UGM (~90%) [38]. A putative explanation for this fact would be that these alternative residues reduce the flexibility of the backbone chain, hindering its ability to locate the carbonyl group in an appropriate position for the H-bond interaction. However, the evaluation of this hypothesis requires further MD computations that are outside the reach of this work.

Stage 2: Formation of the iminium ion

This stage consists of two steps. The first one involves an intramolecular proton transfer from $N5_{FADH}$ to $O4_{FADH}$ (see Fig. 2, step 2). The calculated $\Delta G^{\ddagger}_r$ was 19.8 $\pm$ 0.2 kcal/mol while ΔG^0_r was -4.5 ± 0.3 kcal/mol. The corresponding results of Huang *et. al.* for $\Delta E^{\ddagger}_r$ and ΔE^0_r were 8.22 kcal/mol and -10.42 kcal/mol, respectively. Both calculations agree to indicate that the products of this step are more stable than the reactants. However, the energy difference is smaller in our computations and our estimated barrier is sensibly higher. The large barrier is not surprising since the initial distance between the H atom being transferred and the acceptor is rather long, 2.46 $\pm$ 0.8 Å, while the donor-hydrogen-acceptor angle is far from collinear, 98.7 $\pm$ 7.1 ^{0}. Because of this, a large configurational change needs to take place to enable the transference. The probability of H-tunneling for this step was found to be negligibly, with a transmission coefficient of 0.002 calculated at the zero point energy of the N-H bond. This results is not surprising since the barrier width is rather large.

In Fig. S2a we show the variations of the distances involved in the definition of the reaction coordinate $z_2 = d_3 - d_4$ during step 2. At the transition state TS2 the distance between $O4_{FADH}$ and the H atom has decreased by ~1.3 Å but the $N5_{FADH}$-H distance has not increased substantially. On the other hand the donor-acceptor distance, presented in panel (b), decreases by 0.36 Å in going from reactants to TS2. The analysis of the structures reveals that this shortening is mainly caused by the twisting of the C5X-N5-C4X-C4 torsion of FADH, also shown panel (b). This observation is confirmed by a correlation coefficient of 0.87 between the donor-acceptor distance and the corresponding dihedral angle. Thus, the ability of the isoalloxazine ring to bear this distortion contributes to make possible the transfer. In addition, Fig. S2b shows the distance between the H atom being transferred and $O5X_{GAL}$. This distance reaches a minimum of 1.5 Å when the transference is complete generating a new H-bond between $O4_{FADH}$ and $O5X_{GAL}$. As can be seen in Table 1, the partial charge of $O5X_{GAL}$ decreases along this step while that of the H atom increases. Both, the new H-bond and the variations in the partial charges make the scenario more prone to the attack of the proton onto $O5X_{GAL}$ that initiates the ring opening at the following step. The partial charges of the $N5_{FADH}$, $C4X_{FADH}$, $C5X_{FADH}$, $C4_{FADH}$ and $O4_{FADH}$ also suffer noticeable variations along this step. Basically, the donor and the acceptor of the proton gain electronic density in going from reactants to TS2, while $C4X_{FADH}$, $C5X_{FADH}$ and $C4_{FADH}$ loss it.

Residue His62 lies pretty close to the isoalloxazine ring of FADH (See Fig. 4). They attract to each other since His62 has a positive charge while FADH has a negative charge. Along step 2 His62 moves closer to FADH and its side chain rotates so that the

Table 2. $\Delta E^{i}_{R\rightarrow TS}$ values for key residues at every step of the reaction mechanism shown in Fig. 2.

Residue	ΔΔE step$_1$ (kcal/mol)	ΔΔE step$_2$ (kcal/mol)	ΔΔE step$_3$ (kcal/mol)	ΔΔE step$_4$ (kcal/mol)	ΔΔE step$_5$ (kcal/mol)	ΔΔE step$_6$ (kcal/mol)
His62	0.74 (1.42)	−16.35 (0.77)	2.09 (1.13)	0.50 (0.90)	5.98 (1.54)	−3.37 (1.39)
Val63	0.01 (0.05)	−0.15 (0.04)	−0.05 (0.05)	−0.20 (0.05)	0.10 (0.04)	0.03 (0.05)
Arg176	−2.12 (0.80)	−1.76 (0.84)	3.47 (1.26)	1.06 (1.29)	−8.97 (2.04)	5.36 (1.24)
Asn201	−2.32 (0.45)	−1.37 (0.46)	0.12 (0.63)	1.58 (0.48)	−6.40 (0.61)	2.40 (0.94)
Tyr317	−0.19 (0.61)	3.18 (0.64)	−1.05 (0.63)	−0.49 (0.98)	−0.97 (0.54)	−0.82 (0.59)
Arg327	−16.15 (1.92)	4.72 (2.00)	−1.19 (1.75)	5.31 (1.82)	3.67 (2.39)	−8.10 (1.54)
Tyr395	−4.29 (0.68)	3.81 (0.51)	−0.41 (0.71)	−1.23 (1.04)	1.76 (0.84)	−2.27 (1.63)
Arg423	2.78 (0.65)	−3.18 (1.21)	5.01 (0.88)	2.86 (0.88)	6.53 (1.51)	1.00 (1.60)
Tyr429	−2.35 (0.65)	1.28 (0.93)	0.45 (0.96)	1.31 (1.06)	−1.15 (0.80)	0.60 (0.72)
Asn433	0.95 (0.42)	−1.26 (0.69)	−1.47 (0.41)	1.88 (0.58)	−1.23 (0.47)	−0.30 (0.87)

Numbers within parenthesis indicate the standard deviation of the energies.

NE2 atom establishes a new H-bond with the hydroxyl group at position 2 of the ribitol chain of the cofactor. This movement strengthens the interaction between the two molecules. The importance of this variation can be appreciated from the values of $\Delta E^{i}_{R\rightarrow TS}$ presented in Table 2. It is observed that His62 exerts the largest stabilizing effect on TS2, −16.35 kcal/mol, being the main responsible for the acceleration of this step. The involvement of His62 in the catalysis of step 2 could explain why the substitution of this residue by Ala diminishes the catalytic activity of *Tc*UGM in 98% [38].

The following step (step 3 of Fig. 2) involves the transfer of the H atom from $O4_{FADH}$ to $O5X_{GAL}$ to open the sugar ring and form the iminium ion. The calculated $\Delta G^{\ddagger}_{r}$ and ΔG^{0}_{r} for this process are 13.7 ± 0.2 and −1.1 ± 0.3 kcal/mol, respectively. Experimental results suggested that the opening of the ring, detected as the appearance of the iminium ion, is fast. [28] In agreement with the experimental evidence we found that this step presents a relatively small barrier. Huang *et. al.* estimated an energy barrier rather similar to our $\Delta G^{\ddagger}_{r}$, but their ΔE^{0}_{r} (8.12 kcal/mol) is quite distinct to our ΔG^{0}_{r} estimation. Fig. S3 shows the evolution of the distances involved in the definition of the reaction coordinate $z_3 = d_5 + d_6 - d_7$. It is observed that at TS3 the proton is halfway between the donor and the acceptor, while the bond between $O5X_{GAL}$ and $C1X_{GAL}$ is partially broken. When this bond gets completely broken and the ring opens, the iminium ion is formed. During this process the distance between $C1X_{GAL}$ and $N5_{FADH}$ decreases from 1.50 ± 0.04 Å to 1.34 ± 0.03 Å because the bond between them acquires a partial double bond character. In agreement with experimental findings [46], it was found that $N5_{FADH}$ losses considerably electron density during this step (See Table 1). Besides two out the three atoms bound to this nitrogen ($C1X_{GAL}$ and $C4X_{FADH}$) gain electron density. The analysis of the TS3 stabilization pattern, presented in Table 2, shows that none of the active site residues has a significant influence on the energetic of this step. However, it should be kept in mind that the values of $\Delta E^{i}_{R\rightarrow TS}$ only provide information on static contributions to the interaction energy between a given residue and the quantum subsystem. Any effect of the residue on the conformational freedom of the active site will no be spotted by this analysis.

In that regard, it is interesting to analyse the movement of the hydroxyl groups of the sugar moiety once the chain is open. Fig. S4 shows the probability distributions for the dihedral angles $C2X_{GAL}$-$C3X_{GAL}$-$C4X_{GAL}$-$O4X_{GAL}$, $C3X_{GAL}$-$C4X_{GAL}$-$C5X_{GAL}$-$O5X_{GAL}$ and $C4X_{GAL}$-$C5X_{GAL}$-$C6X_{GAL}$-$O6X_{GAL}$. They describe the ability to rotate of the hydroxyl groups formed by $O4X_{GAL}$, $O5X_{GAL}$ and $O6X_{GAL}$, respectively. The widest distribution corresponds to the hydroxyl group formed by $O6X_{GAL}$, which rotates almost freely. On the contrary, the groups involving $O4X_{GAL}$ and $O5X_{GAL}$ liberate around to their average values. These two atoms participate in the bonds that close the ring in Gal*f* and Gal*p*, respectively. In both cases the impediment to rotate is mainly caused by the strong H-bonds that these hydroxyl groups maintain with $O4_{FADH}$. The $O4X_{GAL}$-$O4_{FADH}$ H-bond is present in 95% of the configurations. The $O5X_{GAL}$-$O4_{FADH}$ H-bond is present in 96%. While the sugar chain remains open, these groups do not participate in any other hydrogen bonding interaction. The remaining hydroxyl groups of the sugar also form H-bonds. $O3X_{GAL}$ interacts with Asn201 and $O2X_{GAL}$ with the phosphate group.

Stage 3: Formation of the flavin-Gal*f* adduct

This stage also has two steps. The first one is the cyclization of the sugar into the furanose form (step 4 of Fig. 2). It occurs accompanied of the proton transfer from $O4X_{GAL}$ to $O4_{FADH}$. We found that this step supports the highest barrier of the whole mechanism: 23.4 ± 0.4 kcal/mol. This agrees with the measurements of Sobrado *et. al.* who determined that this formation of the furanose ring is sensible slower than the opening process *Tc*UGM [28]. The free energy change of the step is 2.9 ± 0.2 kcal/mol.

In Fig. S5 panel a) we show the evolution of the distances involved in the definition of the reaction coordinate $z_4 = d_8 - d_9 - d_{10}$, while panel (b) displays the dihedral angles that determine the orientation of the hydroxyl groups at positions 4 and 5. It is observed that, at the beginning of the process, the two hydroxyl groups change their orientation in a concerted way, while $O4X_{GAL}$ and $C1X_{GAL}$ approximate to each other. These movements destroy the H-bond interaction between $O5X_{GAL}$ and $O4_{FADH}$ and, initially, also drive the H atom to be transferred away from $O4_{FADH}$. However, once the $O4X_{GAL}$-$C1X_{GAL}$ distance gets short enough, a fine tuning in the orientation of the hydroxyl group at position 4 is observed. This reorientation takes its H atom closer to $O4_{FADH}$. Initially, while all these rearrangements take place, the H-$O4X_{GAL}$ bond hardly stretches. Only when the $O4X_{GAL}$-$C1X_{GAL}$ distance gets smaller than 2.3 Å, the H-$O4X_{GAL}$ bond starts to weaken. At the transition state, which appears rather late, the H-O4X$_{GAL}$ and H-O4$_{FADH}$ distances are almost the same. Finally, when the products configuration is reached and the sugar is in the furanose form, O5X$_{GAL}$ and O6X$_{GAL}$ present no H-bond interactions. In contrast, O3X$_{GAL}$ and O2X$_{GAL}$ keep their interactions with Asn201 and the phosphate group, respectively.

The stabilization pattern presented in Table 2 shows that none of the active site residues lowers the energy of TS4 with respect to the reactants in a significant amount. However we note that entropy could play an important role for this step. In general, the conformational freedom of a sugar moiety is larger when the chain is open than when it is closed. Accordingly, it is expected that the cyclization process occurs accompanied by a continuous loss of entropy in the sugar molecule. This would make the reaction slower and the equilibrium position more favourable to the open form. In water solution this effect is compensated by an increase in the entropy of the solvent, but the situation is different within an enzyme. The results of the MD simulation for the open form, discussed at the end of the previous section, suggest that for the reaction under analysis this deleterious entropic effect is ameliorated by the interactions between the carbonyl oxygen of the cofactor and the -OH groups at positions 4 and 5 of the sugar moiety. These interactions attenuate the mobility of the open form and therefore reduce its entropy. Consequently, the entropy change in going from reactants to TS4 or to products is not so adverse. We note, however, that our MD runs are not long enough to allow for an accurate estimation of entropic effects. Thus, the hypothesis put forth in this paragraph needs to be evaluated by additional simulations, especially tailored to that end.

A direct comparison with the results of Huang *et. al.* cannot be done for this step because the authors split the process in three parts. First, a rotation around the C4X$_{GAL}$-C5X$_{GAL}$ bond to take O5X$_{GAL}$ away from O4$_{FADH}$; second a rotation around the C4X$_{GAL}$-C3X$_{GAL}$ bond to place O4X$_{GAL}$ close to C1X$_{GAL}$; third, the attack of O4X$_{GAL}$ onto C1X$_{GAL}$. We found that the computed free energy curve along reaction coordinate z_4 presents no intermediate minimum, indicating that the ring closure takes place in a single concerted step. In other words, the energy minima corresponding to the intermediates detected by Huang *et. al.* do not appear in our free energy computations. This difference could be attributed to dynamical effects which, as explained above, are expected to be large for this cyclization process but are not considered in electronic structure computations.

In the second step of this stage (step 5 of Fig. 2) the H atom is transferred back from O4$_{FADH}$ to N5$_{FADH}$. The calculated $\Delta G_r^{\ddagger}$ and ΔG_r^0 for this process are 19.8 ± 0.4 and 1.0 ± 0.3 kcal/mol, respectively. These results are similar to those of Huang *et. al.* It should also be noted that the barrier and free energy change of this step are pretty similar to those of the reverse of step 2. The result is not surprising since the only difference between them is whether the sugar moiety is in the furanose or pyranose form. The transmission coefficient for H-tunneling for this step is more than fifteen times smaller than that of step 2. This is due to the fact that this step is slightly endothermic while step 2 is exothermic. Nevertheless, in both cases, the probability for H-tunneling is negligibly.

In Fig. S6a we present the evolution of the distances involved in the definition of reaction coordinate $z_5 = d_4 - d_3$, while in panel (b) we show the evolution of the C1X$_{GAL}$-N5$_{FADH}$ distance and the Cremer-Pople angle θ of the furanose ring. The curves for the H-O4$_{FADH}$ and H-N5$_{FADH}$ distances evolve according to the expectations for a direct proton transfer. However, the curve of the C1X$_{GAL}$-N5$_{FADH}$ distance shows an unexpected increase after TS5. The final value, ~1.85 Å, is significantly larger than a typical C-N single bond. Our results thus indicate that the adduct between Gal*f* and FADH becomes rather weak when N5$_{FADH}$ adopts the sp3 hybridization. The enlargement of the C1X$_{GAL}$-N5$_{FADH}$ distance was also described by Huang *et. al.* However, it that case, the final value was somewhat smaller than in our calculations (1.70 Å). In order to check the final distance between C1X$_{GAL}$ and N5$_{FADH}$ we re-simulated the transference using longer simulation times for each window, as well as employing larger QM subsystems. However, we consistently obtained the same result. Moreover, the H transference was simulated applying a restriction on the C1X$_{GAL}$-N5$_{FADH}$ distance, so that it was forced to get values smaller than 1.65 Å. These calculations provided higher $\Delta G_r^{\ddagger}$ and ΔG_r^0 than those obtained without the restriction. Besides, when an unrestricted MD was performed on the products of the restricted transfer, the system spontaneously relaxed to a stable conformation with a C1X$_{GAL}$-N5$_{FADH}$ distance of ~1.85 Å. Fig. S6b indicates that the increase in the C1X$_{GAL}$-N5$_{FADH}$ distance is accompanied with an increase in the Cremer-Pople angle θ. This takes the configuration of the sugar ring from $^2T_3$, for reactants, to E_3 for products. It has to be noted that both, the enlargement of the C1X$_{GAL}$-N5$_{FADH}$ distance and the change in the conformation of the furanose ring, are required to avoid the steric clash between the cofactor and the substrate. The values of $\Delta E^i_{R\to TS}$ presented in Table 2 show that His62 and Arg423 destabilize TS5, but the effect is more than compensated by the stabilization produced by Arg176 and Asn201. Site directed mutagenesis experiments determined that the mutation of Arg176 by Ala produces an impressive reduction in k_{cat} [29]. The involvement of this residue in the catalysis of step 5 could be one of the reasons of this finding.

Stage 4: Formation of UDP-Gal*f*

At first sight, this stage could be considered as the reverse of stage 1, except for the fact that the sugar is now in the funarose form. However, as stated in the previous section, the flavin-Gal*f* bond is already very weak when the transference of the proton from O4$_{FADH}$ to N5$_{FADH}$ is completed. Because of this, the barrier for this step is pretty low, 5.8 ± 0.2 kcal/mol, and ΔG_r^0 is quite negative, −14.8 ± 0.1 kcal/mol. In Fig. S7 we show the evolution of the distances involved in the definition of reaction coordinate $z_6 = d_1 - d_2$. At the transition state, the N5$_{FADH}$ -C1X$_{GAL}$ and C1X$_{GAL}$-O3B$_{UDP}$ distances are both ~ 2.18 Å. This corresponds to a bond order of ~0.32 for the two bonds. As in the case of the flavin-Gal*p* adduct formation, this is consistent with a dissociative S_N2 mechanism. The evolution of the Cremer-Pople angle θ of the furanose ring is also shown in Fig. S7. It changes from 255^0 to 240^0, indicating that the ring conformation returns from E_3 to $^2T_3$. Once the products are formed, the O5X$_{GAL}$ hydroxyl group establishes a new H-bond with the phosphate group. Table 2 shows that the stabilization pattern of this step is similar to that of step 1, with Arg327 being the most stabilizing residue of the TS.

Conclusions

We have presented a detailed description of the energetic and structural changes that take place during the entire mechanism of the reaction catalysed by *Tc*UGM. The results confirm and explain several previous experimental findings, and they also provide new insights on the dynamics of the active site along the reaction.

In agreement with experiments, our results confirm that the first stage of the reaction (formation of the Gal*p*-flavin adduct) proceeds on a single step via a S_N2 dissociative mechanism [28, 34, 46]. Moreover, the computations indicate that Arg327 is the main responsible for the selective stabilization of the TS of this step. This could explain why the substitution of this Arg by Ala reduces the k_{cat} of *Tc*UGM by 69% [29]. The carbonyl oxygen of Gly61 also plays a role in that regard. However, the stabilization energy of this residue could not be quantified.

The second stage of the reaction (formation of the iminium ion) occurs in two steps, as predicted by the electronic structure calculations of Huang *et. al.* First, a proton passes from $N5_{FADH}$ to $O4_{FADH}$. This is followed by the transfer of the proton from $O4_{FADH}$ to $O5X_{GAL}$ that triggers the opening of the ring. The energy decomposition analysis indicated that residue His62 is the main responsible for the catalysis of the first transference. This result could explain why mutations introduced on that position produced a high detrimental effect on the activity of *Tc*UGM. The transfer of the proton from $O4_{FADH}$ to the cyclic oxygen, on the other hand, presents a relatively low barrier and none of the active site residues is particularly relevant to stabilize or destabilize its TS with respect to the reactants.

At the end of stage 2, galactose is in the open-chain form. We analysed, the interactions and dynamics of the hydroxyl groups of the sugar in such situation. We found that the group at position 6 moves freely, without interacting with any active site residue, the one at position 3 forms a H-bond with Asn201 while the one at position 2 forms a H-bond with the phosphate group of UDP. More importantly, the hydroxyl groups at positions 4 and 5 strongly interact with $O4_{FADH}$ via H-bonds. These interactions significantly reduce the conformational freedom of the sugar moiety.

During the closing of the sugar ring to form Gal*f*, the dihedral angles around the $C2X_{GAL}$-$C3X_{GAL}$ and $C3X_{GAL}$-$C4X_{GAL}$ bonds rotate in a concerted way. These rotations take $O5X_{GAL}$ away from $O4_{FADH}$ and locate $O4X_{GAL}$ close to $C1X_{GAL}$. It was found that this step presents the highest free energy barrier of the whole mechanism, in agreement with the proposal of Sobrado *et. al.* [28]. The energy decomposition analysis showed that none of the active site residues is particularly important to reduce the energy of the TS with respect to the reactants, but the MD results discussed in the previous paragraph suggested that entropy could play an important role. In general, the cyclization of a sugar chain occurs with a reduction of its entropy, a fact that hampers the reaction. In this case, by reducing the conformational freedom of the open-chain form, the active site of *Tc*UGM could make the entropy change and the activation entropy of this step less adverse. Unfortunately, the characteristics of our simulations do not allow to quantify this effect. We note, however, that since this step has the largest free energy barrier, any small reduction on that barrier can be significant.

Once Gal*f* is formed, the next step involves the transference of the proton bound to $O4_{FADH}$ towards $N5_{FADH}$. We observed that something unexpected occurs during this process. Once the system has passed over the TS, the furanose ring changes its conformation from $^{2}T_{3}$ to E_3 while the distance between $C1X_{GAL}$ and $N5_{FADH}$ increases to get a final value of ~1.85 Å. The visual inspection of the structures reveals that these modifications are required to avoid the steric clash between the substrate and the cofactor. Huang *et. al.*, who used a different level of theory, different quantum subsystem and different model for the active site, also found a rather long $C1X_{GAL}$-$N5_{FADH}$ distance at the end of this transference. Residues Arg176 and Asn201 make the main contributions to the lowering of the barrier. This role of Arg176 is in line with recent experiments which found that the mutation of this residue by Ala reduce the k_{cat} of *Tc*UGM [29]. During the last step of the reaction, the sugar in the furanose form re-binds to UDP as it detaches from the cofactor. Since the $C1X_{GAL}$-$N5_{FADH}$ bond is already rather weak at the end of the previous step, this last transformation presents a small barrier and a very negative energy change.

Tyr395 and Tyr429 also play an important role in the reaction. Both residues bear strong H-bond interactions with the phosphate group of the cofactor. These bonds are stable throughout the whole catalysed mechanism. Since these interactions are always present, they do not modify the energy of the barriers found along the reaction. Instead, they facilitate the process by keeping the phosphate group at a relatively fixed position, close to the sugar moiety. Thus, UDP is ready to re-bind to the sugar once it adopts the furanose form. Not surprisingly, experiments determined that the substitution of any of these tyrosines by phenylalanine reduced the k_{cat} of *Tc*UGM [29].

Summarizing, the QM/MM molecular dynamics computations presented in this article determined that residues His62, Arg176, Asn201 and Arg327 contribute to the catalytic activity of *Tc*UGM by reducing the barriers of different steps of the mechanism. Tyr385 and Tyr429, on the other hand, play a role by keeping UDP always close to the sugar moiety. Also, the results highlight the participation of the carbonylic oxygen at position 4 of the cofactor. As predicted by Huang *et. al.* this atom provides an alternative route for the transference of the proton between $N5_{FADH}$ and the cyclic oxygen of the substrate. Without this route the barrier for the transference would be prohibitively high. Besides this oxygen restricts the mobility of the open-chain form of the sugar facilitating the ciclyzation process. We hope that the insights obtained from this computational study can contribute to the design of efficient inhibitors of *Tc*UGM.

Methods

Initial settings

The crystallographic structure of reduced *Tc*UGM with UDP was taken from the Protein Data Bank, entry 4DSH. To determine the coordinates of Gal*p* within UGM we superimposed the UDP-Gal*p* molecule, taken from the crystal structure of *Asparragilus fumigatus* UGM (PDB code 3UKF), with the crystallographic UDP of *Tc*UGM. The resultant coordinates of UDP-Gal*p*, together with those of *Tc*UGM, were used as the starting geometry of *Tc*UGM in its holo form. In the initial configuration the nucleophilic group and the leaving group laid on opposite sides of the sugar ring. The distance between $C1X_{GAL}$ and $N5_{FADH}$ was 3.78 Å. The angle between $N5_{FADH}$, $C1X_{GAL}$ and the oxygen atom of UDP, $O3B_{UDP}$, was 144.2°. The flavin cofactor was set in the reduced deprotonated state since it was recently shown that this form augments the nucleophilic character of $N5_{FADH}$ [40]. Besides, since experiments indicate that the pKa of $N1_{FADH}$ is ~ 6.7 while that of $N5_{FADH}$ is > 20, the proton of the reduced flavin was located on $N5_{FADH}$ [49]. The protonation state of the enzyme residues was assigned according with the standard rules except for His62, since recent experiments showed that this residue is protonated when the cofactor is in the reduced state [38]. The resulting file was fed into the Leap module of AMBER and the system was solvated in a 10.0 Å truncated octahedral cell of TIP3P explicit water molecules [50], including the crystallographic water molecules.

The QM/MM molecular dynamics and free energy simulations were performed with the AMBER12 package [51], using periodic boundary conditions with a cutoff distance of 10.0 Å and a time step of 1.0 fs. The potential energy of the classical region was computed with the Amber99SB force field [52] while the self-consistent charge Density Functional Tight Binding method (scc-DFTB) [51] was employed for the QM subsystem. The DFTB method has proved to be appropriate to describe the energetics of many chemical [53] and biochemical reactions [54–56]. More recently, it was shown to provide the best semiempirical description for six-membered carbohydrate rings deformation [57, 58]. The QM subsystem was formed with the flavin cofactor, the substrate, Gly61, His62, Val63, as well as the lateral chains of Arg176, Arg327 and Arg423. This adds up to 232 atoms with a net charge of -1.

The initial structure was first minimized at constant volume and then heated at NVT conditions from 0 K to 310 K by a simulated annealing technique. A weak harmonic restraint on the C_α atoms was implemented during this period. This was followed by 200 ps of equilibration at NPT conditions at 310 K and 1 bar. No restrains were applied in this case. The Pauling Bond Orders, n_x, were determined when galactose either attaches or detaches from the flavin cofactor (process a→b and f→g of Fig. 2). In both cases, the bonds involved are C-O and C-N. The equation used to calculate the orders was,

$$n_x = n_0 e^{(r_x - r_0)/0.6}. \quad (1)$$

Here n_0 denotes the bond order of the fully formed bond while r_0 is the equilibrium distance, which was considered equal to 1.5 Å for the two bonds involved in these reactions. The value of r_x was computed as the average distance among the structures sampled in the umbrella simulations at the transition state. The presence of H-bonds was monitored considering that a H-bond exists if the distance between the donor and the acceptor is < 3.15 Å and the donor-H-acceptor angle is $> 145°$.

When relevant (steps 2 and 5), the probability of H-tunneling was estimated employing the expression for the microcanonical transmission coefficient given at equation 14a of reference [59]. This expression corresponds to tunneling through a one-dimensional barrier whose shape, height and exothermicity are determined by three adjustable parameters. In our estimations these parameters were obtained by fitting the free energy curves of the corresponding proton transfer steps. Since the coordinate in these curves is not the proton coordinate but a difference between the distances of the bond being broken and formed, the effective mass of the particle being transferred was set as $m_H/4$, where m_H is the mass of the proton [60]. The energies employed in these estimations were 1700 cm^{-1} for step 2 and 1800 cm^{-1} for step 5. These are approximately the zero point energies of the N-H and O-H bonds.

Umbrella sampling calculations

The umbrella sampling technique was employed to analyse all the steps involved in the conversion between Gal*p* and Gal*f* within *Tc*UGM (see Fig. 2). Free energy profiles were computed along different reaction coordinates, conveniently defined for each transformation. Harmonic restraints were applied in order to force the system to wander around the selected values of the reaction coordinate. A restraining force of 350.0 kcal/molÅ^2 was employed in all cases and the reaction coordinate was sampled considering windows of 0.08 Å wide. Within each window, an equilibration phase of 65 ps was followed by a production phase of 0.2 ns. The actual values of the reaction coordinate were recorded every 2 fs. Snapshots of the structures were downloaded every 3 ps. The last 30000 data of each window were used to compute the unbiased probability by means of the weighted histogram analysis method (WHAM) [61]. When following each reaction coordinate, the last structure of a given window was used as the starting point for the next one. Simulations of 0.5 ns without any restraint were performed for reactants, products and for each intermediate species in order to check their stability. To check the convergence of the free energy computations, several tests were performed. First, for each step, we compared the free energy profiles obtained using the first half of the data (i.e. the first 15000 values selected at each specific value of the reaction coordinate) with the second half. Besides, every reaction coordinate was sampled both, forward and backwards. Finally, computations for steps 3 and 4 (the opening and closure of the sugar ring) were repeated three times using different initial configurations. Summing all the steps of the reaction, with their corresponding convergence tests, the total length of the QMMM-MD simulations was 187.5 ns. Below we provide the numerical details of the umbrella sampling calculations for each stage.

Stage 1: Formation of the flavin-Gal*p* adduct. This stage consists of just one concerted step in which the bond between Gal*p* and UDP breaks while Gal*p* joins $FADH^-$ (step 1 in Fig. 2). Accordingly, the reaction coordinate was defined as $z_1 = d_2 - d_1$, where d_1 is the distance between $C1X_{GAL}$ and $N5_{FADH}$, while d_2 is the distance between $C1X_{GAL}$ and $O3B_{UDP}$. This coordinate was sampled from -2.03 to 1.89 Å.

Stage 2: Formation of the iminium ion. This stage was proposed to occur in two consecutive steps. The first one is the tautomerization of $FADH^-$ via the transfer of the H atom bonded to $N5_{FADH}$ towards $O4_{FADH}$ (step 2 in Fig. 2). The reaction coordinate for this tautomerization was defined as $z_2 = d_3 - d_4$, where d_3 denotes the $N5_{FADH}$-H distance while d_4 is the $O4_{FADH}$-H distance. Coordinate z_2 was sampled from -1.63 to 1.73 Å.

During the following step, the opening of the ring is initiated by the transfer of the proton linked to $O4_{FADH}$ towards the oxygen atom of the Gal*p* ring, $O5X_{GAL}$, and proceeds with the breakage of the bond between $C1X_{GAL}$ and $O5X_{GAL}$ (step 3 in Fig. 2). We found that a correct description of this process required a reaction coordinate defined as $z_3 = d_5 + d_6 - d_7$, where d_5 is the distance between $O4_{FADH}$ and the proton being transferred, d_6 is the distance between $C1X_{GAL}$ and $O5X_{GAL}$ and d_7 is the distance between the proton and $O5X_{GAL}$. Coordinate z_3 was sampled from 0.82 to 4.18 Å.

We also analysed the possibility of a direct proton transfer from $N5_{FADH}$ to $O5X_{GAL}$. In order to do that we defined a reaction coordinate $z'_3 = d'_5 + d'_6 - d'_7$, where d'_5 is the $N5_{FADH}$-H distance, d'_6 the one between $O5X_{GAL}$ and $C1X_{GAL}$ and d'_7 the $O5X_{GAL}$-H distance. This coordinate was sampled from -1.85 to 0.71 Å. In agreement with previous results of Huang *et. al.* [41] we found that this direct proton transfer is very unlikely since it has a barrier significantly higher than the alternative path.

Stage 3: Formation of the flavin-Gal*f* adduct. This stage also occurs in two steps. First, the hydrogen attached to $O4X_{GAL}$ is transferred to $O4_{FADH}$ while a bond between $O4X_{GAL}$ and $C1X_{GAL}$ is formed (step 4 in Fig. 2). The reaction coordinate for this step was defined as $z_4 = d_8 - d_9 - d_{10}$, where d_8 is the distance between the H atom being transferred and $O4X_{GAL}$, d_9 the distance between the H atom and $O4_{FADH}$ and d_{10} is the distance between $O4X_{GAL}$ and the $C1X_{GAL}$. Coordinate z_4 was sampled from -4.85 to -1.01 Å.

The following step consists of a proton transfer from $O4_{FADH}$ to $N5_{FADH}$ (step 5 in Fig. 2). This can be seen as the reverse of step 2, except for the fact that galactose is now in the furanose form. Therefore, the reaction coordinate was defined as the reverse of step 2 ($z_5 = d_4 - d_3$) and it was scanned from -1.63 to 1.65 Å.

Stage 4: Formation of UDP-Gal*f*. This last step corresponds to the breakage of the bond between $FADH^-$ and Gal*f* along with the formation of a bond between Gal*f* and UDP (step 6 of Fig. 2).

Since this process is analogous to step 1 but occurs in reverse sense we defined $z_6 = d_1 - d_2 = -z_1$ and scanned it from −1.98 to 1.38 Å.

Energy decomposition

An energy decomposition analysis was performed to evaluate how the active site residues stabilize or destabilize the transition states of the successive steps with respect to their correspondent reactants. Different variations of this idea have been implemented to study enzymatic reactions [62–72]. In this case we followed the approach recently employed to compare the catalytic mechanisms of *T. cruzi* transialidase and *T. rangeli* sialidase [54]. Since the approach has been discussed in detail elsewhere we only present here the most relevant equations.

In the QM/MM study of an enzymatic reaction the influence of the classical environment on the activation energy of a given step, $\Delta\Delta E^{env}_{R\rightarrow TS}$, can be evaluated as,

$$\Delta\Delta E^{env}_{R\rightarrow TS} = \Delta E^{QM/MM}_{R\rightarrow TS} - \Delta E^{QM}_{R\rightarrow TS},$$

$$\approx \sum_i \Delta E^i_{R\rightarrow TS}. \qquad (2)$$

Here $\Delta E^{QM/MM}_{R\rightarrow TS}$ is the activation energy within the enzyme (computed with the QM/MM approach) while $\Delta E^{QM}_{R\rightarrow TS}$ is the activation energy of the isolated quantum subsystem (computed at the same QM level). The terms appearing in the summation, $\Delta E^i_{R\rightarrow TS}$, measure the influence of each individual residue on the reaction barrier. They are strictly given by,

$$\Delta E^i_{R\rightarrow TS} = \langle \Psi_{TS} | V_i | \Psi_{TS} \rangle - \langle \Psi_R | V_i | \Psi_R \rangle, \qquad (3)$$

where $|\Psi_{TS}\rangle$ and $|\Psi_R\rangle$ are the wave functions of the quantum subsystem at the transition state and reactants configurations, respectively, while V_i is the non-bonded interaction energy of classical residue i with the quantum subsystem. The evaluation of $\langle \Psi_X | V_i | \Psi_X \rangle$, with X = TS or R, is not trivial since the AMBER code does not compute these values. Instead it provides the energy of the whole system, which accounts for the quantum hamiltonian, H_{QM}, plus the sum of all the non-bonded interactions between the QM subsystem and the classical environment. Thus we estimated each $\langle \Psi_X | V_i | \Psi_X \rangle$ as,

$$\langle \Psi_X | V_i | \Psi_X \rangle \approx$$
$$\langle \Psi_X | H_{QM} + \sum_i V_i | \Psi_X \rangle - \langle \Psi_X | H_{QM} + \sum_{j \neq i} V_j | \Psi_X \rangle. \qquad (4)$$

Here the first term on the right side gives the actual energy of the system at the given configuration. The second one is a fictitious energy calculated with the same wave function by setting the classical environment at exactly the same configuration except for the i-th residue which is transformed into Gly. Average values of $\langle \Psi_X | V_i | \Psi_X \rangle$, with X = TS or R, were computed employing 100 snapshots taken from the umbrella sampling calculations with the reaction coordinate set at the TS or reactants configurations, respectively. For these calculations we defined the QM subsystem as the substrate plus the cofactor, while the active site residues under analysis were His62, Val63, Arg176, Asn201, Tyr317, Arg327, Tyr395, Arg423, Tyr429 and Asn433.

The $\Delta E^i_{R\rightarrow TS}$ computed in this way measures the difference between the actual barrier to reaction and the barrier that would be observed if the interaction between the side chain of residue i and the QM subsystem were turned off. Because of this, neither can it provide information about the effect of the backbone atoms or the effect of Gly residues. Moreover, since no dynamics is run when the i-th residue is replaced by Gly, $\Delta E^i_{R\rightarrow TS}$ does not take into account dynamic effects arising from changes in the conformational freedom of the enzyme upon replacement. Finally we note that positive/negative values of $\Delta E^i_{R\rightarrow TS}$ provide a strong indication about a deleterious/beneficial effect of residue i for the reaction step under consideration. However, they cannot be used to quantitatively estimate changes in k_{cat} produced by the mutation of the i-th residue by Gly because such changes depend on variations in the activation free energy, $\Delta G_{R\rightarrow TS}$.

Supporting Information

Figure S1 Evolution of important distances and angles along step 1. Panel (a): evolution of the distances involved in the definition of the reaction coordinate z_1. Panel (b): evolution of the Cremer-Pople angles θ and ϕ along with the distance between the H atom bonded to $N5_{FADH}$ the carbonyl oxygen of Gly61. The location of the transition state is indicated with an arrow.

Figure S2 Evolution of important distances and angles along step 2. Panel (a): evolution of the distances involved in the definition of the reaction coordinate z_2. Panel (b): evolution of the distance between the donor and the acceptor of the proton, the distance between the proton and the cyclic oxygen, and the torsional angle defined by C5X-N5-C4X-C4 of FADH. The location of the transition state is indicated with an arrow.

Figure S3 Evolution of important distances along step 3. Evolution of the distances involved in the definition of the reaction coordinate z_3. The location of the transition state is indicated with an arrow.

Figure S4 Probability distributions of the torsional angles defining the orientation of the hydroxyl groups at positions 4, 5 and 6 of the sugar moiety in the open form. The bars of the histogram are scaled so that the most likely angles of each distribution have a unitary height.

Figure S5 Evolution of important distances and angles along step 4. Panel (a): evolution of the distances involved in the definition of the reaction coordinate z_4. Panel (b): torsional angles that define the orientation of the hydroxyl groups at positions 4 and 5 of galactose. The location of the transition state is indicated with an arrow.

Figure S6 Evolution of important distances and angles along step 5. Panel (a) evolution of the distances involved in the definition of the reaction coordinate z_5. Panel (b) distance between $N5_{FADH}$ and $C1X_{GAL}$ along with the θ Cremer-Pople angle of furanose. The location of the transition state is indicated with an arrow.

Figure S7 Evolution of important distances and angles along step 6. Evolution of the distances involved in the definition of the reaction coordinate z_6 along with the θ Cremer-Pople angle of the furanose ring. The location of the transition state is indicated with an arrow.

Text S1 PDB file for UDP-Gal*p* bound to UGM (Michaelis complex). This species is labelled as a in Fig. 2.

Text S2 PDB file for the flavin-Gal*p* adduct in UGM. This species if labelled as b in Fig. 2.

Text S3 PDB file for the third species of the mechanism proposed for the reaction catalysed by UGM. This species is labelled as c in Fig. 2.

Text S4 PDB file for the iminium ion in UGM. This species is labelled as d in Fig. 2.

Text S5 PDB file for the fifth species of the mechanism proposed for the reaction catalysed by UGM. This species is labelled as e in Fig. 2.

Text S6 PDB file for the flavin-Gal*f* adduct in UGM. This species is labelled as f in Fig. 2.

Text S7 PDB file for UDP-Gal*f* bound to UGM. This species is labelled as g in Fig. 2.

Acknowledgments

We grateful acknowledge the computational support from Universidad Nacional de Quilmes.

Author Contributions

Conceived and designed the experiments: GPS JP. Performed the experiments: GPS JFG RCP. Analyzed the data: GPS JP JFG RCP. Contributed reagents/materials/analysis tools: JFG RCP. Wrote the paper: GPS JP.

References

1. World Health Organization website. Available: http://www.who.int/mediacentre/factsheets/fs340/en/. Accessed 2014 Sep 16.
2. Bern C, Montgomery SP, Herwaldt BL, Rassi AJ, Marin-Neto JA, et al. (2007) Evaluation and treatment of chagas disease in the united states: A systematic review. J Am Med Assoc 298: 2171–2181.
3. Teixeira ARL, Córdoba JC, Maior IS, Solrzano E (1990) Chagas' disease: Lymphoma growth in rabbits treated with benznidazole. Am J Trop Med Hyg 43: 146–158.
4. Issa VS, Bocchi EA (2010) Antitrypanosomal agents: treatment or threat? The Lancet 376: 768.
5. Berst M, Hellerqvist CG, Lindberg B, Lüderitz O, Svensson S, et al. (1969) Structural investigations on t1 lipopolysaccharides. Eur J Biochem 11: 353–359.
6. Crick DC, Mahapatra S, Brennan PJ (2001) Biosynthesis of the arabinogalactan-peptidoglycan complex of mycobacterium tuberculosis. Glycobiology 11: 107R–118R.
7. Whitfield C, Richards JC, Perry MB, Clarke BR, MacLean LL (1991) Expression of two structurally distinct d-galactan o antigens in the lipopolysaccharide of klebsiella pneumoniae serotype o1. J Bacteriol 173: 1420–1431.
8. Stevenson G, Neal B, Liu D, Hobbs M, Packer NH, et al. (1994) Structure of the o antigen of escherichia coli k-12 and the sequence of its rfb gene cluster. J Bacteriol 176: 4144–4156.
9. Dmitriev BA, Lvov VL, Kochetkov NK (1977) Complete structure of the repeating unit of the o-specific polysaccharide chain of shigella dysenteriae type 3 lipopolysaccharide. Carbohydr Res 56: 207–209.
10. Jann B, Shashkov AS, Kochanowski H, Jann K (1994) Structure of the {O16} polysaccharide from escherichia coli o16:k1: An {NMR} investigation. Carbohydr Res 264: 305–311.
11. Kol O, Wieruszeski JM, Strecker G, Montreuil J, Fournet B, et al. (1991) Structure of the o-specific polysaccharide chain from klebsiella pneumoniae {O1K2} (nctc 5055) lipopolysaccharide. Carbohydr Res 217: 117–125.
12. McConville MJ, Homans SW, Thomas-Oates JE, Dell A, Bacic A (1990) Structures of the glycoinositolphospholipids from leishmania major. a family of novel galactofuranose-containing glycolipids. J Biol Chem 265: 7385–7394.
13. Oppenheimer M, Valenciano AL, Sobrado P (2011) Biosynthesis of galactofuranose in kinetoplastids: Novel therapeutic targets for treating leishmaniasis and chagas' disease. Enzyme Res 2011.
14. Pedersen LL, Turco SJ (2003) Galactofuranose metabolism: a potential target for antimicrobial chemotherapy. Cell Mol Life Sci 60: 259–266.
15. Nikaido H, Sarvas M (1971) Biosynthesis of t1 antigen in salmonella: Biosynthesis in a cell-free system. J Bacteriol 105: 1073–1082.
16. Turco S, Orlandi Jr P, Homans S, Ferguson M, Dwek R, et al. (1989) Structure of the phosphosaccharide-inositol core of the leishmania donovani lipophosphoglycan. J Biol Chem 264: 6711–6715.
17. Weston A, Stern RJ, Lee RE, Nassau PM, Monsey D, et al. (1998) Biosynthetic origin of mycobacterial cell wall galactofuranosyl residues. Tubercle and Lung Dis 78: 123–131.
18. Beverley SM, Owens KL, Showalter M, Griffith CL, Doering TL, et al. (2005) Eukaryotic udp-galactopyranose mutase (glf gene) in microbial and metazoal pathogens. Eukaryotic Cell 4: 1147–1154.
19. Tefsen B, Ram AF, van Die I, Routier FH (2012) Galactofuranose in eukaryotes: aspects of biosynthesis and functional impact. Glycobiology 22: 456–469.
20. Schmalhorst PS, Krappmann S, Vervecken W, Rohde M, Müller M, et al. (2008) Contribution of galactofuranose to the virulence of the opportunistic pathogen aspergillus fumigatus. Eukaryotic Cell 7: 1268–1277.
21. Damveld RA, Franken A, Arentshorst M, Punt PJ, Klis FM, et al. (2008) A novel screening method for cell wall mutants in aspergillus niger identifies udp-galactopyranose mutase as an important protein in fungal cell wall biosynthesis. Genetics 178: 873–881.
22. Dykhuizen EC, May JF, Tongpenyai A, Kiessling LL (2008) Inhibitors of udp-galactopyranose mutase thwart mycobacterial growth. J Am Chem Soc 130: 6706–6707.
23. Almeida IC, Ferguson MA, Schenkman S, Travassos LR (1994) Gpi-anchored glycoconjugates from trypanosoma cruzi trypomastigotes are recognized by lytic anti-alpha-galactosyl antibodies isolated from patients with chronic chagas' disease. Braz J Med Biol Res 27: 443–447.
24. Ralton JE, Milne KG, Güther ML, Field RA, Ferguson MA (1993) The mechanism of inhibition of glycosylphosphatidylinositol anchor biosynthesis in trypanosoma brucei by mannosamine. J Biol Chem 268: 24183–24189.
25. Ferguson MAJ (1997) The surface glycoconjugates of trypanosomatid parasites. Philos T Roy Soc B 352: 1295–1302.
26. Turnock DC, Ferguson MAJ (2007) Sugar nucleotide pools of trypanosoma brucei, trypanosoma cruzi, and leishmania major. Eukaryotic Cell 6: 1450–1463.
27. MacRae JI, Obado SO, Turnock DC, Roper JR, Kierans M, et al. (2006) The suppression of galactose metabolism in *Trypanosoma cruzi* epimastigotes causes changes in cell surface molecular architecture and cell morphology. Mol Biochem Parasit 147: 126–136.
28. Oppenheimer M, Valenciano AL, Kizjakina K, Qi J, Sobrado P (2012) Chemical mechanism of udp-galactopyranose mutase from trypanosoma cruzi: a potential drug target against chagas' disease. PloS one 7: e32918.
29. Boechi L, Oliveira CAF, Fonseca I, Kizjakina K, Sobrado P, et al. (2013) Substrate-dependent dynamics of udp-galactopyranose mutase: Implications for drug design. Protein Science 22: 1490–1501.
30. Tanner JJ, Boechi L, Andrew McCammon J, Sobrado P (2013) Structure, mechanism, and dynamics of udp-galactopyranose mutase. Arch Biochem Biophys 544: 128–141.
31. Sanders DA, Staines AG, McMahon SA, McNeil MR, Whitfield C, et al. (2001) Udp-galactopyranose mutase has a novel structure and mechanism. Nat Struct Mol Biol 8: 858–863.
32. Beis K, Srikannathasan V, Liu H, Fullerton SW, Bamford VA, et al. (2005) Crystal structures of *Mycobacteria tuberculosis* and *Klebsiella pneumoniae* udp-galactopyranose mutase in the oxidised state and *Klebsiella pneumoniae* udp-galactopyranose mutase in the (active) reduced state. J Mol Biol 348: 971–982.
33. Gruber TD, Borrok MJ, Westler WM, Forest KT, Kiessling LL (2009) Ligand binding and substrate discrimination by udp-galactopyranose mutase. J Mol Biol 391: 327–340.
34. Gruber TD, Westler WM, Kiessling LL, Forest KT (2009) X-ray crystallography reveals a reduced substrate complex of udp-galactopyranose mutase poised for covalent catalysis by flavin. Biochemistry 48: 9171–9173.
35. Karunan Partha S, van Straaten KE, Sanders DA (2009) Structural basis of substrate binding to udp-galactopyranose mutase: crystal structures in the reduced and oxidized state complexed with udp-galactopyranose and udp. J Mol Biol 394: 864–877.

36. Partha SK, Sadeghi-Khomami A, Slowski K, Kotake T, Thomas NR, et al. (2010) Chemoenzymatic synthesis, inhibition studies, and x-ray crystallographic analysis of the phosphono analog of udp-gal*p* as an inhibitor and mechanistic probe for udp-galactopyranose mutase. J Mol Biol 403: 578–590.
37. Dhatwalia R, Singh H, Oppenheimer M, Karr DB, Nix JC, et al. (2012) Crystal structures and small-angle x-ray scattering analysis of udp-galactopyranose mutase from the pathogenic fungus aspergillus fumigatus. J Biol Chem 287: 9041–9051.
38. Dhatwalia R, Singh H, Oppenheimer M, Sobrado P, Tanner JJ (2012) Crystal structures of trypanosoma cruzi udp-galactopyranose mutase implicate flexibility of the histidine loop in enzyme activation. Biochemistry 51: 4968–4979.
39. Kizjakina K, Tanner JJ, Sobrado P (2013) Targeting udp-galactopyranose mutases from eukaryotic human pathogens. Current pharmaceutical design 19: 2561.
40. Soltero-Higgin M, Carlson EE, Gruber TD, Kiessling LL (2004) A unique catalytic mechanism for udp-galactopyranose mutase. Nat Struct Mol Biol 11: 539–543.
41. Huang W, Gauld JW (2012) Tautomerization in the udp-galactopyranose mutase mechanism: A dft-cluster and qm/mm investigation. J Phys Chem B 116: 14040–14050.
42. Barlow JN, Girvin ME, Blanchard JS (1999) Positional isotope exchange catalyzed by udp-galactopyranose mutase. J Am Chem Soc 121: 6968–6969.
43. van Straaten KE, Routier FH, Sanders DAR (2012) Towards the crystal structure elucidation of eukaryotic udp-galactopyranose mutase. Acta Crystallogr Sect F Struct Biol Cryst Commun 68: 455–459.
44. Yuan Y, Bleile DW, Wen X, Sanders DAR, Itoh K, et al. (2008) Investigation of binding of udp-gal f and udp-[3-f] gal f to udp-galactopyranose mutase by std-nmr spectroscopy, molecular dynamics, and corcema-st calculations. J Am Chem Soc 130: 3157–3168.
45. Huang Z, Zhang Q, Liu Hw (2003) Reconstitution of udp-galactopyranose mutase with 1-deaza-fad and 5-deaza-fad: analysis and mechanistic implications. Bioorg Chem 31: 494–502.
46. Sun HG, Ruszczycky MW, Chang Wc, Thibodeaux CJ, Liu Hw (2012) Nucleophilic participation of reduced flavin coenzyme in mechanism of udp-galactopyranose mutase. J Biol Chem 287: 4602–4608.
47. Cremer D, Pople JA (1975) General definition of ring puckering coordinates. J Am Chem Soc 97: 1354–1358.
48. Chad JM, Sarathy KP, Gruber TD, Addala E, Kiessling LL, et al. (2007) Site-directed mutagenesis of udp-galactopyranose mutase reveals a critical role for the active-site, conserved arginine residues. Biochemistry 46: 6723–6732.
49. Himo F (2006) Quantum chemical modeling of enzyme active sites and reaction mechanisms. Theor Chem Acc 116: 232–240.
50. Jorgensen WL, Chandrasekhar J, Madura JD, Impey RW, Klein ML (1983) Comparison of simple potential functions for simulating liquid water. J Chem Phys 79: 926–935.
51. Case DA, Babin V, Berryman JT, Betz RM, Cai Q, et al. (2012) Amber 12. University of California, San Francisco 142.
52. Cornell WD, Cieplak P, Bayly CI, Gould IR, Merz KM, et al. (1995) A second generation force field for the simulation of proteins, nucleic acids, and organic molecules. J Am Chem Soc 117: 5179–5197.
53. Krüger T, Elstner M, Schiffels P, Frauenheim T (2005) Validation of the density-functional based tight-binding approximation method for the calculation of reaction energies and other data. J Chem Phys 122: 114110.
54. Pierdominici-Sottile G, Palma J, Roitberg AE (2014) Free-energy computations identify the mutations required to confer trans-sialidase activity into trypanosoma rangeli sialidase. Proteins: Structure, Function, and Bioinformatics 82: 424–435.
55. Pierdominici-Sottile G, Horenstein NA, Roitberg AE (2011) Free energy study of the catalytic mechanism of trypanosoma cruzi trans-sialidase. from the michaelis complex to the covalent intermediate. Biochemistry 50: 10150–10158.
56. Woodcock HL, Hodoscek M, Brooks BR (2007) Exploring scc-dftb paths for mapping qm/mm reaction mechanisms. J Phys Chem A 111: 5720–5728.
57. Barnett CB, Naidoo KJ (2010) Ring puckering: a metric for evaluating the accuracy of am1, pm3, pm3carb-1, and scc-dftb carbohydrate qm/mm simulations. J Phys Chem B 114: 17142–17154.
58. Biarns X, Ardevol A, Planas A, Rovira C, Laio A, et al. (2007) The conformational free energy landscape of β-d-glucopyranose. implications for substrate preactivation in β-glucoside hydrolases. J Am Chem Soc 129: 10686–10693.
59. Ahmed Z (1993) Tunneling through a one-dimensional potential barrier. Phys Rev A 47: 4761–4767.
60. Pierdominici-Sottile G, Martí MA, Palma J (2008) The role of residue thr122 of methylamine dehydrogenase on the proton transfer from the iminoquinone intermediate to residue asp76. Chem Phys Lett 456: 243–246.
61. Kumar S, Rosenberg JM, Bouzida D, Swendsen RH, Kollman PA (1992) The weighted histogram analysis method for free-energy calculations on biomolecules. i. the method. J Comp Chem 13: 1011–1021.
62. Lee FS, Chu ZT, Bolger MB, Warshel A (1992) Calculations of antibody-antigen interactions: microscopic and semi-microscopic evaluation of the free energies of binding of phosphorylcholine analogs to mcpc603. Protein Engineering 5: 215–228.
63. Torrie GM, Valleau JP (1977) Nonphysical sampling distributions in monte carlo free-energy estimation: Umbrella sampling. J Comp Phys 23: 187–199.
64. Lin Yl, Gao J (2009) Internal proton transfer in the external pyridoxal 5?-phosphate schiff base in dopa decarboxylase. Biochemistry 49: 84–94.
65. Pierdominici-Sottile G, Roitberg AE (2011) Proton transfer facilitated by ligand binding. an energetic analysis of the catalytic mechanism of trypanosoma cruzi trans-sialidase. Biochemistry 50: 836–842.
66. Hensen C, Hermann JC, Nam K, Ma S, Gao J, et al. (2004) A combined qm/mm approach to protein-ligand interactions: polarization effects of the hiv-1 protease on selected high affinity inhibitors. J Med Chem 47: 6673–6680.
67. Chatfield DC, P Eurenius K, Brooks BR (1998) Hiv-1 protease cleavage mechanism: A theoretical investigation based on classical md simulation and reaction path calculations using a hybrid qm/mm potential. J Mol Struc: THEOCHEM 423: 79–92.
68. Cunningham MA, Ho LL, Nguyen DT, Gillilan RE, Bash PA (1997) Simulation of the enzyme reaction mechanism of malate dehydrogenase. Biochemistry 36: 4800–4816.
69. Davenport RC, Bash PA, Seaton BA, Karplus M, Petsko GA, et al. (1991) Structure of the triosephosphate isomerase-phosphoglycolohydroxamate complex: an analog of the intermediate on the reaction pathway. Biochemistry 30: 5821–5826.
70. Dinner AR, Blackburn GM, Karplus M (2001) Uracil-dna glycosylase acts by substrate autocatalysis. Nature 413: 752–755.
71. Garcia-Viloca M, Truhlar DG, Gao J (2003) Reaction-path energetics and kinetics of the hydride transfer reaction catalyzed by dihydrofolate reductase. Biochemistry 42: 13558–13575.
72. Major DT, Gao J (2006) A combined quantum mechanical and molecular mechanical study of the reaction mechanism and α-amino acidity in alanine racemase. J Am Chem Soc 128: 16345–16357.

13

Transcriptome Analysis of Yellow Horn (*Xanthoceras sorbifolia* Bunge): A Potential Oil-Rich Seed Tree for Biodiesel in China

Yulin Liu[1], Zhedong Huang[2], Yan Ao[3], Wei Li[1]*, Zhixiang Zhang[2]*

1 College of Biological Science and Biotechnology, Beijing Forest University, Beijing, China, **2** Laboratory of Systematic Evolution and Biogeography of Woody Plants, College of Nature Conservation, Beijing Forest University, Beijing, China, **3** Academy of Forest, Beijing Forest University, Beijing, China

Abstract

Background: Yellow horn (*Xanthoceras sorbifolia* Bunge) is an oil-rich seed shrub that grows well in cold, barren environments and has great potential for biodiesel production in China. However, the limited genetic data means that little information about the key genes involved in oil biosynthesis is available, which limits further improvement of this species. In this study, we describe sequencing and *de novo* transcriptome assembly to produce the first comprehensive and integrated genomic resource for yellow horn and identify the pathways and key genes related to oil accumulation. In addition, potential molecular markers were identified and compiled.

Methodology/Principal Findings: Total RNA was isolated from 30 plants from two regions, including buds, leaves, flowers and seeds. Equal quantities of RNA from these tissues were pooled to construct a cDNA library for 454 pyrosequencing. A total of 1,147,624 high-quality reads with total and average lengths of 530.6 Mb and 462 bp, respectively, were generated. These reads were assembled into 51,867 unigenes, corresponding to a total of 36.1 Mb with a mean length, N50 and median of 696, 928 and 570 bp, respectively. Of the unigenes, 17,541 (33.82%) were unmatched in any public protein databases. We identified 281 unigenes that may be involved in *de novo* fatty acid (FA) and triacylglycerol (TAG) biosynthesis and metabolism. Furthermore, 6,707 SSRs, 16,925 SNPs and 6,201 InDels with high-confidence were also identified in this study.

Conclusions: This transcriptome represents a new functional genomics resource and a foundation for further studies on the metabolic engineering of yellow horn to increase oil content and modify oil composition. The potential molecular markers identified in this study provide a basis for polymorphism analysis of Xanthoceras, and even Sapindaceae; they will also accelerate the process of breeding new varieties with better agronomic characteristics.

Editor: Turgay Unver, Cankiri Karatekin University, Turkey

Funding: This research was financially supported by the Chinese Key Technology Research and Development Program of Twelfth Five-Year Plan (2011BAD22B08), and the Forestry Public Benefit Research Program (201004001). The funders had no role in study design, data collection and analysis, decision to publish, or preparation of the manuscript.

Competing Interests: The authors have declared that no competing interests exist.

* E-mail: bjfuliwei@bjfu.edu.cn (WL); zxzhang@bjfu.edu.cn (ZXZ)

Introduction

Due to the crises of fossil fuel depletion and worsening global environmental conditions, oil-rich seed plants that can be used to produce renewable and environmentally friendly biodiesel have received much attention [1–3]. A variety of vegetable oils that are obtained from rapeseed (canola), soybean, sunflower, peanut, safflower, palm and Jatropha, among others, have been used to produce biodiesel. However, the great majority of these plants are grown on farmland and are used as cooking oil. Under these circumstances, in some developing countries that have limited per capita arable land, the use of food crops to produce biodiesel is not realistic. Therefore, biodiesel production from non-food crops that can be planted in areas that are unsuitable for traditional crops is an ideal solution to this problem [4–6].

Yellow horn (*Xanthoceras sorbifolia* Bunge.) is an oil-rich seed shrub that belongs to the Sapindaceae family and has a life span of more than 200 years [7,8]. The seeds of yellow horn contain abundant oil (55–70%), of which 85–93% is unsaturated fatty acids [9,10]. According to previous studies, the molecular composition of yellow horn oil is similar to the ideal fatty ester structure for biodiesel [11,12]. Therefore, it has been identified as a major biodiesel tree species and the Chinese Government provided special support to aid its development because it can produce over 800 gallons of oil per acre of cultivation [13,14]. Unlike other energy-resource trees, such as palm and Jatropha, that cannot survive low temperatures, it can not only grow well in barren, salty and drought soil, it can also survive temperatures as low as −30 to −41°C. In addition, the yellow horn tree has many other uses, including multiple entries in the Chinese Pharmacopoeia, it can assist in eliminating desertification and erosion, and it is grown as an ornamental tree and used as a source of high-level woody natural oil for cooking [15].

During the last decade many studies have examined yellow horn seed oil; however, these focused on the extraction of oil and methods of biodiesel production from the seed oil [10,16–18]. Unlike other oil crops, such as rapeseed, soybean and Jatropha, no genome-level studies have attempted to determine the oil synthesis metabolic pathway, which could be used to improve the seed yield and oil content. Although conventional breeding strategies continue to play an important role in crop improvement, genetic engineering methods are more rapid and precise, and allow the specific redesigning of crops for target characteristics [19]. For non-model plants, such as yellow horn, for which little or no molecular information is available, next-generation sequencing (NGS) technologies provide a ready means of obtaining genetic information [20–22]. The advent of NGS, such as RNA-Seq, in recent years has created unprecedented opportunities for generating genomic information in previously uncharacterised systems. NGS facilitates rapid, inexpensive and comprehensive analyses of complex genomes due to the collection of large-scale sequence data that can be used for gene discovery [23], expression profiling [24], molecular marker development [25] and functional, comparative and evolutionary genomics studies [26]. To date, the transcriptomes of a large number of plants, including many oil crops such as palm [27], peanut [28], sesame [29], safflower [30], rape [31] and Jatropha [32], have been analysed using NGS.

In this study, we report the results of using Roche 454 RNA-seq technology, which can generate sufficiently long sequence reads [33], to analyse the yellow horn transcriptome, which was derived from a pooled sample of DNA from multiple tissue types (buds, leaves, flowers and seeds). The analysis included functional annotation of the transcripts, identification of unigenes that are involved in oil biosynthesis and metabolism, and the discovery of a series of molecular markers (SSRs, SNPs and InDels). These transcripts represent the first yellow horn sequence dataset. We believe the data will open new perspectives for improving and selecting elite yellow horn varieties to produce a greater quantity of high-quality biodiesel.

Results and Discussion

Transcriptome Assembly

One and a quarter plates of pyrosequencing reactions were conducted using a 454 GS FLX titanium platform. Approximately 600 Mb of data from 1,221,677 raw reads with a GC content of 43.7% were produced; the read lengths ranged from 23 to 1,478 bp with an average length of 491.1 bp and a median length of 537 bp. After SeqClean was used to cut the adaptors and SMART primers and LUCY2 were used to remove low-quality regions and bases, 88.43% of bases were retained and 1,147,624 trimmed reads (GC content = 43.6%) were generated, with total and average lengths of 530.6 Mb and 462 bp, respectively. Then, 32,165 isotigs (total length = 28,133,950 bp, average length = 874.7 bp, N50 length = 1,116 bp, median length = 732 bp, GC content = 41.40%) and 42,787 singlets (total length = 17,190,382 bp, average length = 401.8 bp, N50 length = 549 bp, median length = 484, GC content = 42.70%) were assembled using Newbler2.6. Finally, after a last assembly using CD-HIT for isotigs and singlets, a dataset of 51,867 unigenes (45 to 10,088 bp with a GC content of 41.37%) was obtained, corresponding to 36.1 Mb with mean, N50 and median lengths of 696, 928 and 570 bp, respectively. Of the unigenes 7,127 (13.74%) were equal to or shorter than 200 bp and 10,960 (21.13%) were longer than 1,000 bp (Table 1, Figure 1C).

Characterisation of Non-redundant Unigenes

To understand their functions, the 51,867 yellow horn unigenes were annotated using BLASTx alignment with an E-value cut-off of 10^{-5} against the following protein databases: NR, Swiss-Prot, CDD, Pfam, TrEMBL, COG, GO, KEGG and TAIR. A total of 33,924 (65.41%), 33,872 (65.31%), 30,504 (58.81%), 28,460 (54.87%), 26,643 (51.36%), 24,412 (47.07%), 24,258 (46.77%), 11,181 (21.56%) and 7,415 (14.30%) unigenes had significant matches with sequences in the NR, TrEMBL, Pfam, TAIR, GO, Swiss-Prot, CDD, COG and KEGG databases, respectively (Table 2). Of the unigenes, 34,326 (66.18%) were described in at least one database with high homology (Figure 1A) with unigenes from *Vitis vinifera* (9,580, 18.47%), *Ricinus communis* (7,900, 15.23%), *Populus trichocarpa* (7,337, 14.15%), *Glycine max* (2,052, 3.96%), *Medicago truncatula* (755, 1.46%), *Arabidopsis thaliana* (635, 1.22%), and other species (6,067, 11.70%) (Figure 1B). However, the remaining 17,541 (33.82%) were unmatched (70.55% of unigenes <500 bp and 1.90% of unigenes >1,000 bp) (Figure 1C), suggesting that longer sequences were more likely to have BLAST hits and shorter sequences may have been either too short to get hits or lacked a characterised protein domain, which resulted in false-negative results. Because public databases contain little genomic and transcriptomic information for yellow horn, these unmatched unigenes may represent putative tissue-specific novel genes or non-coding regions.

Functional Classification of Unigenes by COG, GO and KEGG

To further evaluate the completeness of our transcriptome library and the effectiveness of our annotation process, we used the annotated unigene sequences to search for genes involved in COG classifications, GO assignments and KEGG pathway assignments to predict and classify their functions.

Overall, 11,181 unigenes were assigned into 24 COG function categories. Among them, the cluster for "general function prediction only" represented the largest group (4,227, 37.81%), followed by "posttranslational modification, protein turnover, chaperones" (2,086, 18.66%), "replication, recombination repair" (1,526, 13.65%), "transcription" (1,497, 13.39%) and "translation, ribosomal structure and biogenesis" (1,459, 13.05%). However, few unigenes were assigned into "cell motility" (36, 0.32%) and "nuclear structure" (3, 0.03%). Additionally, 967 (8.65%) unigenes were assigned into the category "lipid transport and metabolism" (Figure 2).

Of the unigenes, 26,643 were assigned into the three main GO functional categories and then into 50 sub-categories (Figure 3). For the three main categories, 23,978 unigenes (90.03%) were assigned into the largest category, molecular function, followed by biological process (22,187, 83.28%) and cellular component (15,012, 56.35%). The biological process category was assigned into 25 sub-categories; the most abundant was "metabolic process", which contained 18,891 unigenes (36.42% of the total), indicating that these genes were enriched in the yellow horn transcriptome libraries. The cellular components category was divided into 11 small groups; the largest sub-category was "cell", which included 11,855 (22.86% of the total) unigenes. For molecular function, 23,978 unigenes were categorised into 14 GO terms; the majority fell into "binding" and "catalytic activity", which contained 17,084 and 15,045 unigenes, respectively.

In addition, a total of 7,415 unigenes were assigned to 182 pathways through KEGG pathway assignment. Among them, the largest represented category was "metabolism", containing 3,862 unigenes, followed by "genetic information processing" and

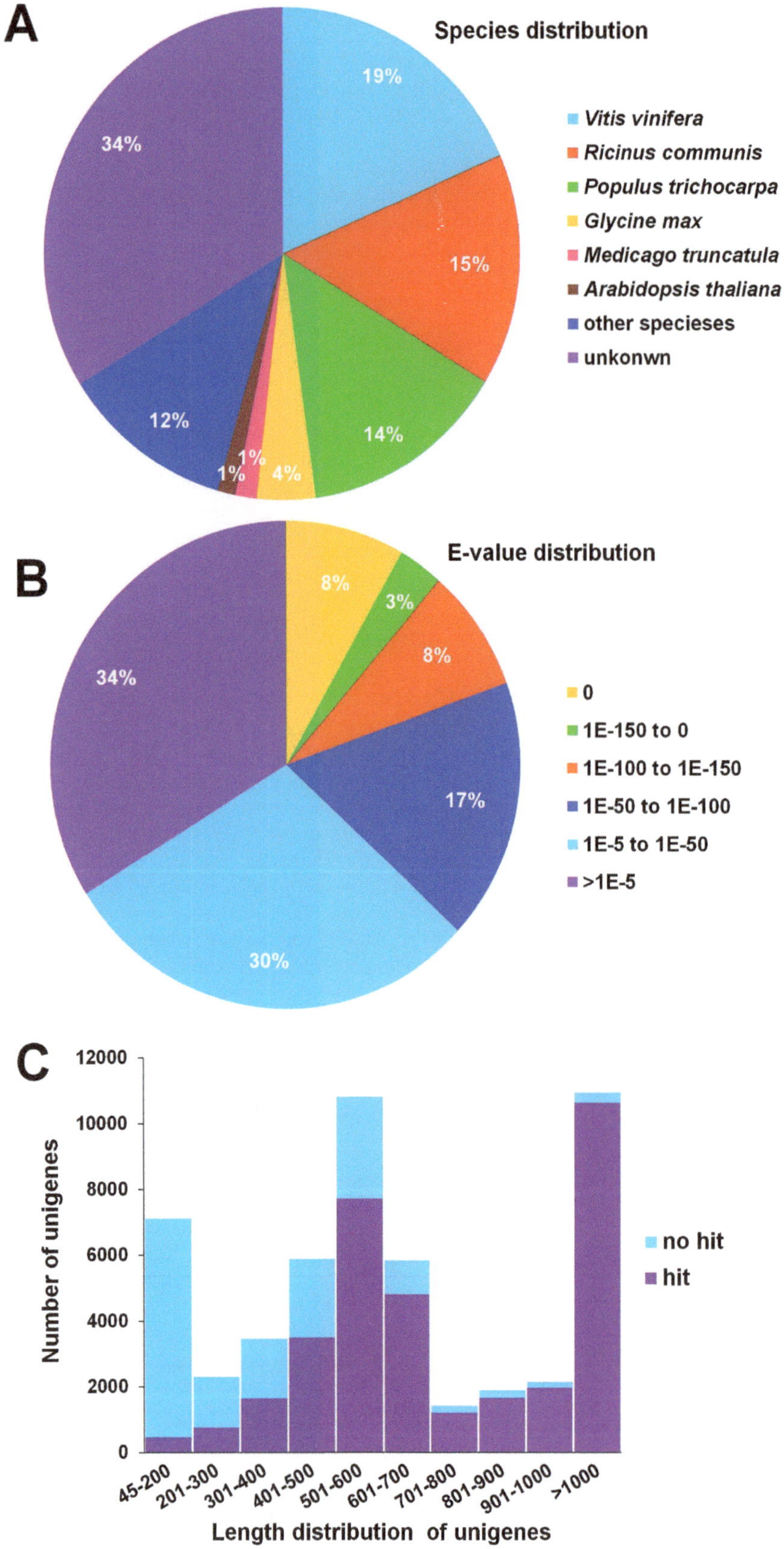

Figure 1. Unigenes functional annotation results. (A) Top-hit species distribution for BLASTx matches for yellow horn unigenes using the following order of priority: NR, TrEMBL and Swiss-Prot. (B) E-value distribution of top BLASTx hits for each unigene. (C) Distribution of unigenes in length with BLASTx hits compared with those without hits.

Table 1. Summary of yellow horn 454 sequencing and assembly.

Items	Raw reads	Trimmed Reads	Isotigs	Singlets	Unigenes
Total number	1,221,677	1,147,624	32,165	42,787	51,867
Total bases (bp)	599,995,187	530,556,412	28,133,950	17,190,382	36,093,173
Average length (bp)	491.1	462.3	874.7	401.8	695.9
N50 (bp)	-	-	1,116	549	928
Median (bp)	537	511	732	484	570
Range of length (bp)	23–1,478	45–965	45–10,088	45–823	45–10,088
GC content (%)	43.70%	43.64%	41.40%	42.70%	41.37%

"cellular processes", which contained 2,736 and 1,070 unigenes, respectively. The pathway "ribosome" involved the largest number of unigenes (475), but both the "primary bile acid biosynthesis" and "nitrotoluene degradation" pathways contained only one unigene. In addition, 548 unigenes were mapped to 15 pathways in the sub-category "lipid metabolism" (Table S1).

Unigenes Related to FA Biosynthesis

In oil plants, fatty acids are stored as a form of TAG and their biosynthesis pathway can be divided into three steps [34]. The first step is *de novo* biosynthesis of fatty acids; this process occurs in plastids and is catalysed mainly by the fatty acid synthase complex (FAS). The second step is the synthesis of triacylglycerol (TAG), which occurs in the endoplasmic reticulum (ER), and the last is the formation of oil bodies (OBs), where TAG is combined with oleosin to form an oil body, which is released from the ER into the cytoplasm [35,36].

According to the KEGG pathway assignment and functional annotation of the unigenes, 40 unigenes were annotated as encoding ten key enzymes involved in FAs biosynthesis (Table 3). The reconstructed pathway of FAs biosynthesis was based on these identified enzymes (Figure 4). First, acetyl-CoA carboxylase (ACCase, EC: 6.4.1.2), as a rate-limiting enzyme in the FAs biosynthesis pathway, catalyses acetyl-CoA to form malonyl-CoA [37]; 16 unigenes that encode its four subunits were identified (four for α-carboxyltransferase, three for β-carboxyltransferase, six for biotin carboxylase and three for biotin carboxyl carrier protein). Next, a series of condensation reactions of malonyl-CoA with a growing ACP-bound acyl chain are catalysed by FAS, consecutively adding two carbon units per cycle over six or seven cycles to form 16:0-ACP or 18:0-ACP, which can then be catalysed by acyl-ACP desaturase (AAD, EC: 1.14.19.2) to form 16:0-ACP or 18:1-ACP [38]. Nineteen unigenes that encoded the five components of FAS were found, including one that encoded malonyl-CoA-ACP transacylase (MAT, EC: 2.3.1.39), eight that encoded 3-ketoacyl ACP synthase (KAS; seven for KAS II (EC: 2.3.1.179) and one for KAS III (EC: 2.3.1.180)), six that encoded 3-ketoacyl ACP reductase (KAR, EC: 1.1.1.100), two that encoded 3-hydroxymyristoyl ACP dehydrase (HAD, EC: 4.2.1.-) and two that encoded enoyl-ACP reductase (EAR, EC:1.3.1.9). Moreover, nine unigenes coded the acyl carrier protein (ACP), an essential cofactor of FAS. After this, under the control of acyl-ACP thioesterase (FAT, EC: 3.1.2.14 3.1.2.-) and palmitoyl-CoA hydrolase (PCH, EC: 3.1.2.2), free FAs are released from the acyl carrier protein (ACP). Two unigenes that encoded FATA, three that encoded FATB and one that encoded PCH which is also involved in FA elongation were identified in this study.

Table 2. Functional annotation of yellow horn unigenes in public protein databases.

Database	Number of annotated unigenes	Percentage (%)
NR	33,924	65.41%
TrEMBL	33,872	65.31%
Pfam	30,504	58.81%
TAIR	28,460	54.87%
GO	26,643	51.36%
Swiss-Prot	24,412	47.07%
CDD	24,258	46.77%
COG	11,181	21.56%
KEGG	7,415	14.30%
Total	34,326	66.18%

In addition, 16 unigenes that encode long-chain acyl-CoA synthetases (LACS), which catalyse the esterification of free FAs to CoA upon arrival in the cytoplasm [39], and seven that encoded acyl CoA binding protein (ACBP), which binds medium- and long-chain acyl-CoA esters with very high affinity and may function as an intracellular carrier of acyl-CoA esters [40], were also identified based on the functional annotation of the transcriptome.

As has been reported previously, overexpression of ACCase, a crucial enzyme in fatty acid synthesis, can alter the fatty acid composition of seeds and increase the fatty acid content, which would lead to an increased oleic acid content and seed yield [41,42]. Most research on FAS has concentrated on ACP and KAS. Functional expression of an ACP from *Azospirillum brasilense* in *Brassica juncea* can improve the content of 18:1 and 18:2 in seeds, enhance the ratio of monounsaturated (18:1) to saturated fatty acids, increase the ratio of 18:2 to 18:3 and reduce the erucic acid content (22:1) [43]. In plastids, three types of KAS are found: KAS I, KAS II and KAS III. KAS II catalyses 16:0-ACP to elongate to 18:0-ACP and KAS III condenses acetyl-CoA with malonyl-ACP to form 4:0-ACP. Using hairpin RNAi to reduce the activity of KAS II can lead to an increase in 16:0 accumulation, up to about 53% of the total, but some transgenic offspring are deformed during early embryonic development [44]. Overexpression of KAS III can also improve 16:0 accumulation, but the rate of lipid synthesis is reduced [45]. In our study, we did not find any unigenes that encoded KAS I, which is highly active with acyl-ACP with chain lengths from C2 to C14, is far less effective for 16:0-ACP and almost inactive for 18:0-ACP [46].

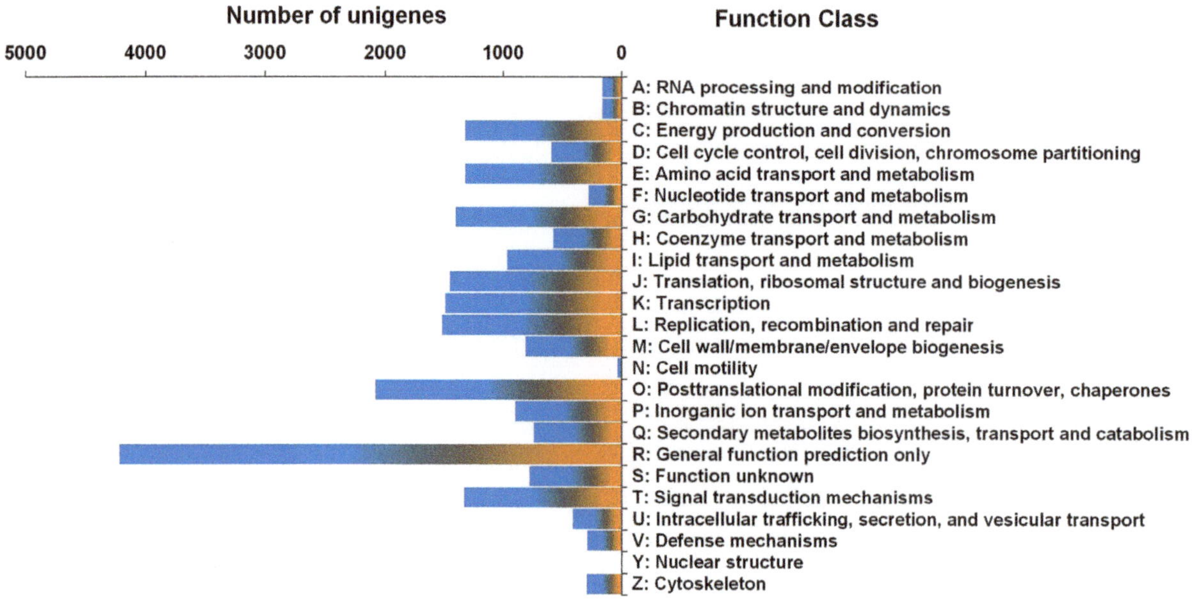

Figure 2. Clusters of orthologous groups (COG) classifications of yellow horn unigenes.

Unigenes Related to TAG and OBs Biosynthesis

In the suggested pathway for TAG biosynthesis [47,48], a total of 33 unigenes that encode six enzymes were found. As the data in Table 4 and Figure 4 show, initially, one unigene that encodes glycerol kinase (GK, EC: 2.7.1.30), which catalyses glycerol to form glycerol-3-phosphate (G-3-P), an initial substrate in the Kennedy pathway, was detected. Then, 12 unigenes that encode the key component of TAG biosynthesis, glycerol-3-phosphate acyltransferase (ATS1 and GPAT, EC: 2.3.1.15) (three for ATS1, one for GPAT5, three for GPAT6, two for GPAT8 and three for GPAT9), which catalyses the first step of the Kennedy pathway, and 11 unigenes that encode lysophosphatidyl acyltransferase (LPAT, EC: 2.3.1.51) were identified (four for LPAT1, two for LPAT2, three for LPAT4 and two for LPAT5). Under catalysis by these two enzymes, sequential esterification of acyl chains from acyl-CoA to the positions of *sn-1* and *sn-2* of G-3-P occur to form lysophosphatidic acid (Lyso-PA) and phosphatidic acid (PA), respectively. The next reaction is catalysed by phosphatidate phosphatase (PP, EC: 3.1.3.4), a key regulator of lipid homeostasis, which was encoded by four unigenes; it plays a role in the removal of the phosphate group from PA and forms diacylglycerol (DAG), which is an essential intermediate in the biosynthesis of phosphatidylcholine (PC). Finally, two enzymes, diacylglycerol *O*-acyltransferase (DGAT, EC: 2.3.1.20) and phospholipid: diacylglycerol acyltransferase (PDAT, EC: 2.3.1.158), which use acyl-CoA and phospholipids as acyl-donors, respectively, transfer an acyl group to the *sn-3* position of DAG to produce TAG. Only one unigene that encoded DGAT1 and four that encoded PDAT1 were identified in the yellow horn transcriptome. In addition, two enzymes, phospholipase A2 (PLA2, EC:3.1.1.4) and lysophosphatidylcholine acyltransferase (LPCAT, EC:2.3.1.23 2.3.1.67) which may regulate the acyl editing were also identified (Table 4 and Figure 4).

Overexpression of a plastidial safflower GPAT and an *Escherichia coli* GPAT in *Arabidopsis* can improve the seed oil content, with average increases of 22 and 15%, respectively [49]. Similarly, substantial increases of 8 to 48% in seed oil content and increases in both overall proportions and amounts of very-long-chain fatty acids in seed TAGs were obtained by overexpression of a mutant form of yeast LPAT in *Arabidopsis* and *Brassica napus* [50]. DGAT is a key enzyme regulating the rate of the Kennedy pathway. It has four types, and we detected DGAT1 in this study. Ectopic expression of DGAT will improve the oil content in seeds, which has been confirmed in *Arabidopsis*, soybean and maize [51–53]. In addition, compared to the single unigene that encoded GPAT1, the finding of four unigenes that encoded PDAT1 in our study provides further evidence that yellow horn has the potential to channel fatty acids that are incorporated in membrane lipids, such as PC, into TAG biosynthesis.

After biosynthesis, pools of TAGs can be stored as a form of OB surrounded by a membrane composed of a layer of phospholipids embedded with several proteins: oleosin, caleosin and steroleosin in mature seeds [35,54]. According to the annotations of the unigenes, nine encoded oleosin, seven encoded caleosin and four encoded steroleosin (Table S2). Oleosin is the most abundant structural protein in OBs; it helps stabilise OBs through increased space bit resistance and charge repulsion, preventing fusion of OBs [35,55]. Caleosin is not only involved in the synthesis and metabolism of OBs, but may also be associated with plant drought tolerance [55,56]. Steroleosin-like proteins may represent a class of dehydrogenases/reductases that are involved in plant signal transduction regulated by various sterols [57]. In conclusion, the detection of unigenes that are involved in oleosin, caleosin and steroleosin biosynthesis will contribute to future functional studies and improvements in production levels by metabolic engineering of yellow horn.

Unigenes Related to FA Desaturation

Many types of enzyme participate in fatty acid desaturation in plants, and can be divided into two types. One type catalyses the formation of monounsaturated fatty acids from saturated fatty acids in plastids (16:0 to 16:1, 18:0 to 18:1); these contain only a soluble enzyme, acyl-ACP desaturase (AAD, EC: 1.14.19.2) [58]. The other type is located on the membranes of the endoplasmic reticulum and chloroplast and introduces double-unsaturated bonds at specifically defined positions (Δ12,

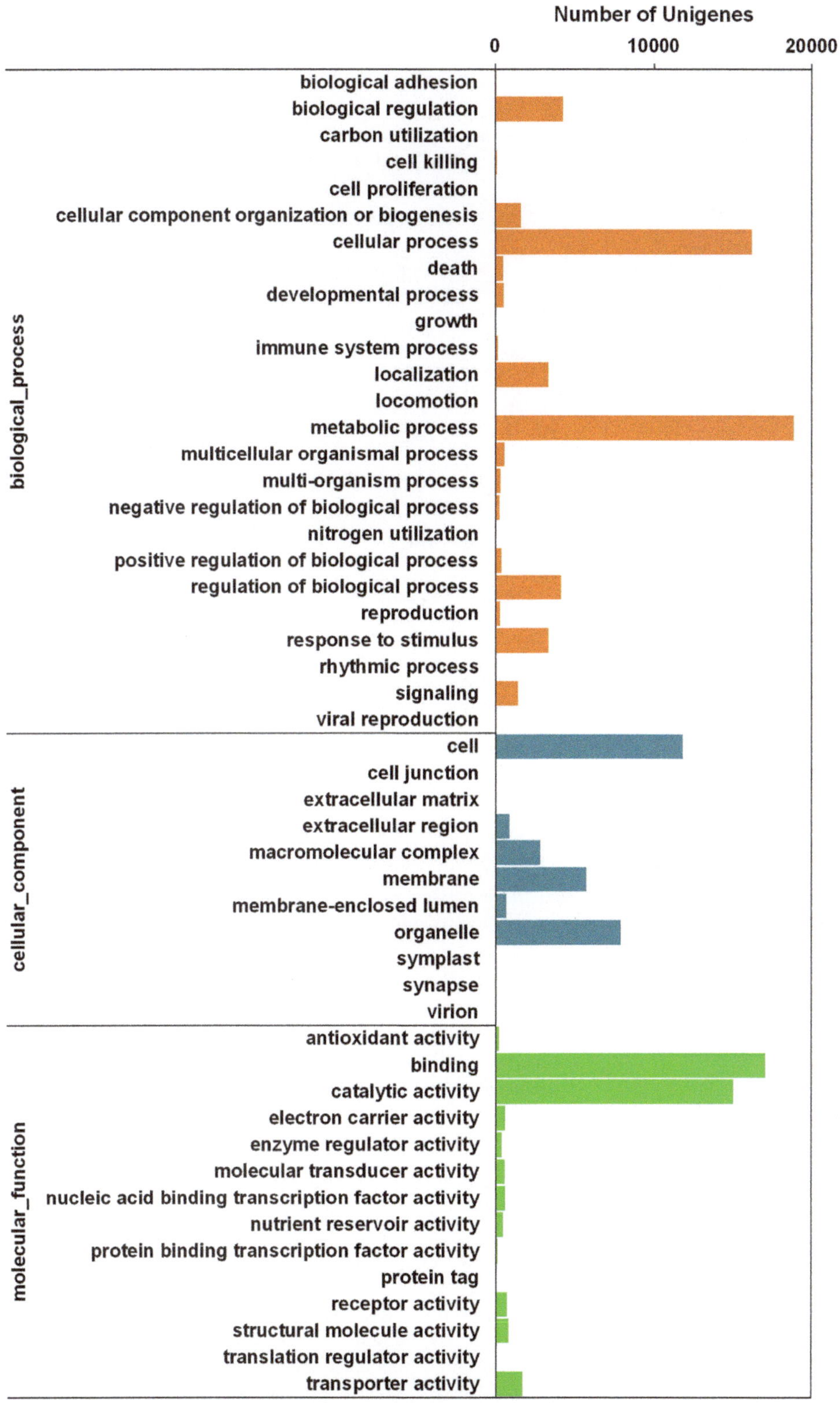

Figure 3. Gene Ontology (GO) categories assigined to the yellow horn unigenes.

Δ15 or Δ6) in fatty acids that are esterified to a glycerol backbone [59], including Δ12(ω6)-Desaturase (FAD2 and FAD6, EC:1.14.19.-), which desaturates oleic acid (18:1) to form linoleic acid (18:2), Δ15(ω3)-Desaturase (FAD3, FAD7 and FAD8, EC:1.14.19.-), which further desaturates linoleic acid (18:2) to form α-linolenic acid (18:3), *etc.* In total, six, five, two, one, five and one unigenes that encode AAD, FAD2, FAD3, FAD6, FAD7 and FAD8 were found (Table 3). Oleic (18:1) and

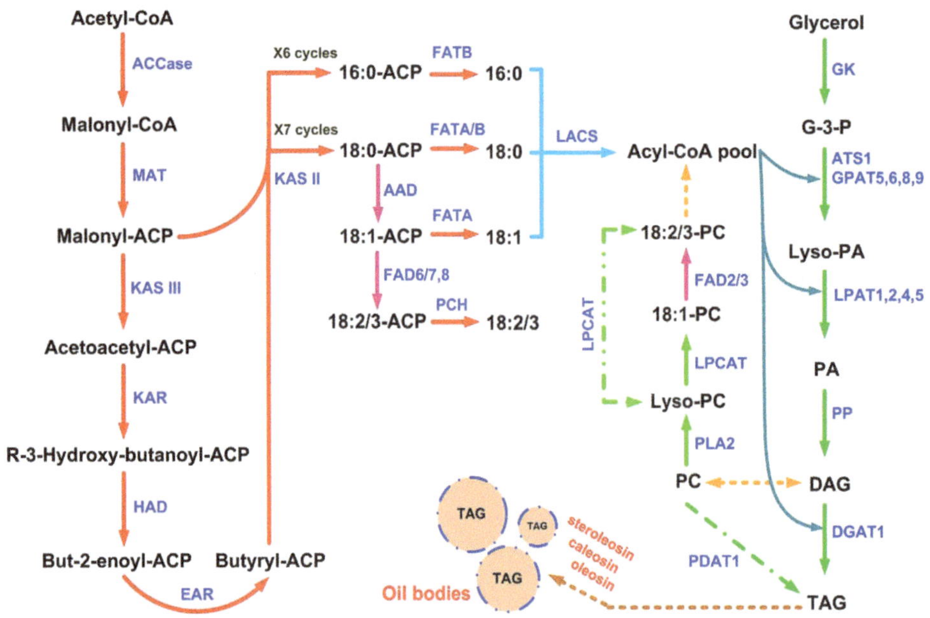

Figure 4. Overview of de novo fatty acid (FA) and triacylglycerol(TAG) biosynthesis pathways. Indentified enzymes include: ACCase, α-acetyl-CoA carboxylase carboxyl transferase (EC:6.4.1.2); MAT, Malonyl-CoA-ACP transacylase (EC:2.3.1.39); KAS, 3-Ketoacyl ACP synthase (KASII, EC: 2.3.1.179; KAS III, EC: 2.3.1.180); KAR, 3-Ketoacyl ACP reductase (EC:1.1.1.100); HAD, 3R-hydroxymyristoyl ACP dehydrase (EC:4.2.1.-); EAR, enoyl-ACP reductase I (EC:1.3.1.9); FATA/B, fatty acyl-ACP thioesterase A/B (EC:3.1.2.14 3.1.2.-); AAD, acyl-ACP desaturase (EC:1.14.19.2); PCH, palmitoyl-CoA hydrolase (EC:3.1.2.2); ACSL, long-chain acyl-CoA synthetase (EC:6.2.1.3); FAD2/6, Δ12(ω6)-Desaturase (EC:1.14.19.-); FAD3/7/8, Δ15(ω3)-Desaturase (EC:1.14.19.-); GK, glycerol kinase (EC:2.7.1.30); ATS1/GPAT, glycerol-3-phosphate acyltransferase (EC:2.3.1.15); LPAT, lysophosphatidyl acyltransferase (EC:2.3.1.51); PP, phosphatidate phosphatase (EC:3.1.3.4); DGAT1, diacylglycerol *O*-acyltransferase 1 (EC:2.3.1.20); PDAT1, phospholipid: diacylglycerol acyltransferase 1 (EC:2.3.1.158); LPCAT, lysophosphatidylcholine acyltransferase (EC:2.3.1.23 2.3.1.67); PLA2, Phospholipase A2 (EC:3.1.1.4). Lipid substrates are abbreviated: 16:0, palmitic acid; 18:0, stearic acid; 18:1, oleic acid; 18:2, linoleic acid.

linoleic acids (18:2) are major constituents of yellow horn oil, according to a previous study, and ideal biodiesel should contain at least one double bond [19]; therefore, AAD, FAD2 and FAD6 are potential biotechnological targets for adjusting yellow horn oil composition.

Unigenes Related to Catabolism Pathways for TAGs and FAs

The complete breakdown of TAGs can be divided into two steps [60]. First, TAGs are metabolised to free FAs; in other words, lipases catalyse the hydrolysis of ester bonds that link fatty acyl chains to the glycerol backbone. During this research, three unigenes that encode triacylglycerol lipase (TGL, EC: 3.1.1.3), which releases fatty acids and intermediate products (DAG or monoacylglycerol) from TAG or DAG, were identified in the transcriptome library (Table 4). In the second step, fatty acids are catabolised to acetyl-CoA, allowing them to be further broken down by oxidation or to follow other metabolic pathways, including re-esterification with glycerol, to form new acylglycerols [61]. Based on the KEGG pathway assignment, we identified 83 unigenes that code for enzymes related to fatty acid catabolism; three key enzymes were acyl-CoA oxidase (ACOX, EC: 1.3.3.6), enoyl-CoA hydratase/3-hydroxyacyl-CoA dehydrogenase (MFP2, EC: 4.2.1.17 1.1.1.35 1.1.1.211) and acetyl-CoA acyltransferase (ACAA, EC: 2.3.1.16), which were encoded by 11, 4 and 8 unigenes, respectively (Table 3). Acetyl-CoA generated through fatty acid catabolism is then used to produce energy for the cell via the citrate cycle or may participate in the synthesis of TAG.

TAG and FA catabolism proceeds in a direction opposite that of their synthesis. Therefore, identifying ways to suppress enzymes involved in TAG and FA catabolism may be another method of increasing the accumulation of lipids under conditions that do not affect plant growth. However, suppressing the expression of TGL increases TAG levels but results in severely stunted growth [62].

Detection of Transcription Factors (TFs) Involved in Oil Synthesis

In previous studies, a set of TFs, including LEC1, LEC2, ABI3, FUS3 and WRI1 which play key roles in seed oil synthesis and deposition, were identified [63–65]. These TFs were reviewed by Fobert and made available online (http://lipidlibrary.aocs.org/plantbio/transfactors/index.htm).

To identify TFs that regulate seed oil synthesis, BLASTx was used to search against the AGRIS (*Arabidopsis* Gene Regulatory Information Server) database with an e-value cut off 10^{-5} [66]. The results showed that 3,341 unigenes were annotated with 905 independent coding sequences of *Arabidopsis* TFs belonging to 49 known TF families. However, none of the unigenes were annotated to the AtRKD family (Table S3). The largest number of unigenes (817) was annotated to the Trihelix family, followed by the C2H2 family (519 unigenes). TF genes in the PHD, CAMTA and JUMONJI families were identified in the yellow horn transcriptome.

Among the 3,347 unigenes, 25 encoding eight TFs involved in oil biosynthesis were detected (Table 5), including ABI3, L1L, ADOF1, EMF2, HSI2, HSI2-L1, AP2 and GL2. However, none of the unigenes showed homology to LEC1, LEC2, FUS3, WRI1, PKL, FIE or SWN.

Table 3. Enzymes/protein related to FA biosynthesis and metabolism identified by annotation of the yellow horn unigenes.

Symbol	Enzymes/Protein	EC Number	Number of unigenes
Fatty acid biosynthesis			
accA	α-acetyl-CoA carboxylase carboxyl transferase	EC:6.4.1.2	4
accB, bccP	acetyl-CoA carboxylase biotin carboxyl carrier protein	–	3
accC	acetyl-CoA carboxylase/biotin carboxylase	EC:6.4.1.2	6
accD	β-acetyl-CoA carboxylase carboxyl transferase	EC:6.4.1.2	3
ACP	acyl-carrier protein		9
MAT	Malonyl-CoA-ACP transacylase	EC:2.3.1.39	1
KASII	3-Ketoacyl ACP synthase II,	EC:2.3.1.179	7
KAR	3-Ketoacyl ACP reductase	EC:1.1.1.100	6
KASIII	3-Ketoacyl ACP synthase III	EC:2.3.1.180	1
EAR	enoyl-ACP reductase I	EC:1.3.1.9	2
HAD	3R-hydroxymyristoyl ACP dehydrase	EC:4.2.1.-	2
FATA	fatty acyl-ACP thioesterase A	EC:3.1.2.14 3.1.2.-	2
FATB	fatty acyl-ACP thioesterase B	EC:3.1.2.14 3.1.2.-	3
Fatty acid elongation			
KCS	3-ketoacyl-CoA synthase	EC:2.3.1.-	20
KR	β-keto reductase	EC:1.1.1.-	3
PHS1	3-hydroxy acyl-CoA dehydratase	EC:4.2.1.-	6
MECR	mitochondrial trans-2-enoyl-CoA reductase	EC:1.3.1.38	1
TER	enoyl reductase	EC:1.3.1.-	7
PCH	palmitoyl-CoA hydrolase	EC:3.1.2.2	1
PPT	palmitoyl-protein thioesterase	EC:3.1.2.22	3
Fatty acid desaturation			
AAD	acyl-ACP desaturase	EC:1.14.19.2	6
FAD2	Δ12(ω6)-Desaturase (endoplasmic reticulum)	EC:1.14.19.-	5
FAD3	Δ15(ω3)-Desaturase (microsomal)	EC:1.14.19.-	2
FAD6	Δ12(ω6)-Desaturase (chloroplast)	EC:1.14.19.-	1
FAD7	Δ15(ω3)-Desaturase (chloroplast)	EC:1.14.19.-	5
FAD8	Δ15(ω3)-Desaturase (chloroplast)	EC:1.14.19.-	1
Fatty acid metabolism			
ACAA1	acetyl-CoA acyltransferase 1	EC:2.3.1.16	8
atoB	acetyl-CoA C-acetyltransferase	EC:2.3.1.9	6
ACADM	acyl-CoA dehydrogenase	EC:1.3.99.3	7
ACOX	acyl-CoA oxidase	EC:1.3.3.6	11
ACSL	long-chain acyl-CoA synthetase	EC:6.2.1.3	16
ADH	alcohol dehydrogenase	EC:1.1.1.1	15
ALDH	aldehyde dehydrogenase (NAD+)	EC:1.2.1.3	16
MFP2	enoyl-CoA hydratase/3-hydroxyacyl-CoA dehydrogenase	EC:4.2.1.17 1.1.1.35 1.1.1.2114	

Identification of Molecular Markers Located in Unigenes Related to Oil Biosynthesis and Metabolism

Simple sequence repeats (SSRs), single nucleotide polymorphisms (SNPs) and insertions and deletions (InDels) are valuable tools for genetic analysis because they are highly polymorphic within species. They can be used for molecular marker-assisted selection (MAS), which is a rapid approach to the development of new crop varieties (particularly in perennials and trees) that reduces the assessment time considerably [67,68], association mapping to find genes related to good agronomic characteristics [69], and polymorphism analysis, among other techniques. Molecular markers located in genes will likely be related to the functions of those genes. Some markers that are located in coding regions related to oil synthesis have been developed and applied [70,71].

In this study, a total of 6,707 SSRs distributed in 5,631 unigenes (4.61 Mb) were identified as potential molecular markers, of which 887 sequences contained more than one SSR. The most common SSRs were dinucleotide repeats, occurring at 3,598 loci (53.65%), followed by trinucleotide repeats (2,976, 44.37%). However, tetranucleotide, pentanucleotide and hexanucleotide repeats were found at lower frequencies, occurring only in 72 (1.15%), 17

Table 4. Enzymes related to TAG biosynthesis and metabolism identified by annotation of the yellow horn unigenes.

Symbol	Enzymes/Protein	EC Number	Number of unigenes
TAG biosynthesis			
GK	glycerol kinase	EC:2.7.1.30	1
ATS1	glycerol-3-phosphate acyltransferase (chloroplast)	EC:2.3.1.15	3
GPAT5	glycerol-3-phosphate acyltransferase 5	EC:2.3.1.15	1
GPAT6	glycerol-3-phosphate acyltransferase 6	EC:2.3.1.15	3
GPAT8	glycerol-3-phosphate acyltransferase 8	EC:2.3.1.15	2
GPAT9	glycerol-3-phosphate acyltransferase 9	EC:2.3.1.15	3
LPAT 1	lysophosphatidyl acyltransferase 1 (chloroplast)	EC:2.3.1.51	4
LPAT 2	lysophosphatidyl acyltransferase 2	EC:2.3.1.51	2
LPAT 4	lysophosphatidyl acyltransferase 4	EC:2.3.1.51	3
LPAT 5	lysophosphatidyl acyltransferase 5	EC:2.3.1.51	2
PP	phosphatidate phosphatase	EC:3.1.3.4	4
DGAT1	diacylglycerol *O*-acyltransferase 1	EC:2.3.1.20	1
PDAT1	phospholipid:diacylglycerol acyltransferase 1	EC:2.3.1.158	4
ACBP	acyl CoA binding protein		7
Acyl editing			
PLA2	Phospholipase A2	EC:3.1.1.4	1
LPCAT	lysophosphatidylcholine acyltransferase	EC:2.3.1.23 2.3.1.67	5
TAG metabolism			
TGL4	TAG lipase	EC:3.1.1.3	3

(0.25%) and 44 (0.66%) SSRs, respectively (Table 6). Polymorphisms in these potential markers will the focus of future research.

To increase the authenticity of SNP and InDel identification, we also filtered the results based on stricter multiple criteria, including read depth and allele frequency, compared with previous studies (see materials and methods). In total, 16,925 SNPs distributed across 4,401 different isotig groups that corresponded to 4,234 different unigenes had a total length of 6.10 Mb (Figure 5), resulting in an SNP occurrence rate of 0.003 per base and four SNPs per unigene. Transitions contained 9,225 (54.51%) SNPs and were primarily transversions (7,700 SNPs, 45.49%). The A/G and C/T transition genotypes had similar percentages, but among the four transversion genotypes, C/G transversions were less frequent than the other three (A/T, G/T, A/C). A total of 6,201 InDels were identified from 3,161 isotig groups that corresponded to 3,094 unigenes, with a total length of 4.41 Mb (Figure 5), which indicates there were two variations per unigene. Insertions (2,862, 46.15%) were slightly less common than deletions (3,339, 53.85%). The length of insertions ranged from 1 to 24 bp, and deletions were 1 to 22 bp in length.

We summed the SSRs, SNPs and InDels that were located in the 281 unigenes involved in the FA and TAG biosynthesis and metabolism pathways, which encoded oleosin, caleosin, steroleosin and TFs that regulated seed oil deposition. As a result, 26 SSRs, 194 SNPs and 60 InDels were found in 74 unigenes, which covered 31 enzymes or proteins and two TFs (ABI3 and HSI2-L1) (Table S4, Table S5 and Table S6). The remaining enzymes, such as MAT, EAR, FATA, ADH, ALDH, PCH, FAD6, PP and TGL4, and TFs, such as L1L, ADOF1, EMF2, HSI2, AP2 and GL2, contained no SSRs, SNPs or InDels.

Because we used larger samples from different plant regions and 1.25 runs to obtain the transcriptome, we found 29,833 (SSRs,

Table 5. Putative transcription factors related to the oil biosynthesis in yellow horn.

TF Name	TF Family	Unigene number
ABI3 (Abscisic Acid Insensitive 3)	ABI3VP1	2
L1L (Leafy Cotyledon 1-Like)	CCAAT-HAP3	1
ADOF1 (Arabidopsis Dof Zinc Finger Protein 1)	C2C2-Dof	1
EMF2 (Embryonic Flower 2)	C2H2	5
HSI2 (High-Level Expression of Sugar-Inducible Gene 2)	ABI3VP1	5
HSI2-L1/HSL1 (HSI2-Like 1)	ABI3VP2	7
AP2 (APETALA2)	AP2-EREBP	1
GL2 (GLABRA2)	Homeobox	3

Table 6. Number of SSRs in yellow horn.

Repeat motif	Repeat numbers							Number of SSRs	Percentage (%)
	5	6	7	8	9	10	>10		
Di-	–	1,052	709	535	395	257	650	3,598	53.65
Tri-	1,687	736	281	154	56	26	36	2,976	44.37
Tetra-	53	15	3	–	–	–	1	72	1.07
Penta-	15	2	–	–	–	–	–	17	0.25
Hexa-	32	8	1	–	1	2	–	44	0.66
Total	1,787	1,813	992	687	449	285	794	6,707	
Percentage (%)	26.64	27.03	14.82	10.27	6.74	4.25	10.24		

SNPs and InDels) potential molecular markers in 9,720 unigenes (18.74%) with a total length of 10.37 Mb, for an average of 3.07 markers per unigene and spaces of 347.58 bp between markers. These markers will play significant roles not only in the production of genetically improved varieties of yellow horn with different oil compositions, increased oil yield and improved agronomic characteristics, but also for studying the evolution and origin of Xanthoceras, and even the Sapindaceae.

Conclusions

In this study, we report the first comprehensive yellow horn sequencing effort using 454 GS FLX. Transcriptome analysis using four tissues (buds, leaves, flowers and developing seeds) of yellow horn found 51,867 unigenes (45 to 10,088 bp), which corresponded to 36.1 Mb and mean, N50 and median lengths of 696, 928 and 570 bp, respectively. These unigenes provide a strong basis for future genomic research to develop microarrays for gene expression assays and can serve as a reference transcriptome for future yellow horn RNA-seq experiments. In addition, 281 unigenes that code for key enzymes and TFs that are involved in reconstructed metabolic pathways for FA and TAG biosynthesis and metabolism were identified. Moreover, a large number of potential molecular markers (6,707 SSRs, 16,925 SNPs and 6,201 InDels) were predicted. Among them, 26 SSRs, 194 SNPs and 60 InDels were identified in 74 unigenes that are related to oil biosynthesis and metabolism. These findings will make a substantial contribution to efforts to improve crop characteristics and will accelerate the breeding of new yellow horn varieties.

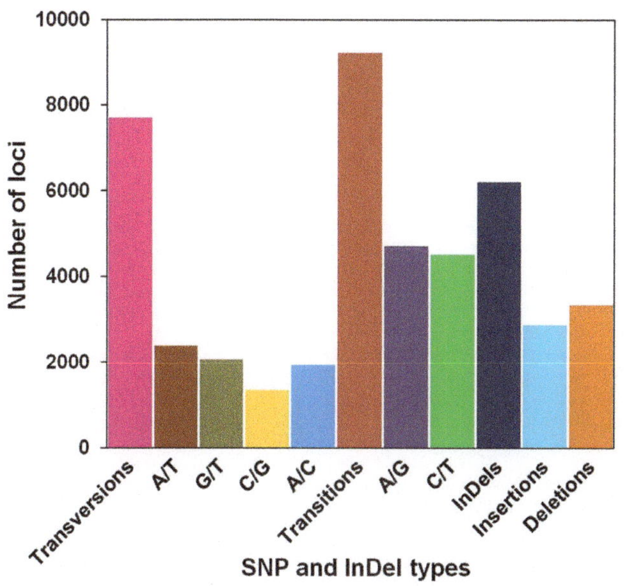

Figure 5. Distribution of putative single nucleotide polymorphisms (SNPs) and insertions and deletions (InDels) in the yellow horn transcriptome.

Materials and Methods

Collection of Tissues for RNA Extraction

Yellow horn is widely distributed in China, so it has not been listed as an endangered or protected species. In this study, we used yellow horn buds, leaves (young and mature leaves), flowers, and developing seeds (10, 20, 30, 40, 50, 60, and 70 days after pollination) which were collected from 30 plants at the two locations. One is located in Xishan forest farm in Haidian, Beijing, China (E116°04′, N40°03′). This forest farm belongs to Beijing Forestry University and the yellow horn trees in this farm have been used for scientific research for several years. The other one is located in a small hill in Chengde city, Hebei province, China (E117°55′, N40°59′). The yellow horn trees which growing in this location were planted for scientific research by Dr. Ao Yan, a co-authors of this research. These samples were immediately frozen in liquid nitrogen and stored at −80°C. Because our research team is also engaged in the work of conservation and utilization of wild plant resources, we confirm that the field studies did not involve any endangered or protected species.

cDNA Library Construction and 454 Sequencing

Total RNA was extracted separately from the buds, leaves, flowers and developing seeds using RNeasy Plant Mini Kits (Qiagen, Inc., Valencia, CA, USA) following the manufacturer's protocol. Extracted RNA was qualified and quantified using a Nanodrop ND-1000 Spectrophotometer (Nanodrop Technologies, Wilmington, DE, USA) and all the samples showed a 260/280 nm ratio from 1.9 to 2.1. Poly(A)$^+$ RNA was purified from total RNA by Oligotex mRNA Mini Kit (Qiagen, Inc., Valencia, CA, USA) following the manufacturer's protocol. After that, equal quantities of total RNA from buds, leaves, flowers and seeds were mixed together. A total of 10 µg of total RNA was used for cDNA library construction. cDNA library construction and normalisation were performed using protocols described previously [72]. The resulting library was sequenced by means of one and a quarter 454 plate runs on a GS-FLX Titanium platform (Roche, USA).

Analysis of 454 Transcriptome Sequencing Results

The raw reads were trimmed before assembly. First, adaptors and SMART primers that were used in the pyrosequencing reactions were cut using the SeqClean software. Then, we used the LUCY2 software [73] to remove low-quality regions and bases. Trimmed reads that were shorter than 45 bp were discarded and the remaining reads were assembled into isotigs and singlets using Newbler (version 2.6) [40]. Finally, after clustering the isotigs and singlets using CD-HIT (version 4.5.6) [74], the obtained unigenes were used in further analyses.

To understand their functions, the yellow horn unigenes were annotated using BLASTx alignment with an E-value cut-off of 10^{-5} against the following protein databases: NCBI non-redundant (NR), Swiss-Prot, Conserved Domain Database (CDD), Pfam protein families database (Pfam), UniProtKB/TrEMBL Protein Database (TrEMBL), Clusters of Orthologous Groups of proteins (COG), Gene Ontology (GO), Kyoto Encyclopaedia of Genes and Genomes (KEGG), and The Arabidopsis Information Resource (TAIR). GO functional classifications and KEGG pathway assignments were performed, as was described previously [75].

Detection of SSRs, SNPs, InDels and TFs

To detect simple sequence repeats (SSRs) in the yellow horn transcriptome, the MISA (http://pgrc.ipk-gatersleben.de/misa/) software was used to identify all 2–6-bp motifs in the unigenes. The minimum repeat unit size was set at six for di-nucleotides and five for tri-, tetra-, penta-, and hexa-nucleotides.

The ssahaSNP software tool [76] was used to identify SNPs and InDels (1 to 100 bp in size) that had high coverage depths. To add the polymorphism and reliability of the SNPs and InDels, as in previous studies [26,77,78], we kept only SNPs and InDels that met the following strict quality criteria: (1) For SNP detection, we identified SNPs that had coverage depths of at least 20 and an alternate allele that was present at a minimum frequency of 20% in all isotigs that contained at least 20 reads. (2) For InDel detection, similar to the standard for SNPs, for insertions or deletions of one base, the coverage depth of the isotig and the InDel allele frequency were set at 20 and 20%, respectively. For insertions or deletions of two or more bases, the coverage depth of the isotigs and the InDel allele frequency were set at 10 and 10%, respectively.

To identify TFs, a BLAST search for all unigenes was conducted using AtTFDB (*Arabidopsis* transcription factor database) with shared identities >77%, as described previously [23].

Data Deposition

The Roche 454 reads of yellow horn were deposited in the NCBI and can be accessed in the Short Read Archive (SRA) under accession number SRP026671.

Supporting Information

Table S1 Pathway annotation of unignenes from yellow horn.

Tabel S2 Unigenes annotated as oleosin, caleosin and steroleosin.

Table S3 Putative transcription factors encoding unigenes in yellow horn.

Table S4 SSRs located in unigenes related to oil biosynthesis and metabolism.

Table S5 SNPs located in unigenes related to oil biosynthesis and metabolism.

Table S6 InDels located in unigenes related to oil biosynthesis and metabolism.

Acknowledgments

We thank Dr. Shihui Niu (College of Biological Science and Biotechnology, Beijing Forest University) for assistance in drawing some figures.

Author Contributions

Conceived and designed the experiments: YLL WL ZXZ. Performed the experiments: YLL. Analyzed the data: YLL. Contributed reagents/materials/analysis tools: YLL ZDH YA. Wrote the paper: YLL WL ZXZ.

References

1. Durrett TP, Benning C, Ohlrogge J (2008) Plant triacylglycerols as feedstocks for the production of biofuels. The Plant Journal 54: 593–607.
2. Pinzi S, Garcia I, Lopez-Gimenez F, Luque de Castro M, Dorado G, et al. (2009) The ideal vegetable oil-based biodiesel composition: a review of social, economical and technical implications. Energy & Fuels 23: 2325–2341.
3. Dorado M, Cruz F, Palomar J, Lopez F (2006) An approach to the economics of two vegetable oil based biofuels in Spain. Renewable Energy 31: 1231–1237.
4. Fairless D (2007) Biofuel: the little shrub that could-maybe. Nature 449: 652–655.
5. Wang F, Xiong XR, Liu CZ (2009) Biofuels in China: opportunities and challenges. In Vitro Cellular & Developmental Biology-Plant 45: 342–349.
6. Huang YH, Wu JH (2008) Analysis of biodiesel promotion in Taiwan. Renewable and Sustainable Energy Reviews 12: 1176–1186.
7. Zu Y, Zhang S, Fu Y, Liu W, Liu Z, et al. (2009) Rapid microwave-assisted transesterification for the preparation of fatty acid methyl esters from the oil of yellow horn (*Xanthoceras sorbifolia* Bunge.). European food research and technology 229: 43–49.
8. Fu H, Guo Y, Li W, Dou D, Kang T, et al. (2010) A new angeloylated triterpenoid saponin from the husks of *Xanthoceras sorbifolia* Bunge. Journal of natural medicines 64: 80–84.
9. Yu H, Zhou S (2009) Preparation of biodiesel from *Xanthoceras sorblfolia* Bunge. seed oil. China Oils and Fats 3: 43–45.
10. Zhang S, Zu YG, Fu YJ, Luo M, Liu W, et al. (2010) Supercritical carbon dioxide extraction of seed oil from yellow horn (*Xanthoceras sorbifolia* Bunge.) and its anti-oxidant activity. Bioresource technology 101: 2537–2544.
11. Kong WB, Liang JY, Ma ZX, Zhang J (2011) Rasearch advance of *xanthoceras sorbifolia* Bunge oil. China Oils and Fats. 36 67–72.
12. Harrington KJ (1986) Chemical and physical properties of vegetable oil esters and their effect on diesel fuel performance. Biomass 9: 1–17.
13. Yao Z, Wang L, Qi J (2009) Biosorption of methylene blue from aqueous solution using a bioenergy forest waste: *Xanthoceras sorbifolia* seed coat. CLEAN–Soil, Air, Water 37: 642–648.
14. Zhang J, Chen G, Sun Q, Li Z, Wang Y (2012) Forest biomass resources and utilization in China. African Journal of Biotechnology 11: 9302–9307.
15. Yao ZY, Qi JH, Yin LM (2013) Biodiesel production from *Xanthoceras sorbifolia* in China: Opportunities and challenges. Renewable and Sustainable Energy Reviews 24: 57–65.
16. Zhang S, Zu YG, Fu YJ, Luo M, Zhang DY, et al. (2010) Rapid microwave-assisted transesterification of yellow horn oil to biodiesel using a heteropolyacid solid catalyst. Bioresource technology 101: 931–936.
17. Fu YJ, Zu YG, Wang LL, Zhang NJ, Liu W, et al. (2008) Determination of fatty acid methyl esters in biodiesel produced from yellow horn oil by LC. Chromatographia 67: 9–14.
18. Li J, Fu YJ, Qu XJ, Wang W, Luo M, et al. (2012) Biodiesel production from yellow horn (*Xanthoceras sorbifolia* Bunge.) seed oil using ion exchange resin as heterogeneous catalyst. Bioresource technology 108: 112–118.

19. Li X, Hou S, Su M, Yang M, Shen S, et al. (2010) Major energy plants and their potential for bioenergy development in China. Environmental management 46: 579–589.
20. Gibbons JG, Janson EM, Hittinger CT, Johnston M, Abbot P, et al. (2009) Benchmarking next generation transcriptome sequencing for functional and evolutionary genomics. Molecular biology and evolution 26: 2731–2744.
21. Nookaew I, Papini M, Pornputtapong N, Scalcinati G, Fagerberg L, et al. (2012) A comprehensive comparison of RNA-Seq-based transcriptome analysis from reads to differential gene expression and cross-comparison with microarrays: a case study in *Saccharomyces cerevisiae*. Nucleic Acids Research 40: 10084–10097.
22. Metzker ML (2009) Sequencing technologies-the next generation. Nature Reviews Genetics 11: 31–46.
23. He M, Wang Y, Hua W, Zhang Y, Wang Z (2012) *De novo* sequencing of *Hypericum perforatum* transcriptome to identify potential genes involved in the biosynthesis of active metabolites. PloS one 7: e42081.
24. Tao X, Gu YH, Wang HY, Zheng W, Li X, et al. (2012) Digital Gene Expression Analysis Based on Integrated *De Novo* Transcriptome Assembly of Sweet Potato [*Ipomoea batatas* (L.) Lam.]. PloS one 7: e36234.
25. Zhang J, Liang S, Duan J, Wang J, Chen S, et al. (2012) *De novo* assembly and Characterisation of the Transcriptome during seed development, and generation of genic-SSR markers in Peanut (*Arachis hypogaea* L.). BMC genomics 13: 90.
26. Sloan DB, Keller SR, Berardi AE, Sanderson BJ, Karpovich JF, et al. (2012) *De novo* transcriptome assembly and polymorphism detection in the flowering plant *Silene vulgaris* (Caryophyllaceae). Molecular Ecology Resources 12: 333–343.
27. Beulé T, Camps C, Debiesse S, Tranchant C, Dussert S, et al. (2011) Transcriptome analysis reveals differentially expressed genes associated with the mantled homeotic flowering abnormality in oil palm (*Elaeis guineensis*). Tree Genetics & Genomes 7: 169–182.
28. Guimarães P, Brasileiro A, Morgante C, Martins A, Pappas G, et al. (2012) Global transcriptome analysis of two wild relatives of peanut under drought and fungi infection. BMC genomics 13: 387.
29. Wei W, Qi X, Wang L, Zhang Y, Hua W, et al. (2011) Characterization of the sesame (*Sesamum indicum* L.) global transcriptome using Illumina paired-end sequencing and development of EST-SSR markers. BMC genomics 12: 451.
30. Li H, Dong Y, Yang J, Liu X, Wang Y, et al. (2012) *De novo* transcriptome of safflower and the identification of putative genes for oleosin and the biosynthesis of flavonoids. PloS one 7: e30987.
31. Trick M, Long Y, Meng J, Bancroft I (2009) Single nucleotide polymorphism (SNP) discovery in the polyploid *Brassica napus* using Solexa transcriptome sequencing. Plant Biotechnology Journal 7: 334–346.
32. Natarajan P, Parani M (2011) *De novo* assembly and transcriptome analysis of five major tissues of *Jatropha curcas* L. using GS FLX titanium platform of 454 pyrosequencing. BMC genomics 12: 191.
33. Mundry M, Bornberg-Bauer E, Sammeth M, Feulner PG (2012) Evaluating characteristics of *de novo* assembly software on 454 transcriptome data: a simulation approach. PloS one 7: e31410.
34. Hills MJ (2004) Control of storage-product synthesis in seeds. Current opinion in plant biology 7: 302–308.
35. Huang AH (1992) Oil bodies and oleosins in seeds. Annual review of plant biology 43: 177–200.
36. Voelker T, Kinney AJ (2001) Variations in the biosynthesis of seed-storage lipids. Annual review of plant biology 52: 335–361.
37. Slabas AR, Fawcett T (1992) The biochemistry and molecular biology of plant lipid biosynthesis. 10 Years Plant Molecular Biology: Springer. 169–191.
38. Ohlrogge J, Chapman K (2011) The seeds of green energy: expanding the contribution of plant oils as biofuels. The Biochemist 33: 34–38.
39. Faergeman NJ, Knudsen J (1997) Role of long-chain fatty acyl-CoA esters in the regulation of metabolism and in cell signalling. Biochemical Journal 323: 1–12.
40. Xiao S, Chye ML (2011) New roles for acyl-CoA-binding proteins (ACBPs) in plant development, stress responses and lipid metabolism. Progress in lipid research 50: 141–151.
41. Roesler K, Shintani D, Savage L, Boddupalli S, Ohlrogge J (1997) Targeting of the Arabidopsis homomeric acetyl-coenzyme A carboxylase to plastids of rapeseeds. Plant Physiology 113: 75–81.
42. Madoka Y, Tomizawa KI, Mizoi J, Nishida I, Nagano Y, et al. (2002) Chloroplast transformation with modified *accD* operon increases acetyl-CoA carboxylase and causes extension of leaf longevity and increase in seed yield in tobacco. Plant and cell physiology 43: 1518–1525.
43. Jha JK, Sinha S, Maiti MK, Basu A, Mukhopadhyay UK, et al. (2007) Functional expression of an acyl carrier protein (ACP) from *Azospirillum brasilense* alters fatty acid profiles in *Escherichia coli* and *Brassica juncea*. Plant Physiology and Biochemistry 45: 490–500.
44. Pidkowich MS, Nguyen HT, Heilmann I, Ischebeck T, Shanklin J (2007) Modulating seed β-ketoacyl-acyl carrier protein synthase II level converts the composition of a temperate seed oil to that of a palm-like tropical oil. Proceedings of the National Academy of Sciences 104: 4742–4747.
45. Dehesh K, Tai H, Edwards P, Byrne J, Jaworski JG (2001) Overexpression of 3-ketoacyl-acyl carrier protein synthase IIIs in plants reduces the rate of lipid synthesis. Plant Physiology 125: 1103–1114.
46. Shimakata T, Stumpf PK (1983) Purification and characterization of β-ketoacyl-ACP synthetase I from *Spinacia oleracea* leaves. Archives of biochemistry and biophysics 220: 39–45.
47. Li-Beisson Y, Shorrosh B, Beisson F, Andersson MX, Arondel V, et al. (2013) Acyl-lipid metabolism. The Arabidopsis book. Published By: American Society of Plant Biologists. Available: http://arabidopsisacyllipids.plantbiology.msu.edu/data/tab_article.pdf.
48. Bates PD, Stymne S, Ohlrogge J (2013) Biochemical pathways in seed oil synthesis. Current opinion in plant biology 16: 358–364.
49. Jain R, Coffey M, Lai K, Kumar A, MacKenzie S (2000) Enhancement of seed oil content by expression of glycerol-3-phosphate acyltransferase genes. Biochemical Society Transactions 28: 959–960.
50. Zou J, Katavic V, Giblin EM, Barton DL, MacKenzie SL, et al. (1997) Modification of seed oil content and acyl composition in the brassicaceae by expression of a yeast *sn-2* acyltransferase gene. The Plant Cell 9: 909–923.
51. Lardizabal K, Effertz R, Levering C, Mai J, Pedroso M, et al. (2008) Expression of *Umbelopsis ramanniana DGAT2A* in seed increases oil in soybean. Plant Physiology 148: 89–96.
52. Jako C, Kumar A, Wei Y, Zou J, Barton DL, et al. (2001) Seed-specific over-expression of an Arabidopsis cDNA encoding a diacylglycerol acyltransferase enhances seed oil content and seed weight. Plant Physiology 126: 861–874.
53. Zheng P, Allen WB, Roesler K, Williams ME, Zhang S, et al. (2008) A phenylalanine in DGAT is a key determinant of oil content and composition in maize. Nature genetics 40: 367–372.
54. Shimada TL, Hara-Nishimura I (2010) Oil-body-membrane proteins and their physiological functions in plants. Biological and Pharmaceutical Bulletin 33: 360–363.
55. Frandsen GI, Mundy J, Tzen JT (2001) Oil bodies and their associated proteins, oleosin and caleosin. Physiologia Plantarum 112: 301–307.
56. Næsted H, Frandsen GI, Jauh GY, Hernandez-Pinzon I, Nielsen HB, et al. (2000) Caleosins: $Ca2^{+}$-binding proteins associated with lipid bodies. Plant molecular biology 44: 463–476.
57. Lin LJ, Tai SS, Peng CC, Tzen JT (2002) Steroleosin, a sterol-binding dehydrogenase in seed oil bodies. Plant Physiology 128: 1200–1211.
58. Murata N, Wada H, Gombos Z (1992) Modes of fatty-acid desaturation in cyanobacteria. Plant and cell physiology 33: 933–941.
59. Tasaka Y, Gombos Z, Nishiyama Y, Mohanty P, Ohba T, et al. (1996) Targeted mutagenesis of acyl-lipid desaturases in *Synechocystis*: evidence for the important roles of polyunsaturated membrane lipids in growth, respiration and photosynthesis. The EMBO journal 15: 6416.
60. Rismani-Yazdi H, Haznedaroglu B, Bibby K, Peccia J (2011) Transcriptome sequencing and annotation of the microalgae *Dunaliella tertiolecta*: Pathway description and gene discovery for production of next-generation biofuels. BMC genomics 12: 148.
61. Jaworski JG, Clough RC, Barnum SR (1989) A cerulenin insensitive short chain 3-ketoacyl-acyl carrier protein synthase in *Spinacia oleracea* leaves. Plant Physiology 90: 41–44.
62. Padham AK, Hopkins MT, Wang TW, McNamara LM, Lo M, et al. (2007) Characterization of a plastid triacylglycerol lipase from Arabidopsis. Plant Physiology 143: 1372–1384.
63. Wang H, Guo J, Lambert KN, Lin Y (2007) Developmental control of Arabidopsis seed oil biosynthesis. Planta 226: 773–783.
64. Baud S, Wuillème S, To A, Rochat C, Lepiniec L (2009) Role of WRINKLED1 in the transcriptional regulation of glycolytic and fatty acid biosynthetic genes in Arabidopsis. The Plant Journal 60: 933–947.
65. Tan H, Yang X, Zhang F, Zheng X, Qu C, et al. (2011) Enhanced seed oil production in canola by conditional expression of *Brassica napus LEAFY COTYLEDON1* and *LEC1-LIKE* in developing seeds. Plant physiology 156: 1577–1588.
66. Palaniswamy SK, James S, Sun H, Lamb RS, Davuluri RV, et al. (2006) AGRIS and AtRegNet. A platform to link cis-regulatory elements and transcription factors into regulatory networks. Plant Physiology 140: 818–829.
67. Mazur B, Krebbers E, Tingey S (1999) Gene discovery and product development for grain quality traits. Science 285: 372–375.
68. O'Malley DM, McKeand SE (1994) Marker assisted selection for breeding value in forest trees. Forest Genetics 1: 207–218.
69. Neale DB, Kremer A (2011) Forest tree genomics: growing resources and applications. Nature Reviews Genetics 12: 111–122.
70. Hu X, Sullivan-Gilbert M, Gupta M, Thompson SA (2006) Mapping of the loci controlling oleic and linolenic acid contents and development of *fad2* and *fad3* allele-specific markers in canola (*Brassica napus* L.). Theoretical and Applied Genetics 113: 497–507.
71. Gupta V, Mukhopadhyay A, Arumugam N, Sodhi Y, Pental D, et al. (2004) Molecular tagging of erucic acid trait in oilseed mustard (*Brassica juncea*) by QTL mapping and single nucleotide polymorphisms in *FAE1* gene. Theoretical and Applied Genetics 108: 743–749.
72. Wang R, Xu S, Jiang Y, Jiang J, Li X, et al. (2013) De novo Sequence Assembly and Characterization of *Lycoris aurea* Transcriptome Using GS FLX Titanium Platform of 454 Pyrosequencing. PloS one 8: e60449.
73. Li S, Chou HH (2004) LUCY2: an interactive DNA sequence quality trimming and vector removal tool. Bioinformatics 20: 2865–2866.
74. Huang Y, Niu B, Gao Y, Fu L, Li W (2010) CD-HIT Suite: a web server for clustering and comparing biological sequences. Bioinformatics 26: 680–682.
75. Wang R, Xu S, Jiang Y, Jiang J, Li X, et al. (2013) *De novo* Sequence Assembly and Characterization of *Lycoris aurea* Transcriptome Using GS FLX Titanium Platform of 454 Pyrosequencing. PloS one 8: e60449.
76. Ning Z, Cox AJ, Mullikin JC (2001) SSAHA: a fast search method for large DNA databases. Genome research 11: 1725–1729.

77. Edwards CE, Parchman TL, Weekley CW (2012) Assembly, gene annotation and marker development using 454 floral transcriptome sequences in *Ziziphus celata* (Rhamnaceae), a highly endangered, Florida endemic plant. DNA research 19: 1–9.

78. Blanca J, Cañizares J, Roig C, Ziarsolo P, Nuez F, et al. (2011) Transcriptome characterization and high throughput SSRs and SNPs discovery in *Cucurbita pepo* (Cucurbitaceae). BMC genomics 12: 104.

14

Efficient Production of Pyruvate from DL-Lactate by the Lactate-Utilizing Strain *Pseudomonas stutzeri* SDM

Chao Gao[1], Jianhua Qiu[1], Cuiqing Ma[1]*, Ping Xu[2,1]

1 State Key Laboratory of Microbial Technology, Shandong University, Jinan, People's Republic of China, 2 State Key Laboratory of Microbial Metabolism, Shanghai Jiao Tong University, Shanghai, People's Republic of China

Abstract

Background: The platform chemical lactate is currently produced mainly through the fermentation of sugars presented in biomass. Besides the synthesis of biodegradable polylactate, lactate is also viewed as a feedstock for the green chemistry of the future. Pyruvate, another important platform chemical, can be produced from lactate through biocatalysis.

Methodology/Principal Findings: It was established that whole cells of *Pseudomonas stutzeri* SDM catalyze lactate oxidation with lactate-induced NAD-independent lactate dehydrogenases (iLDHs) through the inherent electron transfer chain. Unlike the lactate oxidation processes observed in previous reports, the mechanism underlying lactate oxidation described in the present study excluded the costliness of the cofactor regeneration step and production of the byproduct hydrogen peroxide.

Conclusions/Significance: Biocatalysis conditions were optimized by using the cheap DL-lactate as the substrate and whole cells of the lactate-utilizing *P. stutzeri* SDM as catalyst. Under optimal conditions, the biocatalytic process produced pyruvate at a high concentration (48.4 g l^{-1}) and a high yield (98%). The bioconversion system provides a promising alternative for the green production of pyruvate.

Editor: Shuguang Zhang, Massachusetts Institute of Technology, United States of America

Funding: The work was supported by the National Natural Science Foundation of China (31000014 and 31170052), and by the Research Fund for the Doctoral Program of Higher Education of China (Grant no. 20090131110036). The funders had no role in study design, data collection and analysis, decision to publish, or preparation of the manuscript.

Competing Interests: The authors have declared that no competing interests exist.

* E-mail: macq@sdu.edu.cn

Introduction

Lactic acid, the most important hydroxycarboxylic acid, is currently commercially produced by the fermentation of sugars present in biomass [1–3]. In addition to its use in the synthesis of biodegradable polymers [4], lactic acid can also be regarded as a feedstock for the green chemistry of the future [1]. For example, pyruvate, another important platform chemical, can be produced from lactate through direct oxidation [5–7]. Considering the significant difference in their prices, the production of pyruvate from lactate by catalysis is a valuable process.

Most of the reported chemical catalysts convert a major part of lactate to acetaldehyde and CO_2 rather than pyruvate [7]. Indeed, there have been very few attempts to bring about the oxidative dehydrogenation of lactate through chemical catalysis [5,6]. Biocatalysts could catalyze lactate to pyruvate under relatively mild conditions [1,6]. Different enzymes, such as NAD-dependent lactate dehydrogenases (nLDHs) and lactate oxidase, have been employed in the biotechnological production of pyruvate from lactate. However, the costliness of cofactor NAD or the production of the byproduct hydrogen peroxide restricted the industrial application of nLDHs and lactate oxidase, respectively [1,7].

In a previous report, *Pseudomonas stutzeri* SDM was reported to have the ability to produce pyruvate from lactate with oxygen as the terminal electron acceptor [8]. No hydrogen peroxide was produced during *P. stutzeri* SDM catalyzed lactate oxidation, which made the strain a promising biocatalyst for the commercial pyruvate production. NAD-independent lactate dehydrogenases (iLDHs) were reported to be involved in the lactate oxidation process [8–12]. However, iLDHs could not directly use the oxygen as the electron acceptor, which made the lactate oxidation process in strain SDM rather confusing. To further explore the strain with regard to pyruvate production, the aim of the present study is to determine how iLDHs are involved in the oxidation of lactate. After illustrating the pyruvate-producing mechanism, optimal conditions for the production of pyruvate from a cheap substrate, DL-lactate, by the lactate-utilizing strain SDM were also developed.

Results

Pyruvate-producing Mechanism in *P. stutzeri* SDM

P. stutzeri SDM possesses 2 inducible iLDHs that made the strain a good biocatalyst for the 2-oxo-carboxylate production [9,13–15]. Unlike lactate oxidase, iLDHs could not oxidize lactate with oxygen as the directly electron acceptor. There should be an electron transfer system between iLDHs and oxygen in the lactate-utilizing strain SDM. To elucidate the electron transfer system in this case, the effects of different electron transfer inhibitors such as diphenylamine, antimycin A, NaN_3, and salicylhydroxamic acid on the pyruvate-producing activity of the *P. stutzeri* SDM crude

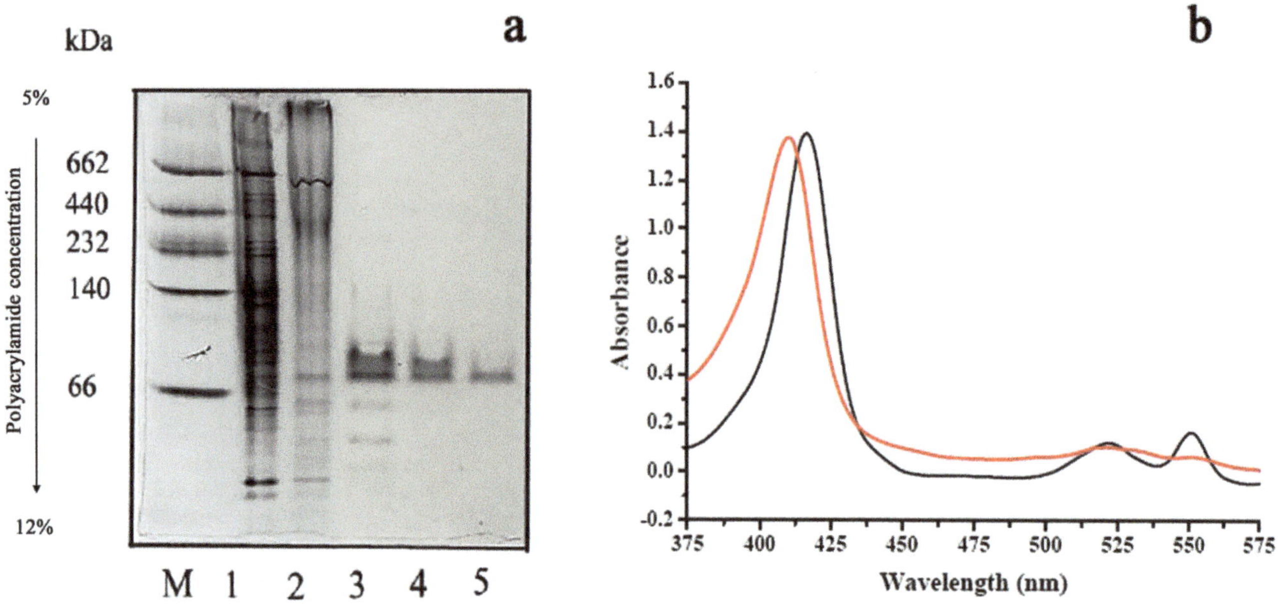

Figure 1. Involvement of cytochrome *c* in lactate oxidation of *P. stutzeri* SDM. (a) Native-PAGE of cytochrome *c* in *P. stutzeri* SDM. Lane M: molecular mass standards in kilodaltons (kDa) (GE Healthcare); lane 1: cell extract of *P. stutzeri* SDM; lane 2: membrane fractions of *P. stutzeri* SDM; lane 3: pooled fraction containing cytochrome *c* after DEAE-Sepharose; lanes 4 and 5: pooled fraction containing cytochrome *c* after DEAE-A25. (b) The absorption spectrum of cytochrome *c* in *P. stutzeri* SDM. Black line: reaction mixtures consisting of cytochrome *c*, crude extract of *P. stutzeri* SDM and DL-lactate. Red line: reaction mixtures consisting of cytochrome *c* and crude extract of *P. stutzeri* SDM.

extract were studied. As shown in Table 1, the cytochrome *c* reductase inhibitor antimycin A distinctly inhibited the pyruvate-producing activity. This implied that iLDHs in *P. stutzeri* SDM might use quinone as the natural electron acceptor. The electron acquired by quinone would be transferred to cytochrome (by cytochrome *c* reductase) and then terminally transferred to oxygen (by cytochrome oxidase). In addition to the traditionally cytochrome oxidase, *Pseudonomas* strains also have some alternative cytochrome oxidase, such as cytochrome cbb_3 oxidase [16]. Although the cytochrome oxidase inhibitors (diphenylamine, NaN_3, and salicylhydroxamic acid) exhibited different inhibition effects on the L- and D-lactate oxidation activities of *P. stutzeri* SDM due to the complexity of the cytochrome oxidase system in *Pseudonomas* strains, those results implied the roles of cytochrome in the lactate oxidation process.

Next, a cytochrome *c* fraction was purified from *P. stutzeri* SDM (Figure 1a). After oxidizing cytochrome *c* with ferricyanide, a crude extract of *P. stutzeri* SDM and DL-lactate were added. As shown in Figure 1b, the addition of DL-lactate to the reaction system produced a characteristic of reduced cytochrome *c*. The reduction of cytochrome *c* by DL-lactate further supported the participation of cytochrome in the lactate oxidation process. On the basis of the results mentioned above, it is concluded that iLDHs and the electron transport chain account for the observed oxidation of lactate in *P. stutzeri* SDM.

Effect of Dissolved Oxygen on Biotransformation Activity

In the case of the lactate oxidation mechanism in *P. stutzeri* SDM, the components of the respiratory chain may influence biotransformation activity distinctly. Because the components of the respiratory chain are different under various dissolved oxygen (DO) concentrations, the effect of DO on the biotransformation activity of *P. stutzeri* SDM was studied in a 5-l reactor. As shown in Figure 2, when 15% DO was employed in the culture, the highest biotransformation activity was obtained.

D-Lactate or L-lactate has been employed in pyruvate production [1,5,7]. Because of the low price and large sources of racemic lactate compared to optical lactate, DL-lactate was used as the substrate for pyruvate production in this work.

Table 1. Effects of different respiratory chain depressors on the lactate oxidation activities.

Depressor	L-Lactate oxidation activity (U mg^{-1})	D-Lactate oxidation activity (U mg^{-1})
No inhibitor	0.170±0.004	0.091±0.004
Salicylhydroxamic acid (0.5 mM)	0.112±0.007	0.069±0.001
Diphenylamine (0.5 mM)	0.149±0.020	0.088±0.002
NaN_3 (0.5 mM)	0.110±0.008	0.086±0.001
Antimycin A (0.1 mM)	0.022±0.002	0.036±0.001

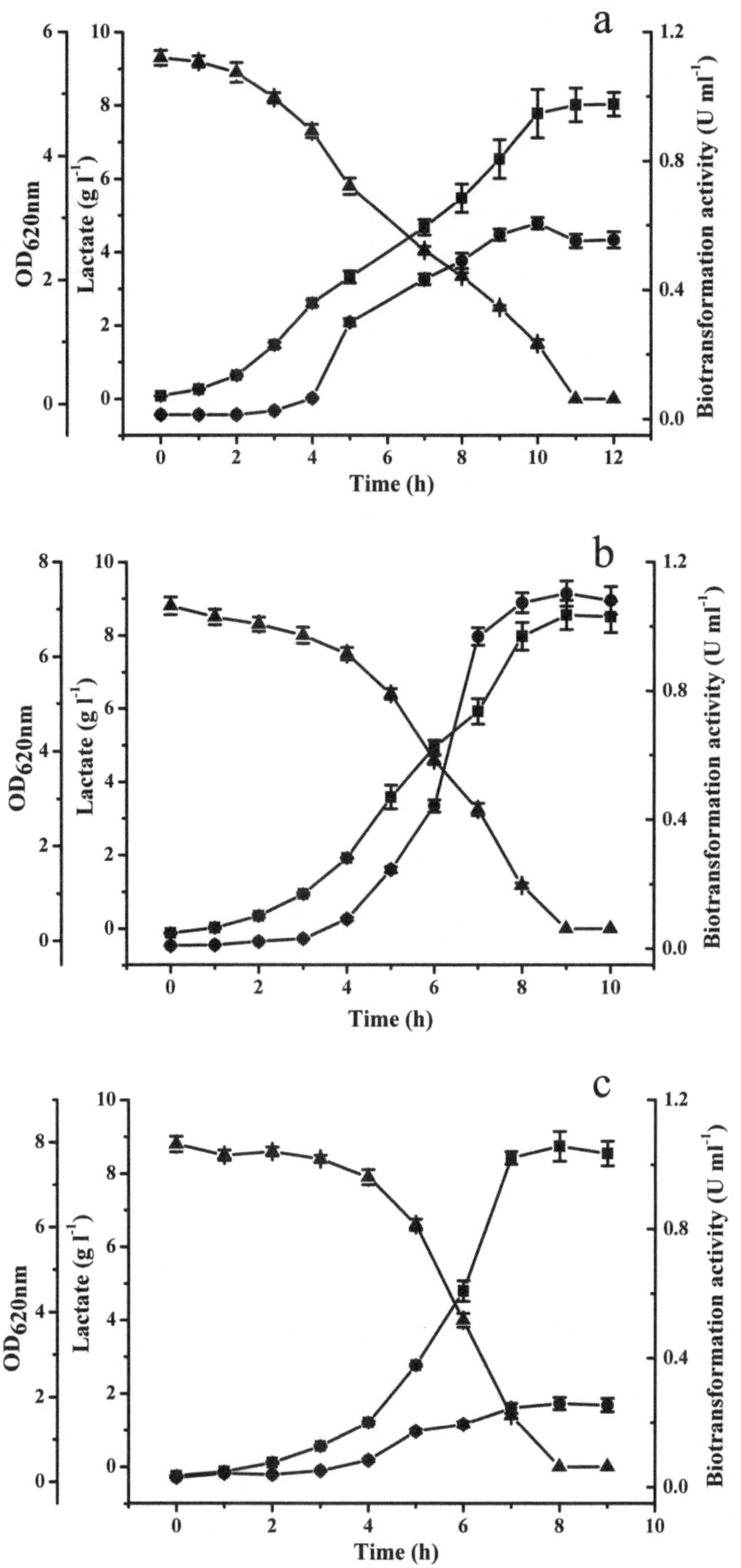
a
OD620nm
Lactate (g l-1)
Biotransformation activity (U ml-1)
Time (h)
b
OD620nm
Lactate (g l-1)
Biotransformation activity (U ml-1)
Time (h)
c
OD620nm
Lactate (g l-1)
Biotransformation activity (U ml-1)
Time (h)

Figure 2. Time course of *P. stutzeri* SDM growth in the media with different dissolved oxygen. (a) 5%; (b) 15%; (c) 30%. (■) OD_{620nm}; (●) Biotransformation activity; (▲) DL-Lactate. Data are the average ± SD of three separate experiments.

Optimal Range of pH

Influence of reaction pH on pyruvate production was determined in 100 mM phosphate buffer containing 6 g dry cell weight (DCW) l^{-1} of cell biomass of *P. stutzeri* SDM and 0.4 M DL-lactate at 30°C. As shown in Figure 3a, after biocatalysis for 24 h, the highest pyruvate production was detected at pH 8.0.

Effect of DL-lactate Concentration on Biocatalysis

Substrate concentration was also important for the conversion of DL-lactate to pyruvate. Effect of the concentration of DL-lactate on biocatalysis was investigated to determine its optimal range. The DL-lactate concentration was varied from 0.15 to 0.75 M. As shown in Figure 3b, after 2 h of biocatalysis, the highest pyruvate production was detected when the concentration of DL-lactate was 0.45 M. High DL-lactate concentration (more than 0.6 M) would lead to substrate inhibition and resulted in lower pyruvate concentrations.

Optimal Reaction Temperature

To investigate the influence of reaction temperature on pyruvate production, the reaction was carried out at different temperatures at pH 8.0. As shown in Figure 4, the highest pyruvate production was detected at 30°C after 24 h. Although high pyruvate production could be detected at 42°C after 2 h, because of biocatalyst instability under high temperature, the optimal reaction temperature was determined to be 30°C.

Optimal Range of Cell Biomass

Biocatalysis was carried out with 2–16 g DCW l^{-1} of *P. stutzeri* SDM as the biocatalyst. As shown in Figure 5, 10 g DCW l^{-1} of *P. stutzeri* SDM was optimal and produced the highest pyruvate concentration and relatively high specific biotransformation activity.

Optimal Range of DO

The biocatalysis production of pyruvate from lactate was a bio-oxidation process. Oxygen would be the terminal electron acceptor. Thus, oxygen was also the substrate of the biocatalysis process and the effect of DO on pyruvate production should be investigated. As shown in Table 2, 0.39 M pyruvate was produced after 39 h with a low amount of DO (5%). Pyruvate production was improved by increasing the DO content to 15%. However, with higher DO (30%), the lactate oxidation decreased, and only 0.19 M pyruvate was produced. This might be due to the substrate inhibition effect of oxygen. The optimal DO content was then determined to be 15%. This result was consistent with our previous report which related to the production of pyruvate from L-lactate [17].

Combining these results, an optimal biotransformation system for the production of pyruvate from DL-lactate was developed. The biocatalysis was conducted at 30°C in distilled water (pH was adjusted to 8.0) with 10 g DCW l^{-1} of *P. stutzeri* SDM as the biocatalyst. The DO saturation was controlled at 15%. The DL-lactate concentration in the 5-l reactor was about 0.45 M. As shown in Figure 6, 0.44 M pyruvate was obtained from 0.45 M DL-lactate after 29 h of biotransformation.

Discussion

Pyruvate is an important starting material widely applied in chemical, pharmaceutical, and agrochemical industries [6,18]. Commercial pyruvate is produced by dehydration and decarboxylation of tartaric acid. This classical chemical route is energy-intensive and controversial with issues of environmental protection and sustainable process development [6,7]. Novel systems which produce pyruvate by biotechnological methods have been the research focus [19–22]. Microbial fermentation currently plays a dominant role in biotechnological production of pyruvate [23–26]. High concentration (135 g l^{-1}) and high volumetric productivity (6 g l^{-1} h^{-1}) of pyruvate have been obtained through fermenta-

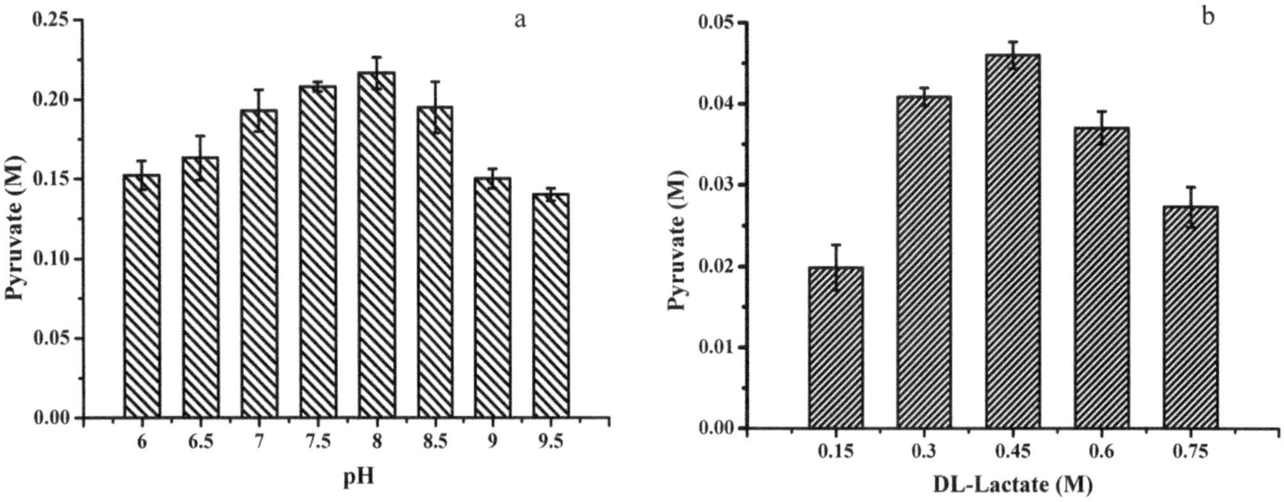

Figure 3. Optimization of biocatalysis conditions. (a) Effect of pH on pyruvate production; (b) Effect of lactate concentrations on pyruvate production.

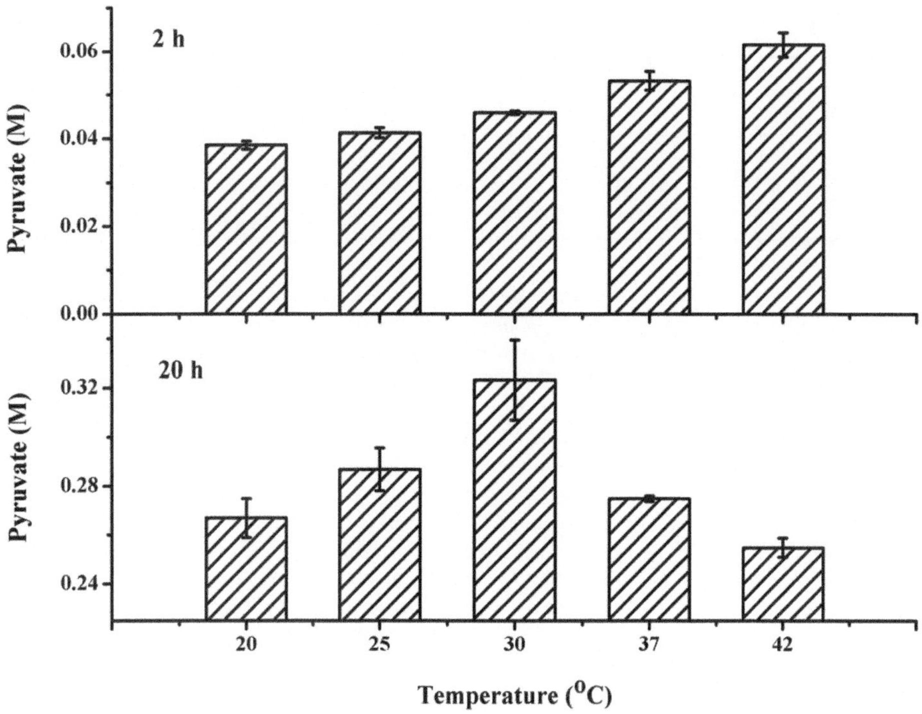

Figure 4. Effect of temperature on the production of pyruvate from DL-lactate.

tion [25,26]. However, biocatalysis processes, due to their simple composition of reaction mixture, high conversion rate of substrate, and convenience of recovery, are also promising in the biotechnological production of pyruvate [6].

Lactate, which can be easily produced from biomass, is the most promising substrate in the biocatalysis production of pyruvate [1,7]. Of the enzymes employed in the lactate-based pyruvate production, lactate oxidase has been studied extensively [1,7].

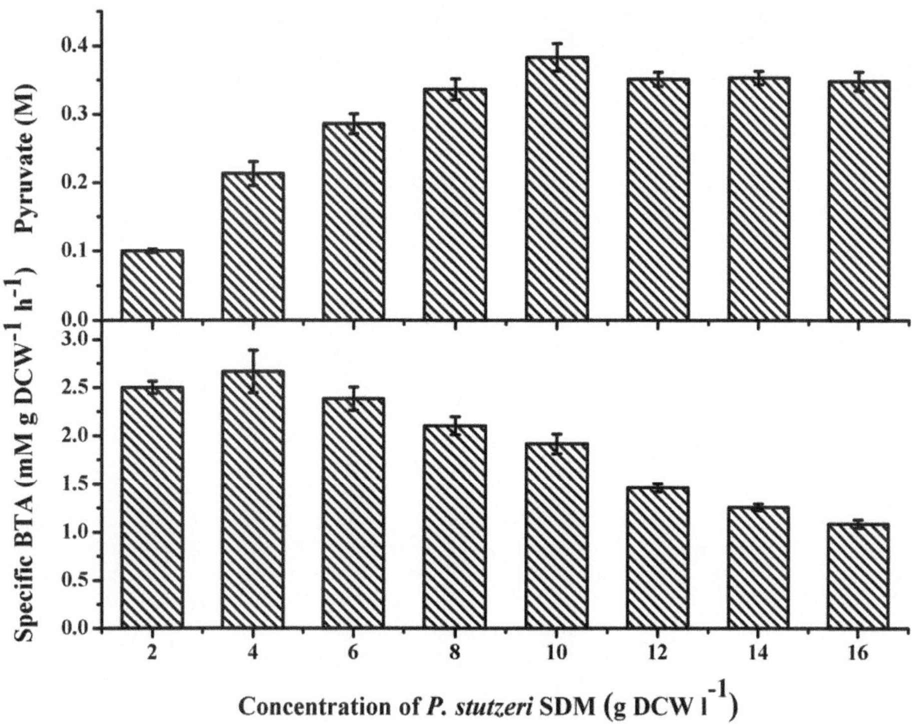

Figure 5. Effect of concentration of *P. stutzeri* SDM on the production of pyruvate from DL-lactate. BTA: biotransformation activity.

Table 2. Effect of DO on the pyruvate production.

DO	Pyruvate concentration (M)	Reaction time (h)	Productivity (mM h^{-1})
5%	0.387	39	9.92
15%	0.44	29	15.17
30%	0.186	25	7.44

Lactate oxidase catalyzes pyruvate formation from L-lactate with oxygen as the second substrate, giving hydrogen peroxide as a byproduct. Hydrogen peroxide decomposes pyruvate to acetate, carbon dioxide, and water, lowering the production yield. If the problem of further oxidation of pyruvate by hydrogen peroxide can be solved, the biocatalysis production of pyruvate from lactate has the potential to be commercialized because of the low price of lactate [1,6,7]. *P. stutzeri* SDM was confirmed to have the ability to produce pyruvate from lactate with the merit of no hydrogen peroxide production. iLDHs in *P. stutzeri* SDM firstly acquired the electron from the lactate and then the electron would be terminally transferred to oxygen. However, the electron transfer process has never been clarified in previous works.

Pyruvate production by (2*R*)-hydroxycarboxylate-viologen-oxidoreductase (HVOR) in *Proteus vulgaris* or *Proteus mirabilis* with the addition of artificial redox mediators was studied in previous reports [27,28]. The regeneration of artificial redox mediators was achieved by chemical or electrochemical methods. However, in the case of *P. stutzeri* SDM, pyruvate was produced without the addition of an artificial redox mediator. The reduction of cytochrome *c* in *P. stutzeri* SDM by DL-lactate implied that the electron transfer components of respiratory chain might play an important role in lactate oxidation. Effects of different respiratory chain inhibitors on the pyruvate production further identified the involvement of the electron transfer chain in the lactate oxidation process. Unlike *Proteus* strains, the cofactor regeneration of iLDHs in *P. stutzeri* SDM may employ the inherent electron transfer system of the strain. This process, which excludes the expensive cofactor regeneration step, makes *P. stutzeri* SDM a rather practicable alternative for the biocatalytic production of pyruvate. Under the optimal conditions, with biocatalyst prepared from 10 g DCW l^{-1} of *P. stutzeri* SDM, 0.44 M pyruvate was obtained after 29 h.

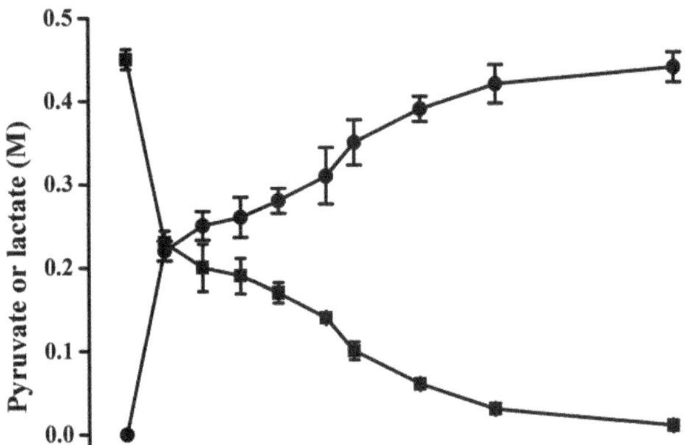

Figure 6. Time course of pyruvate production by *P. stutzeri* SDM. 10 mM EDTA was added in the biocatalysis system to inhibit pyruvate degradation. (■) Pyruvate; (●) DL-lactate. Data are the average ± SD of three separate experiments.

As shown in Table 3, pyruvate production from D-lactate, L-lactate, and DL-lactate has been studied in previous works [29–32]. Genetically modified *Hansenula polymorpha* and *Pichia pastoris* cells expressing glycolate oxidase could catalyze L-lactate into pyruvate with rather high yield [29]. For the decomposition of the byproduct hydrogen peroxide, co-expression of catalase with glycolate oxidase in genetically modified yeasts was needed [1,7]. Whole cells of *P. stutzeri* SDM catalyzed lactate oxidation without the production of hydrogen peroxide. On the other hand, the glycolate oxidase could only utilize L-lactate as the substrate but a large amount of lactate produced today is a racemic mixture of both stereospecific forms. *P. stutzeri* SDM possesses 2 inducible iLDHs which give the strain the ability to use DL-lactate, a much cheaper substrate than the optical lactate, as the substrate to produce pyruvate.

Recently, *Serratia marcescens* ZJB-07166 has been applied in the biotransformation of DL-lactate to pyruvate. The pyruvate concentration of 0.21 M was achieved under an optimum condition [32]. The newly isolated *S. marcescens* ZJB-07166 was regarded as a promising strain for pyruvate production at an industrial scale [32]. Under the conditions optimized in the present study, a much higher concentration of pyruvate (0.44 M) was obtained from the cheap substrate DL-lactate than in previous reports.

In conclusion, the lactate-utilizing *P. stutzeri* strain SDM was confirmed to catalyze lactate oxidation through iLDHs and the inherent electron transfer chain. Preparation of pyruvate from DL-lactate was carried out under the following optimal conditions: DCW, 10 g l^{-1}; pH 8.0; temperature, 30°C; DO saturation, 15%; and DL-lactate concentration, 0.45 M. After 29 h of biotransformation, pyruvate was obtained at a high concentration (48.4 g l^{-1}) and a high yield (98%). The biocatalysis process introduced in this study provides not only a rather feasible method for pyruvate production but also a green pathway for the utilization of lactate produced from biomass.

Materials and Methods

Chemicals

L-Lactate and bovine serum albumin were purchased from Sigma. D-Lactate was purchased from Fluka. DL-Lactate was purchased from Wujiang Ciyun Flavor & Fragrance Co. Ltd. (P. R. China). Diphenylamine, antimycin A, and salicylhydroxamic acid were purchased from Sigma. All other chemicals were of reagent grade.

Microorganism and Biocatalyst Preparation

P. stutzeri SDM (China Center for Type Culture Collection No. M 206010) isolated from soil was used [9]. The minimal salt

Table 3. Overview of literature on the enzymatic synthesis of pyruvate from lactate.

Substrate	Biocatalyst	Pyruvate (g l^{-1})	Yield (%)	Reference
L-Lactate	Whole cells of *Acinetobacter* sp. WLIS with lactate oxidase component	44.6	72	17
L-Lactate	Permeabilized cells of *Hansenula Polymorpha* or *Pichia pastoris* expressing both glycolate oxidase and catalase	111.9	96	29
D-Lactate	*Proteus vulgaris* or *Proteus mirabilis* with HVOR and artificial mediator regeneration system	53.8	94	27, 28
D-Lactate	Whole cells of *Acetobacter* sp. ATCC 21409	10	95	30
DL-Lactate	Whole cells of *P. stutzeri* SDM with iLDHs	22.6	88.6	8
DL-Lactate	Crude extract of *Pseudomonas putida* SM-6 with lactate oxidase and catalase	10.5	82	31
DL-Lactate	Whole cells of *S. marcescens* ZJB-07166	23.1		32
DL-Lactate	Whole cells of *P. stutzeri* SDM with iLDHs	48.4	98	This study

medium (MSM) supplemented with 10.0 g l^{-1} DL-lactate was used as the fermentation medium [9]. For biocatalyst preparation, cells of the SDM strain cultivated at 30°C were harvested from MSM by centrifugation, washed twice with 0.85% (w/v) sterile salt water, and then resuspended in various concentrations with distilled water.

Preparation of Crude Cell Extract

Cells of the SDM strain grown in MSM containing DL-lactate as the sole carbon source were resuspended in 50 mM Tris-HCl (pH 8.0) and disrupted by sonication (Sonics 500 W/20 KHz, USA) in an ice bath. The disrupted cells were subjected to centrifugation for 20 min at 12,000 *g*, and the supernatant was used as crude cell extract.

Purification of Cytochrome in *P. stutzeri* SDM

The membrane fraction of *P. stutzeri* SDM was prepared from the crude cell extract by centrifugation at 147,000 *g* for 180 min. Triton X-100 (10%, w/v) was added to the membrane fraction to a final concentration of 1 mg mg^{-1} of protein. The supernatant (detergent extract) was applied to a column of DEAE Sepharose Fast Flow equilibrated with Buffer A: 50 mM Tris-HCl (pH 8.0) containing 1 mM dithiothreitol (DTT), 5 mM $MgSO_4$, 0.1% Triton X-100, and 1 mM EDTA. The column was washed with Buffer B: 50 mM Tris-HCl (pH 8.0) containing 1 mM DTT, 200 mM KCl, 5 mM $MgSO_4$, 0.1% Triton X-100, and 1 mM EDTA at a flow rate of 5 ml min^{-1}. Cytochrome from column effluents was monitored by measuring the absorbance at 410 nm. The fractions containing cytochrome were concentrated by ultrafiltration and desalted with gel G-25. The cytochrome pool after desalting was then applied to a column of DEAE A-25 pre-equilibrated with Buffer A. The column was washed with a linear gradient of 0–100% Buffer B at a flow rate of 0.5 ml min^{-1}. The fractions containing cytochrome were concentrated and stored at 0 to 4°C. Native polyacrylamide gel electrophoresis (native-PAGE) was performed on a 5–12% native polyacrylamide gradient gel with a Mini-Protean III system (Bio-Rad) according to the protocol by Davis [33]. After electrophoresis, the gel was stained for the protein with Coomassie Brilliant Blue R-250.

Pyruvate-producing Activity Assay

Pyruvate-producing activity was determined in 1 ml of 67 mM phosphate buffer (pH 7.4) containing 20 mM L-lactate or D-lactate. The reaction was initiated by the addition of whole cells or crude cell extract of SDM at 37°C for 10 min. The reaction was stopped by adding 0.05 ml of 1 M NaOH. The mixture was subjected to centrifugation for 1 min at 16,000 *g*, and a 0.5-ml portion of the supernatant was used for pyruvate detection. One unit was defined as the amount of enzyme that converted lactate to 1.0 μM pyruvate per minute under the test conditions.

Analytical Methods

Accurate concentrations of lactate and pyruvate were analyzed by HPLC (Agilent 1100 series, Hewlett-Packard, USA) using an Aminex HPX-87H column (Bio-Rad), which was running at 0.4 ml min^{-1} with 10 mM H_2SO_4 as eluent at 55°C [34]. UV and visible absorption spectra of the cytochrome were measured by an Ultrospec™ 2100 *pro* UV/visible spectrophotometer (GE Healthcare). Protein was determined by the Markwell variation of the Lowry method, with bovine serum albumin as the standard [35].

Author Contributions

Conceived and designed the experiments: PX CM. Performed the experiments: CG JQ. Analyzed the data: CG JQ. Contributed reagents/materials/analysis tools: PX CM. Wrote the paper: CG CM PX.

References

1. Gao C, Ma C, Xu P (2011) Biotechnological routes based on lactic acid production from biomass. Biotechnol Adv 29: 930–939.
2. John RP, Nampoothiri KM, Pandey A (2007) Fermentative production of lactic acid from biomass: an overview on process developments and future perspectives. Appl Microbiol Biotechnol 74: 524–534.
3. Qin J, Wang X, Zheng Z, Ma C, Tang H, et al. (2010) Production of L-lactic acid by a thermophilic *Bacillus* mutant using sodium hydroxide as neutralizing agent. Bioresour Technol 101: 7570–7576.
4. Distel KA, Zhu G, Wang P (2005) Biocatalysis using an organic-soluble enzyme for the preparation of poly(lactic acid) in organic solvents. Bioresour Technol 96: 617–623.
5. Ai M, Ohdan K (1997) Oxidative dehydrogenation of lactic acid to pyruvic acid over iron phosphate catalyst. Appl Catal A: Gen 150: 13–20.
6. Li Y, Chen J, Lun SY (2001) Biotechnological production of pyruvic acid. Appl Microbiol Biotechnol 57: 451–459.
7. Xu P, Qiu JH, Gao C, Ma CQ (2008) Biotechnological routes to pyruvate production. J Biosci Bioeng 105: 169–175.
8. Hao J, Ma C, Gao C, Qiu J, Wang M, et al. (2007) *Pseudomonas stutzeri* as a novel biocatalyst for pyruvate production from DL-lactate, Biotechnol Lett 29: 105–110.
9. Ma C, Gao C, Qiu J, Hao J, Liu W, et al. (2007) Membrane-bound L- and D-lactate dehydrogenase activities of a newly isolated *Pseudomonas stutzeri* strain. Appl Microbiol Biotechnol 77: 91–98.
10. Gao C, Hu CH, Zheng ZJ, Ma CQ, Jiang TY, et al. (2012) Lactate utilization is regulated by the FadR-type regulator LldR in *Pseudomonas aeruginosa*. J Bacteriol 194: 2687–2692.

11. Gao C, Jiang TY, Dou PP, Ma CQ, Li LX, et al. (2012) NAD-Independent L-lactate dehydrogenase is required for L-lactate utilization in *Pseudomonas stutzeri* SDM PLoS ONE 7: e36519.
12. Jiang TY, Gao C, Su F, Zhang W, Hu CH, et al. (2012) Genome sequence of *Pseudomonas stutzeri* SDM-LAC, a typical strain for studying the molecular mechanism of lactate utilization. J Bacteriol 194: 894–895.
13. Gao C, Qiu JH, Li JC, Ma CQ, Tang HZ, et al. (2009) Enantioselective oxidation of racemic lactic acid to D-lactic acid and pyruvic acid by *Pseudomonas stutzeri* SDM. Bioresour. Technol 100: 1878–1880.
14. Gao C, Zhang W, Lv CJ, Li LX, Ma CQ, et al. (2010) Efficient production of 2-oxobutyrate from 2-hydroxybutyrate by using whole cells of *Pseudomonas stutzeri* strain SDM. Appl Environ Microbiol 76: 1679–1682.
15. Gao C, Zhang W, Ma CQ, Liu P, Xu P (2011) Kinetic resolution of 2-hydroxybutanoate racemic mixtures by NAD-independent L-lactate dehydrogenase. Bioresour Technol 102: 4595–4599.
16. Buschmann S, Warkentin E, Xie H, Langer JD, Ermler U, et al. (2010) The structure of *cbb*$_3$ cytochrome oxidase provides insights into proton pumping. Science 329: 327–330.
17. Ma CQ, Xu P, Qiu JH, Zhang ZJ, Wang KW, et al. (2004) An enzymatic route to produce pyruvate from lactate. Appl Microbiol Biotechnol 66: 34–39.
18. Stottmeister U, Aurich A, Wilde H, Andersch J, Schmidt S, et al. (2005) White biotechnology for green chemistry: fermentative 2-oxocarboxylic acids as novel building blocks for subsequent chemical syntheses. J Ind Microbiol Biotechnol 32: 651–664.
19. Miyazaki M, Shibue M, Ogino K, Nakamura H, Maeda H (2001) Enzymatic synthesis of pyruvic acid from acetaldehyde and carbon dioxide. Chem Commun 18: 1800–1801.
20. Ogawa J, Soong CL, Masashi I, Shimizu S (2001) Enzymatic production of pyruvate from fumarate–an application of microbial cyclic–imide transforming pathway. J Mol Catal B: Enzym 11: 355–359.
21. Buto S, Pollegioni L, Angiurl L, Pilone MS (1994) Evaluation of D-amino-acid oxidase from *Rhodotorula gracilis* for the production of alpha-keto acids. Biotechnol Bioeng 44: 1288–1294.
22. Dixon NM, James EW, Lovitt RW, Kell DB (1989) Electromicrobial transformation using the pyruvate synthase system of *Clostridium sporagenes*. Bioelectrochem Bioenerg 21: 245–259.
23. Causey TB, Shanmugam KT, Yomano LP, Ingram LO (2003) Engineering *Escherichia coli* for efficient conversion of glucose to pyruvate. Proc Nat Acad Sci USA 101: 2235–2240.
24. Li Y, Hugenholtz J, Chen J, Lun SY (2002) Enhancement of pyruvate production by *Torulopsis glabrata* using a two-stage oxygen supply control strategy. Appl Microbiol Biotechnol 60: 101–106.
25. van Maris AJ, Geertman JM, Vermeulen A, Groothuizen MK, Winkler AA, et al. (2004) Directed evolution of pyruvate decarboxylase-negative *Saccharomyces cerevisiae*, yielding a C2-independent, glucose-tolerant, and pyruvate-hyperproducing yeast. Appl Environ Microbiol 70: 159–166.
26. Zelić B, Gostovic S, Vuorilehto K, Vasić-Racki D, Takors R (2004) Process strategies to enhance pyruvate production with recombinant *Escherichia coli*: from repetitive fed-batch to in situ product recovery with fully integrated electrodialysis. Biotechnol Bioeng 85: 638–646.
27. Hekmat D, Danninger J, Simon H, Vortmeyer D (1999) Production of pyruvate from (*R*)-lactate in an enzyme-membrane reactor with coupled electrochemical regeneration of the artificial mediator anthraquinone-2,6-disulfonate. Enzyme Microb Technol 24: 471–479.
28. Schinschel C, Simon H (1993) Preparation of pyruvate from (*R*)-lactate with *Proteus* species. J Biotechnol 31: 191–203.
29. Eisenberg A, Seip JE, Gavagan JE, Payne MS, Anton DL, et al. (1997) Pyruvic acid production using methylotrophic yeast transformants as catalyst. J Mol Catal B: Enzym 2: 223–232.
30. Cooper B (1989) Microbial manufacture of pyruvic acid from D-(–)-lactic acid. DE patent 3733157.
31. Gu JS, Xu P, Qu YB (2002) A biocatalyst for pyruvate preparation from DL-lactate: lactate oxidase in a *Pseudomonas* sp. J Mol Catal B: Enzym 18: 299–305.
32. Liu ZQ, Jia LZ, Zheng YG (2010) Biotransformation of DL-lactate to pyruvate by a newly isolated *Serratia marcescens* ZJB-07166. Process Biochem 45: 1632–1637.
33. Davis BJ (1964) Disc electrophoresis II. Method and application to human serum proteins. Ann NY Acad Sci 121: 404–427.
34. Gao C, Xu XX, Hu CH, Zhang W, Zhang Y, et al. (2010) Pyruvate producing biocatalyst with constitutive NAD-independent lactate dehydrogenases. Process Biochem 45: 1912–1915.
35. Markwell MAK, Haas SM, Bieber LL, Tolbert NE (1978) A modification of the Lowry procedure to simplify protein determination in membrane and lipoprotein samples. Anal Biochem 87: 206–210.

15

Earthworm Is a Versatile and Sustainable Biocatalyst for Organic Synthesis

Zhi Guan, Yan-Li Chen, Yi Yuan, Jian Song, Da-Cheng Yang, Yang Xue, Yan-Hong He*

School of Chemistry and Chemical Engineering, Southwest University, Chongqing, P. R. China

Abstract

A crude extract of earthworms was used as an eco-friendly, environmentally benign, and easily accessible biocatalyst for various organic synthesis including the asymmetric direct aldol and Mannich reactions, Henry and Biginelli reactions, direct three-component aza-Diels-Alder reactions for the synthesis of isoquinuclidines, and domino reactions for the synthesis of coumarins. Most of these reactions have never before seen in nature, and moderate to good enantioselectivities in aldol and Mannich reactions were obtained with this earthworm catalyst. The products can be obtained in preparatively useful yields, and the procedure does not require any additional cofactors or special equipment. This work provides an example of a practical way to use sustainable catalysts from nature.

Editor: Matthew H. Todd, University of Sydney, Australia

Funding: This work was financially supported by the National Natural Science Foundation of China (No 21276211), http://www.nsfc.gov.cn/. The funders had no role in study design, data collection and analysis, decision to publish, or preparation of the manuscript.

Competing Interests: The authors have declared that no competing interests exist.

* Email: heyh@swu.edu.cn

Introduction

The exploitation of new catalysts in an environmentally benign manner has become crucially important in recent years [1]. Nature is an extraordinary chemist who evolves its catalysts over millions of years [2]. Such catalysts are models of energy-efficient, environmentally benign chemical agents, as virtually all do their work under mild conditions and generate few waste products [3]. However, the development and use of these natural catalysts in organic chemistry are very limited. To date, the most common biocatalytic methods use either whole-cells or isolated enzymes for a given chemical transformation. Each method has advantages and challenges for both laboratory-scale and industrial-scale chemistry [4]. The use of whole-cell systems benefits from the ability to use low-cost and renewable feedstocks, and co-factor addition and regeneration are not necessary. However, whole-cell systems require expensive equipment and tedious work-up because of large volumes, and have low productivity. More importantly, the accumulation of products and by-products may be toxic to the cell. Besides, the cell membrane may act as a mass transport barrier between the substrates and the enzymes [5]. Compared to whole cells, isolated enzymes offer several benefits, including simpler reaction apparatus, higher productivity owing to higher catalyst concentration, and simpler product purification [6]. Thus, isolated enzymes are usually more efficient in bioconversion than whole cells. However, enzyme purification involves expensive strategies, usually employing costly and polluting chemicals, being also laborious and time-consuming, which restricts the use of isolated enzymes, making it difficult to scale-up [7]. Therefore, it is obligatory to find easy methods to supply the gaps.

Using crude extracts of organisms as biocatalysts may be the right solution for they are environmentally-benign, inexpensive and simple in preparation. Moreover, generally, a relatively crude preparation is far more stable than a highly purified enzyme [8]. Some efforts have been made to use crude extracts of organisms as biocatalysts for organic synthesis. For example, a crude extract of the button mushroom (*Agaricus bisporus*) was employed to catalyse the domino reaction between phenol and cyclic 1,3-dicarbonyls using atmospheric oxygen as the oxidizing agent and yielding annulated benzofuranes [9]. A crude horse liver homogenate was used as a biocatalyst to catalyse the acylglucuronide of mycophenolic acid [10]. Crude preparations of various plants were used to catalyse bioreduction of carbonyl compounds and alcohol biooxidation [11]. However, the reported reaction types catalysed by crude extracts of organisms are very limited.

Earthworms are harmless creatures that live in the soil. They are eco-friendly playing the significant role in decomposing organic wastes, enhancing soil fertility, and improving soil drainage. Earthworms have been used from ancient times in oriental countries as drugs for prevention and treatment of various diseases, and have found applications such as intracystic calculus-contraction and releasing-stimulating agent, anti-choloplania agent, parturifacient, hair growth tonic, antifebrile, spasm-treating agent, hemiplegia-treating agent, urination improving agent, anti-bronchial asthma agent, anti-hypertension agent, therapeutic medicament for thrombosis and others [12].

The most thoroughly studied enzymes from earthworms are proteases that are secreted by alimentary tract of earthworm [13]. For example, an earthworm, *Lumbricus rubellus*, produces alkaline serine proteases that show higher activity and stability than trypsins. These enzymes are stable at temperature below 60°C over a wide range of pH 2–11, and are strongly resistant to organic solvents and detergents. Moreover, they retain full activity for long years at room temperature. They act on various proteins, such as elastin as well as fibrin, and some peptides, such as β-

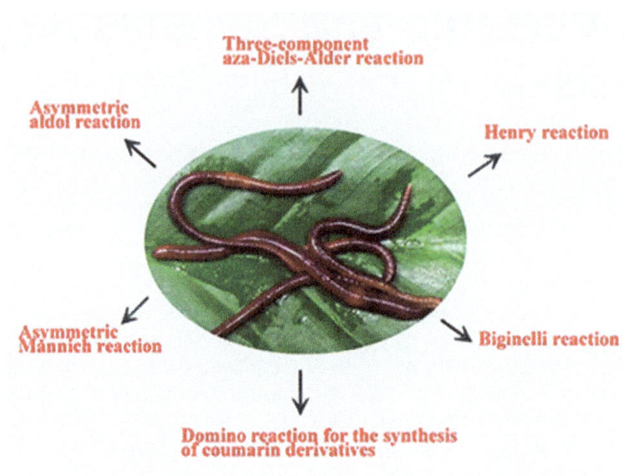

Figure 1. The earthworm-catalysed organic reactions.

amyloid 1–40 and solubilized actual fibrin clots of whole blood in a rat's vena cava [14].

However, the use of earthworms as a source of biocatalysts for biotransformation is very limited. Nakajima et al. reported that one of the earthworm serine protease acts on the hydrolysis of triacylglycerols [15]. They also reported the stereoselective reduction of carbonyl compounds using the cell-free extract from earthworms (*Lumbricus rubellus*) in the presence of NADH or NADPH as a coenzyme [16]. Recently, Santos et al. describe β-carboline imine reductions in high yields and enantiomeric excesses employing the cell-free extract from earthworms (*Eisenia foetida*) in the presence of NADPH [17]. Herein, we report that a crude extract of earthworms is a versatile biocatalyst for the asymmetric direct aldol and Mannich reactions, Henry and Biginelli reactions, direct three-component aza-Diels-Alder reactions for the synthesis of isoquinuclidines, and domino reactions for the synthesis of coumarins (**Fig. 1**). The earthworm species we used is *Eisenia foetida* (Annelida, Oligochaeta), known as red worm.

Results and Discussion

First, we developed a very simple procedure for the preparation of a crude extract from earthworms. The fresh earthworms were homogenized with deionized water. The supernatant obtained after centrifugation was concentrated by dialysis against solid sucrose, dried, and ground to get an extract powder, which was used directly to catalyse the different reactions.

The crude earthworm extract catalysed direct asymmetric aldol reactions

The catalytic asymmetric aldol reaction is one of the most powerful methods for the construction of chiral β-hydroxy carbonyl compounds. The crude earthworm extract could catalyse the direct asymmetric aldol reactions of aromatic aldehydes with cyclic and heterocyclic ketones in MeCN in the presence of a small amount of water (**Fig. 2**). The yields of 53–96%, enantioselectivities of 80–91% ee (for *anti* isomers), and diastereoselectivities of 81:19->99:1 (*anti/syn*) were achieved under the optimized conditions (**Table 1**, entries 1–8). The reaction showed good substrate adaptability to different aromatic aldehydes with electron-withdrawing and electron-donating groups, as well as heterocyclic ketones containing oxygen and sulfur. Moreover, in the absence of the crude earthworm extract, only a trace amount of product was observed in the model aldol reaction of 4-cyanobenzaldehyde and cyclohexanone even after 120 h (**Table 1**, entry 10), indicating that the crude earthworm extract indeed catalysed the aldol reaction in asymmetric manner.

To verify whether the enzyme(s) in the earthworm extract catalysed the aldol reaction, some control experiments were performed (**Table 1**, entries 11–13). Firstly, the high temperature (at 100°C for 24 h) pretreated earthworm extract was used to catalyse the model reaction, which gave the product in a low yield of 39% with 80:20 dr and 85% ee (**Table 1**, entry 11), demonstrating that high temperature treatment caused a decrease of the catalytic ability of the earthworm extract on the aldol reaction. Secondly, the Cu^{2+} pretreated crude earthworm extract

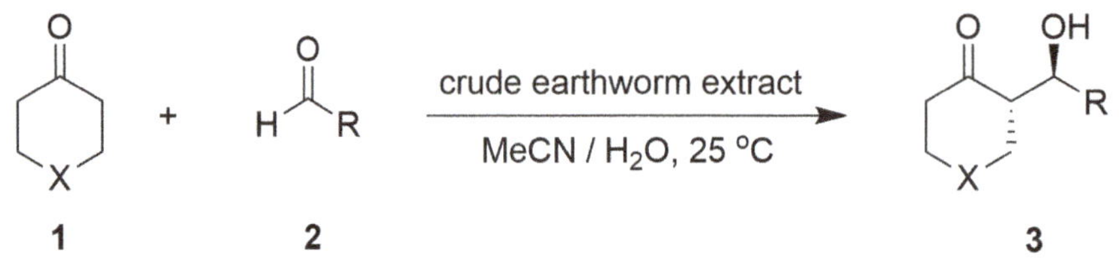

Figure 2. The crude earthworm extract catalysed direct asymmetric aldol reactions.

Table 1. The crude earthworm extract catalysed direct asymmetric aldol reactions[a].

Entry	X	R	Product	Time (h)	Yield (%)[b]	dr (*anti*:*syn*)[c]	ee (*anti*) (%)[c]
1	CH_2	3-CNC_6H_4	**3a**	144	96	91:9	91
2	CH_2	4-ClC_6H_4	**3b**	139	60	92:8	91
3	CH_2	4-$CF_3C_6H_4$	**3c**	120	84	81:19	90
4	CH_2	2,4-$Cl_2C_6H_3$	**3d**	192	82	>99:1	90
5	CH_2	3-$MeOC_6H_4$	**3e**	192	53	88:12	91
6	O	4-$NO_2C_6H_4$	**3f**	103	75	83:17	80
7	S	4-CNC_6H_4	**3g**	117	65	93:7	90
8	CH_2	4-CNC_6H_4	**3h**	120	92	87:13	88
9 (Un-optimized conditions)[d]	CH_2	4-CNC_6H_4	**3h**	120	78	78:22	82
10 (No catalyst)[e]	CH_2	4-CNC_6H_4	**3h**	120	Trace	—	—
11 (100°C pretreated crude earthworm extract)[f]	CH_2	4-CNC_6H_4	**3h**	120	39	80:20	85
12 (Cu^{2+}pretreated crude earthworm extract)[g]	CH_2	4-CNC_6H_4	**3h**	120	11	78:22	86
13 (Cu^{2+} as catalyst)[h]	CH_2	4-CNC_6H_4	**3h**	120	trace	—	—

[a]Optimized reaction conditions: aldehyde (0.50 mmol), ketone (7.50 mmol), the crude earthworm extract (75 mg), deionized water (0.05 mL), and MeCN (0.95 mL) at 25°C.
[b]Yield of the isolated product after silica gel chromatography.
[c]Determined by chiral HPLC analysis performed using Chiralpak AD-H, AS-H, or Chiralcel OD-H column, by comparing the retention time with those of known compounds (For details, please see the Supplementary material).
[d]Un-optimized conditions: 4-cyanobenzaldehyde (0.50 mmol), cyclohexanone (2.50 mmol), the crude earthworm extract (50 mg), deionized water (0.05 mL), and MeCN (0.95 mL) at 30°C.
[e]The reaction was conducted in the absence of the crude earthworm extract, otherwise under the same conditions as entry 9.
[f]The crude earthworm extract (50 mg) in deionized water (3 mL) was stirred at 100°C for 24 h, and then water was removed under reduced pressure before use. The reaction was conducted under the same conditions as entry 9.
[g]The crude earthworm extract (50 mg) in Cu^{2+} solution (0.25 M) [anhydrous $CuSO_4$ (120 mg) in deionized water (3 mL)] was stirred at 30°C for 24 h, and then water was removed under reduced pressure before use. The reaction was conducted under the same conditions as entry 9.
[h]The reaction was conducted using anhydrous $CuSO_4$ (120 mg) as a catalyst instead of the crude earthworm extract, otherwise under the same conditions as entry 9.

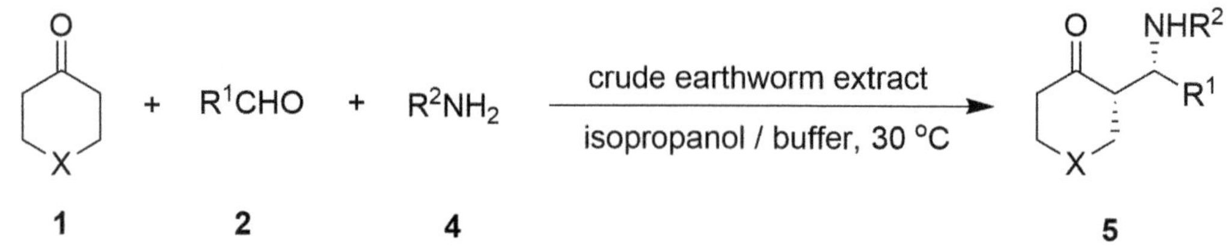

Figure 3. The crude earthworm extract catalysed direct asymmetric Mannich reactions.

was also used to catalyse the model aldol reaction, and only a low yield of 11% was obtained with78:22 dr and 86% ee (**Table 1**, entry 12). Meanwhile, to exclude the effect of Cu^{2+} on the reaction, Cu^{2+} alone was used to catalyse the model aldol reaction, and only a trace amount of product was observed on TLC (**Table 1**, entry 13), proving that Cu^{2+} did not catalyse this transformation. The above control experiments showed that high temperature and metal ion Cu^{2+}, as deactivation factors of enzyme, could greatly decrease the catalytic ability of the crude earthworm extract on the model aldol reaction. Thus, it can be inferred that the enzyme(s) in the crude earthworm extract catalysed the aldol reaction.

The crude earthworm extract catalysed direct asymmetric Mannich reactions

The asymmetric Mannich reaction is a powerful synthetic strategy to prepare chiral β-amino ketones and aldehydes with perfect atom economy through the loss of a molecule of water, and the reaction products are versatile intermediates in the synthesis of chiral amines [18]. The crude earthworm extract could catalyse the direct asymmetric Mannich reactions in isopropanol/buffer (**Fig. 3**). Various substituted aromatic aldehydes, arylamines and cyclic or heterocyclic ketones participated in the reaction smoothly. Mannich products were obtained in yields of 42–83% with diastereoselectivities of 43:57–87:13 (*syn/anti*) and enantioselectivities of 44–76% ee (for *syn* isomers) under the optimized conditions (**Table 2**, entries 1–6). In the absence of the crude earthworm extract, the product was obtained only in 29% yield with 38:62 dr (*syn/anti*) for the model Mannich reaction of 4-nitrobenzaldehyde, aniline and cyclohexanone after 72 h (**Table 2**, entry 8), indicating that the crude earthworm extract indeed catalysed the Mannich reaction in asymmetric manner.

The crude earthworm extract catalysed Henry reactions

The Henry or nitroaldol reaction is an important atom-economical methodology to furnish a β-nitro alcohol. The crude earthworm extract could catalyse Henry reactions in water (**Fig. 4**). Various aromatic aldehydes participated in the reactions with nitromethane, nitroethane and nitropropane giving corresponding products in yields of 41–92% (**Table 3**, entries 1–6). In the absence of the crude earthworm extract, only a trace amount of product was observed (**Table 3**, entry 7), indicating that the crude earthworm extract indeed catalysed the Henry reaction. Unfortunately, there was no obvious enantiomeric excess of the products observed by the chiral HPLC analysis.

Table 2. The crude earthworm extract catalysed direct asymmetric Mannich reactions[a].

Entry	X	R^1	R^2	Product	Time (h)	Yield (%)[b]	dr (*syn:anti*)[c]	ee (*syn*) (%)[c]
1	CH_2	$4\text{-}NO_2C_6H_4$	$3\text{-}BrC_6H_4$	**5a**	96	42	87:13	76
2	CH_2	$4\text{-}NO_2C_6H_4$	$4\text{-}ClC_6H_4$	**5b**	71	66	84:16	74
3	CH_2	$4\text{-}NO_2C_6H_4$	$3\text{-}MeC_6H_4$	**5c**	47	75	79:21	73[d]
4	CH_2	$4\text{-}NO_2C_6H_4$	$4\text{-}MeC_6H_4$	**5d**	47	79	74:26	71[d]
5	S	$4\text{-}ClC_6H_4$	C_6H_5	**5e**	96	70	43:57	44
6	CH_2	$4\text{-}NO_2C_6H_4$	C_6H_5	**5f**	72	83	79:21	75
7 (Un-optimized conditions)[e]	CH_2	$4\text{-}NO_2C_6H_4$	C_6H_5	**5f**	72	82	63:37	62
8 (No catalyst)[f]	CH_2	$4\text{-}NO_2C_6H_4$	C_6H_5	**5f**	72	29	38:62	0

[a]Optimized reaction conditions: aldehyde (0.50 mmol), arylamine (0.55 mmol), cyclohexanone (5.00 mmol) or tetrahydrothiopyran-4-one (1.00 mmol), buffer (NaH_2PO_4-Na_2HPO_4, 0.20 M, pH = 7.53, 0.05 mL), isopropanol (0.95 mL), and the crude earthworm extract (50 mg) at 30°C.
[b]Yield of the isolated product after silica gel chromatography.
[c]Determined by chiral HPLC analysis performed using Chiralpak AD-H or Chiralcel OD-H column, by comparing the retention time with those of known compounds (For details, please see the Supplementary material).
[d]The chiral HPLC did not show baseline separation of peaks.
[e]Un-optimized conditions: 4-nitrobenzaldehyde (0.50 mmol), aniline (0.55 mmol), cyclohexanone (5.00 mmol), deionized water (0.10 mL), isopropanol (0.90 mL) and the crude earthworm extract (50 mg) at 30°C.
[f]The reaction was carried out in the absence of the crude earthworm extract, otherwise under the same conditions as entry 7.

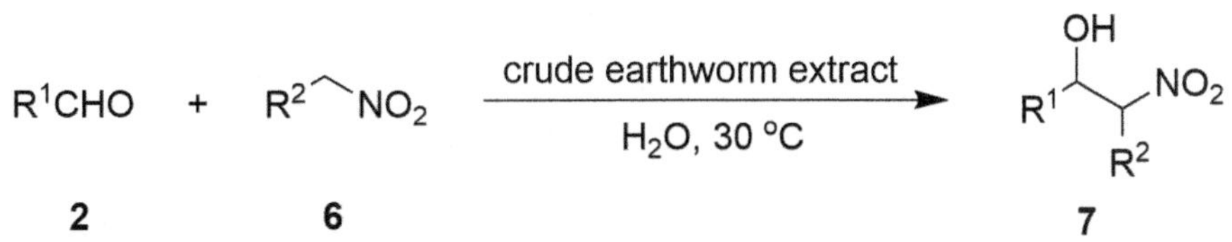

Figure 4. The crude earthworm extract catalysed Henry reactions.

The crude earthworm extract catalysed Biginelli reactions

The Biginelli reaction is an important three-component reaction for construction of dihydropyrimidinones that are widely used in the pharmaceutical industry. The crude earthworm extract could catalyse Biginelli reactions in *n*-butyl acetate/water (**Fig. 5**). Various substituted aromatic aldehydes reacted with urea and acetoacetate giving corresponding products in yields of 14–76% with enantioselectivities of 0–57% ee (**Table 4**, entries 1–7). In the absence of the crude earthworm extract, the product was only obtained in 6% yield (**Table 4**, entry 9), indicating that the crude earthworm extract indeed catalysed the Biginelli reaction.

The crude earthworm extract catalysed domino reactions for the synthesis of coumarin derivatives

Coumarin derivatives are widely used in pharmaceutical and commercial applications. The crude earthworm extract could catalyse the reactions of various salicylaldehyde derivates with different β-keto esters to afford a series of coumarin derivatives (**Fig. 6**). This reaction is a domino process that comprises Knoevenagel and intramolecular transesterification steps. Various coumarin derivatives were obtained in yields of 32–87% in DMSO/water under the optimized conditions (**Table 5**, entries 1–7). In the absence of the crude earthworm extract, only a trace amount of product was observed (**Table 5**, entry 9), indicating that the crude earthworm extract indeed catalysed the domino reaction.

The crude earthworm extract catalysed aza-Diels-Alder reactions for the synthesis of isoquinuclidine derivatives

Isoquinuclidine (azabicyclo[2.2.2]octane) has received considerable attention due to their presence in numerous complex natural products and pharmaceuticals. The crude earthworm extract could catalyse three-component aza-Diels-Alder reactions for the synthesis of isoquinuclidine derivatives (**Fig. 7**). The reactions of aromatic aldehydes, aromatic amines and cyclohexenone in MeCN/H_2O gave corresponding isoquinuclidines in excellent yields of 86–99%, and all reactions afforded the *endo* isomers as the major products with *endo/exo* ratio ranged from 68:32 to 65:35 (**Table 6**, entries 1–5). In the absence of the crude earthworm extract, only a trace amount of product was observed (**Table 6**, entry 6), indicating that the crude earthworm extract indeed catalysed the aza-Diels-Alder reaction. Unfortunately, there was no obvious enantiomeric excess of the products observed by the chiral HPLC analysis.

In addition, we tested the stability of the crude earthworm extract powder by repeating the direct asymmetric aldol reaction of 4-cyanobenzaldehyde and cyclohexanone using the crude earthworm extract that had been stored at 4°C for 6 months. The activity and selectivity did not decrease, demonstrating that the crude earthworm extract is reasonably stable.

Conclusion

Nature is a vital source of catalysts, but the use of natural catalysts in organic synthesis is very limited, mainly due to the difficulties in isolation and purification of enzymes. Herein we demonstrated that using crude extracts of organisms as biocatalysts is a promising choice for development of natural catalysts. The crude earthworm extract is applicable to unusually wide scope of reaction types and substrates, including the asymmetric direct aldol and Mannich reactions, Henry and Biginelli reactions, direct three-component aza-Diels-Alder reactions for the synthesis of isoquinuclidines, and domino reactions for the synthesis of coumarins. Most of these reactions have never before seen in nature. The products can be obtained in preparatively useful

Table 3. The crude earthworm extract catalysed Henry reactions[a].

Entry	R^1	R^2	Product.	Time (h)	Yield (%)[b]	dr (*syn:anti*)[c]
1	3-CNC_6H_4	H	**7a**	160	85	—
2	4-$NO_2C_6H_4$	H	**7b**	93	88	—
3	2-$NO_2C_6H_4$	H	**7c**	93	85	—
4	4-CNC_6H_4	Me	**7d**	118	92	65:35
5	4-$NO_2C_6H_4$	Et	**7e**	118	41	66:34
6	4-CNC_6H_4	H	**7f**	120	88	—
7 (No catalyst)	4-CNC_6H_4	H	**7f**	104	Trace	—

[a]Reaction conditions: aldehyde (0.50 mmol), nitroalkane (2.50 mmol), the crude earthworm extract (50 mg), deionized water (1.00 mL) at 30°C.
[b]Yield of the isolated product after silica gel chromatography.
[c]Determined by ^{1}HNMR in comparison with those of known compounds (For details, please see the Supplementary material).

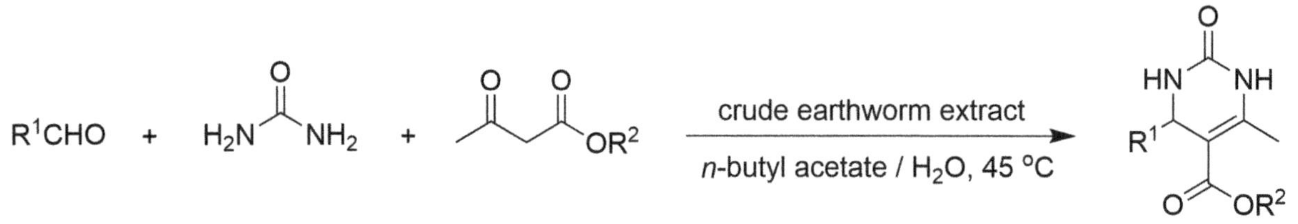

2 **8** **9** **10**

Figure 5. The crude earthworm extract catalysed Biginelli reactions.

yields, and moderate to good enantioselectivities for aldol and Mannich reactions were obtained with this earthworm catalyst. The procedure does not require any additional cofactors or special equipment. The main advantages of using the crude earthworm extract as a catalyst are eco-friendly, environmentally benign, safe, cheap, easily accessible and stable. This work provides an example of a practical way to use sustainable catalysts from nature. It would have a transformative effect on streamlining the practice of organic synthesis.

Methods

Ethics statement

Live earthworms (*Eisenia foetida*, the common Chinese name: "Daping II") were purchased from Tianjin Cheng Gong earthworm farm (Tianjin, China).

Procedure for the preparation of the crude earthworm extract

The live earthworms were rinsed 5 times with tap water to let the internal dirt discharge as much as possible, and then rinsed twice with deionized water. The clean earthworms (200 mL, measured with a graduated beaker) were homogenized with equal volume ice-cold deionized water (200 mL). The homogenate was centrifuged (RCF = 4250 g, for 5 min). The collected supernatants were concentrated by dialysis against solid sucrose at 4°C. The concentrate was placed in a tray and blown with an electric fan at 17–25°C until it turned into a paste. The paste was further dried under vacuum, and then ground into powder with a porcelain mortar to give a gray powder (52.1 g). The extract powder was stored at 4°C for use.

The reaction condition screening program

The reaction conditions for each type of chemical transformation were optimized using corresponding model reactions. These optimizations included: **the solvent optimization** (Different solvents such as CH_2Cl_2, $CHCl_3$, methyl phenyl ether, toluene, *n*-butyl acetate, EtOAc, 1,4-dioxane, MTBE, THF, cyclohexane, DMSO, EtOH, MeOH, *i*-PrOH, H_2O and solvent-free, were screened, respectively, for each type of reaction to find out the optimal solvent); **the water content optimization** [The water contents of 0%, 5%, 10%, 15%, 20%, 25%, 30%, 35%, 40%, 45%, 50% (v/v) in the reaction system were screened, respectively, for each type of reaction to find out the optimal water content];

Table 4. The crude earthworm extract catalysed Biginelli reactions[a].

Entry	R[1]	R[2]	Product	Time (h)	Yield (%)[b]	ee (%)[c]
1	4-MeOC$_6$H$_4$	Et	**10a**	169	44	0
2	3-NO$_2$C$_6$H$_4$	Et	**10b**	118	76	29 (R)
3	4-FC$_6$H$_4$	Et	**10c**	69	53	0
4	3-ClC$_6$H$_4$	Et	**10d**	119	49	10 (R)
5	2-ClC$_6$H$_4$	Me	**10e**	140	69	20[d]
6	3-NO$_2$-C$_6$H$_4$	Me	**10f**	103	14	57[d]
7	C$_6$H$_5$	Et	**10g**	72	65	0
8 (Un-optimized conditions)[e]	C$_6$H$_5$	Et	**10g**	72	44	0
9 (No catalyst)[f]	C$_6$H$_5$	Et	**10g**	72	6	0

[a]Optimized reaction conditions: aldehyde (0.50 mmol), urea (1.00 mmol), acetoacetate (1.00 mmol), deionized water (0.30 mL), *n*-butyl acetate (0.70 mL) and the crude earthworm extract (75 mg) at 45°C.
[b]Yield of isolated product after silica gel chromatography.
[c]Determined by chiral HPLC analysis performed using Chiralpak AD-H column, by comparing the retention time with those of known compounds (For details, please see the Supplementary material).
[d]The absolute configuration was not determined.
[e]Un-optimized conditions: benzaldehyde (0.50 mmol), urea (1.00 mmol), ethyl acetoacetate (1.50 mmol), deionized water (0.05 mL), *n*-butyl acetate (0.95 mL) and the crude earthworm extract (100 mg) at 45°C.
[f]The reaction was carried out in the absence of the crude earthworm extract, otherwise under the same conditions as entry 8.

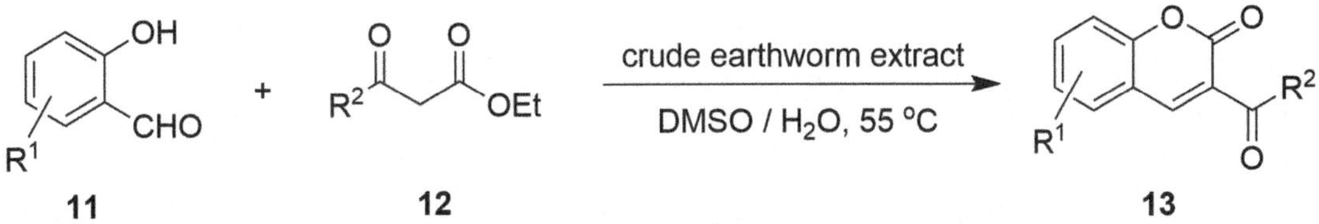

Figure 6. The crude earthworm extract catalysed domino reactions for the synthesis of coumarin derivatives.

the substrate molar ratio optimization (Different molar ratios of substrates were screened, respectively, for each type of reaction to find out the suitable molar ratio of substrates); **the crude earthworm extract loading optimization** (Different amounts of crude earthworm extract were screened, respectively, for each type of reaction to choose the optimal catalyst loading); **the phosphate buffer optimization** [NaH_2PO_4-Na_2HPO_4 buffer, 0.20 M, different pH from 5 to 11 were separately used to replace the optimized water content for each type of chemical transformation to find out the optimal phosphate buffer pH. For the direct asymmetric Mannich reaction, adding buffer (NaH_2PO_4-Na_2HPO_4, 0.20 M, pH = 7.53, 0.05 mL) could improve the yield and enantioselectivity, however, for other types of chemical transformation, addition of phosphate buffer did not lead to obvious improvement of the results. Thus, the reaction medium consisting of phosphate buffer (NaH_2PO_4-Na_2HPO_4, 0.20 M, pH = 7.53, 0.05 mL) and isopropanol (the optimal solvent) (0.95 mL) was used for the Mannich reaction, while deionized water/organic solvent was used for other reactions. For the Henry reaction, the optimal solvent was water]; **the temperature optimization** (The influence of different temperature on each type of reaction was also investigated to find out the optimal temperature).

The selected optimal reaction conditions for each type of chemical transformation were used in the following procedures.

General procedure for the crude earthworm extract catalysed reactions

The aldol reactions. A round-bottom flask was charged with the crude earthworm extract (75 mg), aldehyde (0.50 mmol), ketone (7.50 mmol) and MeCN (0.95 mL), to which deionized water (0.05 mL) was introduced. The resultant mixture was stirred at 25°C for the specified reaction time and monitored by TLC. The reaction was terminated by filtration (with buchner funnel and qualitative filter paper), and ethyl acetate (20 mL) was employed to wash the filter cake. The filtrate was concentrated under reduced pressure. The residue was purified by silica gel flash column chromatography (petroleum ether/ethyl acetate) to give the product.

The Mannich reactions. The crude earthworm extract (50 mg) was added to a round-bottom flask containing an aromatic aldehyde (0.50 mmol), arylamine (0.55 mmol), cyclohexanone (5.00 mmol) or tetrahydrothiopyran-4-one (1.00 mmol), buffer (NaH_2PO_4-Na_2HPO_4, 0.20 M, pH = 7.53, 0.05 mL) and isopropanol (0.95 mL). The resultant mixture was stirred at 30°C for the specified reaction time and monitored by TLC. The reaction was terminated by filtration (with buchner funnel and qualitative filter paper), and ethyl acetate (10 mL) was employed to wash the filter cake. The filtrate was concentrated under reduced pressure. The residue was purified by silica gel flash column chromatography (petroleum ether/ethyl acetate) to give the product.

Table 5. The crude earthworm extract catalysed domino reactions for the synthesis of coumarin derivatives[a].

Entry	R^1	R^2	Product	Time (h)	Yield (%)[b]
1	5-NO_2	Ph	**13a**	71	80
2	5-Cl	Ph	**13b**	72	86
3	5-OMe	Ph	**13c**	102	75
4	1-naphth	Ph	**13d**	96	87
5	H	Ph	**13e**	99	80
6	H	OEt	**13f**	127	32
7	H	Me	**13g**	127	51
8 (Un-optimized conditions)[c]	H	Me	**13g**	100	21
9 (No catalyst)[d]	H	Me	**13g**	100	Trace

[a]Optimized reaction conditions: **11** (4.00 mmol), **12** (1.00 mmol), the crude earthworm extract (150 mg), deionized water (0.15 mL) and DMSO (0.85 mL) at 55°C.
[b]Yield of the isolated product after silica gel chromatography.
[c]Un-optimized conditions: salicylaldehyde (3.00 mmol), ethyl acetoacetate (1.00 mmol), the crude earthworm extract (100 mg), deionized water (0.10 mL) and DMSO (0.90 mL) at 30°C.
[d]The reaction was carried out in the absence of the crude earthworm extract, otherwise under the same conditions as entry 8.

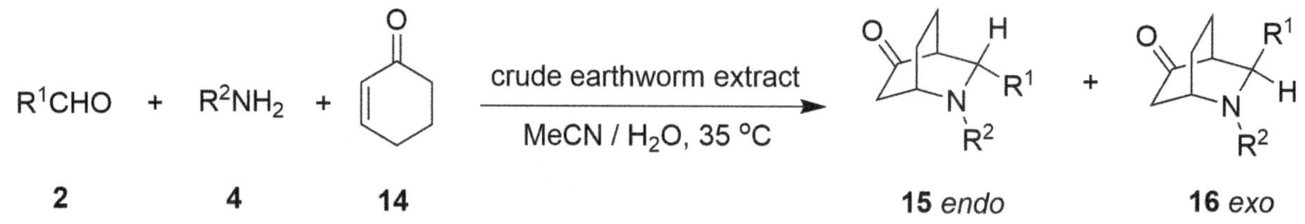

Figure 7. The crude earthworm extract catalysed aza-Diels-Alder reactions.

The Henry reactions. A round-bottom flask was charged with the crude earthworm extract (50 mg), aldehyde (0.50 mmol), nitroalkane (2.50 mmol), and deionized water (1.00 mL). The resultant mixture was stirred at 30°C and monitored by TLC. After a specified reaction time, saturated brine (15 mL) was added to the reaction, and the mixture was extracted four times with ethyl acetate (15 mL). The combined extracts were dried over anhydrous Na_2SO_4, and the solvents were then removed under reduced pressure. The residue was purified by silica gel flash column chromatography (petroleum ether/ethyl acetate) to give the product.

The Biginelli reactions. A round-bottom flask was charged with aldehyde (0.50 mmol), urea (1.00 mmol), acetoacetate (1.00 mmol), *n*-butyl acetate (0.70 mL) and the crude earthworm extract (75 mg), to which deionized water (0.30 mL) was introduced. The resultant mixture was stirred at 45°C and monitored by TLC. After a specified reaction time, ethanol (20 mL) was added to the reaction mixture to dissolve the crude products with stirring. The mixture was then filtrated (with buchner funnel and qualitative filter paper), and ethanol (20 mL) was employed to wash the filter cake. The filtrate was concentrated under reduced pressure. The residue was dissolved in ethyl acetate, and washed with saturated brine to remove the excess urea. The organic phase was dried over anhydrous Na_2SO_4, and the solvents were then removed under reduced pressure. The solid residue was recrystallized from ethyl acetate-petroleum ether to give the first part of the product. The mother liquor was concentrated under reduced pressure, and the residue was then purified by silica gel flash column chromatography (petroleum ether/ethyl acetate) to give the second part of the product. The reaction yield refers to the combination of the two parts of product above.

The domino reactions for the synthesis of coumarin derivatives. A round-bottom flask was charged with the crude earthworm extract (150 mg), salicylaldehyde derivative (4.00 mmol), β-keto ester (1.00 mmol), DMSO (0.85 mL) and deionized water (0.15 mL). The resultant mixture was stirred at 55°C for the specified reaction time and monitored by TLC. The reaction was terminated by filtration (with buchner funnel and qualitative filter paper), and ethyl acetate was employed to wash the filter cake. Saturated brine (15 mL) was then added to the filtrate, and the filtrate was extracted three times with ethyl acetate (15 mL). The combined extracts were dried over anhydrous Na_2SO_4, and the solvents were then removed under reduced pressure. The residue was purified by silica gel flash column chromatography (petroleum ether/ethyl acetate) to give the product.

The aza-Diels-Alder reactions. The crude earthworm extract (100 mg) was added to a round-bottom flask containing aldehyde (0.50 mmol), aromatic amine (1.50 mmol), cyclohexenone (1.50 mmol), MeCN (0.93 mL), and deionized water (0.07 mL). The resultant mixture was stirred at 35°C for the specified reaction time and monitored by TLC. The reaction was terminated by filtration (with buchner funnel and qualitative filter paper), and ethyl acetate (10 mL) was employed to wash the filter cake. The filtrate was then washed twice with water (10 mL), and the aqueous phase was back-extracted twice with ethyl acetate (10 mL). Combined organic phase was dried over anhydrous Na_2SO_4, and the solvents were then removed under reduced pressure. The residue was purified by silica gel flash column

Table 6. The crude earthworm extract catalysed aza-Diels-Alder reactions[a].

Entry	R^1	R^2	Product	Yield (15+16) (%)[b]	*endo* 15/*exo* 16[c]
1	$3\text{-}FC_6H_4$	$4\text{-}MeOC_6H_4$	**15a, 16a**	97	68:32
2	$4\text{-}FC_6H_4$	$4\text{-}MeOC_6H_4$	**15b, 16b**	93	66:34
3	$4\text{-}FC_6H_4$	C_6H_5	**15c, 16c**	99	67:33
4	$3\text{-}ClC_6H_4$	$4\text{-}MeOC_6H_4$	**15d, 16d**	92	65:35
5	$4\text{-}ClC_6H_4$	$4\text{-}MeOC_6H_4$	**15e, 16e**	86	68:32
6 (No catalyst)	$4\text{-}ClC_6H_4$	$4\text{-}MeOC_6H_4$	**15e, 16e**	Trace	—

[a]Reaction conditions: aromatic aldehyde (0.50 mmol), aromatic amine (1.50 mmol), cyclohexenone (1.50 mmol), and the crude earthworm extract (100 mg) in MeCN (0.93 mL) and deionized water (0.07 mL) at 35°C for 96 h.
[b]Yield of the isolated products after silica gel chromatography.
[c]Calculated according to the isolated weights of **15** and **16**. The *endo* or *exo* was determined by ^{1}HNMR in comparison with those of known compounds (For details, please see the Supplementary material).

chromatography (petroleum ether/ethyl acetate) to give the product.

Supporting Information

Table S1 List of the obvious difference between aldol products *syn*-3 and *anti*-3 on ^{1}H NMR and chiral HPLC (Table 1, entries 1–8).

Table S2 List of the difference between Mannich products *syn*-5 and *anti*-5 on chiral HPLC (Table 2, entries 1–6).

Table S3 List of the obvious difference between Henry products *syn*-7 and *anti*-7 on ^{1}H NMR (Table 3, entries 4 and 5).

Table S4 List of the obvious difference between Aza-Diels-Alder products *endo*-15 and *exo*-16 on ^{1}H NMR (Table 6, entries 1–5).

Data S1 ^{1}H NMR, ^{13}C NMR and HPLC data of products.

Materials S1 The materials and general methods.

Spectra S1 ^{1}H and ^{13}C NMR-spectra and HPLC chart.

Acknowledgments

The authors would like to thank the former co-workers Hai-Hong Li, Jian-Feng Cai, Chang-Heng Wang, Ke-Ling Xu, Bang-Hua Xie, Jian-Ping Fu for their enthusiastic preliminary exploration on the earthworm as catalysts.

Author Contributions

Conceived and designed the experiments: ZG YHH. Performed the experiments: YLC YY JS YX. Analyzed the data: ZG YHH DCY YLC YY JS YX. Wrote the paper: YHH ZG.

References

1. Maruoka K, List B, Yamamoto H, Gong LZ (2012) Organocatalysis: a web collection. Chem Commun 48: 10703–10703.
2. Toscano MD, Woycechowsky KJ, Hilvert D (2007) Minimalist active-site redesign: teaching old enzymes new tricks. Angew Chem Int Ed 46: 3212–3236.
3. Arnold FH (2001) Combinatorial and computational challenges for biocatalyst design. Nature 409: 253–257.
4. Bommarius AS, Riebel BR (2004) Biocatalysis: Fundamentals and Applications. Weinheim: Wiley-VCH.
5. Johannes T, Simurdiak MR, Zhao H (2006) Biocatalysis, Encyclopedia of Chemical Processing. Taylor & Francis. DOI: 10.1081/E-ECHP-120017565.
6. Faber K (1997) Biotransformations in Organic Chemistry: A Textbook. Berlin: Springer-Verlag.
7. Boeris V, Spelzini D, Farruggia B, Picó G (2009) Aqueous two-phase extraction and polyelectrolyte precipitation combination: a simple and economically technologies for pepsin isolation from bovine abomasum homogenate. Process Biochem 44: 1260–1264.
8. Messing RA (1975) Immobilized Enzymes for Industrial Reactors. New York: Academic Press.
9. Leutbecher H, Hajdok S, Braunberger C, Neumann M, Mika S, et al. (2009) Combined action of enzymes: the first domino reaction catalyzed by *Agaricus bisporus*. Green Chem 11: 676–679.
10. Kittelmann M, Rheinegger U, Espigat A, Oberer L, Aichholz R, et al. (2003) Preparative enzymatic synthesis of the acylglucuronide of mycophenolic acid. Adv Synth Catal 345: 825–829.
11. Cordell GA, Lemos TLG, Monte FJQ, de Mattos MC (2007) Vegetables as chemical reagents. J. Nat. Prod. 70: 478–492.
12. Ishii Y, Sumi H, Yoshida E (2009) Method of producing a dry earthworm powder. Patent US 20090238891A1.
13. Pan R, Zhang Z-J, He R-Q (2010) Earthworm protease. Applied and Environmental Soil Science, Vol. 2010, Article ID 294258, 13 pages, doi:10.1155/2010/294258
14. Nakajima N, Sugimoto M, Ishihara K (2003) Earthworm-serine protease: characterization, molecular cloning, and application of the catalytic functions. J Mol Catal B-Enzym 23: 191–212.
15. Nakajima N, Sugimoto M, Tsuboi S, Tsuji H, Ishihara K (2005) An isozyme of earthworm seine proteases acts on hydrolysis of triacyllycerol. Biosci Biotechnol Biochem 69: 2009–2011.
16. Ishihara K, Nakajima N (2006) Stereoselective reduction of carbonyl compounds using the cell-free extract of the earthworm, *Lumbricus rebellus*, as a novel source of biocatalyst. Biosci Biotechnol Biochem 70: 3077–3080.
17. Mirabal-Gallardo Y, Soriano M del PC, Santos LS (2013) Stereoselective bioreduction of β-carboline imines through cell-free extracts from earthworms (*Eisenia foetida*). Tetrahedron: Asymmetry 24: 440–443.
18. Arend M, Westermann B, Risch N (1998) Modern variants of the Mannich reaction. Angew Chem Int Ed 37: 1044–1070.

16

Separation, Immobilization, and Biocatalytic Utilization of Proteins by a Supramolecular Membrane

Elisha Krieg[1], Shira Albeck[2], Haim Weissman[1], Eyal Shimoni[3], Boris Rybtchinski[1]*

1 Department of Organic Chemistry, Weizmann Institute of Science, Rehovot, Israel, 2 Israel Structural Proteomics Center (ISPC), Weizmann Institute of Science, Rehovot, Israel, 3 Department of Chemical Research Support, Weizmann Institute of Science, Rehovot, Israel

Abstract

Membrane separation of biomolecules and their application in biocatalysis is becoming increasingly important for biotechnology, demanding the development of new biocompatible materials with novel properties. In the present study, an entirely noncovalent water-based material is used as a membrane for size-selective separation, immobilization, and biocatalytic utilization of proteins. The membrane shows stable performance under physiological conditions, allowing filtration of protein mixtures with a 150 kDa molecular weight cutoff (~8 nm hydrodynamic diameter cutoff). Due to the biocompatibility of the membrane, filtered proteins stay functionally active and retained proteins can be partially recovered. Upon filtration, large enzymes become immobilized within the membrane. They exhibit stable activity when subjected to a constant flux of substrates for prolonged periods of time, which can be used to carry out heterogeneous biocatalysis. The noncovalent membrane material can be easily disassembled, purified, reassembled, and reused, showing reproducible performance after recycling. The robustness, recyclability, versatility, and biocompatibility of the supramolecular membrane may open new avenues for manipulating biological systems.

Editor: Masaya Yamamoto, Institute for Frontier Medical Sciences, Kyoto University, Japan

Funding: This work was supported by grants from the Israel Science Foundation, the Minerva Foundation, the Gerhardt M.J. Schmidt Minerva Center for Supramolecular Architectures, the Helen and Martin Kimmel Center for Molecular Design, and the Yeda Sela Center for Basic Research. E.K. acknowledges the Minerva PhD fellowship. B.R. holds the Abraham and Jennie Fialkow Career Development Chair. The funders had no role in study design, data collection and analysis, decision to publish, or preparation of the manuscript.

Competing Interests: The authors have declared that no competing interests exist.

* E-mail: boris.rybtchinski@weizmann.ac.il

Introduction

Noncovalent materials are adaptive, unlike most conventional covalent systems, rendering them advantageous for a variety of applications. [1–12] They are particularly useful for biotechnology, where biocompatible, stimuli-responsive, and self-healing materials have been utilized as artificial tissues, drug delivery devices, and matrices for encapsulation and controlled release of cells. [13–19] Yet, a number of important biotechnological applications require high material stability. For instance, membrane filtration of biomolecules employs nanoporous materials that should be stable for long periods of time under the pressure-driven flux of solvent, in the presence of buffer, and at physiological pH and salt concentrations. Membrane filtration is an essential tool in the biotechnological industry and appears to be particularly useful for the purification and concentration of proteins. [20–27] Moreover, membranes can be used for immobilization and biocatalytic utilization of enzymes. As enzymes catalyze reactions under very mild conditions, exhibiting efficiency and selectivity largely unmatched by synthetic catalysts, such membrane reactors are emerging components in new, environmentally friendly industrial processes (heterogeneous biocatalysis), which may supplement or replace traditional chemical methods. [28].

Due to high stability demands, virtually all filtration membranes used today are based on polymers or ceramics. [29–32] However, while providing remarkable stability, the irreversible bonding in these systems sets limits to the scope of possible adaptive changes, including self-healing, response to external stimuli, and controlled membrane disassembly. In this respect, functionalization of polymer-based membranes with polymeric hydrogel coatings enabled stimuli-responsiveness, for instance temperature- or pH-induced shrinkage/swelling. [33] Importantly, when dealing with biological samples, adaptive changes in the membrane (e.g., for controlled release of biomolecules) should take place under mild conditions to preserve the biomolecules' structure and function.

Entirely supramolecular membranes can be very advantageous due to the adaptive character of noncovalent bonding. While supramolecular arrays have been used to template the formation of covalent membranes, [34–37] only a few supramolecular membranes have been reported that do not require structural reinforcement by covalent crosslinks in the active layer. [38–43] For instance, Lu et al. have modified polymer membranes with self-assembling amphiphiles in order to achieve unique rejection properties. [39] Dankers et al. have electro-spun self-assembling molecules to form membranes that functioned as matrices for renal epithelial cells; [43] the performance of this system in pressure-driven filtration was not an object of the study. Others have combined the well-established phase inversion membrane fabrication with self-assembly procedures to create block copolymer micelle-based membranes, [40] some of which exhibit tunable porosity [41] and self-healing properties. [42] However, these membranes employ conventional high molecular weight polymers, and those applied to biological systems require elaborate modifications of the self-assembled material prior to use. [35,44] To the best of our knowledge, robust biocompatible filtration membranes based entirely on noncovalent bonds between small amphiphilic molecules and operating under physiological conditions have not been reported.

Hydrophobic interactions [45–47] between large nonpolar groups of amphiphilic molecules in aqueous solution can be remarkably strong, driving self-assembly towards very stable supramolecular systems. [3,48–50] Based on this idea, we have previously designed bolaamphiphile **PP2b** [51] (Figure 1). Its extended hydrophobic core contains two perylene-3,4,9,10-tetracarboxylic acid diimide (PDI) [52] units. **PP2b** self-assembles in aqueous media into a robust three dimensional (3D) fibrous network, resulting in a stable and multiple-stimuli-responsive material. Recently, we have fabricated **PP2b**-based *supramolecular ultrafiltration membranes* capable of size-selective separation of metal and semiconductor nanoparticles. [38] The membranes are readily prepared within 15 minutes via one-step deposition of an aggregated **PP2b** solution on a microfiltration support. This simple, fast and robust method represents a significant advantage over most fabrication techniques for covalent membranes, which are time-consuming and often require meticulously controlled conditions. Owing to its noncovalent nature, the material is easily disassembled by organic solvent (e.g. ethanol), the retained particles are released, and the membrane material itself can be recycled and reused multiple times.

Are hydrophobically stabilized noncovalent materials such as **PP2b** membranes suitable for biotechnological applications that require robustness? In general, hydrophobic groups can interfere with biomolecules by affecting their stability or by forming hydrophobic surfaces that will stick to proteins. Also, presence of high salt concentrations may significantly alter the structure of supramolecular assemblies, [53,54] and weak hydrophobic assemblies in particular. Nevertheless, strong noncovalent bonds are sometimes adequate to produce very robust supramolecular arrays. [55–58] In this respect, the **PP2b** assembly is based on very strong hydrophobic interactions, preventing exposure of the hydrophobic moieties to bulk water. [1,59] It is also enclosed by a shell of polyethylene glycol (PEG) groups (Figure 1), [51] which are known to preserve the native structure of proteins and resist undesired biomolecule adsorption (the latter would result in rapid membrane fouling). [60–62] Thus, in water, the **PP2b**-based network is potentially both robust and biocompatible, while retaining recyclability due to its noncovalent nature.

The present study demonstrates how the **PP2b**-based membranes can be employed for the manipulation of proteins. First, protein filtration experiments were carried out to characterize the membranes' performance in bioseparation of protein mixtures, and to determine its molecular weight cutoff. Next, its biocompatible properties were utilized towards immobilization of enzymes that were used in heterogeneous biocatalysis. Such membrane/immobilized enzyme arrays bear great potential to carry out biocatalytic reactions under continuous flow of reactants (i.e. the enzyme's substrate) that are converted "en passant" into a desired product, combining biocatalytic reaction and catalyst removal in a single step. [63–66] As described herein, the **PP2b**

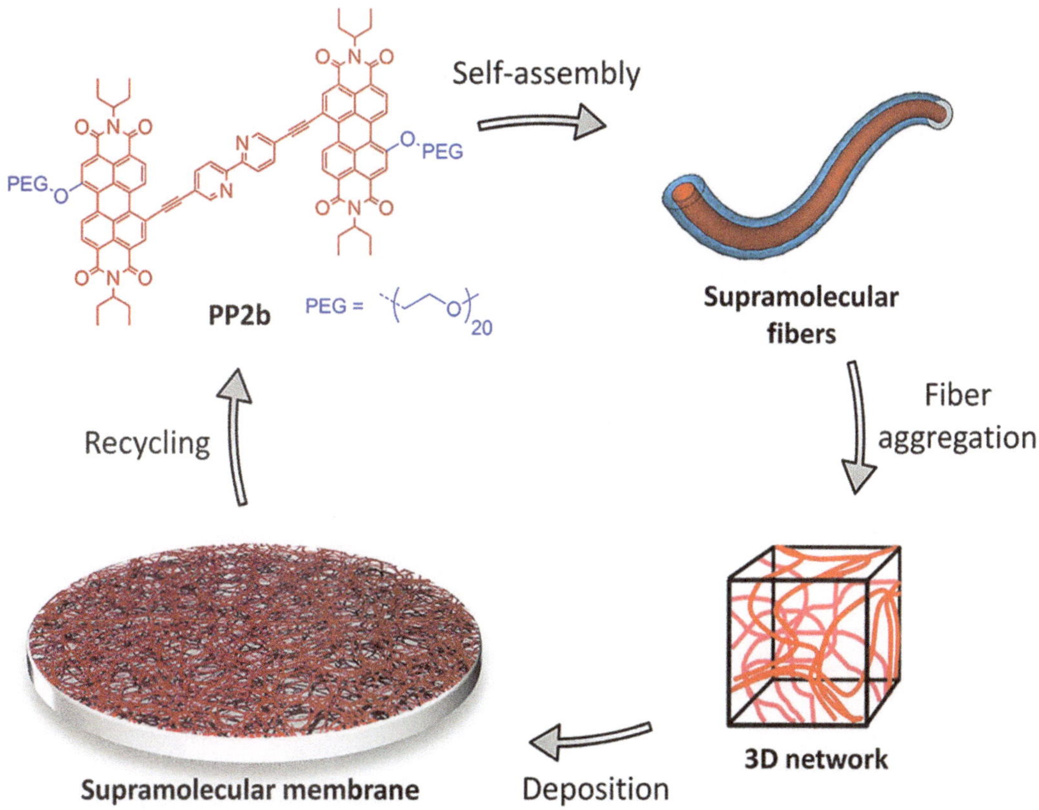

Figure 1. Illustration of membrane fabrication and recycling. Molecular structure of **PP2b**, schematic depiction of supramolecular fibers, 3D network, and membrane. Hydrophobic groups of **PP2b** are located in the fibers' core (red), whereas their hydrophilic PEG shell (blue) provides a biocompatible interface. Recycling of the membrane is achieved by disaggregation or physical removal of the supramolecular layer from the support, followed by purification, and subsequent reassembly in aqueous solution.

membranes are versatile for the manipulation of proteins, allowing recycling of the membrane material and retrieval of proteins. To the best of our knowledge, this is the first report of a noncovalent membrane that performs separation, immobilization and biocatalytic utilization of proteins, thus demonstrating the great potential of hydrophobically stabilized noncovalent materials in biotechnology.

Results and Discussion

1. Fabrication and Characterization of PP2b Supramolecular Membranes

The membranes used in this study were fabricated by depositing 1 ml of an aggregated solution of **PP2b** (0.5 mM) in water on a cellulose acetate (CA) support (0.45 μm pore size; effective filtration area: 5.7 cm^2), analogously to the reported procedure (see Materials and Methods). [38] Hence, each supramolecular membrane was composed of 1.5 mg **PP2b**. Importantly, the membranes remain robust under physiological conditions, e.g. aqueous NaCl (150 mM), and 3-(N-morpholino) propanesulfonic acid (MOPS) buffer solution, as revealed by monitoring the solution flux through the membrane over time (Figure 2). The flux was virtually identical for different aqueous solutions and remained constant during the experiment (30 min), indicating that neither membrane disassembly nor significant compression of the material took place. Fulfilling this first requirement ensured that the membranes are sufficiently stable for typical laboratory-scale filtrations of biological systems. During the course of our experiments, supramolecular membranes showed reproducible performance for filtration and sustained continuous solvent flux for up to 6 hours (see below). The flux of 20 $Lh^{-1}m^{-2}$ at 0.8 bar transmembrane pressure was somewhat lower than usual commercial ultrafiltration membranes with similar rejection properties, which exhibit typical fluxes of 50–100 $Lh^{-1}m^{-2}$. [67] The membrane thickness was approximately 6 μm, as revealed by cryogenic scanning electron microscopy (cryo-SEM) (Figure 3). Higher magnification of its cross-section indicates that the material retains its nanoporous structure under physiological conditions (Figure 3B, inset), which is critical for size-selective bioseparations.

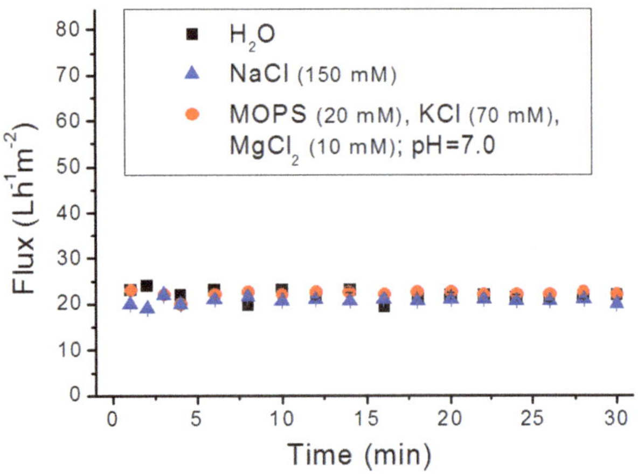

Figure 2. Stable flux of aqueous solutions over time. Flux of water, NaCl(aq), and MOPS buffer solution through a freshly prepared supramolecular membrane.

2. Separation Performance

A mixture of six purified proteins was used to carry out filtration experiments: 1. N-terminal domain of EIIBCA-Bgl residues 2–84 (EIIBCA), 2. *In silico* designed Kemp eliminase (KE70), [68,69] 3. L-carnitine dehydratase (LCD), 4. L-Fuculose-1-Phosphate Aldolase (Aldolase), 5. Citrate Synthase (CS), 6. Bovine Serum Albumin (BSA). This protein mixture represents a broad range of masses, making it suitable for characterization of the filtration cutoff. The proteins were dissolved in MOPS buffer solution (MOPS, 20 mM; KCl, 70 mM; $MgCl_2$, 10 mM; pH = 7.0) with an overall protein concentration of 1.8 mg/ml (0.3 mg/ml for each protein). In order to evaluate the actual size of the individual proteins in solution, they were each analyzed by gel filtration chromatography (GFC) (Figure S3 in Text S1), and their hydrodynamic diameters (D_h) were determined using dynamic light scattering (DLS) (Figure S4 in Text S1). We note that not all proteins in the mixture were monomeric; CS exhibited a molecular weight of approximately 288 kDa (from GFC), corresponding to its well-known hexameric form (301 kDa), [70,71] Aldolase was a tetramer, [72] and the commercially obtained BSA was oligomeric (≥400 kDa, determined by GFC). These findings are consistent with DLS measurements, showing a size increase in the order EIIBCA<KE70< LCD<Aldolase<CS<BSA (Table 1).

Filtration experiments were performed employing a flow of the protein mixture (1.5 ml) in aqueous MOPS buffer solution through a freshly prepared supramolecular membrane at 0.8 bar transmembrane pressure. As the feed solution was passing the membrane, the filtrate was collected (1.5 ml, F1). Subsequently, additional clean buffer solution (6 ml) was filtered to rinse out remaining proteins from the membrane, and another four fractions were collected successively (4×1.5 ml, F2–F5). The dead volume of the filter chamber was 0.8 ml. UV/Vis spectra of all fractions were recorded (Figure 4B, Figure S7 in Text S1). Relative protein concentrations for the collected fractions were quantified via absorbance at 280 nm. The fractions did not exhibit any absorption feature in the visible spectrum, showing that no detectable amount of **PP2b** ($\lambda_{max,1}$ = 393 nm, $\lambda_{max,2}$ = 538 nm) was leaching out of the membrane during filtration. The first two filtrate fractions (F1 and F2) contained considerable amounts of protein; F3 contained only minor amounts, whereas F4 and F5 were virtually protein-free (Figure 4C). Altogether, 38±4% of the proteins passed the membrane. In a control experiment, the protein mixture was also passed through the pristine CA membrane without the **PP2b** layer. As confirmed by UV/Vis spectra of the filtrate, the CA membrane itself did not retain any proteins (Figure 4A), showing that the retention is due to the supramolecular membrane only and not caused by adsorption on the support.

Importantly, proteins that were retained on the supramolecular membrane could be partially recycled by dispersing the used membrane in buffer solution (using a vortex mixer), followed by removing the **PP2b** supramolecular material via centrifugation. The resulting supernatant contained retained proteins while the pellet contained **PP2b** together with some proteins that could not be retrieved. The mechanical dispersion of the supramolecular membrane is enabled by its reversible noncovalent bonding, representing a mild alternative to common release mechanisms in covalent polymers, such as pH and temperature-induced shrinkage/swelling. [32,33] Notably, the enzymatic activity of proteins recycled from the **PP2b** membrane is fully retained (see below).

The filtration experiment results were analyzed by sodium dodecyl sulfate polyacrylamide gel electrophoresis (SDS-PAGE) (Figure 5A). Individual protein concentrations in filtrate and retentate were estimated densitometrically from the gel (Figures

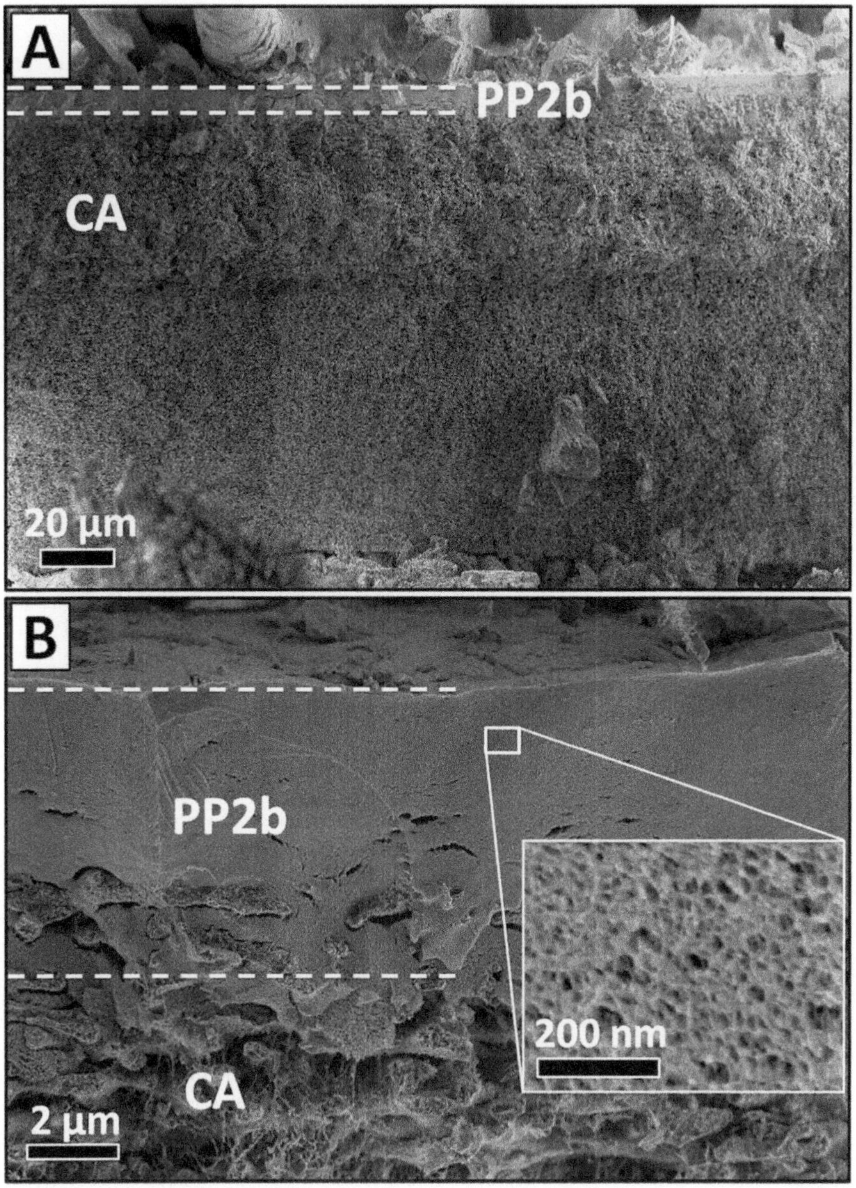

Figure 3. Membrane cross-section. Cryo-SEM images of the supramolecular PP2b membrane on the cellulose acetate (CA) support, which had been permeated by MOPS buffer solution.

Table 1. Size-related values of the proteins: calculated molecular weight (MW), hydrodynamic diameter (D_h), and molecular dimensions estimated from X-ray structures.

Protein	MW [kDa]	D_h [nm]	Protein X-ray structure dimensions [nm] (PDB code)
EIIBCA	8.7	3.4	n.a.
KE70	29	5.4	5.0×4.1×3.8 *(3Q2D)*
LCD	92	7.2	n.a.
Aldolase	158	8.2	7.0×7.0×5.5 *(1DZU)*
CS (hex.)	301	11.4	13.1×12.3×8.2 *(homologous to 1NXG)*
BSA (olig.)[a]	≥400	12.8, 57.8	n.a.

[a]BSA had a bimodal size distribution.

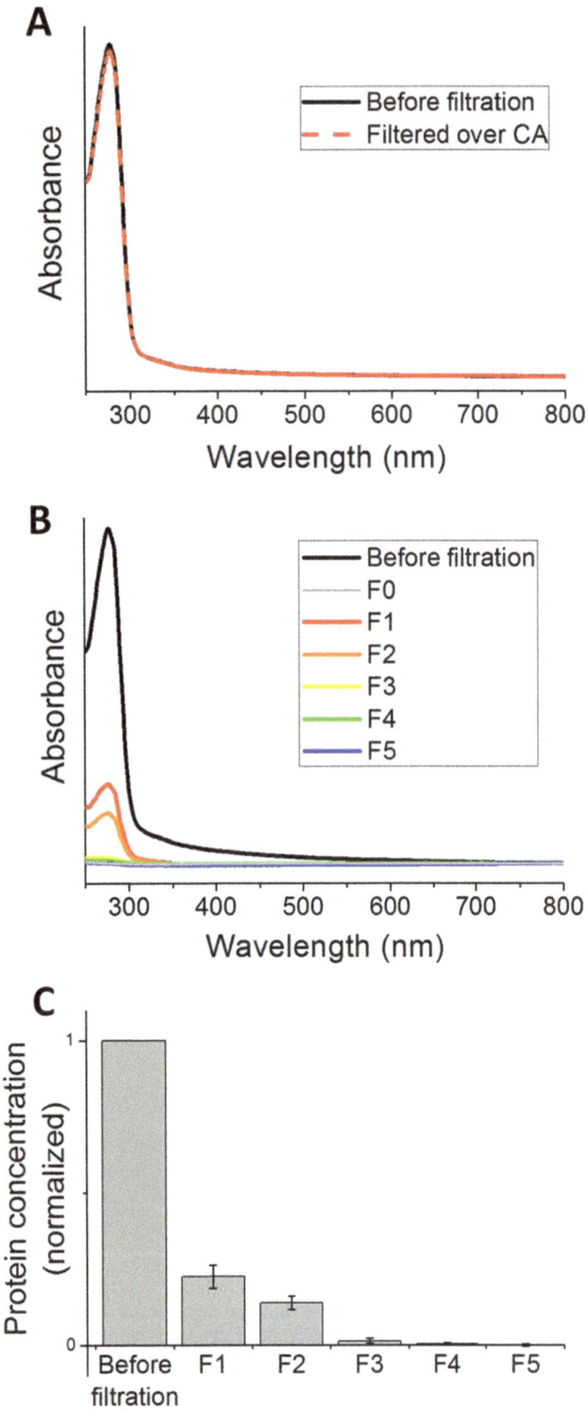

Figure 4. Filtration of protein mixtures. (A) UV/Vis spectra of the protein mixture before and after filtration over the pristine CA membrane. (B) Representative UV/Vis spectra of the protein mixture before filtration through the supramolecular membrane, the filtrate (collected in 5×1.5 ml fractions, F1–F5), and filtered buffer solution as a reference (F0) (C) Total protein concentration in the filtrate fractions F1–F5 as compared to the feed solution, determined from absorbance at 280 nm. Error bars correspond to the standard deviation of 5 independent filtration experiments.

S5, S6 in Text S1). Filtration experiments were repeated five times using independently deposited membranes, which showed consistent separation performance (Figure 5, Figures S6, S7 in Text S1).

Figure 5C shows the retention (the percentage of proteins filtered from the solution) as a function of the molecular weight. It is known for the **PP2b** membrane that retained particles are captured in the interior of the membrane rather than on the surface (depth filtration). [38] In depth filtration, the average observed pore sizes of the membrane are usually an order of magnitude larger than the cutoff value, since particles are captured at constrictions within the membrane or by adsorption on pore surfaces. [73] The retention of proteins in the **PP2b** membrane is clearly size-dependent and follows a typical sigmoid curve, indicating that size-selective capture (e.g. through mechanical sieving) plays a major role in the filtration process, rather than specific (e.g. electrostatic) protein adsorption. However, adsorption might play a minor role, for instance in the retention of small amounts (~10%) of EIIBCA. The molecular weight cutoff, defined as the theoretical weight of a molecule with 90% retention, was determined from the retention curve (Figure 5C), as **150 kDa**. In terms of hydrodynamic diameter, the membrane cutoff is **8 nm** (Figure 5D), which is somewhat larger than the previously reported 5 nm cutoff for spherical gold nanoparticles. [38] Yet, the latter value refers to the nanoparticles' rigid core, which does not include organic capping agent, counter ions, and hydration shell. The available protein crystal structures [69,70,72] reveal that the actual protein dimensions are significantly smaller than the corresponding hydrodynamic diameters (Figure 5C, Table 1). For instance the smallest dimension of KE70 is only 3.8 nm, enabling it to pass the membrane, whereas the smallest dimension of Aldolase is 5.5 nm, resulting in >90% retention. Thus, good agreement with previous nanoparticle filtration experiments is observed. [38] We note that size-selective retention in a membrane can depend on shape and deformability of filtered particles. [73].

Figure 5B details the concentration of the proteins in each of the filtrate fractions and that of the retained proteins recycled from the membrane. Notably, permeation of the three smaller proteins EIIBCA, KE70, and LCD, showed a size-dependent trend: The smallest (EIIBCA) passed the membrane with a short time delay and was completely collected in F1 and F2, whereas the larger (KE70) exhibited a longer delay, and small amounts were detected in F3 as well. Finally, the largest of these three (LCD) was found in considerable amounts in F3, and its traces were detected in F4. For the retained proteins, the efficiency of recycling was not equal. For example, about 66% of CS was regained from the filter, whereas only 25% of retained Aldolase could be recycled, possibly due to aggregation or irreversible entrapment within the supramolecular fibers of the membrane material. The membrane itself can be disassembled in water/ethanol (2:3, v/v) and cleaned from contaminations via chromatography over a short silica column (see Figure S2 in Text S1, showing the ^{1}H-NMR of **PP2b** before and after recycling). It is easily re-assembled and re-used as a membrane, having similar thickness and *identical separation performance* as compared to the membranes prepared from non-recycled **PP2b** (Figure S8 in Text S1).

Facile recycling and reproducible separation performance after recycling can be regarded as a direct result of the noncovalent nature of the membrane material: its nanoscopic structure (i.e. the three dimensional network of supramolecular fibers) [51] is encoded at the molecular level, allowing simple self-assembly and deposition procedures.

The cutoff value of the supramolecular membrane is in the upper range of commonly used ultrafiltration membranes in biotechnology, allowing the retention of large proteins, nucleic acids, lipids and other large lysate components. [20,21,23] We emphasize that the described filtration characteristics can be

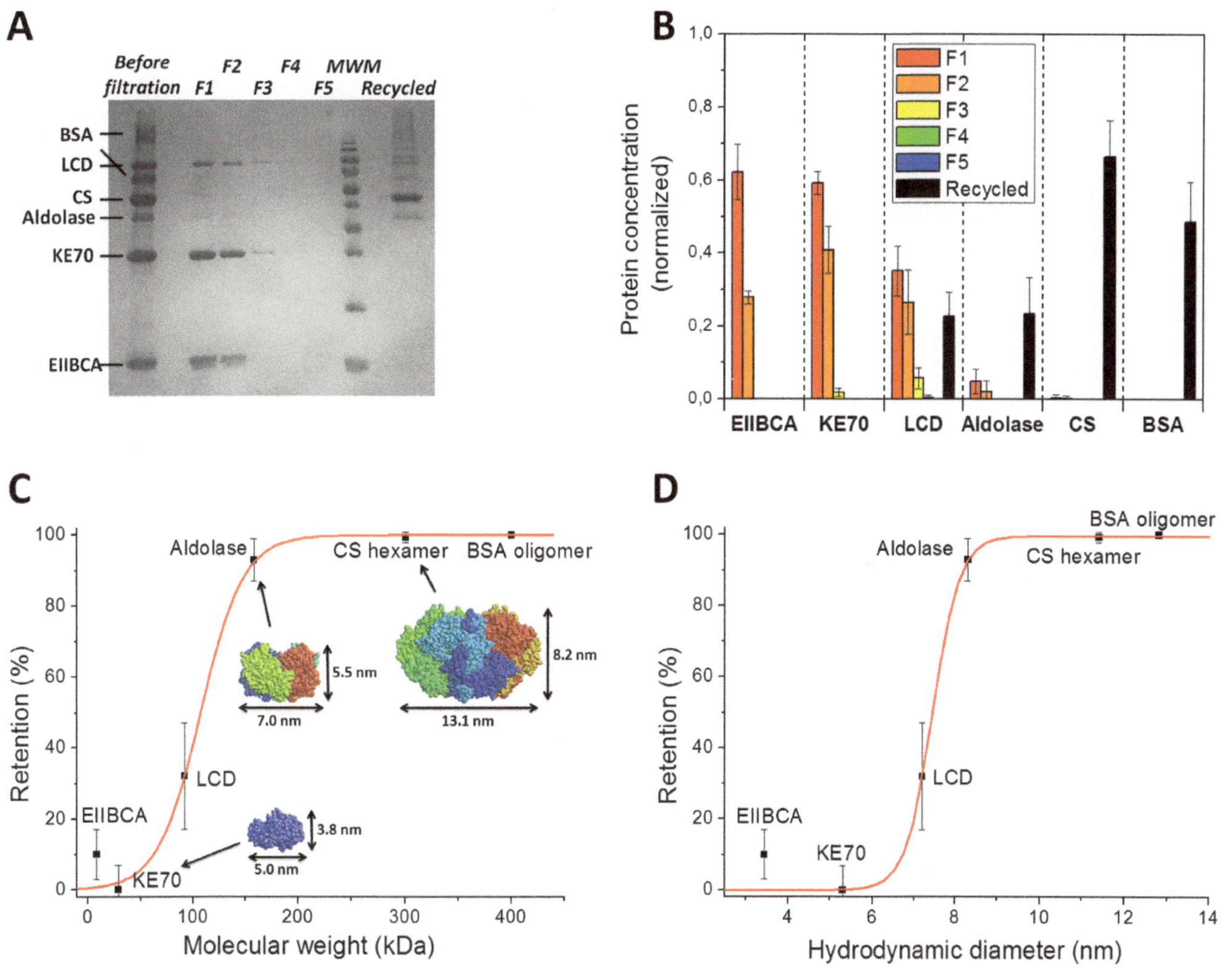

Figure 5. Separation of protein mixtures. (A) Typical SDS-PAGE used for densitometric quantification of individual protein concentrations. MWM = molecular weight marker (170, 130, 95, 72, 55, 43, 34, 26, 17, 11 kDa). (B) Concentrations of individual proteins in fractions F1–F5 (normalized with respect to the non-filtered solution), and recycled proteins. (C) Protein retention against molecular weight (black data points) and sigmoid fit (red curve). Protein structures of KE70 (PDB;3Q2D), Aldolase (PDB;1DZU) and CS hexamer (PDB;1NXG) are shown. (D) Dependence of protein retention on the hydrodynamic diameter (black data points) and sigmoid fit (red curve). All error bars represent the standard deviation of 5 independent filtration experiments.

particularly useful for the rapid separation of protein monomers from protein aggregates. In order to specifically demonstrate monomer/aggregate separation, oligomeric BSA was mixed with specially prepared monomeric BSA and the mixture was filtered over the **PP2b** supramolecular membrane. Due to their small size, the protein oligomers ($D_h \sim 12.8$ nm) cannot be removed by a standard desktop centrifuge (20800 g). However, as revealed by GFC, the supramolecular membrane efficiently removed oligomeric BSA from the mixture, resulting in a filtrate of pure BSA monomer (Figure 6).

3. Biocatalytic Utilization of the Supramolecular Membrane

In order to examine whether the filtration process affected protein function, we tested the enzymatic activity of KE70 at 25°C before and after filtration. Using a standard assay, [68] the activity of the enzyme solution before filtration, in the filtrate, and in neat buffer solution (background reaction) was determined by measuring the kinetics of the enzyme-catalyzed isomerisation of 5-Nitrobezisoxazole via absorbance of the product at 380 nm (OD_{380}) (Figure 7, Figure S10 in Text S1). Due to dilution with residual buffer solution in the filter chamber, the concentration of KE70 in the filtrate was 90% with respect to the feed solution. In comparison, its relative activity was 88%, revealing a near-quantitative (~98%) conservation of activity (see section S10, Figure S9, and Table S1 in Text S1).

In addition to KE70, we also investigated the enzymatic activity of hexameric CS after filtration. In contrast to KE70, CS is completely retained and can be recycled from the membrane. In case of retained enzymes, their structure and function might be affected by the extensive contact with the supramolecular material, and by the subsequent recycling procedure. CS activity before filtration and after recycling from the membrane was quantified at 25°C according to an activity assay based on a literature procedure: [74] measuring the kinetics of Citrate formation from Oxaloacetate and Acetyl Coenzyme A (Acetyl-CoA) by detecting the indicator of the reaction, 2-nitro-5-thiobenzoate (TNB) at 412 nm (OD_{412}) (Figure S11 in Text S1). Taking into account changes in concentration (determined via Bradford test), [75] the

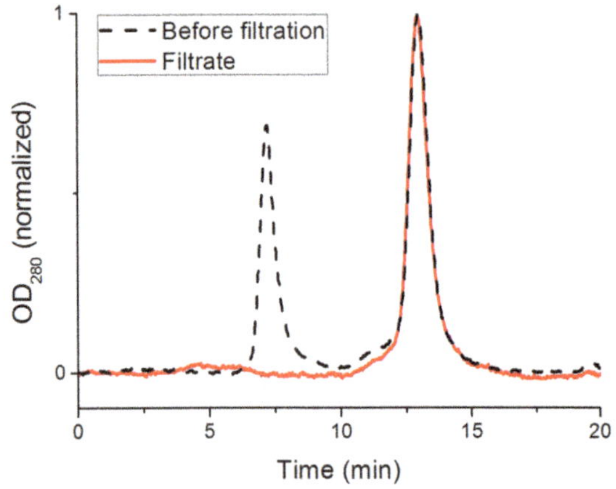

Figure 6. Purification of BSA. Gel filtration chromatogram of a mixture of BSA oligomers and monomers before filtration (black trace), and its filtrate (red trace). Filtration quantitatively removes BSA oligomers (≥ 400 kDa, retention time: 7 min) from the mixture. Smaller BSA aggregates (retention time: 11–12 min) are removed as well. The filtrate contains pure monomeric proteins (~67 kDa, retention time: 13 min).

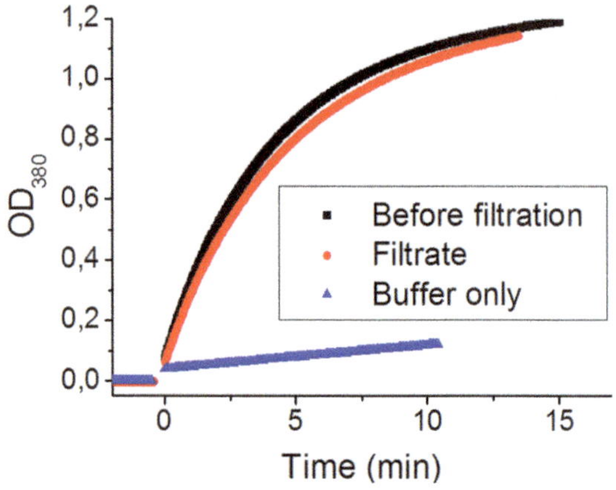

Figure 7. Kinetics of the KE70 activity. Change in absorbance at 380 nm before filtration, after filtration, and of neat buffer solution, following addition of 5-Nitrobezisoxazole (at t = 0 min).

kinetic measurements revealed quantitative retention of enzymatic activity of CS, underscoring the membrane's biocompatibility and the mild conditions of enzyme release from the membrane (see section S11, Tables S2, S3 in Text S1).

Our experiments show that filtration of large enzymes results in their immobilization within the supramolecular network, keeping them exposed to the flux of solvent and solutes. Such an array (membrane/immobilized enzyme) may represent a versatile system to carry out biocatalytic reactions in a heterogeneous manner. Addressing this concept, we investigated the enzymatic activity of membrane-immobilized β-Galactosidase (β-Gal, 465 kDa). β-Gal is significantly larger than the membrane cutoff, resulting in near-quantitative retention. It is ubiquitous in nature and widely utilized in molecular biology, catalyzing the hydrolysis of the glycosidic bond of β-Galactopyranosides. We note that β-Gal and many other hydrolases do not require presence of coenzymes, making them the most relevant enzyme class for biotechnological applications in industry, [28] e.g., β-Gal has wide application in the dairy industry for the production of low-lactose milk. [28].

For immobilization, 1.5 ml of β-Gal (0.2 mg/ml) was filtered over a freshly prepared supramolecular membrane, and rinsed with 7.5 ml of clean buffer solution. Once the enzyme was retained on the membrane, the solution flux slightly dropped from 20 L h^{-1} m^{-2} to 17 L h^{-1} m^{-2}. Subsequently, a typical activity assay solution [76] containing the substrate o-Nitrophenyl β-D-galactopyranoside (ONPG, 0.05 mg/ml) was passed through the membrane. As the colorless ONPG feed solution passed through the membrane it turned yellow, indicating the β-Gal-catalyzed conversion of ONPG into o-Nitrophenol (ONP, $\lambda_{max} = 420$ nm) (Figure 8A). In order to quantify the conversion and to study performance under constant substrate flux for prolonged periods of time, the filter was connected to a UV/Vis flow cuvette, and the absorbance of the filtrate at 420 nm was recorded (Figures S14, S15 in Text S1). Importantly, the reaction showed stable conversion over several hours of uninterrupted substrate flow with an average yield of ~90% ONP (Figure 8B).

Small amounts of enzyme were observed to leach out of the membrane over time, as expected in immobilization that does not involve covalent attachment of the enzyme to the support. [66] However, enzyme leaching was very low (decreasing from 0.24% to 0.046% activity with respect to the original enzyme solution; see section S15, Figure S16, Table S5 in Text S1) and did not influence the overall reaction yield. Emphasizing its remarkable robustness and biocompatibility, the supramolecular membrane/embedded enzyme array sustained stable operation under a constant flux of solution for overall 6 hours (incl. preparation, rinsing steps, and test of enzyme leaching), out of which 3 hours were dedicated to biocatalytic substrate conversion.

In an additional experiment, hexameric CS was immobilized and tested for its biocatalytic activity. While β-Gal facilitates the degradation of a complex molecule (ONPG) into smaller parts (Galactose, ONP), CS represents a distinctly different case: a simple precursor (Oxaloacetate) is converted into a more complex product (Citrate), requiring the presence of a coenzyme (Acetyl-CoA). CS immobilization was performed by simply filtering 2 ml enzyme solution (0.3 mg/ml) over a freshly prepared supramolecular membrane, followed by rinsing with 10.5 ml clean buffer solution. We employed a common assay solution [74] to test activity of immobilized CS. When running the mixed solution of Oxaloacetate, Acetyl-CoA, and 5,5′-Dithiobis(2-nitrobenzoate) (DTNB) through the membrane, biocatalytic formation of Citrate was indicated by the yellow color of the filtrate, which resulted from the subsequent reaction between HS-CoA and DTNB, forming TNB ($\lambda_{max} = 412$ nm) (Figure 9). Thus, conversion of Oxaloacetate into Citrate under constant flow of substrate and coenzyme was readily achieved. Similar to β-Gal, leaching of CS was low (~0.23% activity with respect to the original enzyme solution; see section S13, Figures S12, S13, Table S4 in Text S1).

Overall, immobilization on the supramolecular membranes is versatile and applicable to various enzymes provided that they are large enough to be retained. In order to extend the method to smaller enzymes, one can use standard procedures to enlarge the enzyme without affecting its activity, either by fusing the enzyme to a large inactive protein, or by formation of cross-linked enzyme aggregates (CLEAs) prior to filtration. [64,66,77] Since the design of effective immobilization techniques has been described as one of the main obstacles for industrial-scale biocatalysis, [78] we note that enzyme entrapment in the supramolecular membrane is accomplished in a simple filtration step within 20 minutes. The reactant conversion can be controlled by adjusting the amount of enzyme deposited on the supramolecular membrane. Having a substantial thickness of ~6 μm the noncovalent matrix functions as a depth filter, allowing very high enzyme loading (e.g. 0.4 g

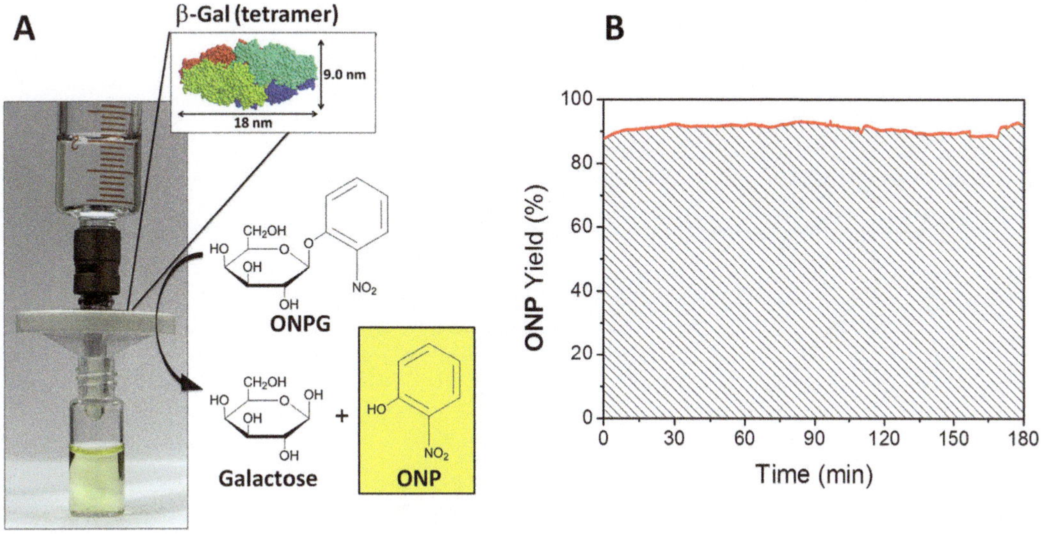

Figure 8. Heterogeneous biocatalysis in the supramolecular membrane using immobilized β-Gal. (A) Hydrolysis of ONPG into Galactose and ONP. (B) Yield of ONP as a function of time during several hours of continuous flux of substrate.

enzyme/1 g **PP2b** membrane) without membrane clogging. In contrast, most commercial ultrafiltration membranes retain particles on their surface (screen filtration), which enables high solution flux but limits the membranes' capacity for loaded enzymes. [73] Since biocatalysis in membrane reactors takes place within the short time of the substrate's passing through the membrane layer, high enzyme loading is important in order to achieve satisfying reaction yields. Heterogeneous biocatalysis also offers an elegant method to facilitate more complex cascade reactions, if two or more enzymes are immobilized in a sequential manner. [66,79] As membrane fabrication and enzyme immobilization are carried out by simple deposition steps, preparation of complex layered structures with alternating membrane/enzyme arrays [79] can be feasible.

Employing a *fully noncovalent* membrane reactor for enzyme biocatalysis has several advantages: no synthetic modification or covalent attachment of the enzyme to a stationary phase is necessary. Moreover, enzymes can be easily retrieved from the membrane by disassembly of the noncovalent membrane material, which is important considering the high cost of enzyme synthesis. These factors represent a significant advantage over the commonly employed covalent attachment of enzymes to a support, which is a slow and complex (multistep) process, can affect enzyme activity, and prohibits simple enzyme retrieval. [28].

Conclusions

In summary, we have demonstrated the application of a noncovalent membrane for separation, immobilization and biocatalytic utilization of proteins. To the best of our knowledge, this is the first example in which these functions were achieved using an entirely noncovalent water-based material. The membrane exhibits stable performance under physiological conditions, allowing size-selective separation with a 150 kDa molecular weight cutoff (~8 nm hydrodynamic diameter cutoff). Importantly, separated proteins - both passing and retained - remain functionally active, underlining the biocompatibility of the noncovalent material and separation process. Although the membrane operates under rather low pressure (0.8 bar) and flow rate (20 $Lh^{-1}m^{-2}$), it has several significant advantages over conventional polymer membranes: 1) Membrane fabrication via deposition of self-assembled fibers is significantly easier and faster than common polymer processing methods. [29] 2) The noncovalent material can be disassembled under benign conditions, e.g. in case of physical damage, clogging, or for intentional protein release. Subsequently, the membrane material can be easily purified, reassembled, and reused multiple times, [38] thus being fully recyclable and cost-efficient. 3) Retained proteins can be retrieved using simple procedures.

Facile membrane fabrication followed by deposition of large proteins represents a simple and fast alternative to covalent enzyme immobilization. Such membrane-immobilized enzymes can be utilized in heterogeneous biocatalysis, which may allow "green" industrial processes and development of enzyme-based sensors. The noncovalent nature of the supramolecular membranes is expected to enable further modifications [14] and realization of more complex tandem reactions. Overall, we demonstrate that self-assembled noncovalent materials based on strong hydrophobic interactions are advantageous for biotechnological applications, where a combination of facile fabrication, robustness, and adaptivity provide new opportunities for manipulating biomolecules.

Materials and Methods

Materials and Methods

Solvents and reagents were purchased from commercial sources and used as received, unless otherwise specified. Synthesis of **PP2b** was reported previously. [51] For all aqueous mixtures double-distilled water was used (Barnstead NANOpure Diamond™ water system). Cellulose acetate (CA) syringe filters were purchased from Whatman (Puradisc FP 30/0.45 CA-S). Protein crystal structures were obtained from the PDB (http://www.rcsb.org) and visualized with *RasMol* (v. 2.7.4.2.). [80,81].

^{1}H-NMR spectra were recorded at room temperature on a 300 MHz spectrometer (Bruker).

UV/Vis absorption measurements were carried out on a Cary-5000 spectrometer (Varian).

MALDI-TOF mass spectrometry was carried out using a REFLEX™ reflector time-of-flight instrument with SCOUT™

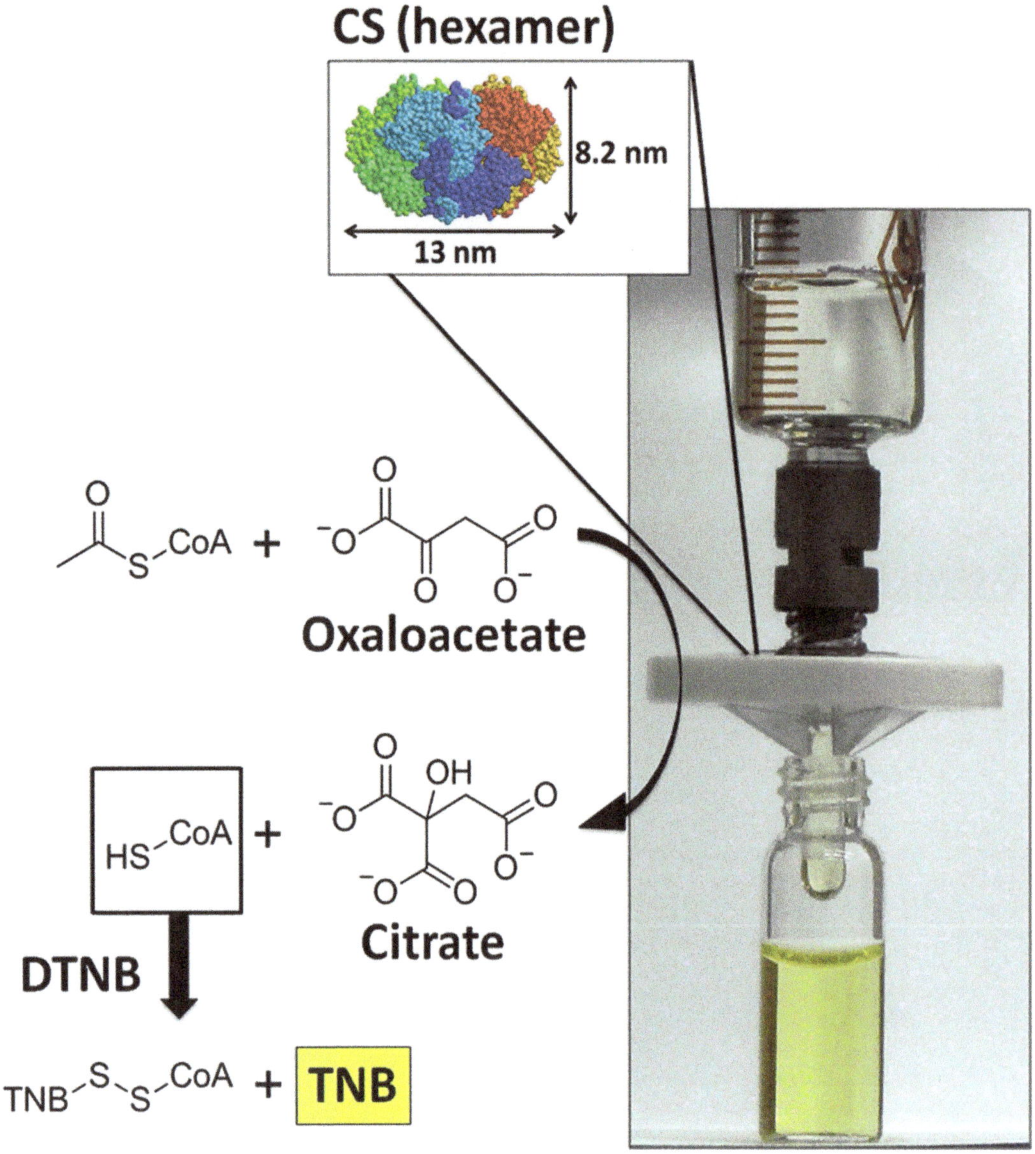

Figure 9. Conversion of Oxaloacetate and Acetyl-CoA into Citrate and HS-CoA over CS immobilized in a PP2b supramolecular membrane. HS-CoA reacts with DTNB in the assay solution to release the indicator of the reaction, TNB (λ_{max} = 412 nm). The color change from clear reactant feed solution to yellow filtrate indicates biocatalytic activity of the immobilized enzymes.

multiprobe (384) inlet. Chloroform was the solvent for all samples analyzed by mass spectrometry.

GFC was performed on an AKTA purifier using an analytical Superdex 200 10/300 GL column (GH Healthcare). The column was equilibrated with buffer containing 20 mM Tris pH 7.5 and 100 mM NaCl. The proteins were injected individually and their migration was compared to commercial molecular weight markers.

DLS was carried out on a Viscotek 802 DLS (Malvern Instruments). Samples were measured in a 12 µl quartz cuvette. Data analysis was performed using OmniSIZE (v. 3,0,0,292). 544 correlator channels were used. The data acquisition time was set to 5–10 seconds. A mass-weighted size distribution was determined from a single correlation function, which was obtained from the average of at least 8 experiments.

SDS-PAGE was performed using 15% gels and stained with Coomassie blue. 5 µl sample buffer was added to 15 µl of each protein sample (0.3 mg/ml) from which 15 µl was applied to each lane. The individual bands were quantified using the QuantityOne volume analysis package (BioRad version 4.4.0). The bands were analyzed densitometrically via the absorbance of UV light integrated over selected band areas. The background absorbance subtraction was set to "local", which reduced the error arising from uneven background brightness.

Cryo-SEM sample preparation involved the high pressure freezing (HPF) technique, which is designed to minimize ice crystal formation. [82–85] For this purpose, a ~1×1 mm small rectangle was cut out from the membrane and placed inside the inner cavity of an aluminium planchette (size = 3.0 mm (diameter)×0.5 mm (thickness); inner cavity = 2.0 mm (diameter)×0.15 mm (depth)). The vacant space in the cavity was filled with 1-hexadecene and it was capped with the flat side of another aluminium planchette. HPF was carried out using a Bal-Tec HPM 010 high pressure freezing machine. Subsequently, the sandwich was transferred into a Bal-Tech BAF 060 freeze fracture system where it was fractured with a pre-cooled razorblade and solvent was allowed to sublime (−105°C, 60 min). Subsequently, it was coated with Pt/C, employing double axis rotary shadowing (DARS). Images were taken in a Zeiss Ultra 55 cryogenic scanning electron microscope operated at 1–5 kV with an aperture size set to 10 μm, utilizing an in-lens secondary electron detector.

Protein Mixture for Filtration Experiments

The protein mixture was prepared by dissolving 0.51 mg of each of the following proteins (from stock solutions) in MOPS buffer solution (1.7 ml): EIIBCA, KE70, LCD, Aldolase, CS, BSA (protein concentrations were determined relative to BSA using Bradford reagent [75]). The solution was vortexed and centrifuged for 5 min at 20800 g, in order to assure complete protein dispersion. 1.5 ml of the clear protein mixture was used for filtration experiments within one hour.

Preparation of PP2b with Optimized PEG Length

PP2b was synthesized following our previously reported procedure. [51] Silica column chromatography (eluent: $CHCl_3$/MeOH, 0–6% MeOH gradient) was used to fractionate the material into two batches with different average PEG lengths (**PP2b(s)** and **PP2b(l)**). As determined by MALDI-TOF mass spectrometry, **PP2b(s)** (m/z[M+Na$^+$] = 2973.46, found: 2973.51) is furnished with PEG chains of 18.5 units length on average (Figure S1A in Text S1) and **PP2b(l)** (m/z[M+Na$^+$] = 3150.57, found: 3150.85) carries PEG chains with 20.75 units on average (Figure S1B in Text S1). While aggregates of pure **PP2b(s)** in water are not stable over time and begin to precipitate from the solution shortly after preparation, **PP2b(l)** forms a stable homogeneous aqueous solution, but its supramolecular structures are not suitable as ultrafiltration membrane, since they are not efficiently retained by the CA support membrane. The optimum average PEG-length in **PP2b** was adjusted by mixing **PP2b(s)** and **PP2b(l)**. The mixture of 0.5 mg (0.17 μmol) **PP2b(s)** and 1.0 mg (0.32 μmol) **PP2b(l)** was found to produce a stable and sufficiently robust supramolecular solution, which was used for fabrication of supramolecular ultrafiltration membranes.

Self-assembly of PP2b

An aqueous **PP2b** supramolecular solution (0.5 mM) was prepared similarly to the previously reported procedure: [38] In a typical experiment, 26 mg (8.8 μmol) of **PP2b(s)** and 53 mg **PP2b(l)** (16.8 μmol) were dissolved in THF (2.56 ml). Then water (48.6 ml) was added, followed by vigorous shaking of the solution and sonication for 5 min. After aging for 2 h, THF was removed by evaporating ~15–20% of the solution at r.t. in high vacuum. The solution was refilled with water to its original volume (51 ml); it was shaken, sonicated for 5 min, and aged for at least 12 h. The mixture was used as a stock solution for preparation of many supramolecular membranes. It was stored at 4°C and could be used for ~2 weeks, producing membranes with consistent separation performance. During storage for prolonged periods of time (>2 weeks), significant amounts of **PP2b** began to precipitate out of the solution, making it necessary to dry and re-assemble the material as described above.

Fabrication of Supramolecular Membranes

Membrane fabrication was similar to previously reported. [38] **PP2b** supramolecular solution (1 ml) was diluted in water (2 ml) and the mixture was filtered over a syringe filter (Whatman Puradisc FP 30, CA; effective filtration area: 5.7 cm^2; pore size: 0.45 μm). During filtration, **PP2b** was deposited on the CA support, as seen by the color change from dark purple feed solution to colorless filtrate, and formation of a dark purple layer on the CA support. The transmembrane pressure was created by opening the gauge to a pressurized nitrogen gas cylinder. It was slowly increased to 0.8 bar during deposition of **PP2b**. Care was taken to avoid air bubbles in the filter chamber, which would hamper the formation of a continuous supramolecular layer. Subsequently, water (2 ml) was filtered through the membrane, which ensures uniform packing of the supramolecular material. [38] In order to adjust pH and ionic strength in the filter chamber to the pH and ionic strength of the protein solution that was to be filtered, neat buffer solution (3 ml) was run through the membrane. The filtrate of that buffer (F0, 1.5 ml) was collected as a reference for UV/Vis spectroscopy. The filter housing was kept filled with buffer solution and the supramolecular membrane was used directly for filtration experiments.

Filtration Experiments

The protein mixture (1.5 ml) was filtered over a freshly prepared supramolecular membrane. Then, 6 ml of protein-free MOPS buffer solution was filtered in order to rinse out residual proteins from membrane and filter housing. Fractions were collected (5×1.5 ml, F1–F5). To test reproducibility of the filtration outcome, the experiments were carried out five times, each time using another freshly prepared membrane. For consistency, the same transmembrane pressure of 0.8 bar was used in all experiments. The collected fractions were analyzed by UV/Vis (Figure S7 in Text S1) and SDS-PAGE (Figure S6 in Text S1). The reported values for protein concentrations and protein retentions (Figure 4, Figure 5) are mean values with error bars representing their standard deviations.

Recycling of Retained Proteins and PP2b

After the filtration experiment, the filter chamber was opened and the **PP2b** supramolecular layer was scratched off the CA support and suspended in MOPS buffer solution (1.5 ml). It was vortexed for 1 h in order to disentangle the **PP2b** supramolecular fibers, thus releasing retained proteins (R). In order to separate **PP2b** from the retained proteins, the solution was centrifuged for 15 minutes at 20800 g (using a desktop centrifuge). The supernatant contained retained proteins while the pellet (P) contained **PP2b** together with some retained proteins that could not be extracted. Both supernatant and pellet were analyzed separately by SDS-PAGE (Figure S6 in Text S1), whereby the pellet was highly concentrated due to its small volume. In order to purify **PP2b** in the pellet from residual proteins and other organic and inorganic contaminations, ten such pellets were dissolved in 20 ml of a water/ethanol mixture (2:3, v/v). Subsequently, **PP2b** was extracted with dichloromethane (40 ml). The colorless aqueous phase was discarded. The organic phase was dried in high vacuum. The solid was washed with hexane (3×50 ml), and with water (1×50 ml). Subsequently, it was purified by silica column chromatography, eluting successively with: 1) 200 ml $CHCl_3$, 2) 200 ml $CHCl_3$/MeOH (98:2, v/v), 3) 200 ml $CHCl_3$/MeOH (92:8, v/v). The product contained pure **PP2b**, as revealed by ^{1}H-NMR (Figure S2 in Text S1). Nearly quantitative recycling was achieved.

Supporting Information

Text S1 Further experimental procedures and specifications, Supporting Figures S1–S16, and Supporting Tables S1–S5.

Acknowledgments

The EM studies were conducted at the Irving and Cherna Moskowitz Center for Nano and Bio-Nano Imaging (Weizmann Institute). EIIBCA, KE70, and CS were produced at the Israel Structural Proteomics Center (ISPC, Weizmann Institute). Monomeric BSA was the gift of Amona Ali and Zeev Gross (Technion). The authors thank U. Alcolombri and Prof. D. Tawfik for providing LCD and useful discussions. Further thanks go to J. Jacobovitch for assistance with protein characterization, and to Dr. E. Wachtel, Dr. N. Shechter, and A. Toth-Petroczy for useful discussions.

Author Contributions

Conceived and designed the experiments: EK SA HW BR. Performed the experiments: EK SA ES HW. Analyzed the data: EK SA HW ES BR. Wrote the paper: EK BR.

References

1. Rybtchinski B (2011) Adaptive Supramolecular Nanomaterials Based on Strong Noncovalent Interactions. ACS Nano 5: 6791–6818.
2. Aida T, Meijer EW, Stupp SI (2012) Functional Supramolecular Polymers. Science 335: 813–817.
3. Krieg E, Rybtchinski B (2011) Noncovalent Water-Based Materials: Robust yet Adaptive. Chem Eur J 17: 9016–9026.
4. Stupp SI, LeBonheur V, Walker K, Li LS, Huggins KE, et al. (1997) Supramolecular Materials: Self-Organized Nanostructures. Science 276: 384–389.
5. Cordier P, Tournilhac F, Soulie-Ziakovic C, Leibler L (2008) Self-healing and thermoreversible rubber from supramolecular assembly. Nature 451: 977–980.
6. Lehn JM (1993) Supramolecular chemistry. Science 260: 1762–1763.
7. Whitesides GM, Mathias JP, Seto CT (1991) Molecular self-assembly and nanochemistry: a chemical strategy for the synthesis of nanostructures. Science 254: 1312–1319.
8. Stoddart JF (2009) Thither supramolecular chemistry? Nature Chem 1: 14–15.
9. Wang Q, Mynar JL, Yoshida M, Lee E, Lee M, et al. (2010) High-water-content mouldable hydrogels by mixing clay and a dendritic molecular binder. Nature 463: 339–343.
10. Fenske T, Korth H-G, Mohr A, Schmuck C (2012) Advances in Switchable Supramolecular Nanoassemblies. Chem Eur J 18: 738–755.
11. Sijbesma RP, Beijer FH, Brunsveld L, Folmer BJB, Hirschberg JHKK, et al. (1997) Reversible polymers formed from self-complementary monomers using quadruple hydrogen bonding. Science 278: 1601–1604.
12. Capito RM, Azevedo HS, Velichko YS, Mata A, Stupp SI (2008) Self-assembly of large and small molecules into hierarchically ordered sacs and membranes. Science 319: 1812–1816.
13. Zhang S (2002) Emerging biological materials through molecular self-assembly. Biotechnology Advances 20: 321–339.
14. Dankers PYW, Meijer EW (2007) Supramolecular Biomaterials. A Modular Approach towards Tissue Engineering. Bull Chem Soc Jpn 80: 2047–2073.
15. Stupp SI (2010) Self-Assembly and Biomaterials. Nano Lett 10: 4783–4786.
16. Lim Y-b, Moon K-S, Lee M (2008) Rod–coil block molecules: their aqueous self-assembly and biomaterials applications. J Mater Chem 18: 2909–2909.
17. Percec V, Wilson DA, Leowanawat P, Wilson CJ, Hughes AD, et al. (2010) Self-Assembly of Janus Dendrimers into Uniform Dendrimersomes and Other Complex Architectures. Science 328: 1009–1014.
18. Huang Z, Lee H, Lee E, Kang S-K, Nam J-M, et al. (2011) Responsive nematic gels from the self-assembly of aqueous nanofibres. Nat Commun 2: 459–459.
19. Shah RN, Shah Na, Del Rosario Lim MM, Hsieh C, Nuber G, et al. (2010) Supramolecular design of self-assembling nanofibers for cartilage regeneration. PNAS 107: 3293–3298.
20. Saxena A, Tripathi BP, Kumar M, Shahi VK (2009) Membrane-based techniques for the separation and purification of proteins: An overview. Adv Coll Int Sci 145: 1–22.
21. van Reis R, Zydney A (2001) Membrane separations in biotechnology. Curr Opin Biotechnol 12: 208–211.
22. van Reis R, Zydney A (2007) Bioprocess membrane technology. J Membr Sci 297: 16–50.
23. Lightfoot EN, Moscariello JS (2004) Bioseparations. Biotechnol Bioeng 87: 259–273.
24. Keller K, Friedmann T, Boxman A (2001) The bioseparation needs for tomorrow. Trends Biotechnol 19: 438–441.
25. Bhattacharjee S, Bhattacharjee C, Datta S (2006) Studies on the fractionation of beta-lactoglobulin from casein whey using ultrafiltration and ion-exchange membrane chromatography. J Membr Sci 275: 141–150.
26. Lu J, Wan Y, Cui Z (2005) Fractionation of Lysozyme and Chicken Egg Albumin Using Ultrafiltration with 30-kDa Commercial Membranes. Ind Eng Chem Res 44: 7610–7616.
27. Peeva PD, Knoche T, Pieper T, Ulbricht M (2012) Performance of Thin-Layer Hydrogel Polyethersulfone Composite Membranes during Dead-End Ultrafiltration of Various Protein Solutions. Ind Eng Chem Res 51: 7231–7241.
28. Illanes As (2008) Enzyme biocatalysis: principles and applications: Springer.
29. Ulbricht M (2006) Advanced functional polymer membranes. Polymer 47: 2217–2262.
30. Vandezande P, Gevers LEM, Vankelecom IFJ (2008) Solvent resistant nanofiltration: separating on a molecular level. Chem Soc Rev 37: 365–405.
31. Benfer S, Árki P, Tomandl G (2004) Ceramic membranes for filtration applications - preparation and characterization. Adv Eng Mater 6: 495–500.
32. Tokarev I, Minko S (2009) Multiresponsive, Hierarchically Structured Membranes: New, Challenging, Biomimetic Materials for Biosensors, Controlled Release, Biochemical Gates, and Nanoreactors. Adv Mater 21: 241–247.
33. Yang Q, Adrus N, Tomicki F, Ulbricht M (2011) Composites of functional polymeric hydrogels and porous membranes. J Mater Chem 21: 2783–2811.
34. Beginn U (1998) Supramolecular templates as porogenes. Adv Mater 10: 1391–1394.
35. Yang SY, Yang J-A, Kim E-S, Jeon G, Oh EJ, et al. (2010) Single-File Diffusion of Protein Drugs through Cylindrical Nanochannels. ACS Nano 4: 3817–3822.
36. Phillip WA, O'Neill B, Rodwogin M, Hillmyer MA, Cussler EL (2010) Self-Assembled Block Copolymer Thin Films as Water Filtration Membranes. ACS Appl Mater Inter 2: 847–853.
37. Beginn U, Zipp G, Möller M (2000) Functional Membranes Containing Ion-Selective Matrix-Fixed Supramolecular Channels. Adv Mater 12: 510–513.
38. Krieg E, Weissman H, Shirman E, Shimoni E, Rybtchinski B (2011) A recyclable supramolecular membrane for size-selective separation of nanoparticles. Nature Nanotech 6: 141–146.
39. Lu Y, Suzuki T, Zhang W, Moore JS, Marinas BJ (2007) Nanofiltration membranes based on rigid star amphiphiles. Chem Mater 19: 3194–3204.
40. Peinemann K-V, Abetz V, Simon PFW (2007) Asymmetric superstructure formed in a block copolymer via phase separation. Nat Mater 6: 992–996.
41. Nunes SP, Behzad AR, Hooghan B, Sougrat R, Karunakaran M, et al. (2011) Switchable pH-Responsive Polymeric Membranes Prepared via Block Copolymer Micelle Assembly. ACS Nano 5: 3516–3522.
42. Tyagi P, Deratani A, Bouyer D, Cot D, Gence V, et al. (2012) Dynamic Interactive Membranes with Pressure-Driven Tunable Porosity and Self-Healing Ability. Angew Chem Int Ed 51: 7166–7170.
43. Dankers PYW, Boomker JM, Huizinga-van der Vlag A, Smedts FMM, Harmsen MC, et al. (2010) The use of fibrous, supramolecular membranes and human tubular cells for renal epithelial tissue engineering: towards a suitable membrane for a bioartificial kidney. Macromol Biosci 10: 1345–1354.
44. Uehara H, Kakiage M, Sekiya M, Sakuma D, Yamonobe T, et al. (2009) Size-Selective Diffusion in Nanoporous but Flexible Membranes for Glucose Sensors. ACS Nano 3: 924–932.
45. Ball P (2008) Water as an active constituent in cell biology. Chem Rev 108: 74–108.
46. Blokzijl W, Engberts JBFN (1993) Hydrophobic effects. Opinions and facts. Angew Chem Int Ed 32: 1545–1579.
47. Chandler D (2005) Interfaces and the driving force of hydrophobic assembly. Nature 437: 640–647.
48. Oshovsky GV, Reinhoudt DN, Verboom W (2007) Supramolecular Chemistry in Water. Angew Chem Int Ed 46: 2366–2393.
49. Ryu J-H, Hong D-J, Lee M (2008) Aqueous self-assembly of aromatic rod building blocks. Chem Commun: 1043–1054.
50. Görl D, Zhang X, Würthner F (2012) Molecular Assemblies of Perylene Bisimide Dyes in Water. Angew Chem Int Ed 51: 6328–6348.
51. Krieg E, Shirman E, Weissman H, Shimoni E, Wolf SG, et al. (2009) Supramolecular gel based on a perylene diimide dye: multiple stimuli responsiveness, robustness, and photofunction. J Am Chem Soc 131: 14365–14373.
52. Würthner F (2004) Perylene bisimide dyes as versatile building blocks for functional supramolecular architectures. Chem Commun: 1564–1579.
53. Lo Nostro P, Ninham BW, Milani S, Lo Nostro A, Pesavento G, et al. (2006) Hofmeister effects in supramolecular and biological systems. Biophys Chem 124: 208–213.
54. Gibb BC (2011) Supramolecular Assembly and Binding in Aqueous Solution: Useful Tips Regarding the Hofmeister and Hydrophobic Effects. Isr J Chem 51: 798–806.
55. Corbin PS, Zimmerman SC (1998) Self-Association without Regard to Prototropy. A Heterocycle That Forms Extremely Stable Quadruply Hydrogen-Bonded Dimers. J Am Chem Soc 120: 9710–9711.
56. Schmuck C, Wienand W (2003) Highly Stable Self-Assembly in Water: Ion Pair Driven Dimerization of a Guanidiniocarbonyl Pyrrole Carboxylate Zwitterion. J Am Chem Soc 125: 452–459.
57. Chen Z, Lohr A, Saha-Möller CR, Würthner F (2009) Self-assembled pi-stacks of functional dyes in solution: structural and thermodynamic features. Chem Soc Rev 38: 564–584.
58. Prins LJ, Reinhoudt DN, Timmerman P (2001) Noncovalent Synthesis Using Hydrogen Bonding. Angew Chem Int Ed 40: 2382–2426.

59. Tidhar Y, Weissman H, Wolf SG, Gulino A, Rybtchinski B (2011) Pathway-Dependent Self-Assembly of Perylene Diimide/Peptide Conjugates in Aqueous Medium. Chem Eur J 17: 6068–6075.
60. Kane RS, Deschatelets P, Whitesides GM (2003) Kosmotropes Form the Basis of Protein-Resistant Surfaces. Langmuir 19: 2388–2391.
61. Kingshott P, Thissen H, Griesser HJ (2002) Effects of cloud-point grafting, chain length, and density of PEG layers on competitive adsorption of ocular proteins. Biomaterials 23: 2043–2056.
62. Dalsin JL, Lin L, Tosatti S, Vörös J, Textor M, et al. (2004) Protein Resistance of Titanium Oxide Surfaces Modified by Biologically Inspired mPEG-DOPA. Langmuir 21: 640–646.
63. Brady D, Jordaan J (2009) Advances in enzyme immobilisation. Biotechnol Lett 31: 1639–1650.
64. Sheldon RA (2007) Enzyme Immobilization: The Quest for Optimum Performance. Adv Synth Catal 349: 1289–1307.
65. Spahn C, Minteer SD (2008) Enzyme Immobilization in Biotechnology. Recent Pat Eng 2: 195–200.
66. Tran DN, Balkus KJ (2011) Perspective of Recent Progress in Immobilization of Enzymes. ACS Catal 1: 956–968.
67. see for example technical specifications of Koch Membrane Systems HFM-180 or Sartorius stedim VF20P4, VF20C4.
68. Röthlisberger D, Khersonsky O, Wollacott AM, Jiang L, DeChancie J, et al. (2008) Kemp elimination catalysts by computational enzyme design. Nature 453: 190–195.
69. Khersonsky O, Röthlisberger D, Wollacott AM, Murphy P, Dym O, et al. (2011) Optimization of the in-silico-designed kemp eliminase KE70 by computational design and directed evolution. J Mol Biol 407: 391–412.
70. Maurus R, Nguyen NT, Stokell DJ, Ayed A, Hultin PG, et al. (2003) Insights into the Evolution of Allosteric Properties. The NADH Binding Site of Hexameric Type II Citrate Synthases. Biochemistry 42: 5555–5565.
71. Nguyen NT, Maurus R, Stokell DJ, Ayed A, Duckworth HW, et al. (2001) Comparative Analysis of Folding and Substrate Binding Sites between Regulated Hexameric Type II Citrate Synthases and Unregulated Dimeric Type I Enzymes. Biochemistry 40: 13177–13187.
72. Joerger AC, Gosse C, Fessner W-D, Schulz GE (2000) Catalytic Action of Fuculose 1-Phosphate Aldolase (Class II) As Derived from Structure-Directed Mutagenesis. Biochemistry 39: 6033–6041.
73. Baker R (2004) Membrane Technology and Applications. Menlo Park: Wiley.
74. Srere PA, Brazil H, Gonen L (1963) The Citrate Condensing Enzyme of Pigeon Breast Muscle and Moth Flight Muscle. Acta Chem Scand 17: S129–S134.
75. Bradford MM (1976) A rapid and sensitive method for the quantitation of microgram quantities of protein utilizing the principle of protein-dye binding. Anal Biochem 72: 248–254.
76. Miller JH (1972) Experiments in molecular genetics. New York: Cold Spring Harbor Laboratory.
77. Cao L, van Rantwijk F, Sheldon RA (2000) Cross-Linked Enzyme Aggregates: A Simple and Effective Method for the Immobilization of Penicillin Acylase. Org Lett 2: 1361–1364.
78. Torres-Salas P, del Monte-Martinez A, Cutiño-Avila B, Rodriguez-Colinas B, Alcalde M, et al. (2011) Immobilized Biocatalysts: Novel Approaches and Tools for Binding Enzymes to Supports. Adv Mater 23: 5275–5282.
79. Onda M, Lvov Y, Ariga K, Kunitake T (1996) Sequential reaction and product separation on molecular films of glucoamylase and glucose oxidase assembled on an ultrafilter. J Ferment Bioeng 82: 502–506.
80. Sayle RA, Milner-White EJ (1995) RASMOL: biomolecular graphics for all. Trends Biochem Sci 20: 374–374.
81. Bernstein HJ (2000) Recent changes to RasMol, recombining the variants. Trends Biochem Sci 25: 453–455.
82. Serp D, Mueller M, Stockar Uv, Marison IW (2002) Low-temperature electron microscopy for the study of polysaccharide ultrastructures in hydrogels. I. Theoretical and technical considerations. Biotechnol Bioeng 79: 243–252.
83. Serp D, Mueller M, Stockar Uv, Marison IW (2002) Low-temperature electron microscopy for the study of polysaccharide ultrastructures in hydrogels. II. Effect of temperature on the structure of Ca^{2+}-alginate beads. Biotechnol Bioeng 79: 253–259.
84. Matzelle T, Reichelt R (2008) Review: Hydro-, micro- and nanogels studied by complementary measurements based on SEM and SFM. Acta Microscopica 17: 45–61.
85. Menger FM, Seredyuk VA, Apkarian RP, Wright ER (2002) Colloidal Assemblies of Branched Geminis Studied by Cryo-etch-HRSEM. J Am Chem Soc 124: 12408–12409.

The Essential Role of the 3′ Terminal Template Base in the First Steps of Protein-Primed DNA Replication

Irene Rodríguez, Elisa Longás¤, Miguel de Vega, Margarita Salas*

Instituto de Biología Molecular "Eladio Viñuela" (Centro Superior de Investigaciones Científicas), Centro de Biología Molecular "Severo Ochoa" (Centro Superior de Investigaciones Científicas-Universidad Autónoma de Madrid), Cantoblanco, Madrid, Spain

Abstract

Bacteriophages ϕ29 and Nf from *Bacillus subtilis* start replication of their linear genomes at both ends using a protein-primed mechanism by means of which the DNA polymerase initiates replication by adding dAMP to the terminal protein, this insertion being directed by the second and third 3′ terminal thymine of the template strand, respectively. In this work, we have obtained evidences about the role of the 3′ terminal base during the initiation steps of ϕ29 and Nf genome replication. The results indicate that the absence of the 3′ terminal base modifies the initiation position carried out by ϕ29 DNA polymerase in such a way that now the third position of the template, instead of the second one, guides the incorporation of the initiating nucleotide. In the case of Nf, although the lack of the 3′ terminal base has no effect on the initiation position, its absence impairs further elongation of the TP-dAMP initiation product. The results show the essential role of the 3′ terminal base in guaranteeing the correct positioning of replication origins at the polymerization active site to allow accurate initiation of replication and further elongation.

Editor: Martin G. Marinus, University of Massachusetts Medical School, United States of America

Funding: This work was supported by the Spanish Ministry of Science and Innovation (grants BFU 2008-00215, and Consolider-Ingenio CSD 2007-00015 to MS); and Autonomous Community of Madrid (grant P-MAT-0283-0505 to MS). Institutional grant from Fundación Ramón Areces to the Centro de Biología Molecular Severo Ochoa. The funders had no role in study design, data collection and analysis, decision to publish, or preparation of the manuscript.

Competing Interests: Margarita Salas is a shareholder of X-Pol Biotech. There are no patents, products in development or marketed products to declare. This does not alter the authors' adherence to all the PLOS ONE policies on sharing data and materials.

* E-mail: msalas@cbm.uam.es

¤ Current address: X-Pol Biotech, Cantoblanco, Madrid, Spain

These authors contributed equally to this work.

Introduction

Some organisms such as bacteriophages and eukaryotic viruses contain replication origins formed by inverted terminal repeats (ITR) and a terminal protein (TP) covalently linked at both 5′ ends of their linear genomes [1], [2]. The location of the replication origins at the ends allows both strands to be replicated continuously from both ends in a symmetric fashion. For the initiation of replication, a free TP molecule forms a heterodimer with the DNA polymerase, providing the OH group of a specific Ser, Thr or Tyr residue to prime DNA replication from the very ends of the chromosome. After elongation, the TP remains covalently linked to the 5′ ends of the DNA, circumventing the problem of replication of linear chromosome ends [3]. The development of an *in vitro* replication system with purified proteins and TP-DNA from bacteriophage ϕ29 has been instrumental for the elucidation of this so-called protein-primed mechanism of DNA replication [1], [2].

The genome of ϕ29 is a linear dsDNA of 19285 bp, containing a TP of 31 kDa covalently linked to each 5′ end [4] that, together with a 6-bp ITR (3′-TTTCAT-5′) [5], [6] forms part of a minimal replication origin. Once the replication origins are specifically recognized by the heterodimer formed by the DNA polymerase and a free TP molecule [7], [8], the DNA polymerase catalyses the formation of a covalent bond between the initiating dAMP and the hydroxyl group of Ser^{232} of the TP, a reaction directed by the second T at the 3′ end of the template strand [9]. Then, to recover the template information corresponding to the terminal 3′ T, the TP-dAMP initiation product translocates backwards one position, the so-called sliding-back mechanism.

Bacteriophage Nf also belongs to the group of phages that infect *Bacillus* [10]. Its linear TP-DNA is 18754 bp long, containing a 6-bp ITR (3′ -TTTCAT-5′) [10], [11] and a 31 kDa TP at both 5′ ends [10]. In this case, the initiation of DNA replication occurs opposite the third 3′ terminal nucleotide of the template [12]. The recovery of the two terminal nucleotides mainly occurs by a step-wise sliding-back mechanism whereby the TP-dAMP formed opposite the third position is translocated one position back to pair with the second T, the third T directing again the formation of the TP-AA product. This product slides-back one position again, recovering the terminal 3′ T of the template. The sliding-back mechanism or variations of it, as the jumping-back that takes place in adenovirus [13] seems to have evolved in protein-priming replication systems as a general process to maintain full-length DNA [14], [2].

Phage ϕ29 DNA polymerase is the only protein-primed DNA-dependent replicase whose 3-D crystal structure has been solved, both as apoenzyme [15] and complexed to the TP [16]. The ϕ29 DNA polymerase/TP complex shows that TP has an extended structure complementary to the DNA polymerase surface. The TP is folded into a N-terminal domain, an intermediate domain that

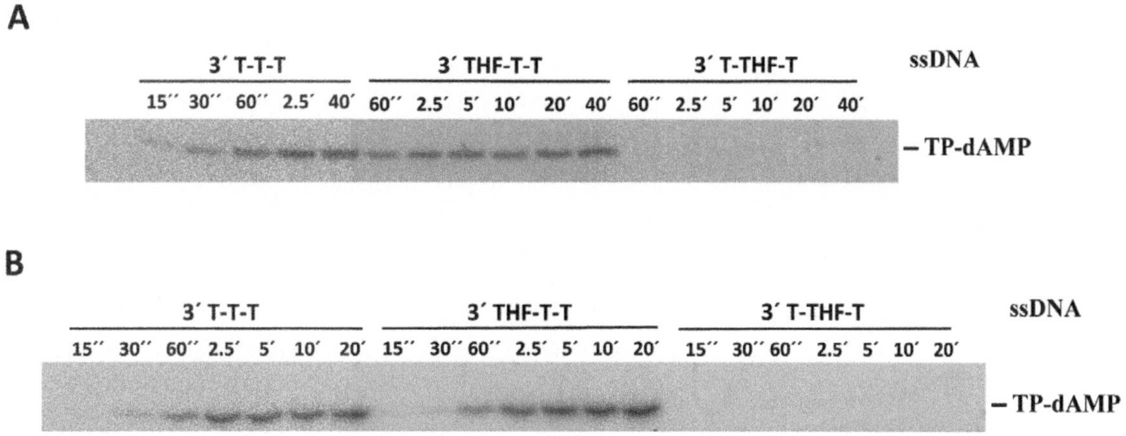

Figure 1. Formation of the TP-dAMP complex catalysed by either ϕ29 (A) or Nf (B) DNA polymerase. Using a single-stranded oligonucleotide with the chemically stable analogue of an abasic site, tetrahydrofuran (THF), at the 3′ end, in the presence of [α-^{32}P]dATP, both ϕ29 and Nf DNA polymerases were able to catalyze the formation of TP-dAMP product. These products were analysed by SDS-PAGE and autoradiography.

interacts with the DNA polymerase subdomain TPR1, and a C-terminal priming domain whose shape and electrostatic potential mimics DNA, occupying the DNA-binding cleft of the polymerase during initiation of DNA replication. This fact precludes the initiation at internal DNA sites, as an upstream 3′ template would sterically clash with the TP, restricting the beginning of DNA synthesis at the ends of the genome [16].

All structural complexes of ϕ29 DNA polymerase with ssDNA show that the 3′ end of the ssDNA molecule reaches the DNA polymerase active site, while its 5′ terminus remains bound to the DNA polymerase in the tunnel that lies downstream of the active site in a non-sequence specific manner, suggesting that it alone cannot establish register of the template [17]. In the context of protein priming, where the polymerase adds the first nucleotide to residue Ser^{232} of the TP by pairing it with the second nucleotide from the 3′ end of the template [9], the priming domain of TP must position the template in the active site by sterically excluding it from the upstream duplex binding region of the polymerase, implying that the TP priming domain acts as a structural barrier. As a result of data obtained with chimerical TPs by swapping the priming domains of ϕ29 and Nf TPs [12], we know that TP and, specifically, its priming domain dictates the template position used to direct the initiation reaction. Thus, once the 3′ end of the template contacts the TP priming domain located at the duplex binding region, the corresponding internal T is placed at the catalytic site to direct formation of the TP-dAMP product.

Table 1. Oligonucleotides used as template.

Oligonucleotide	Bacteriophague	Sequence 3′–5′
3′ T-T-T	*ϕ29*	TTTCATCCCATGTCGCTGTTG TATGTGGT
	Nf	TTTCATTCCAAGTTTCGTTTT AGCTGGGT
3′ THF-T-T	*ϕ29*	(THF)TTCATCCCATGTCGCTGTT GTATGTGGT
	Nf	(THF)TTCATTCCAAGTTTCGTTT TAGCTGGGT
3′ T-THF-T	*ϕ29*	T(THF)TCATCCCATGTCGCTGTT GTATGTGGT
	Nf	T(THF)TCATTCCAAGTTTCGTTT TAGCTGGGT
3′ THF-A-T	*ϕ29*	(THF)ATCATCCCATGTCGCTGTT GTATGTGGT
	Nf	(THF)ATCATTCCAAGTTTCGTTT TAGCTGGGT
3′ THF-T-A	*ϕ29*	(THF)TACATCCCATGTCGCTGTT GTATGTGGT
	Nf	(THF)TACATTCCAAGTTTCGTTT TAGCTGGGT

Single-stranded oligonucleotides containing the wild-type and modified sequences of the ϕ29 or Nf DNA right replication origin were obtained from Sigma. These oligonucleotides were used as template for both the initiation assay (TP-dNMP formation) and the truncated elongation assay. THF indicates the presence of tetrahydrofuran, which is a stable analogue of an abasic site.

To progress in the comprehension of this special mechanism to initiate replication, we have analysed here the influence of the 3′ terminal base of the template strand in the initiation and the first elongation steps of ϕ29 and Nf TP-DNA replication, by using highly purified DNA polymerase and TP of each phage, and template ssDNA oligonucleotides corresponding to either the wild-type sequence of the replication origins or variations of it.

Materials and Methods

Nucleotides and DNAs

Unlabelled nucleotides and dideoxynucleotides, [α-^{32}P]dATP (3000 Ci/mmol), [α-^{32}P]dTTP (3000 Ci/mmol), [α-^{32}P]dGTP (3000 Ci/mmol) and [α-^{32}P]dCTP (3000 Ci/mmol), were supplied by Amersham Biosciences. Single-stranded oligonucleotides containing the wild-type and modified sequences of the ϕ29 or Nf DNA right replication origin, used as template, were obtained from Sigma. Terminal protein-containing ϕ29 DNA (ϕ29 TP-DNA) was obtained as described [18].

Proteins

Wild-type ϕ29 and Nf TPs were expressed in *E. coli* BL21(DE3) cells harboring the gene cloned into plasmid pT7-3 and further purified as described [19]. ϕ29 DNA polymerase mutant D12A/D66A and Nf DNA polymerase mutant D66A, both exonuclease deficient, were purified essentially as described [20].

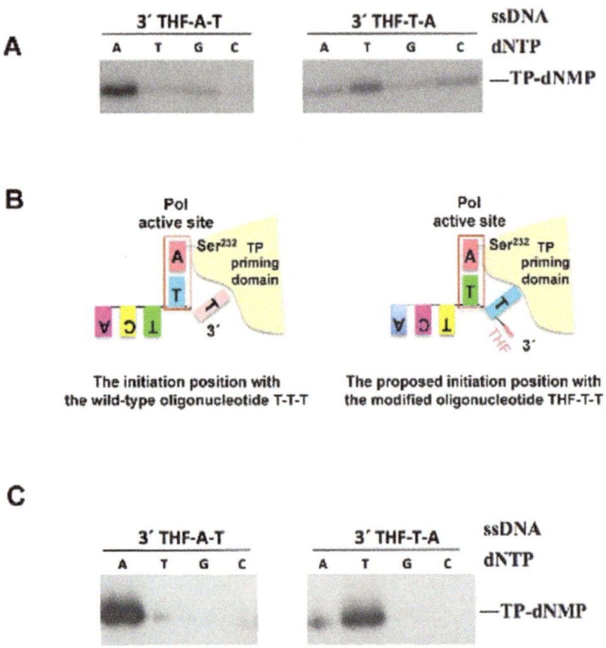

Figure 2. (A) The absence of the 3′ base at the template strand makes ϕ29 DNA polymerase to initiate opposite the third position. The different initiation products were detected by high resolution SDS-PAGE and analysed by autoradiography. ***(B)*** *Schematic representation of the placement of the 3′ end lacking the terminal base at the active site of ϕ29 DNA polymerase.* On the left side, the ϕ29 DNA polymerase catalyses the formation of the covalent bond between the initiating dAMP and the hydroxyl group of Ser^{232} of the TP, using as template the second T at the 3′ end of a wild-type template strand [9]. On the right side, the absence of the terminal base makes the template strand to enter the polymerization domain one position further, leading to the placement of the third 3′ nucleotide at the catalytic site to direct the formation of the initiation complex. **(C)** *The lack of the 3′ terminal base does not alter the initiation position by Nf DNA polymerase.* Bands corresponding to TP-dNMP products synthesized by Nf DNA polymerase were detected by high resolution SDS-PAGE.

Template-Dependent *in Vitro* Formation of the Initiation Complex TP-dNMP

The incubation mixture contained, in 25 µl, 50 mM Tris-HCl (pH 7.5), 1 mM $MnCl_2$, 20 mM ammonium sulphate, 4% (v/v) glycerol, 0.1 mg/ml BSA, 0.1 µM dNTP, 3 µCi of the corresponding labelled dNTP, 350 ng of the specified single-stranded oligonucleotide, 50 ng of ϕ29 or Nf TP and 50 ng of ϕ29 DNA polymerase mutant D12A/D66A or Nf DNA polymerase mutant D66A, and incubated for the indicated times at 30°C. The reactions were stopped by adding 10 mM EDTA and 0.1% SDS, filtered through Sephadex G-50 spin columns, and further analysed by SDS-PAGE as described [18]. The labelled bands corresponding to the TP-dNMP complex were detected by autoradiography.

Truncated Elongation Assays

The incubation mixture contained, in 25 µl, 50 mM Tris-HCl (pH 7.5), 1 mM $MnCl_2$, 20 mM ammonium sulphate, 4% (v/v) glycerol, 0.1 mg/ml BSA, 700 ng of the indicated single-stranded oligonucleotide or 500 ng of ϕ29 TP-DNA as template, 100 ng of purified TP and 100 ng of exonuclease-deficient ϕ29 or Nf DNA polymerase. In the case of the initiation reactions, 5 µM of the indicated dNTP and 4 µCi of the corresponding [α-^{32}P]dNTP were used, and 5 µM dATP, dTTP and dGTP, 4 µCi of [α-^{32}P]dNTP and 100 µM ddCTP in the case of the elongation reaction. In all cases, after incubation for five minutes at 30°C, reactions were stopped by adding EDTA to 10 mM. Samples were filtered through Sephadex G-50 spin columns in the presence of 0.1% SDS, and the excluded volume was analysed by 12% SDS-PAGE gel (360×280×0.5 mm) electrophoresis to obtain enough resolution to distinguish TP bound to the first elongation products. The length of the different products is indicated.

Results and Discussion

ϕ29 DNA polymerase changes the initiation position of replication in the absence of the 3′ terminal base

Previous studies showed that ssDNA containing replication origin sequences can support the template-directed TP-primed initiation reaction using DNA polymerase/TP heterodimers, having become a valuable tool to determine the precise initiation site, as well as to analyse the effect that nucleotide changes at the replication origins have in the first steps of DNA replication [9], [12], [13], [21], [22], [23]. This fact has led us to study the influence of the base of the 3′ terminal nucleotide in the template-directed formation of TP-dAMP (initiation complex). For this, we used a single-stranded oligonucleotide in which the 3′-terminal T was replaced by the chemically stable analogue of an abasic site, tetrahydrofuran (THF; THF-T-T oligonucleotide) (see **Table I**). To prevent degradation of the template oligonucleotides, exonuclease deficient DNA polymerases, ϕ29 D12A/D66A [24] and Nf D66A [25], were used. In the presence of [α-^{32}P]dATP, both ϕ29 (**Figure 1A**) and Nf (**Figure 1B**) DNA polymerases were able to catalyse the formation of TP-dAMP in the absence of the 3′ terminal base of the template, although in the case of ϕ29 DNA polymerase in a lower extent than with the unmodified oligonucleotide (**Figure 1A**). As expected, the absence of the base at the penultimate 3′ nucleotide of the template (T-THF-T oligonucleotide) precluded the initiation reaction by the ϕ29 DNA polymerase/TP heterodimer, as this enzyme has been shown to initiate opposite the second 3′ terminal position of the template [9]. Interestingly, the presence of an abasic site in the penultimate position impaired initiation by Nf DNA polymerase, despite the fact that formation of the TP-dAMP product by this polymerase is guided by the third position of the template (**Figure 1B**). This fact could suggest that proper stability/orientation of the third T at the Nf DNA polymerase depends on the presence of the 3′ neighbour base.

Having established that ϕ29 DNA polymerase is able to initiate using as template the THF-T-T oligonucleotide, the next step was to determine which of the 3′ terminal Ts directed the initiation reaction. For this, single changes into A at the second or third position of the template were introduced and assayed for TP-primed initiation with each of the four [α-^{32}P]dNTPs. Unexpectedly, the substitution of the second T (THF-A-T oligonucleotide) made the polymerase to preferentially catalyse the formation of TP-dAMP product (**Figure 2A**), suggesting that the polymerase was using the third T as template of the initiation reaction instead of the second one. To confirm this possibility, the same kind of experiment was carried out changing the third T on the template into A. As it can be observed in **Figure 2A**, now the ϕ29 DNA polymerase catalysed preferentially the formation of the TP-dTMP product. These results led us to propose a model in which the absence of the terminal base makes the template strand to enter the polymerization domain one position further, making the third 3′ nucleotide to be allocated at the catalytic site to direct the formation of the initiation complex (see **Figure 2B**). As shown in

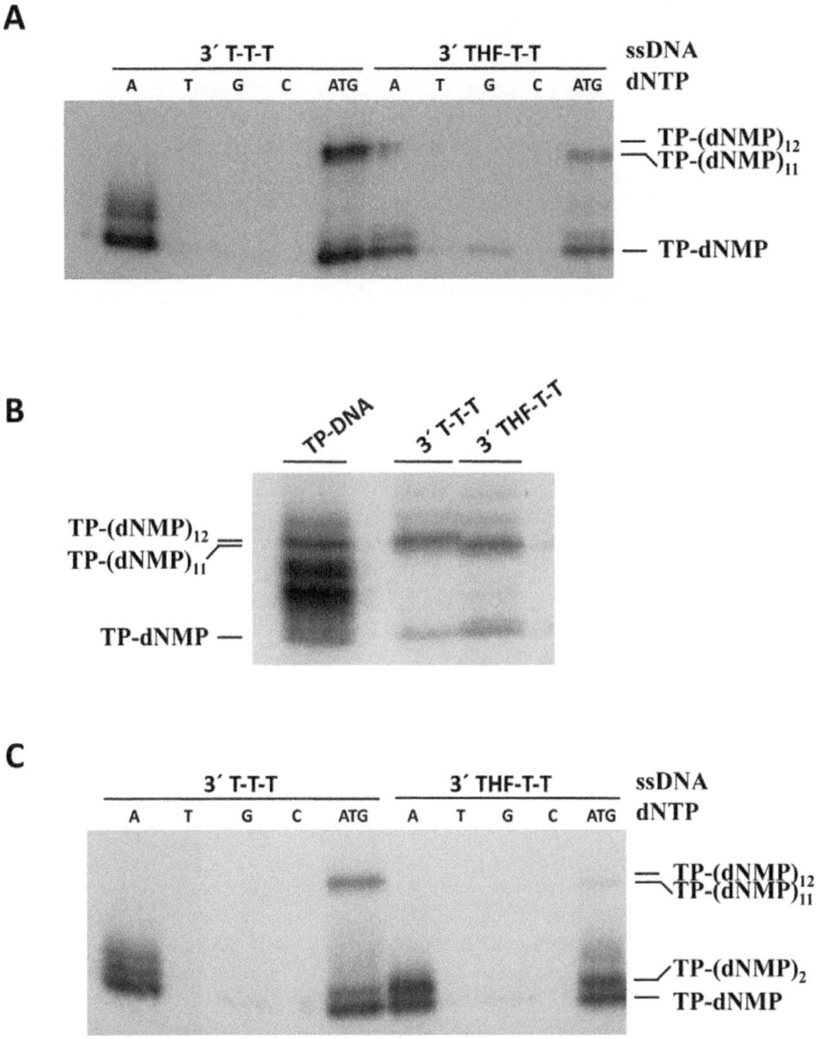

Figure 3. (A) The absence of the 3′ terminal base at the template oligonucleotide prevents the recovery of the 3′ terminal end by ϕ29 DNA polymerase. (**B**) Size of the products synthesized by ϕ29 DNA polymerase with the oligonucleotides 3′ T-T-T and 3′ THF-T-T. The length of the product synthesized by ϕ29 DNA polymerase with the oligonucleotides 3′ T-T-T and 3′ THF-T-T correspond with the intermediate replication products TP-$(dNMP)_{12}$ and TP-$(dNMP)_{11}$, respectively, as deduced from the comparison with the products obtained when the natural ϕ29 TP-DNA is used as template. (**C**) *The lack of the 3′ terminal base at the template strand blocks elongation of the initiation product by Nf DNA polymerase.* In all cases, the different elongation products were analysed by high resolution SDS-PAGE. The length at the different products is indicated.

Figure 2A, the fidelity of ϕ29 DNA polymerase on the THF-T-A oligonucleotide was lower than that obtained with the oligonucleotide THT-A-T. The change of the third T into A (THF-T-A oligonucleotide) gave rise, in addition to the expected initiation product TP-dTMP, to the formation of smaller amounts of TP-dAMP, TP-dGMP and TP-dCMP initiation products. This result could indicate that the stability of the templating base is influenced by the sequence context in such a way that the position previous to the templating base could affect the incorporation of the initiating nucleotide. In this sense, it is possible that the presence of a pyrimidine previous to the initiation position in oligonucleotide THF-T-A decreases the fidelity of ϕ29 DNA polymerase by reducing the proper stabilization of the templating nucleotide that directs the initiation position.

Intriguingly, in spite of the high percentage of sequence similarity between ϕ29 and Nf DNA polymerase/TP heterodimers [26], the latter shows a behaviour quite different from that of the ϕ29 one with respect to the absence of the 3′ terminal base of the template. Nf DNA polymerase does not seem to require the terminal base for the proper localization of the template oligonucleotide since, in the absence of the terminal base, Nf DNA polymerase is still able to use the 3′ third position as template nucleotide, being TP-dAMP and TP-dTMP the main initiation products obtained with THF-A-T and THF-T-A oligonucleotides, respectively (**Figure 2C**). This last result indicates that the presence of THF at the 3′ end does not modify the initiation position by Nf DNA polymerase.

Nf DNA polymerase is not able to elongate the initiation product on oligonucleotides with an abasic site at the 3′ terminus

It has been reported that the TP-dAMP initiation product formed using the second and third terminal nucleotide by the ϕ29 and Nf systems, respectively, is not directly elongated from the initiation site [9], [25]. In ϕ29, the sliding-back mechanism has evolved to recover the terminal T of the template through translocation of the initiation complex one position back, enabling to use again as template the penultimate T to direct the insertion of the second nucleotide [9]. In Nf, the related step-wise sliding-back mechanism allows the recovery of the first two nucleotides by means of two consecutive sliding-back steps [25]. These processes have to take place before definitive elongation of the initiation products occurs.

Once established that the absence of the 3′ terminal base makes the ϕ29 DNA polymerase to initiate opposite the third nucleotide, it was necessary to find out whether the first two nucleotides were recovered during the first steps of replication. To address this question, elongation of the initiation product TP-dAMP was studied in the presence of dATP, dTTP, dGTP and ddCTP, using oligonucleotides containing either the wild-type ϕ29 and Nf replication origins or the variants bearing a 3′ terminal THF. Thus, if the 2 nucleotides are recovered before elongation, the maximum length of the replication product in the presence of ddCTP obtained with ϕ29 DNA polymerase would be TP-$(dNMP)_{12}$ with the oligonucleotide 3′ T-T-T, as previously described [9]. Firstly, to analyse that this elongation is the result of the synthesis onto the TP-dAMP product, each nucleotide was provided separately. As shown in **Figure 3A**, the ϕ29 DNA polymerase gave rise to initiation products only in the presence of dATP, TP-dAMP being the main product, although small amounts of $TP(dAMP)_3$-$TP(dNMP)_4$ complexes were also detectable, the latter as a result of misincorporation in front of the fourth position due to the use of the 3′-5′ exonuclease deficient ϕ29 DNA polymerase mutant D12A/D66A [24]. These results indicate that the longer elongation products that could be yielded in the presence of the four nucleotides would be the result of the direct elongation of TP-dAMP initiation complex (see below). On the other hand, the lack of initiation products when each of the other three [α^{32}P]dNTPs are used indicates that the absence of the 3′ terminal base does not affect the fidelity of the initiation reaction. As seen in **Figure 3A and B**, with the wild-type ϕ29 sequence, and in the presence of dATP, dTTP, dGTP and ddCTP, replication was truncated at $TP(dAMP)_{12}$. The use of the natural ϕ29 TP-DNA as template rendered bands corresponding to intermediate products and allowed us the assignment of the elongation band (**Figure 3B**). On the contrary, the maximum length of the replication product obtained with oligonucleotide THF-T-T was TP-$(dNMP)_{11}$ (**Figure 3A and B**), suggesting that once the TP-dAMP is formed opposite the third position of the template, the initiation complex slides-back one position, recovering the penultimate T, and is then elongated. The lack of a second sliding-back step to recover the 3′ terminal position could be due either to the incapacity to hybridise the TP-$(dAMP)_2$ product opposite the 3′ THF-T (see **Figure 2B**), which would require the breakage of the hydrogen-bond of two base pairs, a situation energetically unfavoured, or because the polymerase recognizes the TP-$(dAMP)_2$ bound to the second and third positions as the functional one, due to the lack of a 3′ terminal base. Therefore, it seems that the THF at the 3′ end does not preclude the sliding-back movement of the TP-dAMP product opposite the penultimate position of the template but does not allow the sliding-back opposite the THF.

In the case of the Nf DNA polymerase, when the initiation with the different nucleotides on the T-T-T oligonucleotide was assayed, bands corresponding to TP-$(dAMP)_{1-3}$ were observed only in the presence of [α-^{32}P]dATP. As it occurred with ϕ29 DNA polymerase, TP-$(dNMP)_{12}$ was the longest replication product observed in the presence of dATP, dTTP, dGTP and ddCTP, indicating the recovery of the two terminal nucleotides (**Figure 3C**). Interestingly, when Nf DNA polymerase had to elongate the THF-T-T oligonucleotide, in contrast to ϕ29 DNA polymerase, replication was blocked mainly at position 2. This result indicates that the TP-dAMP initiation complex slided-back once, but the absence of a 3′ terminal base precluded the second sliding-back step. In addition, instead of elongating directly the TP-$(dAMP)_2$ product, as the Nf DNA polymerase does when it cannot perform the second sliding-back step [12], the absence of the 3′ terminal base hampered elongation. This result could be pointing to either a misalignment of the template at the catalytic site, or to the requirement of a terminal base to allow the DNA polymerase to proceed further.

The present study leads us to propose an essential role for the base of the 3′ terminal nucleotide to allow initiation at the corresponding penultimate nucleotide in the case of ϕ29, and to allow elongation in the case of Nf during the first steps of the protein-primed replication process. In addition, the presence of a base previous to the initiation site in Nf seems to be required to allow proper placement of the templating base for initiation.

Author Contributions

Conceived and designed the experiments: IR EL MdV MS. Performed the experiments: IR EL. Analyzed the data: IR EL MdV MS. Wrote the paper: IR MdV MS.

References

1. Salas M (1991) Protein-priming of DNA replication. Annu Rev Biochem 60, 39–71.
2. Salas M, Miller J, Leis J, DePamphilis M (1996) Mechanisms for priming DNA synthesis. Cold Spring Harbor Laboratory Press, New York.
3. Kornberg A, Baker T (1992) DNA replication. 2nd edition. Freeman (San Francisco).
4. Salas M, Mellado RP, Viñuela E, Sogo JM (1978) Characterization of a protein covalently linked to the 5′ termini of the DNA of *Bacillus subtilis* phage ϕ29. J Mol Biol 119, 269–291.
5. Escarmís C, Salas M (1982) Nucleotide sequence of the early genes 3 and 4 of bacteriophage ϕ29. Nucleic Acids Res 10, 5785–5798.
6. Yoshikawa H, Ito J (1982) Nucleotide sequence of the major early region of bacteriophage ϕ29. Gene 17, 323–335.
7. Blanco L, Prieto I, Gutiérrez J, Bernad A, Lázaro JM et al. (1987) Effect of NH_4^+ ions on ϕ29 DNA-protein p3 replication: Formation of a complex between the terminal protein and the DNApolymerase. J Virol 61, 3983–3991.
8. Gutiérrez J, Vinos J, Prieto I, Méndez E, Hermoso JM et al. (1986) Signals in the ϕ29 DNA-terminal protein template for the initiation of phage ϕ29 DNA replication. Virol 155, 474–483.
9. Méndez J, Blanco L, Esteban JA, Bernad A, Salas M (1992) Initiation of ϕ29 DNA replication occurs at the second 3′ nucleotide of the linear template: a sliding-back mechanism for protein-primed DNA replication. Proc Natl Acad Sci U S A 89, 9579–9583.
10. Yoshikawa H, Ito J (1981) Terminal proteins and short inverted terminal repeats of the small *Bacillus* bacteriophage genomes. Proc Natl Acad Sci USA 78, 2596–2600.
11. Yoshikawa H, Garvey KJ, Ito J (1985) Nucleotide sequence analysis of DNA replication origins of the small *Bacillus* bacteriophages: Evolutionary relationships. Gene 37, 125–130.
12. Longás E, Villar L, Lázaro JM, de Vega M, Salas M (2008) Phage ϕ29 and Nf terminal protein-priming domain specifies the internal template nucleotide to initiate DNA replication. Proc Natl Acad Sci USA 47, 18290–18295.
13. King AJ, van der Vliet PC (1994) A precursor terminal protein-trinucleotide intermediate during initiation of adenovirus DNA replication: Regeneration of molecular ends in vitro by a jumping back mechanism. EMBO J 13, 5786–5792.
14. Salas M (1999) Mechanisms of initiation of linear DNA replication in prokaryotes. Genet Eng (NY) 21, 159–171.
15. Kamtekar S, Berman AJ, Wang J, Lázaro JM, de Vega M et al. (2004) Insights into strand displacement and processivity from the crystal structure of the protein-primed DNA polymerase of bacteriophage ϕ29. Mol Cell 16, 609–618.
16. Kamtekar S, Berman AJ, Wang J, Lázaro JM, de Vega M et al. (2006) The ϕ29 DNA polymerase:protein-primer structure suggests a model for the initiation to elongation transition. EMBO J 25, 1335–1343.
17. Berman AJ, Kamtekar S, Goodman JL, Lázaro JM, de Vega M et al. (2007) Structures of ϕ29 DNA polymerase complexed with substrate: the mechanism of translocation in B-family polymerases. EMBO J. 26, 3494–3505.

18. Peñalva MA, Salas M (1982) Initiation of phage ϕ29 DNA replication in vitro: formation of a covalent complex between the terminal protein, p3, and 5′-dAMP. Proc Natl Acad Sci USA 79, 5522–5526.
19. Mencía M, Gella P, Camacho A, de Vega M, Salas M (2011) Terminal protein-primed amplification of heterologous DNA with a minimal replication system based on phage phi29. Proc Natl Acad Sci USA 46, 18655–18660.
20. Lázaro JM, Blanco L, Salas M (1995) Purification of bacteriophage ϕ29 DNA polymerase. Methods Enzymol 262, 42–49.
21. Illana B, Zaballos A, Blanco L, Salas M (1998) The The RGD sequence in phage ø29 terminal protein is required for interaction with ø29 DNA polymerase. Virol 248(1), 12–19.
22. Martín AC, Blanco L, García P, Salas M, Méndez J (1996) In vitro protein-primed initiation of pneumococcal phage Cp-1 DNA replication occurs at the third 3′ nucleotide of the linear template: A stepwise sliding-back mechanism. J Mol Biol 260, 369–377.
23. Caldentey J, Blanco L, Bamford DH, Salas M (1993) In vitro replication of bacteriophage PRD1 DNA. Characterization of the protein-primed initiation site. Nucleic Acids Res 21, 3725–3730.
24. Bernad A, Blanco L, Lázaro JM, Martín G, Salas M (1989) A conserved 3′-5′ exonuclease active site in prokaryotic and eukaryotic DNA polymerases. Cell 59, 219–228.
25. Longás E, de Vega M, Lázaro JM, Salas M (2006) Functional characterization of highly processive protein-primed DNA polymerases from phages Nf and GA-1, endowed with a potent strand displacement capacity. Nucleic Acids Res 34, 6051–6063.
26. González-Huici V, Salas M, Hermoso JM (2000) Sequence requirements for protein primed initiation and elongation of phage ϕ29 DNA replication. J Biol Chem 275, 40547–40553.

18

Ecogenomic Perspectives on Domains of Unknown Function: Correlation-Based Exploration of Marine Metagenomes

Pier Luigi Buttigieg[1,2]*, Wolfgang Hankeln[1], Ivaylo Kostadinov[3], Renzo Kottmann[1], Pelin Yilmaz[1], Melissa Beth Duhaime[4], Frank Oliver Glöckner[1,2]

1 Microbial Genomics and Bioinformatics Group, Max Planck Institute for Marine Microbiology, Bremen, Germany, **2** Jacobs University gGmbH, Bremen, Germany, **3** Departament de Biologia Marina i Oceanografia, Institut de Ciències del Mar, Barcelona, Spain, **4** Department of Ecology and Evolutionary Biology, University of Michigan, Ann Arbor, Michigan, United States of America

Abstract

Background: The proportion of conserved DNA sequences with no clear function is steadily growing in bioinformatics databases. Studies of sequence and structural homology have indicated that many uncharacterized protein domain sequences are variants of functionally described domains. If these variants promote an organism's ecological fitness, they are likely to be conserved in the genome of its progeny and the population at large. The genetic composition of microbial communities in their native ecosystems is accessible through metagenomics. We hypothesize the co-variation of protein domain sequences across metagenomes from similar ecosystems will provide insights into their potential roles and aid further investigation.

Methodology/Principal findings: We calculated the correlation of Pfam protein domain sequences across the Global Ocean Sampling metagenome collection, employing conservative detection and correlation thresholds to limit results to well-supported hits and associations. We then examined intercorrelations between domains of unknown function (DUFs) and domains involved in known metabolic pathways using network visualization and cluster-detection tools. We used a cautious "guilty-by-association" approach, referencing knowledge-level resources to identify and discuss associations that offer insight into DUF function. We observed numerous DUFs associated to photobiologically active domains and prevalent in the *Cyanobacteria*. Other clusters included DUFs associated with DNA maintenance and repair, inorganic nutrient metabolism, and sodium-translocating transport domains. We also observed a number of clusters reflecting known metabolic associations and cases that predicted functional reclassification of DUFs.

Conclusion/Significance: Critically examining domain covariation across metagenomic datasets can grant new perspectives on the roles and associations of DUFs in an ecological setting. Targeted attempts at DUF characterization in the laboratory or *in silico* may draw from these insights and opportunities to discover new associations and corroborate existing ones will arise as more large-scale metagenomic datasets emerge.

Editor: Iddo Friedberg, Miami University, United States of America

Funding: This study was funded by the Max Planck Society (http://www.mpg.de/en). The funders had no role in study design, data collection and analysis, decision to publish, or preparation of the manuscript.

Competing Interests: The authors have declared that no competing interests exist.

* E-mail: pbuttigi@mpi-bremen.de

Introduction

In recent years, genomic sequencing projects have revealed a large number of novel genes across a wide range of organisms and environments. Many of these have poor sequence-level similarity to genes that have been characterized in a laboratory setting and, consequently, have not been annotated with functional roles. These 'hypothetical' genes are becoming increasingly prevalent in bioinformatics databases. For example, the Pfam 24 database [1] stored some 11,912 protein domain families derived from conserved sequence data with ~26% dubbed "domains of unknown function" (DUFs). This proportion is predicted to soon overtake that of functionally characterized domains [2], and calls for community action [3] and cross-disciplinary efforts [4] towards their identification have been made.

Several groups have employed biochemical and molecular techniques in DUF characterization. Among these, Deng et al. [5] characterized members of a Pfam family, formerly known as DUF62, as S-adenosyl-l-methionine hydroxide adenosyltransferases using a range of enzymatic analyses. Similarly, Weinitschke et al. characterized members of the DUF81 family as the sulfite exporter TauE by examining co-transcription of genes in the metabolism of C_2 sulfonates [6]. Computational approaches have also been used to aid DUF characterization. Goonesekere et al. [7] applied secondary structure analysis and the use of three-dimensional homology models to functionally annotate 8 DUFs.

On a larger scale, Jaroszewski et al. [8] used structural genomics approaches to determine the three-dimensional structures for more than 250 DUF families, granting insights into their potential activities. This latter study reported that the majority of DUFs analyzed were either divergent structural variants of well-characterized families or showed notable substructure similarity to known proteins. The authors inferred that these variations may have been conserved as they extended an organism's functional repertoire in an ecologically adaptive manner. Here, we explored a marine metagenomic dataset to gain an ecological perspective on DUF functionality.

Metagenomics allows insight into an entire ecological community's genetic content [9,10], providing a rich source of *in situ* sequence data from a range of ecosystems. These metagenomic resources are ripe for mining, with intra-ecosystem variation in microbial function cited as a high-value target for exploration [11]. If genomic conservation of DUFs is adaptive in nature, then it is probable that these DUFs are co-selected with other domains linked to that adaptation. We thus hypothesized that the abundances of co-selected domains will co-vary across metagenomes from similar ecosystems. By identifying co-varying groups of domains, it may be possible to speculate on DUF functionality using a 'guilty-by-association' approach [12]. This approach is often used in the identification of novel metabolic modules by relating experimental perturbations to subsequent gene expression patterns [13]. In this model, genes with similar responses are grouped into putative modules to guide more rigorous future investigation. Further, correlative approaches have been employed to explore ecosystem-level interactions between microbial taxa and their environment [14,15].

Here, we propose that correlations of protein domain sequences found in intra-ecosystem metagenomic data may provide insight into the functional roles of DUFs. We critically examined the correlation of protein domains of known and unknown function across the Global Ocean Sampling (GOS) collection [16], comprising a globally distributed set of epipelagic microbial community metagenomes. We found several sets of associations that show promise in guiding DUF characterization in an experimental setting.

Results and Discussion

Pfam versions in analysis and interpretation

Here, we queried raw sequences of selected GOS metagenomes against version 24 of the Pfam database [1]. Of the 3,069 DUFs in Pfam v24, we detected 2,531 in our analysis. In subsequent Pfam database releases, many models and their functional descriptions have been updated to reflect the growing knowledge surrounding them. The associations presented here are best validated by domain-level knowledge. We thus referred to Pfam v26 [17] when discussing our results, allowing several correlative hypotheses to be evaluated against later functional characterizations.

Correlation analysis

Removal of rows (sites) and columns (Pfams) dominated by zero abundances was performed to decrease the sparseness of the dataset and focus on observed variation. The dataset was thus reduced from a matrix of 80 GOS metagenomes and 3,587 Pfam domains to 69 sites and 1,863 Pfam domains. A total of 670 of the domains retained were DUFs. We standardized abundances by their site maxima to mitigate the effect of differing sample size across metagenomes; however, the co-variation of domain abundances due to sample size may still provide insights into their potential associations, assuming proportionality. Thus, we used the unstandardized matrix (UM) and its standardized counterpart (SM) in further analysis. Totals of 975 (UM) and 167 (SM) Pfams were retained following removal of those with no correlations stronger than a rho of 0.80 and Bonferroni-corrected P-values below 1×10^{-6}. These included 225 (UM) and 75 (SM) DUFs. Correlation of domains in the UM resulted in largely Gaussian distribution of rho coefficients (mean≈0.41, s.d.≈0.26); however, a slight second peak around a rho of 0.80 was observed (**Figure S1**). Correlation of domains in the SM displayed a more Gaussian distribution of rho values (mean≈0.03, s.d.≈0.19); however, with a positive skew (**Figure S2**). Based on these distributions, we chose a correlation threshold and conservative significance values to focus on robust associations.

In the UM, a total of 94 DUFs showed biased correlations to Pfam domains assigned to a single metabolic category (Table 1). Of these, 56 were biased towards photobiological processes (**Tables S1 and S2**), many of which were also associated with a septum formation inhibitor domain (MinC_C) involved in the control of cell division. In the analysis of the SM, 48 DUFs showed biased correlations, with 45 in favor of photobiological processes (**Table S3**). Of these 45, 15 showed some bias (≥40% of the maximum correlated category) towards domains in carbohydrate and/or coenzyme metabolism. The three remaining DUFs were associated to domains in inorganic ion transport and metabolism (DUF1008); carbohydrate transport and metabolism (DUF111); and transcription (DUF37). Such biased connectivity of DUFs to domains in a particular functional category is not a guarantee of related function; however, it provides associative context for hypothesis generation.

Network exploration

General network characteristics. The networks generated from the correlation of the UM and SM described above comprised 975 nodes and 166,232 edges and 167 nodes and 1,897 edges, respectively. The network derived from the unstandardized abundances contained two regions of highly interconnected vertices, each with a mesh-like topology (**Figure 1**, Boxes 1 and 2; see **Figure S3** for node labels). The larger of these regions (Box 1) contained a core of highly interconnected nodes with domains from a wide variety of metabolic categories, as well as 'spokes' of less interconnected nodes. The highly-enmeshed topology of this large connected complex (LCC) prevents speculation on DUF function by visual inspection alone. However, a spoke of the UM LCC (**Figure 1**, Box 3) containing DUF2805 and DUF37 was defined by domains linked to protein biosynthesis at the level of translation (**Table 2**).

Network regions with photobiological relevance. The smaller region of this network (**Figure 1**, Box 2) was dominated by DUFs and included ten domains implicated in photobiological processes (**Table S4**). These included a central photosystem II domain, a subunit of the photosystem I reaction center, a domain that targets precursors of light-harvesting prosthetic groups, and a proton-pumping domain essential for photoheterotrophic cell growth. The taxonomic distribution of the DUFs in this region (after Pfam v26) was heavily weighted toward the Cyanobacteria and other photobiologically active taxa, such as the Viridiplantae. It is premature to connect the DUFs in this region directly to phototrophic mechanisms; however, a photoresponsive mode of life may have some bearing on their function and association in marine metagenomes.

Additionally, we noted the domains bridging this putative module to the bulk of the LCC were also photobiologically relevant (**Figure 1**, Box 4; **Table S5**). For example, eight domains were involved in cobalamin metabolism and the large

Table 1. DUFs* exhibiting bias in correlative associations to metabolic categories (unstandardized data).

DUF	Primary category	Secondary category (fraction of primary category)
DUF59	TransR	AA (0.44)
DUF3429		AA (0.44)
DUF805		AA (0.48)
DUF140		CoE (0.45)
DUF354		CoE (0.5)
DUF151		RRR (0.43)
DUF192		RRR (0.44)
DUF87		RRR (0.46)
DUF37		Transcr (0.50)
DUF1730		None
DUF2805		None
DUF3159	AA	Carb, E, Lip, Nuc (0.50)
DUF521		CoE, E, TransR (0.50)
DUF1329		None
DUF208		None
DUF2141		None
DUF2899		None
DUF403		None
DUF407		None
DUF490	CoE	CWME, E, Photo, Sec (0.50)
DUF1499		None
DUF1820		None
DUF92		None
DUF2062	E	AA, Carb, CoE, RRR (0.50)
DUF137		None
DUF74		None
DUF2130	Carb	CoE, Photo (0.50)
DUF897		None
DUF1445	Lip	None
DUF501		None
DUF1732	PostModChaps	Carb, Nuc (0.33)
DUF1385		None
DUF1008	Ion	None
DUF88	Nuc	None
DUF2334	RRR	None
DUF429	Sec	None
DUF1907	Sig	None
DUF2779	Transcr	CWME, PostModChaps (0.50)

*DUFs with primary connectivity to photobiology domains (n = 56) present in **Table S2**.
AA: Amino acid transport and metabolism.
Carb: Carbohydrate transport and metabolism.
CellDiv: Cell cycle control, cell division, chromosome partitioning.
CoE: Coenzyme transport and metabolism.
CWME: Cell wall, membrane, and envelope biogenesis.
Def: Defence mechanisms.
E: Energy production and conversion.
Ion: Inorganic ion transport and metabolism.
Lip: Lipid transport and metabolism.
Nuc: Nucleotide transport and metabolism.
Photo: Photobiology.
PostModChaps: Posttranslational modification, protein turnover, chaperones.
RRR: Replication, recombination, and repair.
Sec: Secondary metabolite biosynthesis, transport, and catabolism.
Sig: Signal transduction mechanisms.
Transcr: Transcription.
TransR: Translation, ribosomal structure and biogenesis.

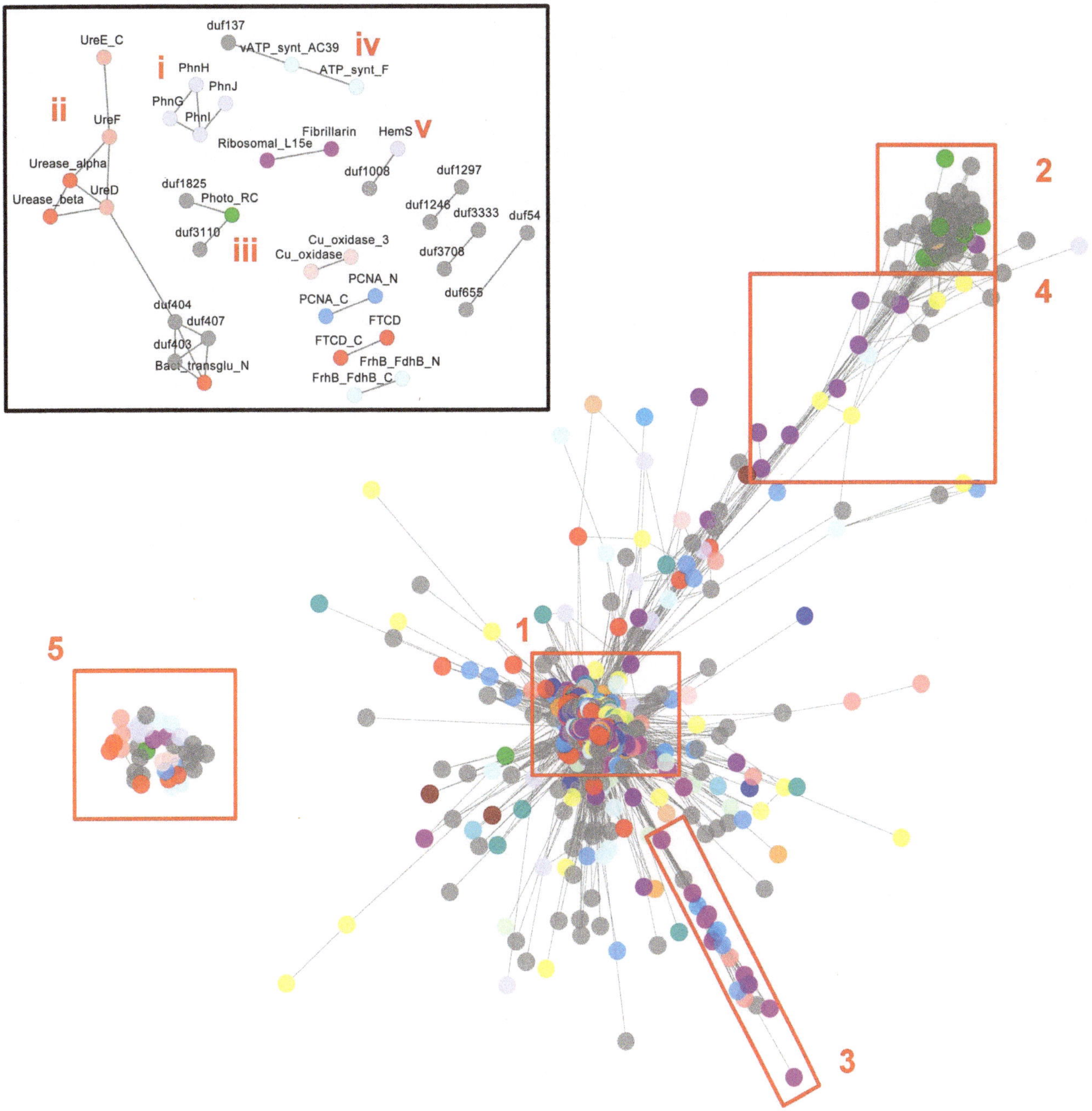

Figure 1. Force-directed, spring-embedded network visualization of pairwise correlations between Pfam domain abundances across selected GOS metagenomes. Nodes represent Pfam domains and edges correlations greater than a Spearman's rho of 0.80. Shorter edges indicate stronger correlations. A large network with two enmeshed regions (Box 1 and 2) bridged by a small number of nodes (Box 4) dominates the graph. Several small networks of functionally related nodes are also present (Box 5, inset). Node colors represent functional categories; refer to **Figure S7** for description. See text for detailed descriptions.

and small subunits of RuBisCO were also present. Cobalamin (vitamin B_{12}) is characterized by a corrin ring, which is chemically similar to the porphyrin ring found in heme, chlorophyll, and cytochrome. Further, cobalamin-dependent methyltransferases are involved in C1 metabolism and CO_2 fixation [18].

We observed an SM-derived network (**Figure 2, i**; see **Figure S4** for node labels) with similar membership to the bridge and photobiological regions discussed above (**Table S6**). This network featured 112 Pfams, which included DUFs (n = 72), photoreactive domains (n = 10), and coenzyme transport and metabolism domains (n = 8). We also observed a network (**Figure 2**, ii) with similar membership to the spoke comprising domains related to protein biosynthesis discussed above (**Figure 1**, Box 4).

Networks with known functional relatedness. A number of comparatively small UM-derived networks were also observed (**Figure 1**, Box 5 & inset), including domain groups known to be functionally related. Detection of correlative associations between such domains is encouraging and lends merit to the potential of metagenomic datasets in functional module detection. For example, we observed a group of phosphonate metabolism

Table 2. Pfam domains contained in a prominent spoke of the UM-derived association network (Figure 1, Box 3).

Category	Pfam ID	Pfam comment (abridged)
CWME	LpxD	UDP-3-O-[3-hydroxymyristoyl] glucosamine N-acyltransferase catalyses an early step in lipid A biosynthesis. Members of this family also contain a hexapeptide repeat (Pfam:PF00132). This family constitutes the non-repeating region of LPXD proteins.
CoE	Porphobil_deamC	–
NA	DUF2805	This is a bacterial family of proteins with unknown function.
	DUF37	This domain is found in short (70 amino acid) hypothetical proteins from various bacteria. The domain contains three conserved cysteine residues. Swiss:Q44066 from Aeromonas hydrophila has been found to have hemolytic activity (unpublished).
PostModChaps	NifU	This is an alignment of the carboxy-terminal domain. This is the only common region between the NifU protein from nitrogen-fixing bacteria and rhodobacterial species. The biochemical function of NifU is unknown.
	SmpB	–
	Bac_DnaA_C	–
	DNA_gyraseB_C	The amino terminus of eukaryotic and prokaryotic DNA topoisomerase II are similar, but they have a different carboxyl terminus. The amino-terminal portion of the DNA gyrase B protein is thought to catalyse the ATP-dependent super-coiling of DNA. See Pfam:PF00204. The carboxyl-terminal end supports the complexation with the DNA gyrase A protein and the ATP-independent relaxation. This family also contains Topoisomerase IV. This is a bacterial enzyme that is closely related to DNA gyrase.
Transcr	RNA_pol_Rpb2_3	RNA polymerases catalyse the DNA dependent polymerisation of RNA. Prokaryotes contain a single RNA polymerase compared to three in eukaryotes (not including mitochondrial. and chloroplast polymerases). Domain 3, s also known as the fork domain and is proximal to catalytic site.
	Sigma70_r2	Region 2 of sigma-70 is the most conserved region of the entire protein. All members of this class of sigma-factor contain region 2. The high conservation is due to region 2 containing both the −10 promoter recognition helix and the primary core RNA polymerase binding determinant. The core binding helix, interacts with the clamp domain of the largest polymerase subunit, beta prime. The aromatic residues of the recognition helix, found at the C-terminus of this domain are though to mediate strand separation, thereby allowing transcription initiation.
TransR	B5	This domain is found in phenylalanine-tRNA synthetase beta subunits.
	Glu-tRNAGln	This is a family of Glu-tRNAGln amidotransferase C subunits. The Glu-tRNA Gln amidotransferase enzyme itself is an important translational fidelity mechanism replacing incorrectly charged Glu-tRNAGln with the correct Gln-tRANGln via transmidation of the misacylated Glu-tRNAGln. This activity supplements the lack of glutaminyl-tRNA synthetase activity in gram-positive eubacterteria, cyanobacteria, Archaea, and organelles.
	Phe_tRNA-synt_N	–
	Ribosomal_L11	–
	Ribosomal_L12	–
	Ribosomal_S17	–
	Ribosomal_S5	–
	RNase_PH_C	This family includes 3′-5′ exoribonucleases. Ribonuclease PH contains a single copy of this domain, and removes nucleotide residues following the -CCA terminus of tRNA. Polyribonucleotide nucleotidyltransferase (PNPase) contains two tandem copies of the domain. PNPase is involved in mRNA degradation in a 3′-5′ direction. The exosome is a 3′-5′ exoribonuclease complex that is required for 3′ processing of the 5.8S rRNA. Three of its five protein components, Swiss:P46948 Swiss:Q12277 and Swiss:P25359 contain a copy of this domain. Swiss:Q10205, a hypothetical protein from S. pombe appears to belong to an uncharacterised subfamily. This subfamily is found in both eukaryotes and archaebacteria.

Refer to Table 1, footnote for list of abbreviations.

domains as well as a group of urea metabolism domains in the UM-derived networks (**Figure 1**, inset, i and ii resp.). Analysis of the standardized data also revealed the phosphonate network (**Figure 2**, iii), as well as some of the urea metabolic domains noted above, embedded in the LCC (**Figure 2**, hollow arrowheads). Phosphonate and urea are significant components of the marine dissolved organic matter pool and numerous microbial genera, such as *Prochlorococcus*, possess corresponding uptake and utilization capacities [19–21]. The urea metabolism domains were associated with a bacterial transglutaminase-like domain (Bact_transglu_N), DUF403, DUF404, and DUF407 in the UM-derived networks. These latter domains occurred on the periphery of the SM-derived LCC (**Figure 2**, solid arrowhead). In Pfam 26, DUF407 has been reclassified as a circularly permuted ATP-grasp family (CP_ATPgrasp_1) and DUF404 has been merged with this family. Further, Pfam architectures place instances of DUF403, now referred to as Alpha-E, in proximity to both Bact_transglu_N and CP_ATPgrasp_1 in microbial genomes. Goonesekere et al. [7] also observed that DUFs 404 and 407 co-occur. Together, these domains function in a peptide synthesis/modification system [22].

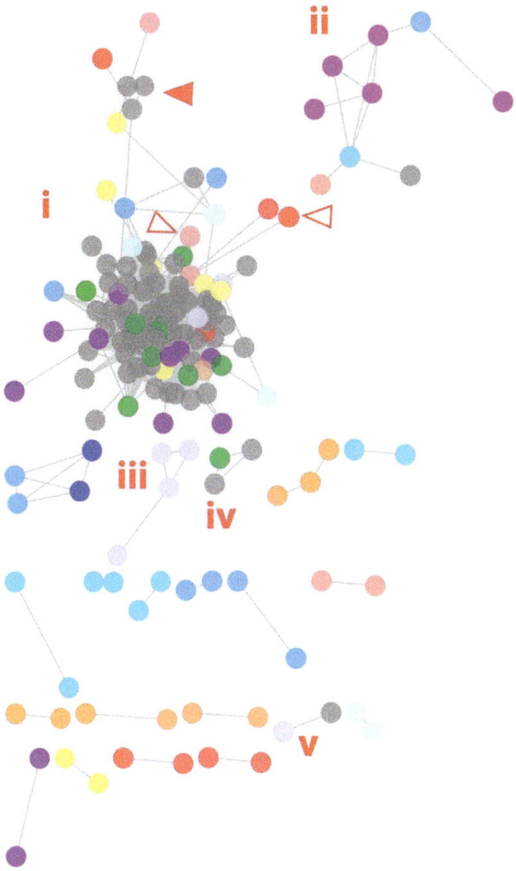

Figure 2. Force-directed, spring-embedded network visualization of pairwise correlations between standardized Pfam domain abundances across selected GOS metagenomes. Abundances were standardized by site maxima. Nodes represent Pfam domains and edges correlations greater than a Spearman's rho of 0.80. Shorter edges indicate stronger correlations. The largest network (i) is dominated by DUFs and domains linked to photobiological processes. Domains linked to urea metabolism were also present in this network (hollow arrowheads). Smaller networks featured domains linked to translation (ii), phosphonate metabolism (iii), and cyanophage activity (iv). Numerous pairs of functionally related domains were also present. Node colors represent functional categories; refer to **Figure S7** for description. See text for detailed descriptions.

We observed a number of associations between N- and C-terminal domain abundances in analysis of both the SM and UM. Examples include arginine-tRNA-protein transferase (ATE_N; ATE_C), coenzyme F420 hydrogenase (FrhB_FdhB_N; FrhB_FdhB_C), and domains involved in eukaryotic DNA polymerase processivity (PCNA_N; PCNA_C).

We also observed several small networks with DUFs associated to domains that may shed light on their roles. Notably, a haemin-degrading domain was associated with DUF1008 in analysis of both the SM (**Figure 2**, v) and UM (**Figure 1**, inset, v). DUF1008 has been reclassified as a haem utilization ChuX/HutX domain in Pfam v26, which corroborates this correlative association. Further, a photoreaction center domain was associated with DUF1825 and DUF3110 in both the SM (**Figure 2**, iv) and UM analyses (**Figure 1**, inset, iii). These DUFs principally occur in the *Cyanobacteria*; however, their specific correlation with the photoreaction center domain is likely caused by the co-occurrence of these domains in marine cyanophages, which infect the genera *Synechococcus* and *Prochlorococcus*. Previous work has identified a number of cyanophage-borne photosynthetic genes which are transferred to their hosts, including *psbA* and *psbD* [23], both of which bear the Photo_RC domain. Further, DUF3708, annotated as a phosphate ATP-binding cassette transporter, was associated with DUF3333 in the UM-derived networks. Both these DUFs are prevalent in the *Alphaproteobacteria* and Pfam architectures note them in proximity to the inner membrane component of a binding-protein-dependent transport system (BPD_transp_1). DUF137 was associated with two ATP synthase subunits. We observed these three domains to be prevalent in the *Archaea*, thus, attributing DUF137 with ATP synthase activity is premature, as taxonomic restriction may also explain their correlation. A similar explanation can be applied to the association of DUF655 and DUF54, the former domain reportedly resembling a ribosomal protein [24]. DUF1297 and DUF1246, associated as a pair, share several Pfam architectures, are listed as Pfam interaction partners, and are also prominent in the *Archaea*. Similarly, a pair comprising DUF126 and DUF521 shared similar taxonomic distributions, occurring primarily in the *Bacteria* and *Archaea*. The remaining networks, often featuring few domains with broad taxonomic distribution and poor characterization, were not amenable to our 'guilty-by-association' interpretative approach.

Transitivity clustering

Associations between DUFs in highly-enmeshed regions are difficult to evaluate by visual inspection. Transitivity clustering [25] offers a means to detect interconnected substructures in these regions, isolating them by adding and removing edges against a cost function (in this case derived from correlation strength). Our application of this method produced 49 clique-like transitivity clusters (TCs) with 3 or more members from the UM (**Figure 3**; see **Figure S5** for node labels) and 13 such clusters from the SM (**Figure 4** see **Figure S6** for node labels).

Much of the highly-enmeshed core of the UM-derived LCC remained in a single cluster (**Figure 3**: TC1, n = 464) which included 30 DUFs. The density and functional diversity of this region prevented speculation on DUF function. However, domains with known and related functions were clustered, including the urea and phosphonate metabolic components observed above (**Figure 3**: TC40, TC44; **Figure 4**: TC12, TC23). Further, the DNA mismatch repair domains, MutS I–V (**Figure 3**: TC14; **Figure 4**: TC37), which were initially obscured within enmeshed network regions, were clustered.

Photobiology. We observed prominent transitivity clusters with photobiologically relevant domain membership. A UM-derived cluster (**Figure 3**: TC2; **Table S7**) contained most domains from the photobiological region of the original network (**Figure 1**, Box 2). The membership of the largest SM-derived cluster (**Figure 4**: TC1; **Table S8**) was similar and included 28 DUFs, five photobiological domains, two cobalamin metabolism domains, and a fructosamine kinase domain.

Notably, several photobiologically relevant domains occurred in separate clusters. In the UM-derived clusters, a photosystem I reaction center domain was clustered with five DUFs, while a domain that has been implicated in the assembly and stability of photosystem I complex in chloroplasts was clustered with four DUFs (**Figure 3**: TC9 and TC15 resp.). Further, a photosystem I reaction center domain (PsaL) was clustered with DUF2839 (UM) and DUF1824 (SM). Lastly, Photo_RC and DUF1825 were associated in an UM-derived cluster (TC73) while an SM-derived cluster also included DUF3110 (TC5). These DUFs have distinct taxonomic distributions, which may partially account for their clustering (see Taxonomic perspectives, below). Cobalamin

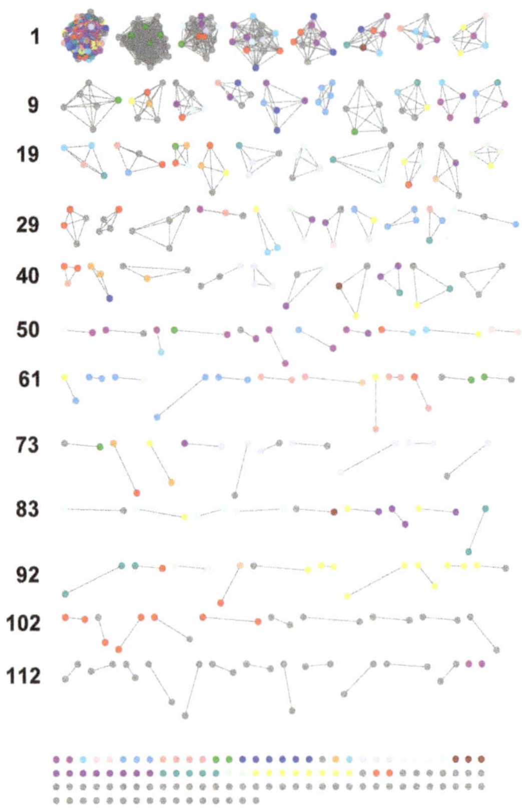

Figure 3. Transitivity clusters derived from correlations of Pfam abundances across selected GOS metagenomes (unstandardized data). Edge-weights (correlations) determine the cost of adding or removing edges during clustering. We observed clusters with domains linked to photobiology; oligotrophic adaptations; DNA maintenance and repair; and iron supply. Node colors represent functional categories; refer to **Figure S7** for description. See text and **Tables 3, 4, and 5** for detailed descriptions.

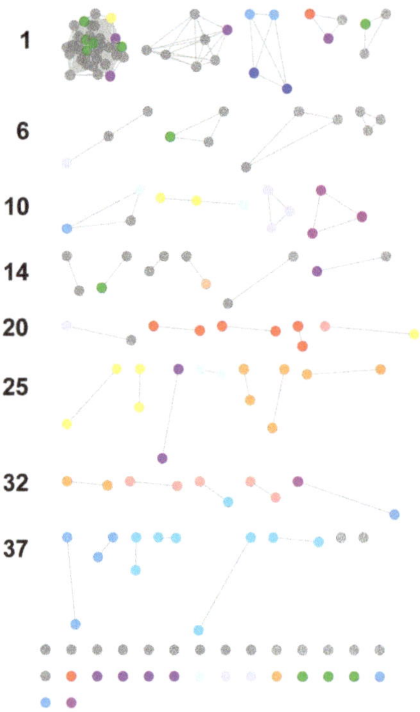

Figure 4. Transitivity clusters derived from correlative associations of Pfam domains across GOS metagenomes (standardized data). Edge-weights (correlations) determine the cost of adding or removing edges during clustering. The largest cluster contained DUFs and domains linked to photobiology (**Table S5**). Node colors represent functional categories; refer to **Figure S7** for description. See text for detailed descriptions.

synthesis domains, which may participate in pigment synthesis, were clustered with the septum formation inhibitor domain, MinC_C. MinC_C shares a Pfam architecture with an amidase domain involved in cobalamin synthesis (CbiA) and was clustered with a domain which catalyzes the conversion of cobalamin into its coenzyme form; DUF3531 which occurs primarily in the *Cyanobacteria* and *Viridiplantae*; and DUF3104 which occurs primarily in the *Cyanobacteria* (**Figure 3**: TC27).

Several UM-derived clusters, discussed below, appeared to have a common functional theme, which suggested putative roles for the DUFs they contain.

Nutrient-limitation. We observed three UM-derived clusters (**Table 3**) featuring domains with special relevance in nutrient-limited environments, including capacities to flexibly metabolize nutrients, repress costly biosynthetic pathways, and cope with metabolic stress. Indeed, many marine microbes are adapted to low or intermittent nutrient availability and such adaptations are reflected in their genomes [26,27]. A cluster comprising 13 members (**Figure 3**: TC4) included domains involved in molybdenum cofactor biosynthesis; nitrate assimilation and nitrate inducible dehydrogenase activity; phosphatase activity; oxygenation of nitrogen, sulfur, phosphorus and selenium atoms in xenobiotics; carbon-sulfur and carbon-nitrogen bond chemistry; and a chorismate mutase domain, whose activity promotes phenylalanine and tyrosine biosynthesis at the expense of tryptophan synthesis. Minimizing the synthesis of biochemically costly amino acids such as tryptophan [28] is a viable strategy to promote ecological competitiveness in oligotrophic conditions. Continuing this theme, a cluster including DUF3047 and DUF2155 (**Figure 3**: TC12) contained domains that catalyze the cleavage and cannibalization of nitrogen-bearing components from polyamines and purines. Lastly, a four-membered cluster (**Figure 3**: TC19) contained an organic solvent tolerance protein; a regulator of the acetate and glycerol operons and glyoxalate shunt; a domain involved in monitoring cellular nitrogen levels, the nitrogen stress response, and with possible roles in iron metabolism; and a fatty acid degradation regulatory domain, also found in regulators of sugar biosynthesis operons and iclR activators. These associations suggest that DUFs in these clusters may contribute to metabolic adaptations to oligotrophic conditions.

DNA metabolism and repair. Two UM-derived clusters were characterized by DNA metabolism and maintenance domains (**Table 4**). The microbial communities of epipelagic waters face sustained solar irradiation, which causes DNA damage [29]. It is likely the following clusters reflect adaptations to cope with this environmental threat. A cluster of 9 nodes (**Figure 3**: TC5) included amino acid dehydrogenase domains; domains involved in the synthesis of purines and aromatic amino acids; domains involved in DNA synthesis and repair; and a tRNA synthetase domain. Milligan et al.'s in vitro finding on the ability of several amino acids to repair oxidative DNA damage by reducing guanyl radicals [30] may offer some insight into the presence of amino acid dehydrogenases in this cluster: epipelagic microbes may similarly couple amino acid oxidation and DNA radical

Table 3. Pfam domains contained in transitivity clusters putatively linked to nutrient-limitation (unstandardized data).

Cluster	Category	Pfam ID	Pfam Comment (abridged)
TC4	AA	Alliinase_C	Allicin is a thiosulphinate that gives rise to dithiines, allyl sulphides and ajoenes, the three groups of active compounds in Allium species. Allicin is synthesised from sulfoxide cysteine derivatives by alliinase, whose C-S lyase activity cleaves C(beta)-S(gamma) bonds. It is thought that this enzyme forms part of a primitive plant defence system.
		CM_2	Chorismate mutase catalyses the conversion of chorismate to prephenate in the pathway of tyrosine and phenylalanine biosynthesis. This enzyme is negatively regulated by tyrosine, tryptophan and phenylalanine.
	CoE	MoaC	Members of this family are involved in molybdenum cofactor biosynthesis. However their molecular function is not known.
		Mob_synth_C	This region contains two iron-sulphur (3Fe-4S) binding sites.
		ThiS	ThiS (thiaminS) is a 66 aa protein involved in sulphur transfer. Thiocarboxylate is formed at the last G in the activation process. Sulphur is transferred from ThiI to ThiS in a reaction catalysed by IscS. MoaD, Swiss:P30748 a protein involved sulphur transfer in molybdopterin synthesis, is about the same length and shows limited sequence similarity to ThiS.
	E	FdhD-NarQ	Nitrate assimilation protein, NarQ, and FdhD are required for formate dehydrogenase activity.
	Ion	FMO-like	This family includes FMO proteins, cyclohexanone monooxygenase Swiss:P12015, and Swiss:Q10532.
	NA	DUF3108	This bacterial family of proteins has no known function.
		DUF328	Members of this family are functionally uncharacterised. They are about 250 amino acids in length.
		DUF3501	This family of proteins is functionally uncharacterised. This protein is found in bacteria and archaea.
	Nuc	Ureidogly_hydro	Ureidoglycolate hydrolase carried out the third step in the degradation of allantoin.
	Transcr	NIF	This family contains a number of NLI interacting factor isoforms and also N-terminal regions of RNA polymerase II CTC phosphatase and FCP1 serine phosphatase. This region has been identified as the minimal phosphatase domain.
		Sigma70_ner	The domain is found in the primary vegetative sigma factor. The function of this domain is unclear and can be removed without loss of function.
TC12	NA	DUF2155	This domain, found in various hypothetical prokaryotic proteins, has no known function.
		DUF3047	This bacterial family of proteins has no known function.
	Nuc	Allantoicase	These proteins allow the use of purines as secondary nitrogen sources in nitrogen-limiting conditions.
	PostModChaps	DS	Eukaryotic initiation factor 5A (eIF-5A) contains an unusual amino acid, hypusine. The first step in the post-translational formation of hypusine is catalysed by the enzyme deoxyhypusine synthase (DS). The modified version of eIF-5A, and DS, are required for eukaryotic cell proliferation.
	Sig	HPP	These proteins are integral membrane proteins with four transmembrane spanning helices. The most conserved region of the alignment is a motif HPP. The function of these proteins is uncertain but they may be transporters.
TC19	CWME	OstA_C	Family involved in organic solvent tolerance in bacteria.
	PostModChaps	GlnD_UR_UTase	This is a family of bifunctional uridylyl-removing enzymes/uridylyltransferases (UR/UTases, GlnD) that are responsible for the modification of the regulatory protein P-II, or GlnB. In response to nitrogen limitation, these transferases catalyse the uridylylation of the PII protein, which in turn stimulates deadenylylation of glutamine synthetase (GlnA). Moreover, uridylylated PII can act together with NtrB and NtrC to increase transcription of genes in the sigma54 regulon, which include glnA and other nitrogen-level controlled genes. It has also been suggested that the product of the glnD gene is involved in other physiological functions such as control of iron metabolism in certain species.
	Transcr	IclR	This family of bacterial transcriptional regulators includes the glycerol operon regulatory protein and acetate operon repressor both of which are members of the iclR family. However this family covers the C-terminal region that may bind to the regulatory substrate (unpublished observation, Bateman A.).
		FCD	This domain is the C-terminal ligand binding domain of many members of the GntR family. This domain probably binds to a range of effector molecules that regulate the transcription of genes through the action of the N-terminal DNA-binding domain. This domain is found in Swiss:P45427 and Swiss:P31460 that are regulators of sugar biosynthesis operons.

Refer to Table 1, footnote for list of abbreviations.

deactivation. Another cluster (**Figure 3**: TC39) comprised DUF836, classified as a glutaredoxin-like domain; an inorganic pyrophosphatase domain; and a domain involved in recombination, repair of double strand DNA breaks, and resistance to irradiative and chemical DNA-damage. As a glutaredoxin-like domain, DUF836 may participate both in DNA metabolism through glutathione-dependent synthesis of deoxyribonucleotides and in antioxidant defense [31]. Glutaredoxins may also serve in the assembly and transfer of iron/sulfur complexes [32] and thus may have particular importance in safeguarding DNA integrity and metabolism in iron-limited marine waters.

Iron-limitation. Seawater iron concentrations have previously been linked to iron metabolism genes in GOS [33], and the importance of iron supply may offer insight into two further clusters (**Table 5**). A four-membered cluster (**Figure 3**: TC23) contained DUF255; domains of an iron/manganese superoxide dismutase; and the Gram-negative bacterial TonB domain. TonB works in conjunction with outer membrane transport proteins in the active uptake of siderophore-bound iron (II) and cyanocobalamin. The ability of the TonB system to divert the proton motive force in aid of active iron transport is likely to be an asset in the typically iron-limited marine water column. Close association of this iron (II) supply-line to iron-dependent superoxide dismutases may be due, once again, to the threat of radicals generated in the irradiated water column. Another association of iron transport and iron-dependent enzyme domains was observed in a cluster (**Figure 3**: TC24) containing DUF58; a pair of pyruvate oxidoreductase domains; and a periplasmic binding protein domain often involved in iron transport. Pyruvate oxidoreductase has multiple iron/sulfur clusters and participates in the metabolism of short carboxylic acids as well as the carbon-fixing, reductive carboxylate cycle.

Miscellaneous clusters. We observed several UM-derived clusters that we could not place in a larger interpretive framework, but which included DUFs in a noteworthy functional context (**Table S9**). For example, DUF484 and DUF2066 were clustered with domains involved in sodium translocation linked to amino acid transport and redox as well as a ligase that initiates the glutathione biosynthesis pathway (**Figure 3**: TC11). The glutathione pathway is implicated in a wide range of cellular functions, including amino acid transport, and has been specifically linked to sodium-dependent transport in eukaryotic systems [34,35]. Next, DUF37, annotated as a haemolytic domain in Pfam v26, was clustered with a number of domains implicated in transcription and ribosomal function (**Figure 3**: TC7). This cluster is similar to a network derived from the SM (**Figure 2**, ii) and a spoke of the UM (**Figure 1**, Box 3; **Table 2**). Another cluster was characterized by domains capable of phosphorylation (**Figure 3**: TC10) and contained a DUF (DUF894) reclassified as a transmembrane secretion effector in a major facilitator superfamily (MFS_3); several kinase domains including a DUF (DUF227) reclassified as a kinase targeting insect hormones (EcKinase); and a selenium transferase domain. Phosphorylation plays a key role in selenocysteine formation through selenophosphate in Archaea and Eukarya. A pair of proline metabolism domains was clustered with DUF525 and DUF461 (**Figure 3**: TC29). Lastly, DUF192, putatively involved with extracellular sugar processing, was clustered with an isomerase domain, which functions in both the

Table 4. Pfam domains contained in transitivity clusters putatively linked to DNA maintenance and repair (unstandardized data).

Cluster	Category	Pfam ID	Pfam Comment
TC5	AA	ELFV_dehydrog_N	-
		Pro_dh	-
	CoE	DHFR_1	-
		ApbE	This prokaryotic family of lipoproteins are related to ApbE from Salmonella typhimurium. ApbE is involved in thiamine synthesis. More specifically is may be involved in the conversion of aminoimidazole ribotide (AIR) to 4-amino-5-hydroxymethyl-2-methyl pyrimidine (HMP).
	NA	DUF1800	This is a family of large bacterial proteins of unknown function.
		DUF177	-
	Nuc	ASL_C	This domain is found at the C-terminus of adenylosuccinate lyase(ASL; PurB in E. coli). It has been identified in bacteria, eukaryotes and archaea and is found together with the lyase domain Pfam:PF00206. ASL catalyses the cleavage of succinylaminoimidazole carboxamide ribotide to aminoimidazole carboxamide ribotide and fumarate and the cleavage of adenylosuccinate to adenylate and fumarate.
		Thymidylat_synt	Swiss:P28176 is not included as a member of this family, Although annotated as such there is no significant sequence similarity to other members.
	TransR	tRNA-synt_1c_C	Other tRNA synthetase sub-families are too dissimilar to be included. This family includes only glutamyl and glutaminyl tRNA synthetases. In some organisms, a single glutamyl-tRNA synthetase aminoacylates both tRNA(Glu) and tRNA(Gln).
TC39	E	Pyrophosphatase	-
	NA	DUF836	-
	RRR	Exonuc_V_gamma	The Exodeoxyribonuclease V enzyme is a multi-subunit enzyme comprised of the proteins RecB, RecC (this family) and RecD. This enzyme plays an important role in homologous genetic recombination, repair of double strand DNA breaks resistance to UV irradiation and chemical DNA-damage. The enzyme (EC:3.1.11.5) catalyses ssDNA or dsDNA-dependent ATP hydrolysis, hydrolysis of ssDNA or dsDNA and unwinding of dsDNA. This family consists of two AAA domains.

Refer to Table 1, footnote for list of abbreviations.

Table 5. Pfam domains contained in transitivity clusters putatively linked to iron supply and utilization (unstandardized data).

Cluster	Category	Pfam ID	Pfam Comment
TC23	CWME	TonB	-
	Ion	Sod_Fe_N	superoxide dismutases (SODs) catalyse the conversion of superoxide radicals to hydrogen peroxide and molecular oxygen. Three evolutionarily distinct families of SODs are known, of which the Mn/Fe-binding family is one.
		Sod_Fe_C	superoxide dismutases (SODs) catalyse the conversion of superoxide radicals to hydrogen peroxide and molecular oxygen. Three evolutionarily distinct families of SODs are known, of which the Mn/Fe-binding family is one.
	NA	DUF255	-
TC24	E	POR	This family includes a region of the large protein pyruvate-flavodoxin oxidoreductase and the whole pyruvate ferredoxin oxidoreductase gamma subunit protein. It is not known whether the gamma subunit has a catalytic or regulatory role. Pyruvate oxidoreductase (POR) catalyses the final step in the fermentation of carbohydrates in anaerobic microorganisms. This involves the oxidative decarboxylation of pyruvate with the participation of thiamine followed by the transfer of an acetyl moiety to coenzyme A for the synthesis of acetyl-CoA. The family also includes pyruvate flavodoxin oxidoreductase as encoded by the nifJ gene in cyanobacterium which is required for growth on molecular nitrogen when iron is limited.
		POR_N	This family includes the N terminal structural domain of the pyruvate ferredoxin oxidoreductase. This domain binds thiamine diphosphate, and along with domains II and IV, is involved in inter subunit contacts. The family also includes pyruvate flavodoxin oxidoreductase as encoded by the nifJ gene in cyanobacterium which is required for growth on molecular nitrogen when iron is limited.
	Ion	Peripla_BP_2	"This family includes bacterial periplasmic binding proteins. Several of which are involved in iron transport."
	NA	DUF58	This family of prokaryotic proteins have no known function. Swiss:P71138 a protein of unknown function in the family has been misannotated as alpha-dextrin 6-glucanohydrolase.

Refer to Table 1, footnote for list of abbreviations.

pentose phosphate pathway and the Calvin cycle, and a Type I restriction modification DNA specificity domain (**Figure 3**: TC46). This latter domain targets and either degrades DNA foreign to bacterial cells (such as viral DNA) or methylates DNA. This cluster may reflect the coupling of viral defense mechanisms to sugar scavenging – a possible adaptation to capitalize on the 'spoils of war' in resource limited environments. Despite their isolation, the above clusters offer interesting perspectives on the involvement of DUFs in epipelagic community metabolism.

Several SM-derived clusters may also grant interesting perspectives on DUF co-occurrence. An Mg^{2+}-dependent acid phosphatase involved in the biosynthesis of several cofactors including cobalamin and heme was clustered with six DUFs (**Figure 4**: TC2). These DUFs are prevalent in the *Cyanobacteria* (see below) with very low representation in the *Proteobacteria*, a restriction that may account for their clustering. However, domains in several smaller clusters showed dissimilar taxonomic distributions. These included a cluster (**Figure 4**: TC6) comprising a voltage-gated chloride channel, DUF2930, and DUF2214, the latter predicted to be a membrane protein. Pairs of domains with dissimilar taxonomic distributions were also observed: DUF3531 and DUF3641; a cobalamin-5-phosphate synthase domain (CobS) and DUF3727; a septum formation inhibitor (MinC_C) and DUF3119; DUF2010, reclassified as a mitochondrial PGP phosphatase, and DUF1823; and a divalent ion tolerance protein (CutA1) and DUF92. Comments on other transitivity clusters are available in the supplementary **Material S1**.

Taxonomic perspectives

The taxonomic distribution of Pfam domains, particularly those present in the highly abundant marine *Cyanobacteria*, will inevitably influence their association and condition any interpretation. We retrieved the taxonomic distribution of the DUF families analyzed above from the Pfam web-portal to qualitatively contextualize the observed associations. However, quantifying the degree to which the taxonomic make-up of microbial communities confounds functional associations in metagenomic samples is a non-trivial task. Such assertions are contingent on the taxonomy and functional annotation of the current genome collection, which is unlikely to reflect the true *in situ* diversity. Further, correct assignment of sequencing reads to known taxa is often problematic. Thus, the taxonomic distributions presented below are intended to provide a tentative context to advise hypothesis generation (as demonstrated above) and are not intended as a basis for phylogenomic profiling.

In the UM-derived clusters (**Figure 3**, **Table 6**), four phyla contained greater than 5% of DUF instances, namely the *Proteobacteria*, *Firmicutes*, *Actinobacteria*, and *Cyanobacteria*. The DUFs of the largest UM-derived transitivity cluster (TC1) had similar distributions to the whole collection, while DUFs whose instances were concentrated in cyanobacterial genomes dominated the second cluster (~55%). The proteobacterial proportion of this latter cluster was dominated by the *Gammaproteobacteria* (~56%) and a proportion of *Alphaproteobacteria* (~23%). The genera *Rhodobacterales* and *Rhodospirillales*, known to possess proteorhodopsin genes encoding light-powered proton pumps, were present in the alphaproteobacterial division. As reviewed by DeLong and Béjà [36], there is increasing evidence that proteorhodopsin pumps and light-powered heterotrophy are more broadly distributed in epipelagic bacteria than previously thought. In fact, proteorhodopsin pumps have recently been identified in marine eukaryotes [37] suggesting the limits of their occurrence in marine microbes has yet to be fully established. This suggests that the correlation of these DUFs may not be entirely due to their restriction to photoautotrophs. Indeed, DUFs with distributions more restricted to photoautotrophs were clustered separately. For example, less than a percentage of organisms bearing DUFs from TC9 were *Proteobacteria* and those from TC15 were almost exclusively *Cyanobacteria* (~95%). Other clusters with conspicuous taxonomic restrictions included TC31, with domains prevalent in actinobacterial genomes, and TC43, with domains prevalent in bacterial, fungal, and plant genomes.

The DUFs in SM-derived clusters (**Figure 4**, **Table 7**) were predominately found in cyanobacterial genomes (≥50% of all occurrences). Exceptions included TC6 and TC7, which contained DUFs comparably distributed between proteobacterial and cyanobacterial genomes, as well as TC14 which featured DUFs distributed in the *Fungi*, *Firmicutes*, and *Cyanobacteria*.

Establishing benchmarks

Assigning the associations discussed above with meaningful measures of confidence is an immediate concern. False positives may lead experimentalists in fruitless directions, while false negatives may limit functional discovery. Similar difficulties are encountered when attempting to benchmark protein interaction networks and attempts to minimize them rely on the reproducibility of interactions across datasets or the use of well-characterized model systems or gold standards [38].

Several aspects of metagenomic data currently hinder the construction of such benchmarks. Firstly, metagenomes are likely to contain genome fragments from organisms with no metabolically well-characterized counterparts. This greatly weakens the credibility of model-systems or –organisms as gold standards. Secondly, studies on the scale of the GOS expedition are, presently, difficult to replicate. The increasing collection of marine metagenomes, fed by initiatives such as TARA Oceans [39], as well as the growing genome collection driven by programs such as the Genomic Encyclopedia of *Bacteria* and *Archaea* [40], may soon offer the opportunity to construct such standards. However, experimental confirmation or falsification of our assertions by bench scientists is perhaps the most conclusive basis for evaluation. In the spirit of initiatives such as the Computational Bridge to Experiments (ComBrEx) [41], groups with the infrastructure and expertise to test *in silico* predictions *in vitro* may, en masse, provide a degree of confidence estimation for studies similar to ours. DUFs which have been described as at least partially characterized in subsequent releases of Pfam may anticipate such estimation. Relative to Pfam v24, we noted 28 DUF families present in our UM and SM datasets which have been renamed or merged into existing families in Pfam v26 (**Table S10**). In the UM-derived data, eight of these DUFs were not clustered while a further eight occurred in the highly-enmeshed TC1 (**Figure 3**). We were thus unable to compare these updates to our results. However, the updated descriptions of six of the remaining DUFs – DUF989, DUF403, DUF404, DUF407, DUF227, and DUF1008 – reflect their clustering in this analysis. For example, a heme-binding domain towards the C-terminus of DUF989, now merged with the Cytochrome-C family, is clustered with families such as CobT, involved in cobalamin synthesis. The chemical similarity of the corrin and porphyrin rings found in cobalamin and heme, respectively, may provide a basis for the co-occurrence of these domains. The remaining DUFs have been compared to their updated descriptions in the sections above.

While the novelty and scale of metagenomic datasets present obstacles in constructing gold standards, these factors are assets in inferring functionality through intra-ecosystemic domain covariation. We attempted to limit false assertions by using conservative detection and correlation thresholds coupled with knowledge-guided curation of our results. These approaches were used to gauge if domain associations offered plausible insights into DUF functionality in an epipelagic setting, and noteworthy associations were discussed above.

Table 6. Phylum-level* taxonomic distribution of DUFs in selected transitivity clusters (unstandardized data).

Sample	n(DUF)	Phylum	Instances	% of total instances
Overall	225	*Proteobacteria*	73177	47.83
		Firmicutes	26904	17.59
		Actinobacteria	15128	9.89
		Cyanobacteria	7798	5.10
TC1	30	*Proteobacteria*	21877	46.11
		Firmicutes	10783	22.73
		Actinobacteria	4646	9.79
TC2	28	*Cyanobacteria*	2206	54.48
		Proteobacteria	820	20.25
		Viridiplantae	588	14.52
		Firmicutes	240	5.93
TC3	8	*Proteobacteria*	2199	77.32
		Actinobacteria	202	7.10
		Fungi	145	5.10
TC4	3	*Proteobacteria*	1590	62.62
		Firmicutes	363	14.30
		Bacteroidetes	289	11.38
		Actinobacteria	160	6.30
TC9	5	*Cyanobacteria*	362	64.07
		Viridiplantae	155	27.43
TC10	3	*Proteobacteria*	2027	35.42
		Metazoa	1140	19.92
		Firmicutes	1019	17.81
		Actinobacteria	971	16.97
TC11	2	*Proteobacteria*	1021	98.74
TC12	2	*Proteobacteria*	317	94.63
TC15	4	*Cyanobacteria*	266	92.68
		Viridiplantae	15	5.23
TC16	2	*Proteobacteria*	355	64.20
		Actinobacteria	163	29.48
		Metazoa	30	5.42
TC29	2	*Proteobacteria*	1614	79.16
		Actinobacteria	120	5.89
TC30	3	*Proteobacteria*	1363	67.24
		Actinobacteria	303	14.95
		Cyanobacteria	126	6.22
TC31	4	*Actinobacteria*	1708	71.73
		Proteobacteria	218	9.16
		Fungi	159	6.68
		Firmicutes	143	6.01
TC49	3	*Proteobacteria*	1480	76.21
		Firmicutes	202	10.40
		Bacteroidetes	193	9.94

*Only Phyla with >5% of DUF instances are shown.

Table 7. Phylum-level* taxonomic distribution of DUFs in selected transitivity clusters (standardized data).

Sample	n(DUF)	Phylum	Instances	% of total instances
Overall	75	*Proteobacteria*	5575	32.31
		Cyanobacteria	4845	28.08
		Firmicutes	1946	11.28
		Viridiplantae	1505	8.72
		Actinobacteria	1026	5.95
TC1	28	*Cyanobacteria*	2290	54.48
		Proteobacteria	760	18.08
		Viridiplantae	600	14.28
		Firmicutes	244	5.81
TC2	6	*Cyanobacteria*	338	66.14
		Viridiplantae	73	14.29
		Actinobacteria	52	10.18
		Firmicutes	29	5.68
TC5	2	*Cyanobacteria*	125	65.79
		dsDNAvirusesnoRNAstage	30	15.79
		Viridiplantae	24	12.63
TC6	2	*Proteobacteria*	109	37.59
		Cyanobacteria	91	31.38
		Viridiplantae	77	26.55
TC7	2	*Proteobacteria*	109	37.59
		Cyanobacteria	91	31.38
		Viridiplantae	77	26.55
TC8	3	*Cyanobacteria*	199	70.32
		Viridiplantae	41	14.49
		Proteobacteria	19	6.71
TC14	2	*Fungi*	89	28.34
		Firmicutes	69	21.97
		Cyanobacteria	63	20.06
		Viridiplantae	34	10.83
		Proteobacteria	24	7.64
TC16	2	*Cyanobacteria*	121	51.05
		Viridiplantae	71	29.96
		Chlorobi	24	10.13
TC18	2	*Cyanobacteria*	113	39.37
		Viridiplantae	53	18.47
		Proteobacteria	42	14.63
		Firmicutes	30	10.45
		Bacteroidetes	16	5.57

*Only Phyla with >5% of DUF instances are shown.

Conclusion

We observed that intercorrelation of protein domain sequences across intra-ecosystem metagenomic datasets can provide perspectives on the potential roles of domains of unknown function. Naturally, even strong correlation across metagenomic datasets cannot provide direct functional annotations, as numerous factors may account for domain covariance in natural systems. However, critically evaluating strongly correlated domains with knowledge-level resources can provide an interpretive context to complement more targeted efforts in DUF characterization. The observed association of domains involved in microbial phosphonate metabolism, urea metabolism, and other ecologically relevant capacities encourage this approach. Of the 225 and 75 DUFs retained following correlation analysis in the UM and SM datasets respectively, we detected 94 (UM) and 48 (SM) DUFs with connectivity biased towards a single metabolic category. Further, the results above list 73 DUFs from the UM-derived and 41 DUFs from SM-derived networks and transitivity clusters whose associations may reflect ecological features of the marine epipelagic zone. While these results represent only a fraction of the DUFs detected in the GOS metagenomes, this analysis is an initial step in using ecogenomic variation to assist functional discovery. The opportunity to routinely perform such exploratory analyses and establish quantitative benchmarks is emerging as data from large-scale metagenomic, metatranscriptomic, and metaproteomic sampling campaigns becomes publically available. The perspectives that can be derived from this data will almost certainly forward efforts to characterize DUFs where homology-based approaches cannot.

Materials and Methods

Detecting domains of unknown function

A collection of 10,133,846 unassembled reads from the Global Ocean Sampling expedition metagenomes GS000a-GS023, GS025-GS051, GS108a-GS117b, GS119-GS123, GS148-GS149, and MOVE858 [16] were downloaded from the CAMERA web-portal [42] and queried against all hidden Markov models (HMMs) present in the Pfam 24 database using the HMMER3 software (version 3.0b3). Hits were deemed significant if their domain independent E-value was less than or equal to 1e-3, their bias composition correction was at least an order of magnitude less than their full score, the length of the query alignment was at least 20% of the query length, and the model alignment was at least 20% of the HMM length. Results were stored in a relational database and cross-tabulated into a "site×Pfam" matrix, wherein the abundance of each Pfam at a given site was enumerated. Pfams assigned to COG functional metabolic categories available from the integrated microbial genomes (IMG) system [43], as well as an additional category for photobiologically active domains (**Table S1**), were used in further analyses.

Correlation analysis and network visualization

The site×Pfam matrix described above was imported into the R statistical computing environment (http://cran.r-project.org/). Pfam categories listed in more than one metabolic category were removed. Distributions of total, non-zero abundances across Pfam categories and sites were used to advise data preparation. Categories with less than 20 non-zero abundances across the 80 sites analyzed and sites with less than 1,000 non-zero abundances across the 3,587 Pfam categories analyzed were removed. A copy of the resulting matrix was subject to row standardization, whereby the Pfam abundances across a given site (row) were divided by the maximum Pfam abundance of that site. The Spearman's rank correlation of Pfam categories in both matrices was determined using the *rcorr()* function from the R package Hmisc. Pfam categories with no correlations greater than a rho of 0.80 and with a P-value less than a Bonferroni corrected cut-off of $\sim 1\times10^{-6}$ were removed. DUFs that showed correlations biased towards Pfams in a single metabolic category were noted. Bias was declared when the number of correlations of a DUF to Pfams in a given category was at least double that of correlations to any other category.

The igraph R package [44] was used to create an adjacency matrix from these correlation results. This was imported into Cytoscape [45] for visualization. Network vertices, each corresponding to a Pfam category, were connected by an edge if their correlation satisfied the thresholds stated above. The Spearman's rho statistic provided weights in an edge-weighted, spring-embedded visualization. Pfam categories were color-coded according to their assigned metabolic category.

Network exploration

Networks were manually inspected for distinct topological features, particularly those where DUFs were associated with characterized Pfams from a narrow range of COG categories. The domains comprising these features were investigated further by retrieving descriptions from Pfam 26 via its webportal (www.pfam.sanger.ac.uk) and further literature where appropriate.

The TransClust [25,46] algorithm was run from the Cytoscape plugin, clusterMaker [47], to detect clique-like clusters. TransClust was run with a maximum sub-cluster size set to 50, a maximum time allowance of 2 seconds to execute each loop in the algorithm, and using edge weight (correlation) as an array source. The resulting clusters were visualized and evaluated as described above.

Taxonomic annotation

The DUFs in several transitivity clusters were annotated with their collective taxonomic distribution. Distributions were retrieved from the Pfam website (www.pfam.sanger.ac.uk; 2012-02-10) and the results stored in a relational database. Only clusters with ≥4 members and ≥2 DUFs were examined and those with notable distributions were discussed.

Supporting Information

Material S1 Comments on selected transitivity clusters not discussed in text.

Table S1 Pfam 24 domains categorized as photobiologically relevant in this analysis.

Table S2 DUFs with correlative bias towards photobiologically relevant domains (unstandardized data).

Table S3 DUFs with correlative bias towards metabolic categories (standardized data). Refer to **Table 1**, footnote for list of abbreviations.

Table S4 Domains present in a network region characterized by photobiologically relevant domains (unstandardized data; Figure 1, Box 2). Refer to **Table 1**, footnote for list of abbreviations.

Table S5 Domains present in a network region bridging two enmeshed regions (unstandardized data; Figure 1, Box 4). Refer to **Table 1**, footnote for list of abbreviations.

Table S6 Domains present in the primary network generated from standardized Pfam abundances across GOS sites (Figure 2, i). Refer to **Table 1**, footnote for list of abbreviations.

Table S7 Pfam domains contained in the second largest transitivity cluster derived from unstandardized domain abundances (Figure 3: TC2). Refer to **Table 1**, footnote for list of abbreviations.

Table S8 Pfam domains contained in the largest transitivity cluster derived from standardized domain abundances (Figure 4: TC1). Refer to **Table 1**, footnote for list of abbreviations.

Table S9 Pfam domains contained in the selected transitivity clusters derived from unstandardized domain abundances (Figure 3). Refer to **Table 1**, footnote for list of abbreviations.

Table S10 DUFs with updated statuses in Pfam v26 and their location in transitivity clusters (TCs) from both the unstandardized (UM; Figure 3) and standardized (SM; Figure 4) datasets.

Figure S1 Histogram of correlation strength between unstandardized abundances of Pfam domains across GOS metagenomes. The distribution's mean and standard deviation were ~0.41 and ~0.26 respectively. A normal distribution with equal mean and standard deviation is indicated by a blue contour.

Figure S2 Histogram of correlation strength between abundances of Pfam domains across GOS metagenomes, standardized by site maxima. The distribution's mean and standard deviation were ~0.03 and ~0.19 respectively. A normal distribution with equal mean and standard deviation is indicated by a blue contour.

Figure S3 High resolution, unannotated version of Figure 1 with individual node labels.

Figure S4 High resolution, unannotated version of Figure 2 with individual node labels.

Figure S5 High resolution, unannotated version of Figure 3 with individual node labels.

Figure S6 High resolution, unannotated version of Figure 4 with individual node labels.

Figure S7 Color key for network nodes.

Author Contributions

Integrated the Pfam domain taxonomic data into the study: WH. Supervised the project: FOG. Reviewed the manuscript: WH IK RK PY MBD FOG PLB. Conceived and designed the experiments: RK PLB. Performed the experiments: IK RK PLB WH. Analyzed the data: PLB WH MBD PY. Wrote the paper: PLB.

References

1. Finn RD, Mistry J, Tate J, Coggill P, Heger A, et al. (2010) The Pfam protein families database. Nucleic Acids Res 38: D211–222.
2. Bateman A, Coggill P, Finn RD (2010) DUFs: families in search of function. Acta Crystallogr Sect F Struct Biol Cryst Commun 66: 1148–1152.
3. Roberts RJ (2004) Identifying protein function–a call for community action. PLoS Biol 2: E42.
4. Galperin MY, Koonin EV (2010) From complete genome sequence to "complete" understanding? Trends Biotechnol 28: 398–406.
5. Deng H, McMahon SA, Eustáquio AS, Moore BS, Naismith JH, et al. (2009) Mechanistic insights into water activation in SAM hydroxide adenosyltransferase (duf-62). Chembiochem 10: 2455–2459.
6. Weinitschke S, Denger K, Cook AM, Smits THM (2007) The DUF81 protein TauE in Cupriavidus necator H16, a sulfite exporter in the metabolism of C2 sulfonates. Microbiology 153: 3055–3060.
7. Goonesekere NCW, Shipely K, O'Connor K (2010) The challenge of annotating protein sequences: The tale of eight domains of unknown function in Pfam. Comput Biol Chem 34: 210–214.
8. Jaroszewski L, Li Z, Krishna SS, Bakolitsa C, Wooley J, et al. (2009) Exploration of uncharted regions of the protein universe. PLoS Biol 7: e1000205.
9. Wooley JC, Godzik A, Friedberg I (2010) A primer on metagenomics. PLoS Comput Biol 6(2): e1000667.
10. Gilbert JA, Dupont CL (2011) Microbial metagenomics: Beyond the genome. Ann Rev Mar Sci 3: 347–371.
11. Delmont TO, Malandain C, Prestat E, Larose C, Monier J-M, et al. (2011) Metagenomic mining for microbiologists. ISME J 5: 1837–1843.
12. Merico D, Gfeller D, Bader GD (2009) How to visually interpret biological data using networks. Nat Biotechnol 27: 921–924.
13. Breitling R, Vitkup D, Barrett MP (2008) New surveyor tools for charting microbial metabolic maps. Nat Rev Microbiol 6: 156–161.
14. Fuhrman J, Steele J (2008) Community structure of marine bacterioplankton: patterns, networks, and relationships to function. Aquat Microb Ecol 53: 69–81.
15. Fuhrman JA (2009) Microbial community structure and its functional implications. Nature 459: 193–199.
16. Rusch DB, Halpern AL, Sutton G, Heidelberg KB, Williamson S, et al. (2007) The Sorcerer II Global Ocean Sampling expedition: northwest Atlantic through eastern tropical Pacific. PLoS Biol 5: 398–431.
17. Punta M, Coggill PC, Eberhardt RY, Mistry J, Tate J, et al. (2012) The Pfam protein families database. Nucleic Acids Res 40: D290–301.
18. Fuchs G (2011) Alternative pathways of carbon dioxide fixation: insights into the early evolution of life? Ann Rev Microbiol 65: 631–658.
19. Martinez A, Tyson GW, Delong EF (2009) Widespread known and novel phosphonate utilization pathways in marine bacteria revealed by functional screening and metagenomic analyses. Environ Microbiol 12: 222–238.
20. Villarreal-Chiu JF, Quinn JP, McGrath JW (2012) The genes and enzymes of phosphonate metabolism by bacteria, and their distribution in the marine environment. Front Microbiol 3: 19.
21. Kujawinski EB (2011) The impact of microbial metabolism on marine dissolved organic matter. Ann Rev Mar Sci 3: 567–599.
22. Iyer LM, Abhiman S, Maxwell Burroughs A, Aravind L (2009) Amidoligases with ATP-grasp, glutamine synthetase-like and acetyltransferase-like domains: synthesis of novel metabolites and peptide modifications of proteins. Mol Biosyst 5: 1636–1660.
23. Sullivan MB, Lindell D, Lee JA, Thompson LR, Bielawski JP, et al. (2006) Prevalence and evolution of core photosystem II genes in marine cyanobacterial viruses and their hosts. PLoS Biol 4: e234.
24. Miyazono K-ichi, Shirokane M, Sawano Y, Tanokura M (2009) Crystal structure of the DUF54 family protein PH1010 from hyperthermophilic archaea Pyrococcus horikoshii OT3. Proteins 74: 256–260.
25. Wittkop T, Emig D, Truss A, Albrecht M, Böcker S, et al. (2011) Comprehensive cluster analysis with Transitivity Clustering. Nat Protoc 6: 285–295.
26. Polz MF, Hunt DE, Preheim SP, Weinreich DM (2006) Patterns and mechanisms of genetic and phenotypic differentiation in marine microbes. Philos Trans R Soc B 361: 2009–2021.
27. Lauro FM, McDougald D, Thomas T, Williams TJ, Egan S, et al. (2009) The genomic basis of trophic strategy in marine bacteria. Proc Natl Acad Sci USA 106: 15527–15533.
28. Seligmann H (2003) Cost-minimization of amino acid usage. J Mol Evol 56: 151–161.
29. Meador JA, Baldwin AJ, Catala P, Jeffrey WH, Joux F, et al. (2009) Sunlight-induced DNA damage in marine micro-organisms collected along a latitudinal gradient from 70°N to 68°S. Photochem Photobiol 85: 412–420.
30. Milligan JR, Aguilera JA, Ly A, Tran NQ, Hoang O, et al. (2003) Repair of oxidative DNA damage by amino acids. Nucleic Acids Res 31: 6258–6263.
31. Fernandes AP, Holmgren A (2004) Glutaredoxins: glutathione-dependent redox enzymes with functions far beyond a simple thioredoxin backup system. Antioxid Redox Signal 6: 63–74.
32. Rouhier N, Lemaire SD, Jacquot J-P (2008) The role of glutathione in photosynthetic organisms: emerging functions for glutaredoxins and glutathionylation. Annu Rev Plant Biol 59: 143–166.
33. Toulza E, Tagliabue A, Blain S, Piganeau G (2012) Analysis of the Global Ocean Sampling (GOS) project for trends in iron uptake by surface ocean microbes. PLoS ONE 7: e30931.
34. Lash LH, Jones P (1984) Renal glutathione transport: Characteristics of the sodium-dependent system in the basal-lateral membrane. J Biol Chem 259: 14508–14514.
35. Kannan R, Yi JR, Tang D, Li Y, Ziokovic BV, et al. (1996) Evidence for the existence of a sodium-dependent glutathione (GSH) transporter. J Biol Chem 271: 9754–9758.
36. DeLong EF, Béjà O (2010) The light-driven proton pump proteorhodopsin enhances bacterial survival during tough times. PLoS Biol 8: e1000359.
37. Slamovits CH, Okamoto N, Burri L, James ER, Keeling PJ (2011) A bacterial proteorhodopsin proton pump in marine eukaryotes. Nat Commun 2: 183.
38. Börnke F (2008) Protein Interaction Networks. In: Junker BH, Schreiber F, editors. Analysis of Biological Networks. Hoboken, NJ, , USA: John Wiley & Sons, Inc. pp. 207–232.
39. Karsenti E, Acinas SG, Bork P, Bowler C, De Vargas C, et al. (2011) A holistic approach to marine eco-systems biology. PLoS Biol 9: e1001177.
40. Wu D, Hugenholtz P, Mavromatis K, Pukall R, Dalin E, et al. (2009) A phylogeny-driven genomic encyclopaedia of Bacteria and Archaea. Nature 462: 1056–1060.
41. Roberts RJ, Chang Y-C, Hu Z, Rachlin JN, Anton BP, et al. (2011) COMBREX: a project to accelerate the functional annotation of prokaryotic genomes. Nucleic Acids Res 39: D11–14.
42. Seshadri R, Kravitz SA, Smarr L, Gilna P, Frazier M (2007) CAMERA: a community resource for metagenomics. PLoS Biol 5: e75.
43. Markowitz VM, Chen I-M, Palaniappan K, Chu K, Szeto E, et al. (2010) The integrated microbial genomes system: an expanding comparative analysis resource. Nucleic Acids Res 38: D382–390.
44. Csardi G, Nepusz T (2006) The igraph software package for complex network research. InterJournal Complex Sy: 1695.
45. Smoot ME, Ono K, Ruscheinski J, Wang P-L, Ideker T (2011) Cytoscape 2.8: new features for data integration and network visualization. Bioinformatics 27: 431–432.
46. Wittkop T, Emig D, Lange S, Rahmann S, Albrecht M, et al. (2010) Partitioning biological data with transitivity clustering. Nat Methods 7: 419–420.
47. Morris JH, Apeltsin L, Newman AM, Baumbach J, Wittkop T, et al. (2011) clusterMaker: a multi-algorithm clustering plugin for Cytoscape. BMC Bioinformatics 12: 436.

A Highly Productive, Whole-Cell DERA Chemoenzymatic Process for Production of Key Lactonized Side-Chain Intermediates in Statin Synthesis

Matej Ošlaj[1], Jérôme Cluzeau[2], Damir Orkić[2], Gregor Kopitar[1], Peter Mrak[1]*, Zdenko Časar[2,3]*

1 Genetics, Anti-Infectives, Lek Pharmaceuticals d.d., Mengeš, Slovenia, **2** API Development, Sandoz Development Center Slovenia, Lek Pharmaceuticals d.d., Mengeš, Slovenia, **3** Faculty of Pharmacy, University of Ljubljana, Ljubljana, Slovenia

Abstract

Employing DERA (2-deoxyribose-5-phosphate aldolase), we developed the first whole-cell biotransformation process for production of chiral lactol intermediates useful for synthesis of optically pure super-statins such as rosuvastatin and pitavastatin. Herein, we report the development of a fed-batch, high-density fermentation with *Escherichia coli* BL21 (DE3) overexpressing the native *E. coli deoC* gene. High activity of this biomass allows direct utilization of the fermentation broth as a whole-cell DERA biocatalyst. We further show a highly productive bioconversion processes with this biocatalyst for conversion of 2-substituted acetaldehydes to the corresponding lactols. The process is evaluated in detail for conversion of acetyloxy-acetaldehyde with the first insight into the dynamics of reaction intermediates, side products and enzyme activity, allowing optimization of the feeding strategy of the aldehyde substrates for improved productivities, yields and purities. The resulting process for production of ((2*S*,4*R*)-4,6-dihydroxytetrahydro-2*H*-pyran-2-yl)methyl acetate (acetyloxymethylene-lactol) has a volumetric productivity exceeding 40 g L^{-1} h^{-1} (up to 50 g L^{-1} h^{-1}) with >80% yield and >80% chromatographic purity with titers reaching 100 g L^{-1}. Stereochemical selectivity of DERA allows excellent enantiomeric purities (*ee* >99.9%), which were demonstrated on downstream advanced intermediates. The presented process is highly cost effective and environmentally friendly. To our knowledge, this is the first asymmetric aldol condensation process achieved with whole-cell DERA catalysis and it simplifies and extends previously developed DERA-catalyzed approaches based on the isolated enzyme. Finally, applicability of the presented process is demonstrated by efficient preparation of a key lactol precursor, which fits directly into the lactone pathway to optically pure super-statins.

Editor: Andrew C. Marr, Queen's University Belfast, United Kingdom

Funding: This work nwas funded by Lek pharmaceuticals d.d. a Sandoz company. The funders had no role in study design, data collection and analysis, decision to publish, or preparation of the manuscript.

Competing Interests: This work was funded by Lek pharmaceuticals d.d., a Sandoz company. This does not alter the authors' adherence to all the PLOS ONE policies on sharing data and materials. The authors of this work are employed by Lek pharmaceuticals d.d., a Sandoz company. The interests of this commercial funder are pursued through the patent applications: WO2008119810 and WO2009092702. The sole purpose of this work is to share the findings obtained during the development of the described process with the scientific community and the publication has no impact on any products; in development or marketed. The funders had no role in study design, data collection and analysis, decision to publish, or preparation of the manuscript.

* E-mail: peter.mrak@sandoz.com (PM); zdenko.casar@sandoz.com (ZC)

Introduction

Statins, inhibitors of 3-hydroxy-3-methylglutaryl-coenzyme A (HMGCoA) reductase [1–3], are one of the most pronounced success stories of modern medicinal chemistry [4]. Beside their cholesterol-lowering capabilities, they have been found also to possess many other beneficial effects [5–9]. Statins consist of a chiral diol side chain, appended to a cyclic fragment. Initially discovered as microbial metabolites [10–13], statins have been rapidly developed into even more efficient synthetic analogues by a partial modification of their structure. The fully synthetic derivatives are frequently addressed as super-statins [14]. Due to the fact that the side chain represents an essential building block in all statins, vast research efforts have been made for its efficient construction and incorporation into the final statin structure. These studies have demonstrated that statins are readily accessible through a variety of different approaches utilizing various types of side-chain precursors. Indeed, statins have been built from open side-chain derivatives, lactol type precursors [14] and very recently from a lactonized side-chain derivative [15,16]. Due to the increased commercial demand for statins and the need for a simple preparation on industrial scale, impetus for even more efficient, environmentally friendly and easily scalable preparation of the side-chain derivatives appeared. Moreover, synthetically demanding chiral diol structure of the side chain and high quality requirements of pharmaceutical industry on its (stereo)chemical purity shifted the research from initially pure chemical synthesis to enzyme-based approaches, which are known to perform with high stereoselectivity [17–32].

Among applications of various enzymes in synthesis of the statin side chain [33–36], utilization of 2-deoxyribose-5-phosphate aldolase (DERA) [37–41] for the synthesis of chiral lactol precursors [42–47] attracted considerable attention recently. This is due to the fact that DERA-catalyzed aldol condensations [48–51] of prochiral aldehyde precursors **(1)** and **(2)** provide lactols **(3)** (Figure 1), containing two chiral centers of the final statin side chain with excellent stereopurity and high yield. DERA (EC

4.1.2.4) is a lysine-mediated aldolase (type I), able to accept acetaldehyde **1** (donor) and glyceraldehyde-3-phosphate (acceptor) as its natural substrates [52–55]. This ubiquitous enzyme is unique among aldolases by being capable of condensing two aldehydes and is well-known for its relaxed substrate specificity; both on the donor and the acceptor side [42–43,56–58]. While acetaldehyde **1**, acetone, flouroacetone and propanal can be used as donors, even broader selection is possible for acceptor molecules. In addition to a remarkable number of various aldehydes, aldose-sugars and their phosphates can also be accepted [42,56–58]. Use of DERA for *in vitro* catalysis with unnatural substrates, involved in single DERA-catalyzed aldol condensation, was first shown by Barbas *et al.* [56], followed by description of sequential aldol condensations catalyzed by *E. coli* DERA [42,57], which opened a new field in chemoenzymatic synthesis. Several 6′-substituted dideoxy-sugars can be prepared using DERA in high enantiomeric purity [42,57]. An immediate application of this reaction was reported by Greenberg *et al.* [44], using an alternative DERA enzyme in synthesis of chiral (halo)lactols, where *ee* of >99.9% and *de* >96.6% are reported in a highly productive and scalable process for production of (4*R*,6*S*)-6-(chloromethyl)tetrahydro-2*H*-pyran-2,4-diol (**3b**). Other processes employing DERA for sequential aldol condensations have been reported since [46,59–60].

These processes, however, have some major drawbacks. One of them, arising from a high enzyme inactivation rate (driven by the aldehyde-lysine coupling), is the high catalyst load necessary for reactions with industrially suitable yields and productivities [61]. Apart from interfering with surface lysine residues (thereby destabilizing the enzyme's tertiary structure) acetaldehyde (**1**) has a detrimental effect on the enzyme's activity when the catalytic lysine residue in the active site is inactivated [38,61–62].

The problem of DERA inactivation in the presence of aldehydes has been addressed before either by employing random (epPCR, DNA shuffling) and/or targeted mutagenesis [61,63] or, alternatively, by screening of environmental DNA libraries [44] and immobilization procedures [64–66]. The immobilization approach aimed at improving stability of DERA was attempted also in our laboratory (unpublished data), however this approach was found to be economically unfavorable. While great efforts have been invested into resolving the inactivation problems in DERA-catalyzed reactions, a high enzyme load is still necessary for high-titer processes, making the biocatalyst a major cost driver. In contrast to the described techniques, which stream toward reduction of the enzyme load or its recycling, a different approach, aiming at reduction of the price impact of the biocatalyst, is to minimize its production costs.

Compared to even the most straightforward enzyme downstream procedures, the use of a whole-cell biocatalyst [67–73] in form of a fermentation broth is by far the most inexpensive approach to biocatalysis unless very stable and recyclable enzymes can be used [73]. Several industrially scalable whole-cell bioconversion processes utilizing *Escherichia coli* as the enzyme expression host are known today [67–77]. The feasibility of such approach, however, is limited by specific properties of each individual case of enzyme and bioconversion process. One major prerequisite for a whole-cell catalyzed process, for example, is the permeability of cellular envelopes for substrates and products when considering a cytoplasmic expression of the enzyme. Another important prerequisite is the absence of secondary enzymatic activities in the cell biomass, which could lower the yields of the desired product by further metabolization (e.g. cleavage by hydrolyses in the case of **2g,** see the results section) or by catabolic utilization of the reaction substrates. Third, but equally important criterion is the absence of impurities either derived from the fermentation medium or produced by the host microorganism, which would influence overall product quality and result in additional, undesired purification steps.

Although several implications for suitability of (halo)lactols (**3**) obtained *via* DERA-catalyzed approaches in statin synthesis have been made, a clear-cut direct application of these intermediates for assembly of statins with heptenoic side-chain residue has remained vague. This is due to the fact that usually lactols (**3**) need to be first oxidized to lactones (**4**), then ring-opened and after several synthetic steps (including protection/deprotection sequences) side-chain derivatives suitable for super-statin assembly are obtained [44]. All these additional chemical steps in the total synthesis devalorize highly effective and straight-forward enzymatic step. Recently, we have developed the most efficient chemical approach to halolactone derivative (**4**) [78–79], which has been successfully hydroxylated *via* S_N2 reaction with acetates to acetyloxymethylene-lactone (**5**) followed by deacetylation with tin catalyst [78], which was latter replaced for a nontoxic, environmentally

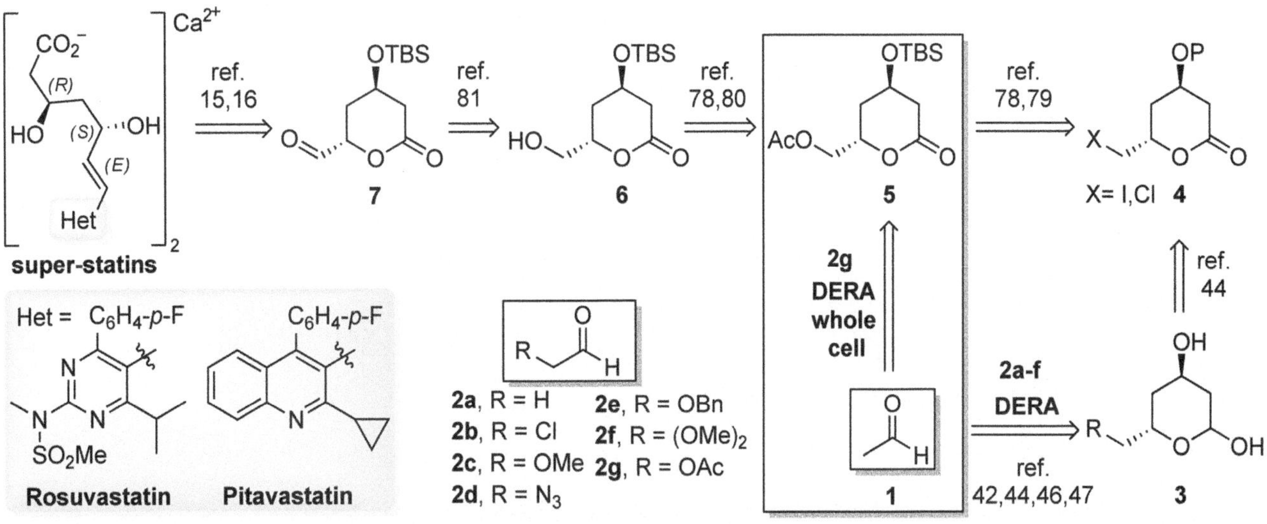

Figure 1. Current art in the direct synthesis of optically pure super-statins from the lactol precursors.

acceptable and inexpensive chemoselective acetyl cleavage [80], catalyzed by the pancreatic lipase powder to give **(6)**. Furthermore, the obtained hydroxymethylene-lactone **(6)** was for the first time successfully transformed to its formylated analogue **(7)** suitable for coupling with the heterocyclic counterpart [81]. Moreover, we have demonstrated for the first time that statins can be directly assembled from the lactonized side-chain precursor **(7)** [15,16].

Having proved the direct lactone pathway to statins, we have opened a further possibility for direct application lactols **(3)**/ lactones **(4)**/**(5)** derived *via* DERA-catalyzed approach without the necessity of their lengthy transformation to dihydroxyl protected open-chain derivatives before the final statin formation. Therefore, we were prompted to prepare a key acetyloxymethylene-lactone **(5)** precursor *via* DERA-catalyzed approach (Figure 2) from acetaldehyde **(1)** [82] and aldehyde **(2g)** [83] in order to take advantage of its efficient assembly into statin structure and thus provide efficient total synthesis of super-statins based on enzymatic technology (Figure 1) [84].

Here, we report development of a high-density *E. coli* fermentation process, yielding a broth with high activity of DERA. By directly utilizing this broth in a highly productive, high-yield, whole-cell, chemo-enzymatic process, we show that all of the above outlined challenges can be successfully overcome. Special focus is given to a process yielding acetyloxymethylene-lactol **3g** wherein we point out the role of reaction kinetics of the two sequential condensation steps for the outcome of the reaction. Measurement of the reaction intermediate (**8**) and side-product (**3a**) accumulation allowed optimization of the feeding strategy, which proved to be a yield-determining parameter. Further conversion to advanced downstream statin intermediates confirms the efficient total synthesis of super-statins based on enzymatic technology and employing our previously described methodology [15–16,78–81,83].

Materials and Methods

Strains and Culture Conditions

E. coli BL21 (DE3) (Invitrogen, USA) was used for the expression of wild type *E. coli* DERA from the pET30a+plasmid (Novagen, USA). The expression plasmid was constructed by PCR amplification of the *deoC* gene from *E. coli* DH5α genome using primers GCCGATATCCGTAGCTGCTGGCGCTCTTACC and CGGCATATGACTGATCTGAAAGCA-AGCAGCC followed by cleavage of the amplified fragment as well as the host plasmid with *Nde*I and *Blp*I restriction endonucleases and assembly of the fragments in a T4 ligation reaction to yield pET30/*deoC*. All shake flask cultivations were performed on a rotary shaker (250 rpm, 5 cm radius) at 28°C. The seed medium (VD) consisted of Yeast extract (10 g L^{-1}), NaCl (5 g L^{-1}), glycerol (5 g L^{-1}) and NaH_2PO_4 (2 g L^{-1}) and was supplemented with Kanamycin (25 μg mL^{-1}). The seed culture was prepared by inoculation of 30 mL of fresh VD medium (250 mL shake flask) with 200 μL of overnight *E. coli* BL21 (DE3) pET30/*deoC* VD culture. After

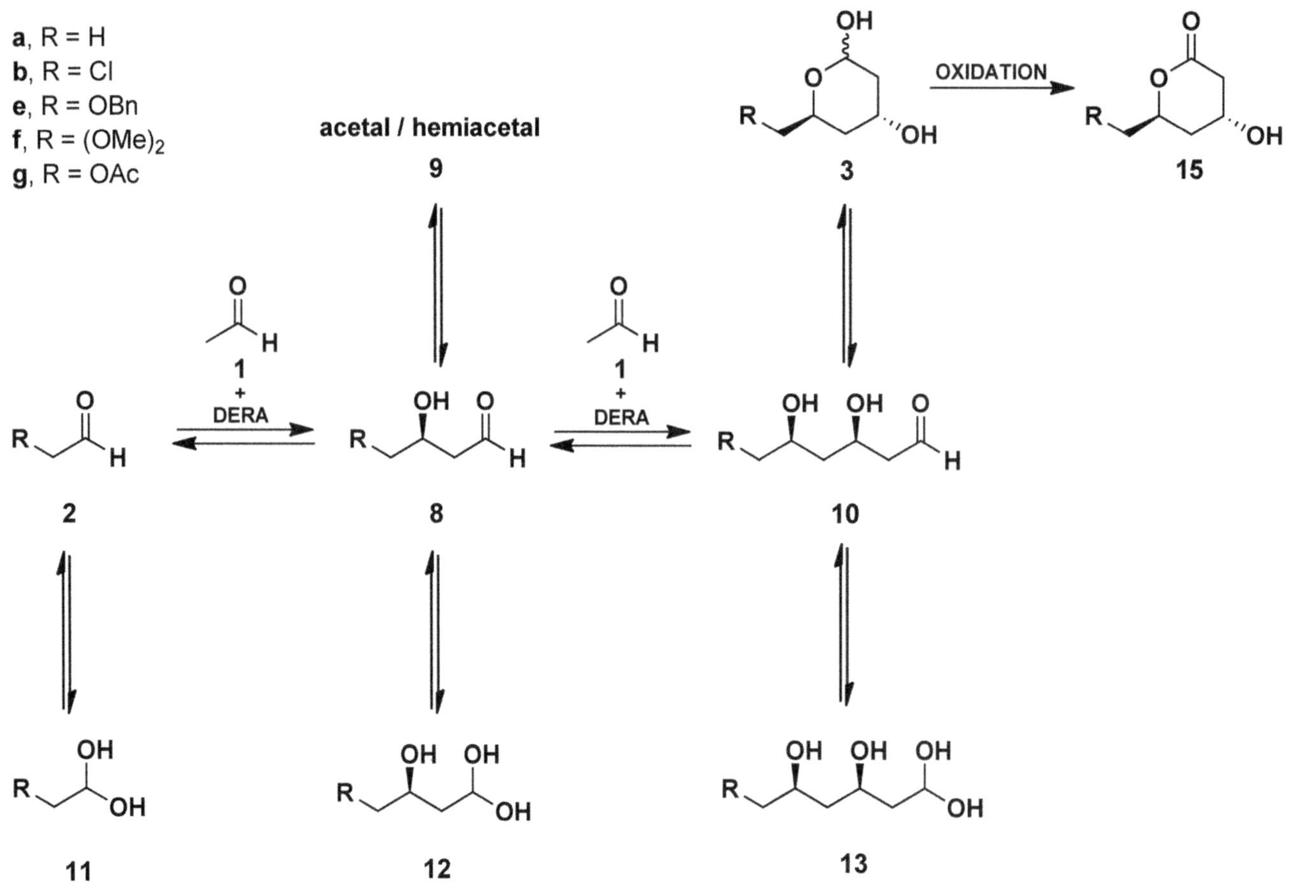

Figure 2. The DERA-catalyzed sequential aldol condensation with 2-substituted acetaldehydes. Several acetal/hemiacetal (**9**) species were found in equilibrium with **8**. These may include monomeric cyclic hemiacetal as well as dimeric and trimeric cyclic acetals/hemiacetals (Information S3). Compounds **15** are obtained by chemical oxidation of the lactols **3**.

approximately 8 h of cultivation at 28°C, OD_{600} reached 1.5–2.0 and the seed culture was used for inoculation of the main culture. Expression of DERA in a high-density fermentation culture was performed in a 2 L bioreactor (ISF-100, Infors HT, CH) The main fermentation medium and the glucose feeding solution were described before [85] and were both supplemented with kanamycin (25 μg mL^{-1}). 1 L of the main medium was inoculated with 15 mL of seed culture and was grown with initial conditions (28°C, pH 7.0, 700 rpm, 1 L min^{-1} aeration, atmospheric pressure) until dO_2 dropped to 30%. After that, rpm and airflow were automatically raised in order to maintain the dO_2 above 30% (1800 rpm and 3 L/min were the maximum settings reached). pH was adjusted automatically to 6.8, using ammonia solution (12.5%). The feeding with glucose solution was initiated upon depletion of the glycerol in the main fermentation medium which was indicated by an instant drop in oxygen consumption (10–12 h into the process). The flow rate for the feeding solution was initially set to 0.08 mL min^{-1} and exponentially increased to 0.27 mL min^{-1} in the course of 24 h. The expression was induced by addition of 0.2 mM IPTG 18 h after inoculation. The resulting culture broth (DERA fermentation broth) was harvested between 34 and 39 h after inoculation and used directly in the DERA reactions. Alternatively, cell-free lysates were used (where indicated), prepared by sonication of the culture broth (Digital Sonifier 450, Branson, USA; std. horn, T <8°C, 5×10 s pulses, 70% amplitude), followed by centrifugation in order to remove cell debris.

DERA Activity Measurements

Activity of DERA was assayed using 7-deoxyribosyl-4-methyl umbelliferone as reported by Greenberg *et al.* [44,60,63]. Fluorescence was measured for 30 min at 28°C using fluorescence spectrometer SpectraMax M2 (Molecular Devices, USA) set to excite at 370 nm and detect emission at 455 nm. The following modifications to the original method were made. 96-well microtiter plates (96w Costar blk/blk bottom), covered with an optically transparent foil (MicroAmp, Applied Biosystems, USA) were used for the measurements. Samples were diluted 1000 fold in buffer (100 mM Bis-Tris propane, pH 8.5, total volume 200 μL). Each of the 7 different loads of properly diluted sample, spanning from 10 μL to 170 μL (2 μg to 40 μg of biomass per assay; OD_{600}: 0.005 to 0.1), was measured in triplicate after addition of 10 μL of BSA (40 mg/mL) and 20 μL of the fluorogenic substrate [44] (~ 1.5 mg/mL in 20% DMSO; corrected by dilution to match $A_{320} = 65.0 \pm 0.5$ at 28°C) for determination of reaction velocity. Wet cell weight (WCW; weight of sediment per sample volume after centrifugation at 16000 G for 10 min) of samples was measured prior to the activity measurement and was used to calculate specific activity of the biomass. In cases where cell-free lysates were used as catalysts, the WCW of the source culture was used for specific activity calculation for practical reasons. The specific DERA activity was determined as slope of the mean fit to at least 5 points in an initial velocity vs. biomass load plot and is given in kRFU s^{-1} g^{-1} as an average of the triplicate samples. Zero value for blank sample was included, but not forced in the linear regression. Acceptance criteria for a valid measurement were: $R^2>0.99$, offset <0.02.

General Reaction Conditions

The substrates acetaldehyde **1**, chloracetaldehyde **2b**, dimetoxyacetaldehyde **2f** and benzyloxyacetaldehyde **2e** used for the reaction were obtained from Sigma and were of p.a. grade. Acetyloxyacetaldehyde **2g** was prepared as described before [83]. All reactions were performed in a 2 L stirred vessel (ISF-100, Infors HT, CH) at 37°C. pH was adjusted automatically using ammonia solution (12.5%) to 6.2. Stirrer speed was set to 700 rpm. The *E. coli* BL21 (DE3) pET30/*deoC* fermentation broth was added first (80% of the final volume of the reaction). Substrates were added directly into the reaction mixture into the vicinity of Rushton impeller using a programmable pump (Costametric 4100, Thermo separation products, US) with the acetyldehyde being diluted in water so that both substrates filled the remaining 20% of the final reaction volume. The condenser of the reactor exhaust was cooled to 2°C in order to minimize the loss of acetaldehyde **1** due to evaporation.

GC-MS and GC-FID Analytics

Reaction-mixture samples were quenched with 4 volumes of acetonitrile. Cells and precipitates were removed by centrifugation and the samples were further diluted by a factor of 10 in acetonitrile. Analyses were performed on an Agilent (USA) 7890A gas chromatographic (GC) system equipped with flame ionization detector (FID). A HP-5MS column with dimensions of 60 m×250 μm (ID) ×0.25 μm was used. 1 μL of samples was injected onto the column by split injection (split ratio 20:1) via inlet, which was held at 200°C. Helium (6.0) was used as a carrier gas at constant pressure of 45 PSI (initial flow 3.4 mL min^{-1}). Temperature program was set as follows: initial temperature 50°C (5 min), gradient 10°C min^{-1} until 250°C (5 min). Analytes were detected on a FID detector. Detector temperature was set to 250°C. Gas flow rates were set to 30 mL min^{-1} for hydrogen gas (fuel), 300 mL min^{-1} for air (oxidizer) and 25 mL min^{-1} for nitrogen gas (make-up).

For the GC-MS analytics, samples were prepared as above and injected into GC-MS system with chemical ionization (CI) for determination of CI mass spectrum. Analysis were performed on an Agilent 7890A gas chromatographic (GC) system coupled with Agilent mass selective detector 5975C inert XL EI/CI MSD using inert chemical ionization (CI) source. Methane gas was used as a chemical ionization reagent to perform a positive chemical ionization (PCI). 1 μL of sample was injected onto a HP-5 Trace Analysis 5% Phenyl column with dimensions of 30 m×250 μm (ID) ×0.25 μm by split injection (split ratio 20:1) via inlet, which was held at 200°C. Helium (6.0) was used as the carrier gas at constant pressure of 10 PSI (initial flow 1.2 mL min^{-1}). Temperature program was set as follows: initial temperature 50°C (5 min), gradient 10°C min^{-1} until 230°C (5 min).

Isolation of the Reaction Products and Intermediates

After the completion of reactions, pH was lowered to 4.5 using phosphoric acid. 200 g L^{-1} of Na_2SO_4 was added in order to facilitate extraction of reaction products and intermediates with EtOAc. Whole-broth extraction with 2 vol. of EtOAc was repeated 4 times. EtOAc phase was collected, dried over $MgSO_4$ and filtered, followed by evaporation yielding a crude product in the form of a yellow-brownish oil. Where pure compounds were needed, the oil was dissolved in water and mixed with 5 vol. of toluene, which was then removed and discarded. Aqueous phase was extracted with 2 vol. of CH_2Cl_2. The first extract was discarded and the extraction was then repeated 6 times. The CH_2Cl_2 phase was collected, dried over $MgSO_4$ and evaporated. The resulting light yellow oil was dissolved in 95/5 water/acetonitrile (500 g L^{-1}) and injected (1 mL) into a reverse phase preparative LC system (Äkta Purifier, Amersham-Biotech, SE) using the preparative C18 column (EVF D17, RP18 25–40 um − 15 g, Merck Chimie SAS). Flow rate of 2.5 mL min^{-1} was used in isocratic mode with 95/5 water/acetonitrile (pH = 5) for the first 55 min, followed by a gradient from 95/5 to 50/50 (water/

acetonitrile) over the next 15 min. The resulting 2 mL fractions were analyzed by GC-FID and the fractions containing the desired compound were collected. Acetonitrile was evaporated and the compound was extracted from the aqueous phase with CH_2Cl_2 to give pure compounds.

Chemical Oxidation of the Lactols

To a whole-cell catalyst reaction mixture with **3g** (250 mL, **3g** concentration 55.0 g/L), NaCl (83.9 g, 1456 mol) was added at r.t. and mixed until dissolved using mechanical stirring. Then, EtOAc was added (83 mL) and cooled to 0–5°C, pH adjusted to 3 by phosphoric acid and to this solution, $Ca(OCl)_2$ (65%, 37.7 g, 171.3 mmol) was added portion-wise over 3 h, keeping the pH between 2–6 (by phosphoric acid) and the temperature range between 5–25°C by cooling with ice bath. After additional 0.5 h, the reaction was finished and chlorine was driven out by bubbling N_2 through the solution (exhaust going to a trap consisting of solution of NaOH and $Na_2S_2O_3$ in water). From the reaction mixture, EtOAc was distilled under reduced pressure, Celite® (10 g) was added and the mixture filtered. The filter cake was washed with 2×80 mL of distilled water and combined filtrate fractions were extracted with CH_2Cl_2 (10×250 mL), The CH_2Cl_2 phases were dried with $CaCl_2$, filtered and evaporated to obtain **15g** (14.6 g, 85.5% assay, 92% crude yield).

Results and Discussion

High-density Fed-batch Fermentation

The fermentation process for production of DERA whole-cell biocatalyst was developed with three targets in mind. First, the volumetric DERA activity should support direct use of the fermentation broth for a highly productive, chemoenzymatic process comparable to the process described by Greenberg *et al.* [44]. Secondly, the use of a defined mineral medium with means of controlling the residual carbon source would minimize contaminating impurities during the work-up of the chemoenzymatic reaction broth. And thirdly, the raw-material and process costs should be as low as possible in order to make the process not only industrially scalable but also economically favorable compared to any other DERA-catalyzed process known. The process outline was based on a mineral-medium, high-density *Escherichia coli* fermentation protocol [85], which, however, is focused on biomass production and not specifically evaluated for protein expression. We have found, nevertheless, that the protocol allows appropriate high-density process for a high-level DERA expression when the time of induction, inducer concentration and carbon source feeding rates are optimized (Information S1). Typical characteristics of the final broth with the optimized procedure, harvested at 36 h after inoculation were: 180–215 g L^{-1} wet cell weight (OD_{600} ~ 100–110), 210–250 kRFU s^{-1} g^{-1} specific DERA activity. The DERA protein was estimated to be present in 4.07 g L^{-1} and represented ~ 50% of the total soluble protein after analysis of cell-free extracts with SDS-PAGE (Information S1). In comparison, shake-flask experiments using standard expression conditions (LB, 37°C, 1 mM IPTG at OD_{600} = 0.6, 6 h induction time) yielded 8 g L^{-1} WCW and 300 kRFU s^{-1} g^{-1}.

The resulting culture broth (DERA whole-cell catalyst) was used directly in the aldol condensation reactions. Raw material cost calculation (using bulk prices of raw material) for this process, which was successfully scaled-up to 70 L and 200 L bioreactors in a linear manner from the 2 L scale protocol, was calculated to be less than 0.2 € per kg of the whole-cell catalyst. According to the advances in high-density *E. coli* fermentation, [86,87] given additional optimization, it should be possible to increase the biomass yields, and consequently, the volumetric DERA activity of the fermentation broth even further.

Activity Assay

Due to the well-known inactivation of DERA by the aldehyde substrates [38,61–62], and the fact that a whole-cell catalyst was to be used for chemoenzymatic reactions, a straightforward and accurate activity assay to support the work was essential and therefore given much attention. The fluorometric method for determination of DERA activity was described before [44,60,63], however it's precision on a whole-cell catalyst remained to be tested. A series of experiments was performed with the aim to evaluate the impact of cellular envelopes on the efficiency of the retro-aldol reaction resulting in release of the fluorescent 4-methyl umbelliferone from the relatively bulky substrate 7-deoxyribosyl-4-methyl umbelliferone the cell-free extracts were always compared to the whole-cell catalyst from which they originated). In addition, the light scattering due to the presence of the cells in the samples may have an impact on the fluorescence readings [88]. The latter was evaluated by measuring activity of a cleared DERA lysate spiked with increasing amounts of DERA-expressing cells in the range between 0.01 and 0.12 g L^{-1} WCW (OD_{600} = 0.005 to 0.012). In this range, independent measurements showed high linearity (R^2>0.99) in velocity vs. biomass-load plot for the appropriately diluted samples of the DERA fermentation broth (Figure 3a and 3b). The spiking of the cell-free lysate had no effect in the range described (Figure 3c). Linearity of specific biomass activity within samples with constant biomass concentration was validated using mixtures of *E. coli* BL21 pET30/*deoC* cultures with high DERA activity and w.t. *E. coli* BL21 cultures. The cell concentration in the samples was kept constant in the aforementioned range, within which, the fraction of the DERA-expressing cells varied. In parallel, measurements were performed with cell-free extracts of identical samples, prepared by thorough sonication and removal of cell debris by centrifugation (Figure 3d). The results show highly linear response both for the whole-cell catalyst and the cell-free extract mixtures. Moreover, in comparison, both catalysts had specific activities at essentially the same level. Data obtained in a similar experiment, where the cell-free lysate was prepared directly from the high-density DERA fermentation broth, and activity of the two was compared after appropriate dilution, were in agreement with the above.

To confirm that the activity detected on the whole cells is not due to potential lysis of the cells and release of DERA extracellularly during the fermentation or the activity assay, the following measurements were made. First, the supernatant of a freshly harvested high-density culture of *E. coli* BL21 pET30/*deoC* was analyzed. The activity of the extracellular DERA contributed 4.5±2% to the total fermentation broth activity, whereas the washed biomass contributed the remainder of the activity (Information S6). Next, using SDS-PAGE, we compared the amount of extracellular enzyme present in the assay mixture at the end of the measurement where either the cleared lysate or the whole broth was used as a catalyst. The results clearly show that lysis of the cells during the assay is minimal, since only traces of the DERA protein were found in the whole-cell assay supernatant compared to the cleared-lysate assay mixture. At the same time, similar activity values were measured for both samples (Information S6). Taken together, the results strongly suggest that the bulky 7-deoxyribosyl-4-methyl umbelliferone is used as a substrate by the intracellularly expressed DERA and that both the substrate and the product, 4-methyl umbelliferone, freely diffuse in and out of the cells during the assay. Whether this is a simple diffusion

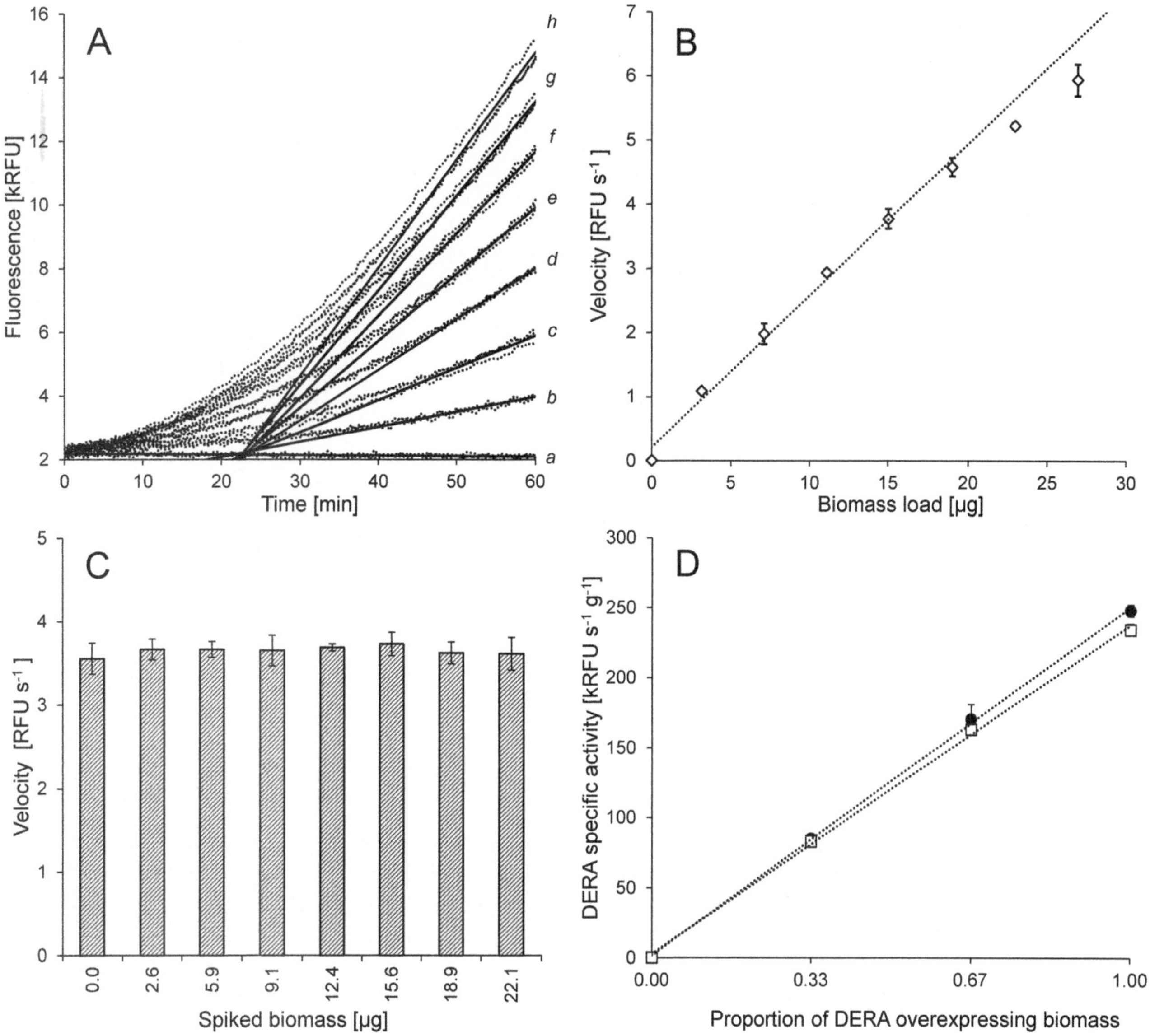

Figure 3. DERA activity measurements of the whole-cell catalyst. A. Fluorescence raw data for a DERA activity assay (dotted lines). Velocities for triplicate samples of the whole-cell catalyst were measured for 7 different loads (*b–h*, 3.16 µg–26.9 µg in 3.96 µg increments) of biomass. After normalization with the blank (*a*), maximum slopes were determined for each sample and averaged (solid lines) to yield velocity for a given biomass load. **B.** Velocity vs. biomass load plot. The first 5 points are taken for the specific activity calculation. Linear regression: $y = 0.2366x+0.2073$ $R^2 = 0.9936$ **C.** Comparison of velocities measured for cell-free lysate spiked with increasing loads of biomass. **D.** Validation of linearity of the activity assay within samples with constant biomass. The whole-cell catalyst *E. coli* BL21 (DE3) pET30/*deoC* was mixed with w.t. *E. coli* BL21 (DE3) biomass (●). Linear regression: $y = 248.94x+1.3840$, $R^2 = 0.9995$. In parallel, sonicated and cleared samples were measured (□). Linear regression: $y = 235.00x+2.6433$, $R^2 = 0.9989$.

process or an assisted transport mechanism is involved, remains beyond our investigations.

Aldehyde-caused inactivation of DERA is a major problem in use of this enzyme for chemoenzymatic processes since it makes the reuse of the catalyst impossible [38,61–62] and demands high loads for highly productive reactions [38,44,61–62]. Therefore, we were curious to see how the DERA whole-cell catalyst handles acetaldehyde **1** inactivation compared to the cell-free extract. In general, one can expect higher stability of proteins in their natural intercellular environment compared to cell-free extracts, due to the mechanisms organisms have, for dealing with missfolding issues [89], aggregation of proteins [90], oxidative stress [91], temperature stress [92], etc. We compared the inactivation rate of diluted DERA whole-cell catalyst and cell-free extract in presence of 75, 150 and 225 mM acetaldehyde **1** by measuring residual activity with the fluorometric method. A small, but consistent stabilization of enzyme activity is apparent from the results (Figure 4), indicating that the cellular environment may indeed have a beneficial effect on DERA stability.

Whole-cell Enzymatic Synthesis of Lactonized Statin Side-chain Precursors

Batch processes (the whole amount of the substrates was added to the reaction at $t = 0$), using DERA whole-cell catalyst and 400 mmol L^{-1} of various, readily accessible, 2-substituted

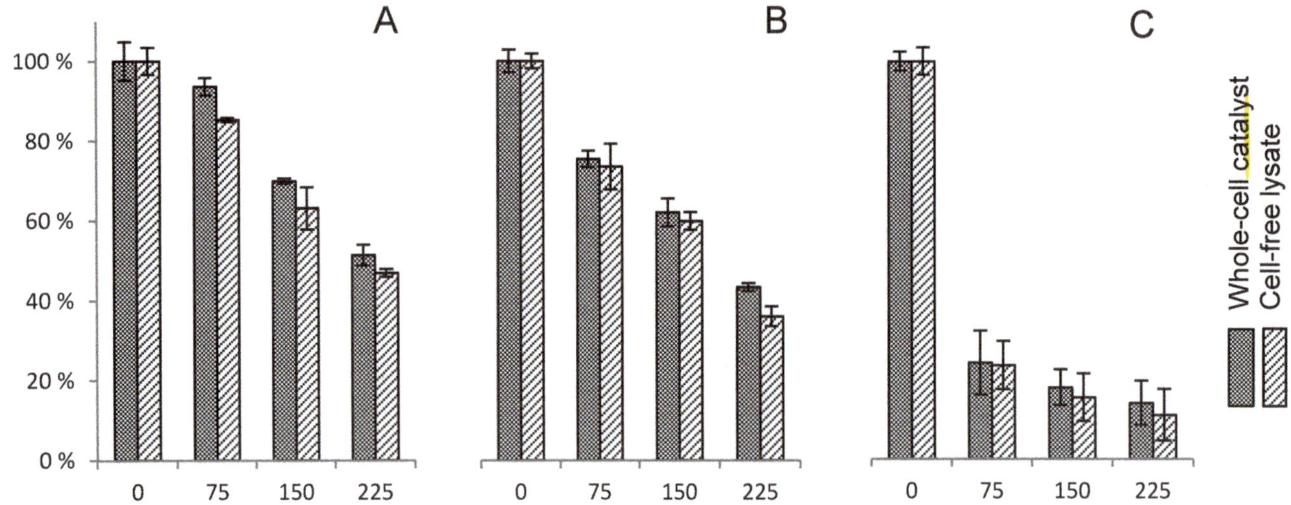

Figure 4. Inactivation of DERA whole-cell catalyst and DERA cell-free lysate with various aldehydes. Samples were treated with 75 mM, 150 mM and 225 mM substrate aldehydes for 15 minutes prior to the activity assay. The specific DERA activity was 226.8 kRFU s^{-1} g^{-1} and 226.6 kRFU s^{-1} g^{-1} for the whole-cell catalyst and for the cell-free lysate, respectively. Residual activities are given relative to non-treated whole-cell catalyst. Aldehydes used were acetaldehyde (A), **2b** (B) and **2g** (C).

acetaldehydes (**2b, 2e, 2f** and **2g**) [83,93–95] with 2.1 molar equivalents of acetaldehyde **1** were performed as described in the methods section. The GC-MS analysis confirmed accumulation of the expected corresponding lactols (**3b, 3e, 3f** and **3g**). The compounds were isolated and characterized with ^{1}H-NMR, ^{13}C-NMR and HRMS (Information S2). The confirmation of the structure of **3f** and **3g,** previously shown to be excellent starting material for super-statin synthesis [15–16,78–81,83], broadens the scope of molecules which can be obtained by DERA-catalyzed reactions, and this is the first time these molecules have been synthesized enzymatically. The successful production of the dimethoxy-substituted lactol **3f** for which unsuccessful synthesis attempts (using DERA) have been reported [57], was especially surprising. In agreement with Gijsen *et al.* [42,57], significant amounts of 2,4,6-trideoxyhexose (**3a**) have been detected when only acetaldehyde **1** was used as a substrate for the reaction. Surprisingly, **3a** was also found accumulating as a rule (albeit in lesser quantities) in all other reactions, where apart from acetaldehyde **1**, 2-substituted acetaldehydes were used (Table 1). This was not emphasized to date, despite the numerous reports on the DERA-catalyzed sequential reactions using acetaldehyde **1** as one of the substrates [37–44,46,57,59–60]. Although there is little value in this compound in terms of starting material for super-statin synthesis (due to lack of appropriate leaving group at the 6′ position, which would allow further transformations toward super-statin assembly), the finding is important. Namely, the chemical and physical properties of **3a** make the removal of this compound from the target cyclic hemiacetal side-chain products, obtained by DERA chemoenzymatic process, difficult during the work-up of the reactions. In addition, the accumulation rate of **3a** is an important indicator of the substrate availability during the reactions and was taken into account for optimization of substrate feeding.

The yields of the described batch reactions in Table 1 are given as non-isolated yields (calculated from the GC-FID analysis of the reaction samples and purified, NMR-evaluated compound samples). The mass balance of the reactions, taking into account the unreacted substrates, was found to be short of the theoretical outcome. This led us to search for the missing substrate balance in either accumulating intermediates or side products. Since both whole broth and cleared lysates are enzymatically complex systems, a side reaction of some kind would not be surprising. The GC-MS analysis showed accumulation of compounds with the mass spectra in agreement with products of single aldol condensation (**8b, 8e, 8f** and **8g**) in the reaction mixture at considerable level (Figure 5a and 5b: the reactions with **2b** and **2g** are shown for illustration). The calculated yields of the reactions with the **2e** and the **2f** were well below those observed with **2b** and **2g** and no further attention was given to the former two. After purification and characterization (^{1}H-NMR, ^{13}C-NMR and HRMS) of **8g** (Information S3), the mass balance of the reaction with acetyloxy-acetaldehyde (**2g**) was re-assessed. More than 90% of the substrate **2g** was accounted for, either in the final product (**3g, 10g**) or in the intermediate (**8g**). Although accumulation of **8b** in certain reaction conditions was indicated before [42], the significant accumulation of **8** during all of the reactions was a surprising observation since no attention to this intermediate is given in the past optimizations of DERA processes [42–46,59–60]. The observation offered another indicator by which the reaction conditions could be optimized.

Interestingly, an additional product of DERA was found accumulating in the reactions using **2b** as one of the substrates. The mass spectra of this compound indicate a product of condensation of two 2-chloro-acetyldehyde molecules with one acetaldehyde (Information S8). This new compound, presumably (4*R*,6*S*)-3-chloro-6-(chloromethyl)tetrahydro-2*H*-pyran-2,4-diol, was found accumulating in higher amounts in the reaction where 1 to 1 ratio between acetaldehyde and **2b** was used (Table 1).

Incidentally, a small amount of acetic acid was detected accumulating during the reactions with **2g** and acetaldehyde **1** in a rate not found in any of the reactions using **2b** as a substrate (Information S10). This finding points toward likely hydrolysis of **2g** to acetic acid and 2-hydroxy-acetaldehyde by a hydrolase present in the whole-cell catalyst. Indeed, incubation of **2g** alone with the whole-cell catalyst results in partial degradation and accumulation of acetic acid. Interestingly, we have also detected accumulation of trace quantities of a side product with mass spectra matching 2,4-dideoxy erythrose (data not shown). This

Table 1. Reaction species in batch reactions using whole-cell catalyst and various substrate aldehydes.

Substrate (mmol L^{-1})	1	2	8	3a	3	10	*	Yield
1 (840)	2.7 (19)	n.a	0.	136.2 (173)	**0**	0.6 (1)		62%
2b (400)+**1** (840)	7.5 (32)	0	4.3 (9)	9.5 (15)	**223.6 (258)**	5 (9)	22.5	65%
2b (400)+**1** (400)	0	11.2 (78)	27.0 (56)	0	**93.2 (108)**	2.8 (5)	40.6	27%
2e (400)+**1** (840)	0	150 (236)	80	91 (116)	**85 (43)**	3 (2)		11%
2f (400)+**1** (840)	0	88.6 (238)	52.2	85.7 (109)	**74.5 (54)**	2.4 (2)		14%
2g (400)+**1** (840)	0	50.0 (103)	76.9 (76)	32.6 (41)	**275.3 (182)**	13.7 (9)		45%
2g (400)+**1** (400)	0	77.8 (190)	98.2 (83)	2 (3)	**103.5 (80)**	4.7 (2)		20%

Whole-cell DERA catalyst was used (DERA specific activity = 232 kRFU s^{-1} g^{-1}, WCW = 207 g L^{-1}). Data are given as GC-FID peak area after 60 minutes after addition of the substrates, and calculated to molar concentration (mmol L^{-1}, in brackets) where isolated reference material was available. There was no considerable change in reaction-species concentration after 60 min (reaction was followed up to 180 min) indicating loss of DERA activity early in the reaction.
*presumably 2,6-chloro-2,4-dideoxyhexose. GC-MS data for this compound are available in Information S8.

molecule could originate from DERA-catalyzed condensation of acetaldehyde **1** and hydroxyl-acetaldehyde [42] arising from hydrolysis of **2g**. The above, were the only indications of a side activity in the whole-cell catalyst we have observed during our work.

An important finding, indicating transfer or diffusion of the reaction substrates, intermediates and products across the cellular envelopes, was made in the observation of practically identical reaction dynamics and yields when using either whole-cell catalyst or cell-free lysate with matched DERA activity for production of **3g** (data not shown). Since large majority of the DERA was found in the cellular fraction during the whole-cell process (Information S6), the reaction must proceed predominately inside the cells in order to match the reaction species dynamics observed with the cell-free extract process. Concentrations of the reaction substrates, intermediates and products in the acetonitrile-quenched reaction supernatant were found equal to intracellular concentrations obtained by sonication of the acetonitrile-quenched biomass pellet (GC-FID analysis). It is worth noting that the tightly packed biomass from the reaction mixtures represents ~ 20% of the reaction volume and that error in the above observation due to presence of compounds in the intercellular space is unlikely. The temporal distribution of intracellular vs. extracellular concentrations shows no apparent effects in the reaction kinetics that could arise from diffusion limitations. The accumulation of the intermediate **8g** both intra- and extra-cellularly in the first part of the process and its later consumption, as well as the observations with the activity assay experiments, further support the finding.

The Dynamics of the Reactions

Due to the complex nature of the reaction wherein the activity of the catalyst is changing over time in a non-constant manner (inactivation of DERA) [38,61–62,96], occurring reactions are reversible (retro-aldol reaction) [54–55], the aldehyde substrates [53], intermediates and products are in equilibrium with their presumably nonreactive forms [42,97–101], (Information S3 and

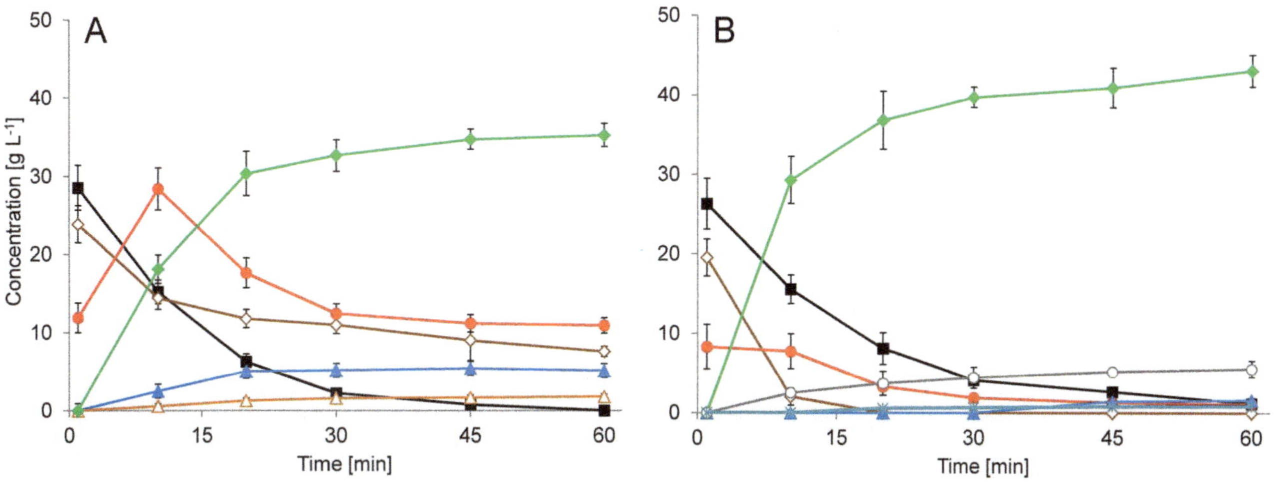

Figure 5. Time course of whole-cell, DERA-catalyzed batch reactions. Reactions were performed using *E. coli* BL21 (DE3) pET30/*deoC* fermentation cultures directly (DERA specific activity = 232 kRFU s^{-1} g^{-1}, WCW = 207 g L^{-1}). Results are given as mass concentrations obtained from GC-FID analysis. The measured quantity of a particular compound, with the exception of the stable 6-ring hemiacetals (**3**), represents the sum of the corresponding equilibrium forms (hydrate, aldehyde and acetal/hemiacetal), which exist under the reaction conditions. **A:** Reaction species data from reactions using 400 mmol L^{-1} of **2g** and 840 mmol L^{-1} of **1** are shown. **1** (■, black), **3a** (▲, blue) **3g** (♦, green), **8g** (●, red), **10g** (Δ, orange) and **2g** (◇, brown). **B:** Reaction species data from reactions using 400 mmol L^{-1} of **2b** and 840 mmol L^{-1} of **1** are shown. **1** (■, black), **3a** (▲, blue) **3b** (♦, green), **8b** (●, red), **10b** (Δ, orange), **2b** (◇, brown), 2,6-chloro-2,4-dideoxyhexose (□, grey). Concentration of the latter (Information S8) is evaluated based on the assumption, that the GC-FID response factor is similar to that of **3b**.

S4), and one of the substrates is involved in a parallel reaction (formation of **3a**), in-depth understanding of the reaction dynamics is extremely difficult. Even more so, the complexity increases when considering existence of alternative double condensation products (Information S8) found in batch reactions with **2b**. A simplified scheme of the reaction species for a process with **2g** is illustrated in Figure 6.

The observed accumulation of the single-aldol-condensation intermediates **8** in the reactions can generally be explained by second-order kinetics for consecutive reactions [102]. The difference in the initial substrate concentrations for the first and second condensation alone can explain the effect, given similar kinetic constants for both steps. The accumulation rate of the intermediate **8**, however, was found to be significantly higher in reactions with **2g** compared to the ones using **2b** (Figure 5). Also, in comparison to the process with **2b**, the reaction yielding **3a** seems more competitive to the condensation converting **8g** to **3g**, in reactions using **2g**. In addition, high accumulation rates of the intermediate **8g** were observed when the substrates were fed continuously at a rate which results in concentration of the substrate **2g** near zero throughout the reaction (Information S9). The availability of **2g** for the enzyme is even lower since it was found predominantly in hydrate form in water (Information S4). Therefore, the rate of the second condensation must be lower, to allow the observed accumulation of the intermediate (**8g**). This can be attributed to markedly different kinetic constant values for the first and the second condensation step due to DERA properties, or to the equilibriums between the aldehyde, acetal/hemiacetal and hydrate forms of the reaction species (Figure 2).

In fact the equilibrium between the reaction product **3** and its open, aldehyde form (**10**), is considered to be the reason why the sequential aldol reaction stops at the second condensation (the equilibrium is strongly shifted toward **3**) [42,57]. Surely, one possibility that could influence the rate of the second condensation, is the role of acetal or hemiacetal (**9**) equilibrium forms originating from the single-condensation intermediate (**8**) under the reaction conditions.

Indeed, although the reaction species (with exception of stable 6-ring hemiacetals **3**) are observed only in the aldehyde form in the GC analysis, the ^{1}H-NMR and ^{13}C-NMR spectra for purified **8g** showed (in addition to the expected hydrate form **12g**) presence of several distinct acetal-specific signals under mild conditions. Due to the number of the acetal signals observed for these equilibrium forms, which we collectively termed **9**, indisputable evidence on their structures remains elusive. According to the published examples, the species **9** may include cyclic lactol [42], dimeric lactol [97], dimeric hemiacetal [98] or cyclic trioxane [99] (Information S3).

Shifts in equilibrium between **8g**, **9g** and **12g** can be observed when various solvents are used (D_2O, $CDCl_3$, DMSO-d_6) for the ^{1}H-NMR and ^{13}C-NMR measurements. The equilibrium is shifted toward **8g** and **9g** in nonpolar $CDCl_3$ whereas the hydrate **12g** is predominant in polar solvents, although aldehyde **8g** and acetal-specific proton signals of **9** can also be observed (Information S3). In contrast, only hydrate and aldehyde species are found for the initial substrates **1** and **2g** in D_2O as indicated by ^{1}H-NMR (Information S4).

Therefore, in analogy with DERA-catalyzed coupling of hydroxyacetaldehyde and acetaldehyde **1**, where the formation of stable hemiacetals both with the initial substrate and the product of the first aldol condensation was found preventing successive reaction steps [42], one can foresee a similar, yet less prominent effect influencing the second condensation to form **3g.**

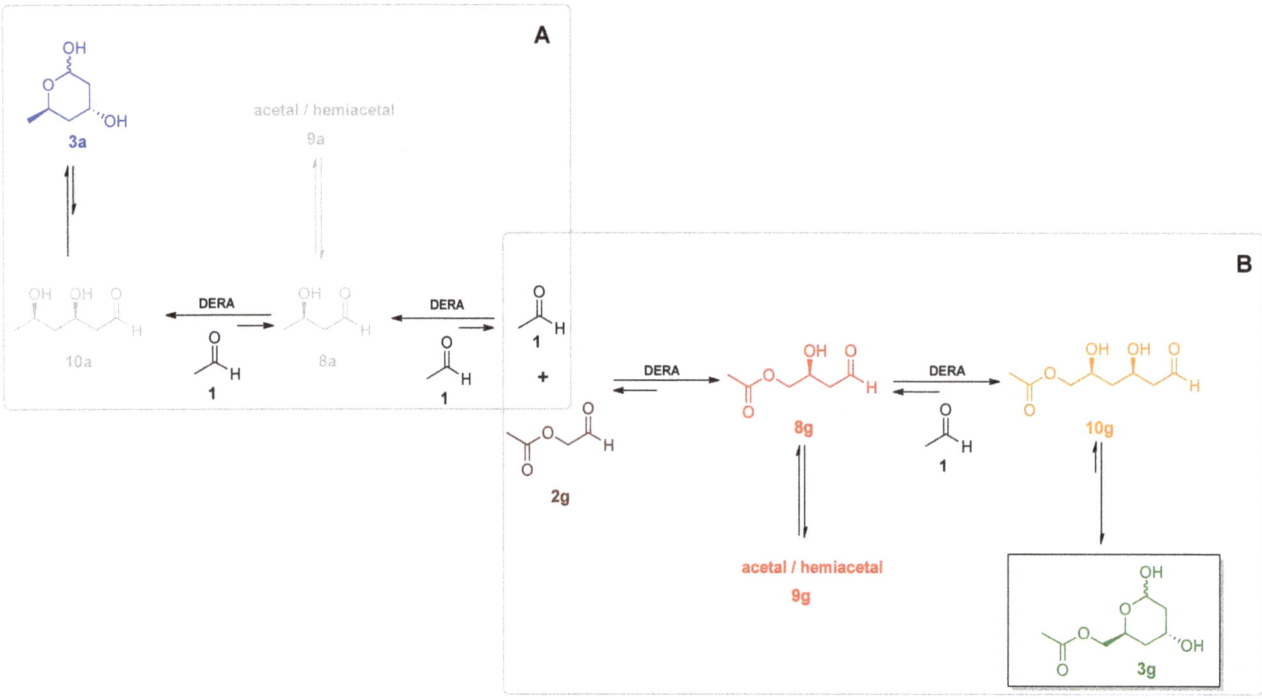

Figure 6. Compounds influencing the production of acetiloxy-lactol (3g). Hydrate forms of the aldehydes (**11, 12 and 13**) are not depicted here. **A:** Reaction species arising from acetaldehyde alone. **B:** Reaction species arising from acetaldehyde **1** and **2g.** Several acetal/hemiacetal (**9g**) species were found in equilibrium with **8g**. These may include monomeric cyclic hemiacetal as well as dimeric and trimeric cyclic acetals/hemiacetals (Information S3).

On the other hand, we can predict but one difference between **8b** and **8g** in the possibility of formation of cyclic acetals/hemiacetals. The presence of the acetyl group in **8g** changes the possible diversity in these equilibria compared to the situation with **8b** (Information S3). Although the reversible rearrangement (transesterification) of **8g**, leading to a five membered lactol, could reduce the availability of the mono-aldol intermediate to the enzyme, we cannot claim that this is the sole reason for the observed lower second condensation rate compared to the reactions with the chloro-substituded counterparts. Still, consideration of the existence of the equilibrium forms of the intermediates offers an insight, complementary to the mechanistic studies published for DERA [52–53,61], when interpreting the dynamics of sequential aldol condensations.

Next, analysis of the residual DERA activity, determined for washed whole-cell samples taken from the batch reaction with 400 mmol L^{-1} **2g** and 840 mmol L^{-1} acetaldehyde **1**, showed rapid inactivation of DERA. Only 28% of the initial activity was detected after the first 10 minutes of the reaction, 18% after 30 minutes and less than 15% after 1 h (Information S7). Reliable residual activity measurements for the parallel cell-free extract reaction could not be performed due to the lack of a convenient method for removal of the inactivating aldehydes present in the assay samples. This rapid loss of enzyme activity was another observation calling for optimization of the feeding strategy, although it was not a surprising find. The issue has already been addressed by Greenberg *et al.* [44]. In their report, chloroacetaldehyde (**2b**) was found to be significantly more inhibiting compound compared to acetaldehyde **1**. In our hands, derived from the results where the residual activity of the whole-cell catalyst in presence of various aldehydes was compared, acetaldehyde **1** was found to be almost as potent inactivator of DERA as **2b**. On the other hand, addition of **2g**, the prime focus of our interest, resulted in significantly higher inactivation rates making the development of a highly productive process with this substrate even more challenging (Figure 4). For this reason, and in light of the accumulating **8** and **3a** in the reaction, as well as the formation of alternative products such as the 2,6-chloro-2,4-dideoxyhexose (Figure 5, Information S8), it is obvious that the feeding strategy described previously (proposing addition of the two reactants in a constant 1:2.1 molar ratio) [44], is not generally applicable to the DERA-catalyzed processes and can be improved by controlling feeding rates of each individual substrate throughout the reaction.

Fed-batch Whole-cell Process

Using a semi-empiric approach, various feeding profiles for each of the two substrates (**1** and **2g**) were tested in fed-batch reactions using DERA whole-cell catalyst (data not shown). When using a constant molar ratio (**2g** : **1** = 1 : 2.1) of the substrates, similarly to the batch process described above (Figure 5a), the intermediate **8g** was formed much faster than was consumed to form **3g**. Consequently, a surplus of acetaldehyde was created, resulting in enhanced accumulation of **3a**. At the end, rapid loss of DERA activity and lack of acetaldehyde (consumed for **3a** formation) lead to unconverted **8g** and low yields. As the next step, with the aim to control the accumulation of **3a**, we set up fed-batch reactions with addition of the total quantity of **2g** at the start of the reaction. Acetaldehyde was fed in a profile that allowed consumption of the accumulated **8g** almost completely. Although this feeding strategy showed high yields, and both **3a** and **8g** levels below 5% each, the inactivation of DERA prevented high yields in processes with **2g** loads higher than 400 mmol L^{-1}.

Finally, a compromise was made in the feeding strategy (Figure 7), taking into account the indications to the reaction dynamics we have obtained. The ratio between **2g** and acetaldehyde was adjusted to 1 : 2.2 in order to compensate for the acetaldehyde consumed in **3a** formation. **2g** was fed faster than acetaldehyde **1** and only in the first 30 minutes of the process. This resulted in a higher rate of **3g** formation in the first part of the process compared to the strategy where all of the **2g** was added at the beginning. At the same time, the formation of **3a** was low. As **2g** was being consumed, a surplus of acetaldehyde was created, driving both conversion of **8g** to **3g** and formation of **3a**. The remaining 25% of the acetaldehyde (fed in the next 30 minutes) allowed efficient conversion of the large part of the mono-aldol intermediate in a reasonable timeframe of the remaining 60 minutes, although some **3a** was formed due to this acetaldehyde surplus.

More than 20 fed batch reactions with **2g** loads between 500 mmol L^{-1} and 700 mmol L^{-1} were made using this feeding profile and titers of 100 g L^{-1} of **3g** were reached (103.2 g L^{-1} including the open form **10g**) in 2 h reaction time using 700 mmol L^{-1} of **2g** (Figure 8). Due to higher process robustness and higher yields (78–86%), we chose to use lower substrate loads (550 mmol L^{-1} of **2g**), which routinely resulted in 85–91 g L^{-1} titers of **3g** in 120 min (Figure 7). The chromatographic purity of the reaction-mixture samples at harvest time were 82%–86% and crude **3g** prepared with whole-broth extraction from these reactions had >70% assay for the **3g** as a rule. The volumetric productivity of the process was over 40 g L^{-1} h^{-1}. Despite the optimization, accumulation of **3a** and unreacted intermediate (**8g**) could not be completely avoided. **3a** and **8g** were found, in average, in 10% (w/w) each, compared to **3g**. The ratio between **3g** and its open, aldehyde form (**10g**) was constant at the reaction conditions (~ 3% w/w) but could be influenced in favor of **10g** by lowering the pH. Residual activity of DERA during the reaction was tested and although <10% of the activity was detected at the completion of the reaction, the inactivation profile shows significant improvement compared to the batch process (Information S7). Using the fed-batch process ~ 45% of the activity remained after 30 minutes and ~ 25% after 60 min of the process even with 550 mmol L^{-1} load of **2g** (Figure 7).

For validation of applicability of the process for preparation of other substituted lactols, chlorolactol (**3b**) reaction was selected due to the challenging precedents in the literature [44,60]. Using the same conditions as used for production of **3g**, with 700 mmol L^{-1} of **2b**, 102 g L^{-1} of **3b** (104.2 g L^{-1} including the open form **10b**) was prepared in 120 min (88% yield, volumetric productivity 51 g $L^{-1}h^{-1}$). Reaction species profiles were found to be similar to the acetyloxylactol (**3g**) process, however - as expected, the accumulation of the mono-aldol intermediate **8b** was less pronounced in the reaction with **2b** resulting in higher production rates in the first part of the process (Figure 8). Compared to the batch process, formation of 2,6-chloro-2,4-dideoxyhexose was at barely detectable levels.

The resulting process not only improves the volumetric productivity reported for the w.t. *E. coli* DERA catalyzed production of **3b** [42,44,60], but also achieves this with a considerably cheaper biocatalyst. The material cost contribution of the whole-cell catalyst is estimated to be in the range of 2.0–3.5 € per assay kilogram of the crude isolate of chiral lactols (depending on the target titer). Furthermore, the fermentation and the chemoenzymatic process were joined into a one-pot telescope process, additionally simplifying the industrialization.

Preparation of Advanced Super-statin Intermediates

The conversion of the isolated lactols **3** to a useful statin side chain can be achieved by oxidation of lactols **3** to the lactones **15,**

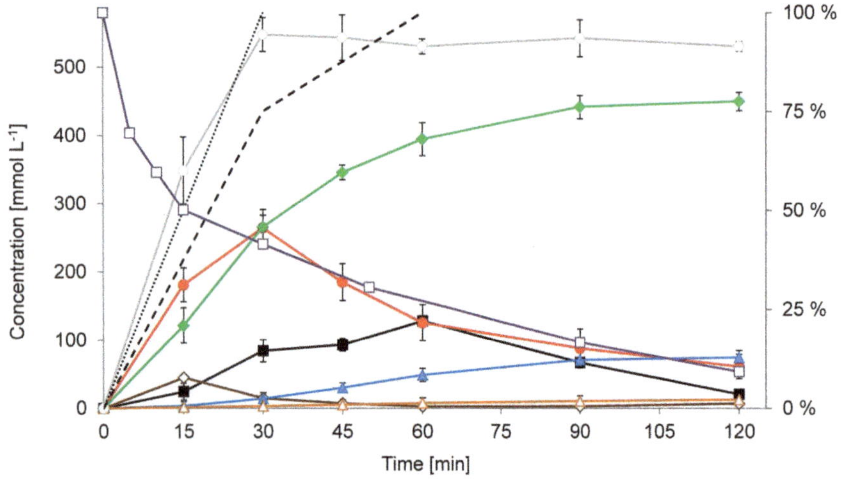

Figure 7. Time course of whole-cell, DERA-catalyzed, fed-batch reactions yielding 3g. Reaction species data from three independent experiments using (in total) 550 mmol L^{-1} of **2g** and 1200 mmol L^{-1} of **1** are shown. Whole-cell catalyst (*E. coli* BL21 (DE3) pET30/*deoC* high-density culture) with 217 kRFU s^{-1} g^{-1} DERA specific activity and 182 g L^{-1} WCW was used. Results are given as molar concentrations obtained from GC-FID analysis. The measured quantity of a particular compound, with the exception of the stable 6-ring hemiacetals (**3**), represents the sum of the corresponding equilibrium forms (hydrate, aldehyde and acetal/hemiacetal) which exist under the reaction conditions. **1** (■, black), **3a** (▲, blue) **3g** (♦, green), **8g** (●, red), **10g** (Δ, orange), **2g** (◇, brown), cumulative molarity of reaction species originating from **2g** (□, grey; sum of **2g**, **8g**, **10g** and **3g**). Secondary vertical axis shows in %: residual DERA activity (□, violet), cumulative feed of **2g** (dotted line), cumulative feed of **1** (dashed line).

followed by silyl protection of 4-hydroxyl substituent resulting in compound **5** [78], an advanced intermediate, recently described *en route* to preparation of rosuvastatin [15,80,81] and pitavastatin [16,80,81]. The lactol oxidation using various oxidation methods such as $Br_2/BaCO_3$ [103], *N*-iodosuccinimide [104], Ag_2CO_3 [105], Pt/C-O_2 [106], $RuCl_2(PPh_3)_3$/cyclohexanone [107], MnO_2 [108], or NaOCl/AcOH [60] was described before. Aiming toward a scalable and sustainable process we tested several process options in order to use the whole-cell DERA reaction mixtures in a straightforward manner (Table 2). Initially, oxidation of **3g** using bromine in the presence of barium carbonate was used to obtain the corresponding lactone **4** in good 78% yield (Table 2, entry 1). Prior to oxidation, addition of acetonitrile to the whole-cell reaction mixture and filtration of the precipitate was used to remove majority of the cells and cell debris. Inconveniently, the remaining proteins, carbohydrates and other organic material from the extract, as well as the presence of undesired compounds **3a** and **8g**, make the oxidation reaction more difficult as it requires 3.3 equivalents of bromine to fully oxidize the lactol **3g**. Even more equivalents were needed when chemoenzymatic reaction mixtures having higher amounts of **3a** and **8g** were used. To avoid the large amount of hazardous and toxic bromine needed, the process using bleach [60] was tested and modified (avoiding the solvent exchange) with the use of a mixture of aqueous NaH_2PO_4 solution and ethylacetate instead of acetic acid as solvent. This reaction also required large excess of

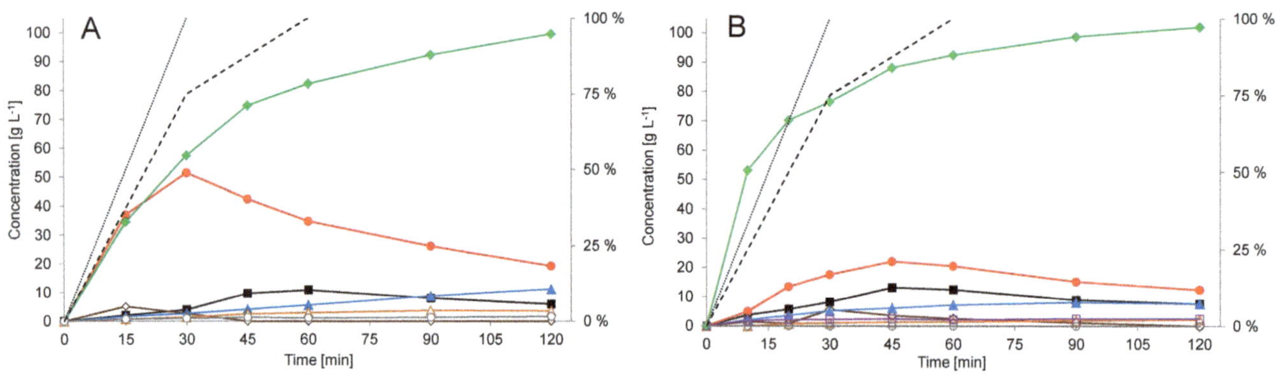

Figure 8. Time course of exemplary whole-cell, DERA-catalyzed, fed-batch reactions with ~50 g L^{-1} h^{-1} volumetric productivity. Whole-cell catalyst (E. coli BL21 (DE3) pET30/deoC high-density culture) with 247 kRFU s^{-1} g^{-1} DERA specific activity and 215 g L^{-1} WCW was used. Results are given as mass concentrations obtained from GC-FID analysis. The measured quantity of a particular compound, with the exception of the stable 6-ring hemiacetals (**3**), represents the sum of the corresponding equilibrium forms (hydrate, aldehyde and acetal/hemiacetal) which exist under the reaction conditions. **A:** Reaction species data from reaction using (in total) 700 mmol L^{-1} of **2g** and 1540 mmol L^{-1} of **1** are shown. **1** (■, black), **3a** (▲, blue) **3g** (♦, green), **8g** (●, red), **10g** (Δ, orange), **2g** (◇, brown), acetic acid (□, grey). Secondary vertical axis shows in %: cumulative feed of **2g** (dotted line), cumulative feed of **1** (dashed line). **B:** Reaction species data from reaction using (in total) 700 mmol L^{-1} of **2b** and 1540 mmol L^{-1} of **1** are shown. **1** (■, black), **3a** (▲, blue) **3b** (♦, green), **8b** (●, red), **10b** (Δ, orange), **2b** (◇, brown), acetic acid (□, grey), 2,6-chloro-2,4-dideoxyhexose (□, purple). Secondary vertical axis shows in %: cumulative feed of **2b** (dotted line), cumulative feed of **1** (dashed line).

Table 2. The optimization of lactol (**3g**) oxidation yielding lactone **15**.

	Cell removal method	Oxidant (eq.)	Additive	Yield
1	MeCN precipitation	Br_2 (3.3)	$BaCO_3$	78%
2	MeCN precipitation	NaOCl (6.5)	NaCl/NaH_2PO_4	77%
3	None	NaOCl (6.5)	NaCl/NaH_2PO_4	84%
4	Centrifugation	NaOCl (6.5)	NaCl/NaH_2PO_4	84%
5	Centrifugation	$Ca(OCl)_2$ (9.5)	NaCl/NaH_2PO_4	64%
6	Centrifugation	$Ca(OCl)_2$ (4.7)	NaCl/H_3PO_4	78%
7	None	$Ca(OCl)_2$ (2.5)	NaCl/H_3PO_4	84%

the oxidant to reach completion of the reaction (6.5 eq.) and led to a similar, 77% yield (Table 2, entry 2).

In order to avoid the use of large amounts of acetonitrile for the removal of the biomass, the oxidation was tested directly on the whole-cell catalyst reaction mixture and on the centrifuged reaction mixture. Both reactions gave comparable and better, 84% yields (Table 2, entries 3 and 4) when the acetonitrile was omitted. An issue with the bleach oxidation process was raised upon scale-up; the large volume of water was introduced to the reaction due to the low bleach concentration. To address this issue, we decided to replace bleach with its solid equivalent calcium hypochlorite. At first, the reaction required larger amounts of the oxidant (Table 2, entry 5) and gave lower yield mostly because of filtration problems during the work-up. The replacement of sodium phosphate with phosphoric acid, while maintaining the pH at 3, led to reduction of oxidant usage and made the work-up filtration easier. This resulted in higher yield (Table 2, entry 6) and surprisingly improvements in both the amount of oxidant used and yield (84%) were achieved when using the whole-cell catalyst reaction mixture directly (entry 7). The lactone **15g** (oil) was isolated from the ethyl-acetate phase and purification of impurities derived mainly from compound **3a** was not possible except by chromatography. None the less, the lactone **15g** was used without purification and protected in standard conditions with TBDMSCl in dichloromethane in presence of imidazole to give **5** and then further converted to the compound **6** using previously reported procedure [80,84], to obtain crystalline material in an overall yield of 62% for the three steps. The crystallization of **6** allowed removal of the impurities originating from the whole-cell chemoenzymatic process and carried over through oxidation and silylation steps. ^{1}H-NMR and GC analysis of **6** showed very high purity (99.8% GC purity). Enantiomeric and diastereoisomeric purity was determined to be similar as previously reported (>99.9% *ee*, >99.8% *de*, Information S5) [80,84]. Combined, the results show an efficient way of converting the lactols **3** derived from the whole-cell DERA catalyzed reactions in further steps toward statins. To this end, we avoided problematic and technically demanding downstream procedures yielding an especially low cost, scalable and green process with good overall yields. Again, the importance of controlling the DERA reaction intermediates and side products such as **8g** and **3a** (respectively) can be stressed out; this time not from the reaction yield point of view, but rather in terms of influence to the following chemical steps, namely the oxidation, silylation and deacetylation. The lactol **3a**, for example, is not problematic *per se* since it is efficiently removed in the late steps. The additional amount of TBDMSCl needed at the silylation step due to its presence, however, brings unnecessary extra costs to the process.

Conclusions

The described process for production of chiral, lactonized super-statin intermediates, using a whole-cell DERA catalyst derived directly from a fed-batch, high-density fermentation with *E. coli* BL21 (DE3) overexpressing native *E. coli deoC*, shows unprecedented productivity and is highly cost effective. Volumetric productivities of 50 g L^{-1} h^{-1}, with >80% yield and >80% chromatographic purity have been achieved with this process for 6′-chloro and 6′-acetyloxy substituted lactols. Moreover, ((2*S*,4*R*)-4,6-dihydroxytetrahydro-2*H*-pyran-2-yl)methyl acetate and (4*R*,6*S*)-6-(dimethoxymethyl)tetrahydro-2*H*-pyran-2,4-diol have been prepared, using DERA catalysis, for the first time. Although new insights into the dynamics of the reaction intermediates and side products are presented in this work, the complexity of the reaction leaves room for additional improvement of understanding of the DERA-catalyzed sequential aldol condensation reactions. This is especially true for the role of equilibria of the reaction aldehydes with their hydrate form on the one side and acetal/hemiacetal forms on the other. In addition, the dynamics of the availability of the substrates, defined by the feeding strategy and equilibria with their non-reactive forms in combination with substrate preference of the DERA enzyme, can lead the reaction to form completely unexpected products such as 2,6-chloro-2,4-dideoxyhexose. Nevertheless, excellent enantiomeric purities (*ee* >99.9%) which were demonstrated on downstream advanced intermediates allow the products obtained by the described process to fit directly into our recently described lactone pathway to optically pure super-statins. The enzymatic synthesis of ((2*S*,4*R*)-4,6-dihydroxytetrahydro-2*H*-pyran-2-yl)methyl acetate is the last piece in design of an efficient and industrially scalable synthesis of super-statins based on the direct coupling of lactonised diol side-chains to the heterocyclic part of the molecule.

Supporting Information

Information S1 Fermentation data.

Information S2 NMR, HR-MS, GC-FID and GC-MS data.

Information S3 NMR, HR-MS, GC-FID and GC-MS identification of the mono-aldol intermediates (8, 9, 12).

Information S4 Proton NMR of aldehyde precursors 1 and 2g in D_2O.

Information S5 Determination of stereochemical purity.

Information S6 Comparison of DERA activity in the whole broth, washed cells and cell-free lysate with the assay mixture supernatant. DERA assay on whole cells.

Information S7 Residual DERA activity measurements and DERA distribution in the whole cell process with 2g.

Information S8 Proposed alternative double condensation product observed in reactions with 2b.

Information S9 A fed-batch process keeping 2g concentration near dynamic zero.

Information S10 Accumulation of acetic acid in reactions with 2g.

Acknowledgments

We thank Irena Babič for her invaluable contribution in chemo-enzymatic experiments and activity assays, Tadeja Žohar for isolation of pure compounds, Dr. Ivan Plantan for the work on the hypochlorite oxidation step, Matej Bizjak for the enthusiastic support with the semi-industrial bioprocess and optimization of fermentation conditions, Dr. Martin Črnugelj for the NMR analytics and Prof. Dr. Janez Košmrlj for the HR-MS analytics.

Author Contributions

Conceived and designed the experiments: PM ZČ GK. Performed the experiments: MO JC. Analyzed the data: MO JC GK ZČ PM. Contributed reagents/materials/analysis tools: MO DO JC PM. Wrote the paper: PM ZČ.

References

1. Tobert JA (2003) Lovastatin and beyond: the history of the HMG-COA reductase inhibitors. Nat Rev Drug Discov 2: 517–526.
2. Istvan ES, Deisenhofer J (2001) Structural mechanism for statin inhibition of HMG-CoA reductase. Science 292: 1160–1164.
3. Brautbar A, Ballantyne CM (2011) Pharmacological strategies for lowering LDL cholesterol: statins and beyond. Nat Rev Cardiol 8: 253–265.
4. Kidd J (2006) Life after statin patent expires. Nat Rev Drug Discov 5: 813–814.
5. Almuti K, Rimawi R, Spevack D, Ostfeld RJ (2006) Effects of statins beyond lipid lowering: potential for clinical benefits. Int J Cardiol 109: 7–15.
6. Switzer JA, Hess DC (2006) Statins in stroke: prevention, protection and recovery. Expert Rev Neurotherapeutics 6: 192–202.
7. Lahera V, Goicoechea M, de Vinuesa SG, Miana M, de las Heras N, et al. (2007) Endothelial dysfunction, oxidative stress and inflammation in atherosclerosis: beneficial effects of statins. Curr Med Chem, 14: 243–248.
8. Paraskevas KI, Tzovaras AA, Briana DD, Mikhailidis DP (2007) Emerging indications for statins: a pluripotent family of agents with several potential applications. Curr Pharm Des, 13: 3622–3636.
9. Bersano A, Ballabio E, Lanfranconi S, Mazzucco S, Candelise L, et al. (2008) Statins and stroke. Curr Med Chem 15: 2380–2392 and references cited therein.
10. Endo A, Kuroda M, Tsujita Y (1976) ML-236A, ML-236B, and ML-236C, new inhibitors of cholesterogenesis produced by *Penicillium citrinum*. J Antibiot 29: 1346–1348.
11. Endo A (1992) The discovery and development of HMG-CoA reductase inhibitors. J Lipid Res 33: 1569–1582.
12. Endo A, Hasumi K, (1993) HMG-CoA reductase inhibitors. Nat Prod Rep 10: 541–550.
13. Endo A (2008) A gift from nature: the birth of the statins. Nature Med 14: 1050–1052.
14. Časar Z (2010) Historic overview and recent advances in the synthesis of superstatins. Curr Org Chem 14: 816–845.
15. Časar Z, Steinbücher M, Košmrlj J (2010) Lactone pathway to statins utilizing the Wittig reaction. The synthesis of rosuvastatin. J Org Chem 75: 6681–6684.
16. Fabris J, Časar Z, Gazić Smilović I (2012) The use of a lactonized statin side-chain precursor in a concise and efficient assembly of pitavastatin. Synthesis 44: 1700–1710.
17. Vasić-Rački Đ (2006) History of industrial biotransformations - Dreams and realities. In: Liese A, Seelbach K, Wandrey C, editors. Industrial Biotransformations 2nd ed. Weinheim: Wiley-VCH. 1–29.
18. Liese A, Villela Filho M (1999) Production of fine chemicals using biocatalysis. Curr Opin Biotechnol 10: 595–603.
19. Schulze B, Wubbolts MG (1999) Biocatalysis for industrial production of fine chemicals. Curr Opin Biotechnol 10: 609–615.
20. Wandrey C, Liese A, Kihumbu D (2000) Industrial biocatalysis: past, present, and future. Org Process Res Dev 4: 286–290.
21. Zaks A (2001) Industrial biocatalysis. Curr Opin Chem Biol 5: 130–136.
22. Koeller KM, Wong C-H (2001) Enzymes for chemical synthesis. Nature 409: 232–240.
23. Schmid A, Dordick JS, Hauer B, Kiener A, Wubbolts M, et al. (2001) Industrial biocatalysis today and tomorrow. Nature 409: 258–268.
24. Straathof AJJ, Panke S, Schmid A (2002) The production of fine chemicals by biotransformations. Curr Opin Biotechnol 13: 548–556.
25. Patel RN (2002) Microbial/enzymatic synthesis of chiral intermediates for pharmaceuticals. Enzyme Microb Technol 31: 804–826.
26. Bertau M (2002) Novel developments in biocatalytic organic chemistry. Curr Org Chem 6: 987–1014.
27. Schoemaker HE, Mink D, Wubbolts MG (2003) Dispelling the myths-biocatalysis in industrial synthesis. Science 299: 1694–1697.
28. Yazbeck DR, Martinez CA, Hu S, Tao J (2004) Challenges in the development of an efficient enzymatic process in the pharmaceutical industry. Tetrahedron: Asymmetry 15: 2757–2763.
29. Panke S, Held M, Wubbolts M (2004) Trends and innovations in industrial biocatalysis for the production of fine chemicals. Curr Opin Biotechnol 15: 272–279.
30. Panke S, Wubbolts M (2005) Advances in biocatalytic synthesis of pharmaceutical intermediates. Curr Opin Chem Biol 9: 188–194.
31. Pollard DJ, Woodley JM (2006) Biocatalysis for pharmaceutical intermediates: the future is now. Trends Biotechnol 25: 66–73.
32. Hudlicky T, Reed JW (2009) Applications of biotransformations and biocatalysis to complexity generation in organic synthesis. Chem Soc Rev 38: 3117–3132.
33. Müller M (2005) Chemoenzymatic Synthesis of Building Blocks for Statin Side Chains. Angew Chem Int Ed 44: 362–365.
34. Liljeblad A, Kallinen A, Kanerva LT (2009) Biocatalysis in the Preparation of the Statin Side Chain. Curr Org Synth 6: 362–379.
35. Ma SK, Gruber J, Davis C, Newman L, Gray D, et al. (2010) A green-by-design biocatalytic process for atorvastatin intermediate. Green Chem 12: 81–86.
36. Schürman M, Wolberg M, Panke S, Kierkels H (2010) The development of short, efficient, economic, and sustainable chemoenzymatic processes for statin side chains. In: Dunn PJ, Wells AS, Williams MT, editors. Green Chemistry in the Pharmaceutical Industry 1st ed. Weinheim: Wiley-VCH. 127–144.
37. Whalen LJ, Wong C-H (2006) Enzymes in organic synthesis: aldolase-mediated synthesis of iminocyclitols and novel heterocycles. Aldrichim Acta 39: 63–71.
38. Dean SM, Greenberg WA, Wong C-H (2007) Recent Advances in Aldolase-Catalyzed Asymmetric Synthesis. Adv Synth Catal 349: 1308–1320.
39. Samland AK, Sprenger GA (2006) Microbial aldolases as C–C bonding enzymes-unknown treasures and new developments. Appl Microbiol Biotechnol 71: 253–264.
40. Bolt A, Berry A, Nelson A (2008) Directed evolution of aldolases for exploitation in synthetic organic chemistry. Arch Biochem Biophys 474: 318–330.
41. Clapés P, Fessner W-D, Sprenger GA, Samland AK (2010) Recent progress in stereoselective synthesis with aldolases. Curr Opin Chem Biol 14: 154–167.
42. Gijsen HJM, Wong C-H (1994) Unprecedented asymmetric aldol reactions with three aldehyde substrates catalyzed by 2-deoxyribose-5-phosphate aldolase. J Am Chem Soc 116: 8422–8423.
43. Gijsen HJM, Qiao L, Fitz W, Wong C-H (1996) Recent advances in the chemoenzymatic synthesis of carbohydrates and carbohydrate mimetics. Chem Rev 96: 443–474.
44. Greenberg W, Varvak A, Hanson SR, Wong K, Huang H, et al. (2004) Development of an efficient, scalable, aldolase-catalyzed process for enantioselective synthesis of statin intermediates. Proc Natl Acad Sci USA 101: 5788–5793.
45. Ikunaka M (2007) Catalytic asymmetric carbon-carbon bond formations: their evolution from biocatalysis to organocatalysis over the millennium. Org Process Res Dev 11: 495–502.
46. Yoshida Y, Noritake T, Watanabe M, Nakamoto Y (2005) Method for producing (4R,6S)-6-benzyloxymethyl-4-hydroxy-tetrahydro-2-pyrone. (Ube Industries, Ltd., Japan) Jap Pat Appl JP 2005229858 A Chem Abstr (2005) 143: 228047.
47. Časar Z, Mesar M, Kopitar G, Mrak P, Ošlaj M (2008) Synthesis of statins. (Lek Pharmaceuticals d.d.) PCT Int Appl WO 2008119810 A2 Chem Abstr (2008) 149: 448104.
48. Machajewski TD, Wong C-H (2000) The catalytic asymmetric aldol reaction. Angew Chem Int Ed 39: 1352–1374.
49. Mestres R (2004) A green look at the aldol reaction. Green Chem 6: 583–603.
50. Sukumaran J, Hanefeld U (2005) Enantioselective C-C bond synthesis catalysed by enzymes. Chem Soc Rev 34: 530–542.
51. Mlynarski J, Paradowska J (2008) Catalytic asymmetric aldol reactions in aqueous media. Chem Soc Rev 37: 1502–1511.

52. Heine A, DeSantis G, Luz JG, Mitchell M, Wong C-H, et al. (2001) Observation of covalent intermediates in an enzyme mechanism at atomic resolution. Science 294: 369–374.
53. Heine A, Luz JG, Wong C-H, Wilson IA (2004) Analysis of the class I aldolase binding site architecture based on the crystal structure of 2-deoxyribose-5-phosphate aldolase at 0.99 A° resolution. J Mol Biol 343: 1019–1034.
54. Racker E (1952) Enzymatic synthesis and breakdown of deoxyribose phosphate. J Biol. Chem. 196: 347–365.
55. Pricer WE, Horecker BL (1960) Deoxyribose aldolase from Lactobacillus plantarum. J Biol Chem 235: 1292–1298.
56. Barbas CF, Wang Y-F, Wong C-H (1990) Deoxyribose-5-phosphate aldolase as a synthetic catalyst. J Am Chem Soc 112: 2013–2014.
57. Gijsen HJM, Wong C-H (1995) Sequential three- and four-substrate aldol reactions catalyzed by aldolases. J Am Chem Soc 117: 7585–7591.
58. Chen L, Dumas PD, Wong CH (1992) Deoxyribose-5-phosphate aldolase as a catalyst in asymmetric aldol condensation. J Am Chem Soc 114: 741–748.
59. Kierkels JGT, Mink D, Panke S, Lommen FAM, Heemskerk D (2003) Process for the preparation of 2,4-dideoxyhexoses and 2,4,6-trideoxyhexoses. (DSM N.V.) PCT Int Appl WO 2003006656 A2 Chem Abstr (2003) 138: 112523.
60. Greenberg W, Wong K, Varvak A, Swanson R (2004) Chemoenzymatic methods for the synthesis of statins and statin intermediates. (Diversa Corporation), PCT Int Appl WO 2004027075 A2 Chem Abstr (2004) 140: 302423.
61. Jennewein S, Schürmann M, Wolberg M, Hilker I, Luiten R, et al. (2006) Directed evolution of an industrial biocatalyst: 2-deoxy-D-ribose 5-phosphate aldolase. Biotechnol J 1: 537–548.
62. Habeeb AFSA, Hiramoto R (1968) Reaction of proteins with glutaraldehyde. Arch Biochem Biophys 126: 16–26.
63. Franke D, Hsu C-C, Wong C-H (2004) Directed evolution of aldolases. Methods Enzymol 388: 224–238.
64. Suau T, Calveras J, Clapés P, Benaiges MD, Álvaro G (2005) Immobilization of fuculose-1-phosphate aldolase from *E. coli* to glyoxal-agarose gels by multipoint covalent attachment. Biocatal Biotransfor 23: 241–250.
65. Suau T, Álvaro G, Benaiges MD, López-Santín J (2009) Performance of an immobilized fuculose-1-phosphate aldolase for stereoselective synthesis. Biocatal Biotransfor 27: 136–142.
66. Wang A, Wang M, Wang Q, Chen F, Zhang F, et al. (2011) Stable and efficient immobilization technique of aldolase under consecutive microwave irradiation at low temperature. Bioresour Technol 102: 469–474.
67. Faber K (2011) Biotransformations in Organic Chemistry: A Textbook 6th ed. Berlin: Springer: 9–10.
68. Ishige T, Honda K, Shimizu S (2005) Whole organism biocatalysis. Curr Opin Chem Biol 9: 174–180.
69. Fukuda H, Hama S, Tamalampudi S, Noda H (2008) Whole-cell biocatalysts for biodiesel fuel production. Trends Biotechnol 26: 668–673.
70. Wohlgemuth R (2009) The locks and keys to industrial biotechnology. New Biotechnol 25: 204–213.
71. Xiao Z, Lv C, Gao C, Qin J, Ma C, et al. (2010) A novel whole-cell biocatalyst with NAD+regeneration for production of chiral chemicals. PLoS ONE 5: e8860.
72. Yuryev R, Liese A (2010) Biocatalysis: the outcast. ChemCatChem 2: 103–107.
73. de Carvalho CCCR (2011) Enzymatic and whole cell catalysis: finding new strategies for old processes. Biotechnol Adv 29: 75–83.
74. Matsuyama A, Yamamoto H, Kobayashi Y (2002) Practical application of recombinant whole-cell biocatalysts for the manufacturing of pharmaceutical intermediates such as chiral alcohols. Org Process Res Dev 6: 558–561.
75. Gröger H, Rollmann C, Chamouleau F, Sebastien I, May O, et al. (2007) Enantioselective reduction of 4-fluoroacetophenone at high substrate concentration using a tailor-made recombinant whole-cell catalyst. Adv Synth Catal 349: 709–712.
76. Domínguez de María P, Stillger T, Pohl M, Kiesel M, Liese A, et al. (2008) Enantioselective C-C bond ligation using recombinant Escherichia coli-whole-cell biocatalysts. Adv Synth Catal 350: 165–173.
77. Ema T, Ide S, Okita N, Sakaia T (2008) Highly efficient chemoenzymatic synthesis of methyl (*R*)-O-chloromandelate, a key intermediate for clopidogrel, via asymmetric reduction with recombinant Escherichia coli. Adv Synth Catal 350: 2039–2044.
78. Časar Z (2008) Straightforward and efficient synthesis of (4R,6S)-4-(tert-butyldimethylsiloxy)-6-(hydroxymethyl)tetrahydropyran-2-one. Synlett: 2036–2040.
79. Časar Z, Tramšek M, Goršek A (2010) Calorimetric insight into coupling between functionalized primary alkyl halide and vinylic organocuprate reagent: experimental determination of reaction enthalpies in the synthesis of (R)-ethyl 3-(tert-butyldimethylsilyloxy)hex-5-enoate – a key lactonized statin side chain precursor. Acta Chim Slov 57: 66–76.
80. Troiani V, Cluzeau J, Časar Z (2011) Application of chemoselective pancreatin powder-catalyzed deacetylation reaction in the synthesis of key statin side chain intermediate (4R,6S)-4-(tert-butyldimethylsilyloxy)-6-(hydroxymethyl)tetrahydropyran-2-one. Org Process Res Dev 15: 622–630.
81. Časar Z, Košmrlj J (2009) The first convenient entry to δ-formyl-δ-valerolactone precursor for the synthesis of statins via lactonized side chain. Synlett: 1144–1148.
82. Alcaide B, Almendros P (2008) Organocatalytic reactions with acetaldehyde. Angew Chem Int Ed 47: 4632–4634.
83. Udovič M, Tramšek M, Plantan I, Cluzeau J (2011) Synthesis of acetoxyacetaldehyde. (Lek Pharmaceuticals d.d.) PCT Int Appl WO 2011064249 A1 Chem Abstr (2011) 154: 617792.
84. Cluzeau J, Časar Z, Mrak P, Ošlaj M, Kopitar G (2009) ((2S,4R)-4,6-Dihydroxytetrahydro-2H-pyran-2-yl)methyl carboxylate and process for the production thereof. (Lek Pharmaceuticals d.d.) PCT Int Appl WO 2009092702 A2 Chem Abstr (2009) 151: 218948.
85. DASGIP Technology (2002) DASGIP Application Note on High Density E. coli Fermentation. (DASGIP Information and Process Technology GmbH, an Eppendorf Company) Available: http://www.dasgip.com/media/content/pages/downloads/Application_DASGIP_E.Coli_2002.pdf Accesed 2012 November 15.
86. Shiloach J, Fass R (2005) Growing E. coli to high cell density–A historical perspective on method development Biotech Adv 23: 345–357.
87. Shojaosadati SA, Kolaei SMV, Babaeipour V, Farnoud AM (2008) Recent advances in high cell density cultivation for production of recombinant protein. Iran J Biotech 6: 63–84.
88. Sharma A, Schulman SG (1999) Introduction to Fluorescence Spectroscopy. New York: Wiley Interscience. 174 p.
89. Craig EA, Eisenman HC, Hundley HA (2003) Ribosome-tethered molecular chaperones: the first line of defense against protein misfolding? Curr Opin Microbiol 6: 157–162.
90. Tyedmers J, Mogk A, Bukau B (2010) Cellular strategies for controlling protein aggregation. Nat. Rev. Mol. Cell Bio.11: 777–788.
91. Chiang SM, Schellhorn HE (2012) Regulators of oxidative stress response genes in Escherichia coli and their functional conservation in bacteria. Arch Biochem Biophys 525: 161–169.
92. Guisbert E, Yura T, Rhodius VA, Gross CA (2008) Convergence of molecular, modeling, and systems approaches for an understanding of the Escherichia coli heat shock response. Microbiol Mol Biol Rev 72: 545–554.
93. Gutsulyak DV, Nikonov GI (2012) Chemoselective Ruthenium-Catalyzed Reduction of Acid Chlorides to Aldehydes with Dimethylphenylsilane. Adv Synth Catal 354: 607–611.
94. Chuang C-Y, Vassar VC, Ma Z, Geney R, Ojima I (2002) Electronic effects on the regio- and enantioselectivity of the asymmetric aminohydroxylation of O-Substituted 4-hydroxy-2-butenoates. Chirality 14: 151–162.
95. Crestia D, Guerard C, Bolte J, Demuynck C (2001) Rabbit muscle aldolase (RAMA) as a catalyst in a new approach for the synthesis of 3-deoxy-D-manno-2-octulosonic acid and analogues. J Mol Cat B: Enz 11: 207–212.
96. Sakuraba H, Yoneda K, Yoshihara K, Satoh K, Kawakami R, et al. (2007) Sequential aldol condensation catalyzed by hyperthermophilic 2-deoxy-D-ribose-5-phosphate aldolase. Appl Environ Microbiol 73: 7427–7434.
97. Schmidt RR, Zimmermann P (1986) Synthesis of D-erythro-sphingosines. Tetrahedron Lett 27: 481–484.
98. Wild R, Schmidt RR (1994) Sphingosine and phytosphingosine from D-threose: Synthesis of a 4-keto-ceramide. Tetrahedron: Asymmetry 5: 2195–2208.
99. lamparuthi E, Ramesh E, Raghunathan R (2012) $InCl_3$ as an efficient catalyst for cyclotrimerization of aldehydes. Synthesis of 1,3,5-trioxane under solvent-free conditions. Synth Commun 35: 2801–2804.
100. Park CP, Lee JH, Yoo KS, Jung KW (2010) Efficient Diacetoxylation of Alkenes via Pd(II)/Pd(IV) Process with Peracetic Acid and Acetic Anhydride. Org Lett 12: 2450–2452.
101. O'Connor CJ, Barton RH (1998) Acyl transfer isomerization of Glycerol 1,2-Dibutyrate and Propane-1,2-diol 1-Butyrate. Aust J Chem 51: 455–459.
102. Steinfeld JI, Francisco JS, Hase WL (1989) Chemical Kinetics and Dynamics. Englewood Cliffs: Prentice Hall. 548 p.
103. Pravdic N, Fletcher HG (1971) The oxidation of 2-acetamido-2-deoxyaldoses with aqueous bromine. Two diastereoisomeric 2-acetamido-2,3-dideoxyhex-2-enono-1,4-lactones from 2-acetamido-2-deoxy-D-glucose, -mannose, and -gallactose. Carbohydr Res 19: 339–352.
104. Lee T-J (1985) An expeditious chiral route to analogs of mevinolin and compactin. Tetrahedron Lett 26: 4995–4996.
105. Clive DLJ, Murthy KSK, Wee AGH, Prasad JS, da Silva GVJ, et al. (1988) Total synthesis of both (+)-compactin and (+)-mevinolin. A general strategy based on the use of a special $TiCl_3/C_8K$ mixture for dicarbonyl coupling. J Am Chem Soc 110: 6914–6916.
106. Sharma M, Bernacki RJ, Paul B, Korytnyk W (1990) Fluorinated carbohydrates as potential plasma membrane modifiers. Synthesis of 4- and 6-fluoro derivatives of 2-acetamido-2-deoxy-D-hexopyranoses. Carbohydr Res 198: 205–221.
107. Rajagopal S, Vancheesan S, Rajaram J, Kuriacose JC (1992) $RuCl_2(PPh_3)_3$-catalyzed transfer hydrogenation of D-glucose. J Mol Cat 75: 199–208.
108. Mori Y, Suzuki M (1990) Synthetic study on 1,3-polyols. An efficient enantioselective synthesis of tarchonanthuslactone. J Chem Soc Perkin Trans 1: 1809–1812.

Fermentation Products of Solvent Tolerant Marine Bacterium *Moraxella* spp. MB1 and Its Biotechnological Applications in Salicylic Acid Bioconversion

Solimabi Wahidullah*, Deepak N. Naik, Prabha Devi

Bioorganic Chemistry Lab, Chemical Oceanography Division, CSIR- National Institute of Oceanography, Dona Paula, Goa, India

Abstract

As part of a proactive approach to environmental protection, emerging issues with potential impact on the environment is the subject of ongoing investigation. One emerging area of environmental research concerns pharmaceuticals like salicylic acid, which is the main metabolite of various analgesics including aspirin. It is a common component of sewage effluent and also an intermediate in the degradation pathway of various aromatic compounds which are introduced in the marine environment as pollutants. In this study, biotransformation products of salicylic acid by seaweed, *Bryopsis plumosa,* associated marine bacterium, *Moraxella* spp. MB1, have been investigated. Phenol, conjugates of phenol and hydroxy cinnamic acid derivatives (coumaroyl, caffeoyl, feruloyl and trihydroxy cinnamyl) with salicylic acid (3–8) were identified as the bioconversion products by electrospray ionization mass spectrometry. These results show that the microorganism do not degrade phenolic acid but catalyses oxygen dependent transformations without ring cleavage. The degradation of salicylic acid is known to proceed either via gentisic acid pathway or catechol pathway but this is the first report of biotransformation of salicylic acid into cinnamates, without ring cleavage. Besides cinnamic acid derivatives (9–12), metabolites produced by the bacterium include antimicrobial indole (13) and β-carbolines, norharman (14), harman (15) and methyl derivative (16), which are beneficial to the host and the environment.

Editor: Stefan Strack, University of Iowa, United States of America

Funding: This work was supported by NIO-CSIR. The funders had no role in study design, data collection and analysis, decision to publish or preparation of the manuscript.

Competing Interests: The authors have declared that no competing interests exist.

* E-mail: solima@nio.org

Introduction

Salicylic acid (SA) is a key intermediate in the catabolism of PAHs, naphthalene, naphtaquinone, phenanthrene and fluorene [1,2]. It is widely produced by plants and some bacterial genera such as *Pseudomonas* and *Vibrio* spp. [3]. In several bacteria salicylate is active as siderophore [3] and also plays a role in gene regulation such as the expression of antibiotic resistance [4]. SA and its derivatives, particularly acetylsalicylic acid, are commonly used as effective analgesics and are available to the public in a wide variety of formulations [5]. Pharmaceuticals are released into the environment through human excretion, agricultural run-off, and wastewater from sewage treatment plants and pharmaceutical manufacturers [6].

The degradation of SA is known to proceed either via gentisic acid pathway or catechol pathway. *Pseudomonas* spp. is known to degrade salicylate by oxidative decarboxylation to produce catechol, the key intermediate being salicylate-1-hydroxylase [7,8]. The hydroxylation of salicylic acid at the C-5 position by salicylate-5-hydroxylase to yield gentisic acid has also been observed in *Rhodococcus* [9], *Lignobacter* [10] and *Micrococcus* spp. [11]. Grund et al. [12] demonstrated the existence of two different routes within the genera *Streptomyces*. *S. olivaceiscleroticus* DSM415595, and *S. niger* DSM40302 converted SA to catechol. However, *S. umbvinus* DSM40278 converted SA to gentisic acid. Salicylate-1-hydroxylase is one of the model enzymes for flavin containing monooxygenase [13] while salicylate-5-hydroxylase which is responsible for the formation of gentisic acid intermediate which requires unusual co-factors CoA and ATP [14]. There is yet another group who opine that *Moraxella* spp. strain VG45 isolated from oil field samples degraded SA via salicylate-5-hydroxylase, gentisate 1,2 dioxygenase and then by a glutathione independent maleyl pyruvate hydrolase [15]. In the current study SA is not degraded by *Moraxella* spp. strain MB1 instead SA undergoes esterification with phenolic acids, metabolites produced by the strain MB1.

The main objective of the present investigation is to study the role of the marine bacterium *Moraxella* MB1 in the bioconversion of salicylic acid in the marine environment, where it is released as pollutant. In our earlier communication [16], we have reported decarboxylation of a nephrotoxin, citrinin, into decarboxycitrinin by the seaweed (*Bryopsis plumosa*) associated marine bacterium *Moraxella* spp. MB1. In the present study, metabolism of SA by the same bacterium has been studied and the extracellular metabolites identified using electrospray ionization (ESI-MS) and tandem mass spectrometry (ESI-MS/MS). Phenol and conjugates of phenol and hydroxy cinnamic acid derivatives (coumaroyl, caffeoyl, feruloyl and trihydroxy cinnamyl) with SA (3–8) were identified as the products of bioconversion. In order to obtain a better understanding of the metabolism of SA, metabolites produced by the bacterium have also been analysed. Besides cinnamic acid derivatives (9–12) the bacterium was found to produce adenine, guanine, and antimicrobial indole (13) and β-carbolines (14–16). Based on the present findings, probable metabolic pathways have been postulated for SA bioconversion products by *Moraxella* spp. strain MB1.

Results

Identification of Fermentation Products of *Moraxella* spp. MB1 with SA (in Biphasic Medium)

NMR analysis. Examination of the proton NMR spectra of the products in the control flask (without culture, Figure 1A) and the bio- transformed products from the experimental flask (inoculated with *Moraxella* spp. MB1, Figure 1B) indicated that the sharp singlet due to carboxyl proton of salicylic acid at δ 10.37 (in control), is replaced by a broad hump in the spectrum of experimental product. The broad signal probably results from the carboxyl protons in different environment in different molecules and as the concentration of these molecules is low, the signal is broad. An additional broad signal at δ 5.37 evident in the spectrum of transformed product (Figure 1B) was attributed to the olefenic protons of the acrylic moiety present in CA derivatives.

ESI-MS analysis. ESI-MS is a very sensitive technique able to identify individual metabolites present in complex mixtures in picogram to fentogram range without prior purification. It provides information about the molecular mass and fragmentation pattern of the analyte. The mass spectra of phenolic acids were compared with those of the standards CA derivatives. Lacking commercial standards, p-coumaroyl, feruloyl, caffeoyl and trihydroxy-cinnamoyl conjugates were assigned by their parent ion and their spectroscopic fragmentation observed in both negative (−) and positive (+) ionization modes (Figure S1: A, C, Figure S2: B–F in File S1). Identification of compounds based on MS^2 and MS^3 analysis and comparison with literature, is discussed below and summarized in Table-1. As evident, positive ESI-MS (Figure 2C) is more sensitive technique than negative ESI-MS (Figure 3A, 3B) in detecting the compounds under the conditions used for analysis (Table-1). Among the identified compounds only three were detected in negative ESI-MS using methanol as solvent (Figure 3A), but, when the negative ESI-MS was recorded in methanol containing 0.1% formic acid it was much richer in deprotonated molecules, exhibiting 10 peaks (Figure 3B). In contrast, all the compounds were detected as protonated $[M+H]^+$ molecular ions or as $[M-H_2O]^+$ ions in positive ESI-MS (Figure 2C).

Identification of Metabolites Under Negative ESI-MS Conditions

As mentioned earlier, negative ESI-MS profile of bioconversion products (Figure 3A) displayed only three signals of which the signal at m/z 137 (Figure S1B in File S1) was due to the unreacted SA (1) and the remaining two anions were seen at m/z 93 and m/z 283. The ion at m/z 93, due to phenol (2) (Figure S1C in File S1),

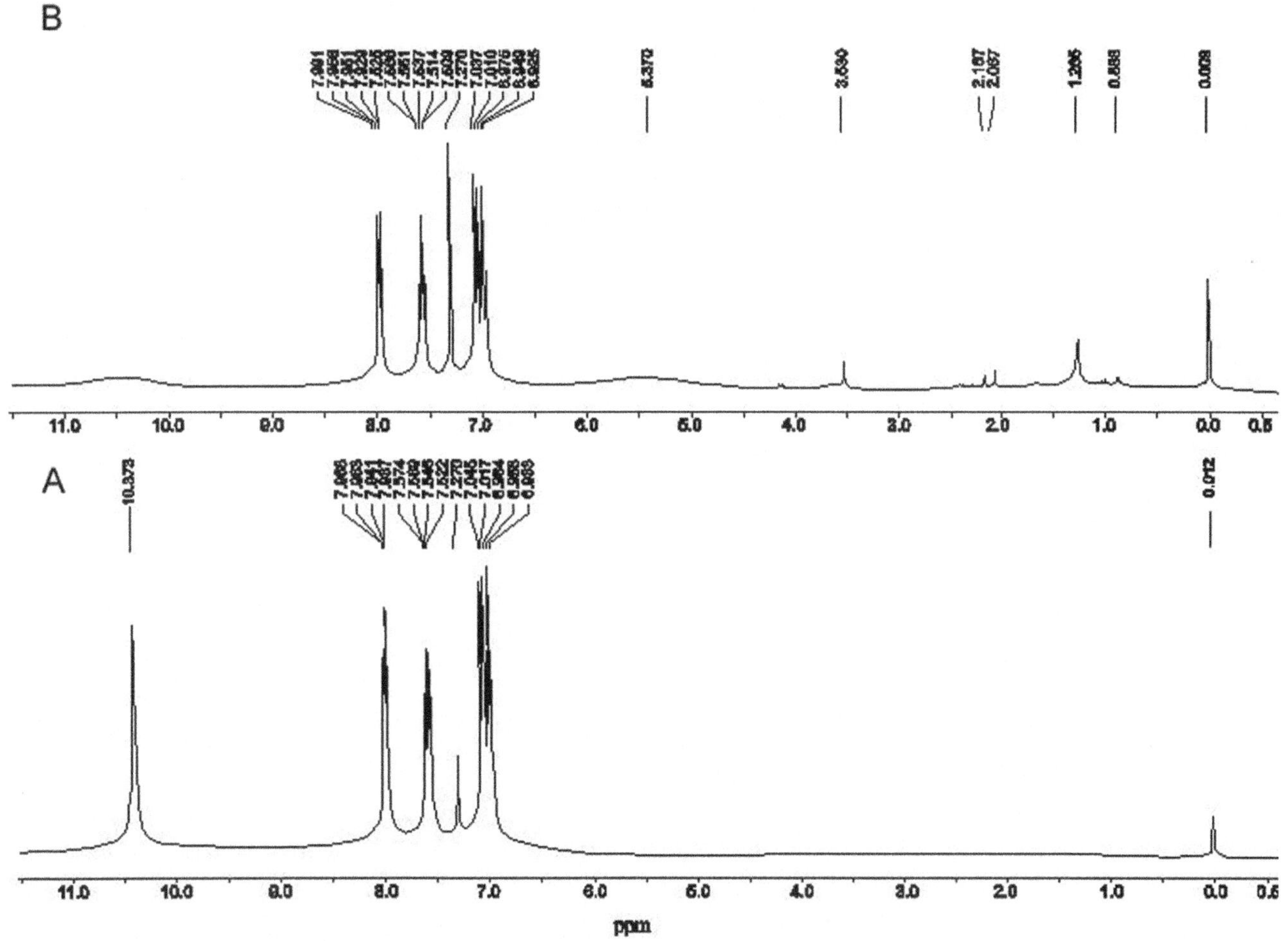

Figure 1. ^{1}H NMR spectra of control salicylic acid (A) and its bioconversion products (B).

Table 1. ESI-MS/MS characterization of metabolites in Ethyl Acetate extract of fermentation of *Moraxella* sp. MB1in biphasic medium containing SA: A-conjugates of SA with phenol and hydroxy cinnamic acids; B- Hydroxycinnamic acids derivatives produced by the bacterium; C-Standard Phenolic acids.

Compound identification	**MW**	**MS (m/z)**	**MS^2 fragmentation observed (m/z)**
A: Conjugates			
SA (1)	138	137$[M-H]^-$	93$[Phenolate]^-$
Phenol (2)	94	93[M−H]	75$[Phenyl]^-$
p-CoBA (3)	284	283$[M-H]^-$	163 $[Coumaratel]^-$
P-tri-OHCi (4)	272	273$[M+H]^+$	179 [Trihydroxy cinnamoyl]$^+$; 161 [Trihydroxy cinnamoyl – $H_2O]^+$; 143 [Trihydroxy cinnamoyl - $2H_2O]^+$; 197[Trihydroxy cinnamic acid +H]$^+$; 237 $[M+H-2H_2O]^+$; 219$[M+H-3H_2O]^+$
2-CBA (5)	300	301$[M+H]^+$	283$[M+H-H_2O]^+$; 257$[M+H-CO_2]^+$; 239$[M+H-(H_2O+CO_2)]^+$163[Caffeoyl]$^+$; 181[Caffeic acid+H]$^+$
2-FBA (6)	314	315$[M+H]^+$	177[Feruloyl]$^+$; 195[Ferulic acid+H]$^+$; 133[Ferulic acid+H$-H_2O-CO_2]^+$; 297$[M+H-H_2O]^+$; 253$[M+H-H_2O-CO2]^+$
2-Tri-OH-CiBA (7)	316	317$[M+H]^+$	179[Trihydroxy cinnamoyl]$^+$; 197[Trihydroxy cinnamic acid +Hl]$^+$; 299$[M+H-H_2O]^+$; 271$[M+H-HCOOH]^+$; 219$[M+H-3H_2O-CO_2]^+$
p-CoBA MeEt (8)	298	299$[M+H]^+$	281$[M+H-H_2O]^+$; 267$[M+H-CH_3OH]^+$; 255$[M+H-CO_2]^+$; 237$[M+H-(H_2O+CO_2)]^+$; 219$[M+H-(2H_2O+CO_2)]^+$161[p-methoxy cinnamoyl]$^+$
B. Hydroxycinnamic acids derivatives			
p-Co-MeEt (9)	178	179$[M+H]^+$	161[p-methoxycinnamoyl]$^+$; 143[p-methoxycinnamoyl-$H_2O]^+$; 135$[M+H-CO_2]^+$; 133$[M+H-HCOOH]^+$;103$[M+H-(CH_3OH+CO_2)]^+$
p-MeEt of Co-MeEs (10)	192	193$[M+H]^+$	161[p-methoxycinnamoyl]$^+$; 149$[M+H-CO_2]^+$;143$[M+H-(CH_3OH+H_2O)]^+$; 133$[M+H-(HCOOCH3)]^+$;103$[M+H-(OCH_3+COOCH_3)]^+$
Methyl caffeate (11)	194	195$[M+H]^+$	181$[M+H-CH_3]^+$;177$[M+H-H_2O]^+$;163[caffeoyl]$^+$;151$[M+H-CO_2]^+$; 149$[M+H-HCOOH]^+$; 133$[M+H-(H_2O+CO_2)]^+$
p-OH-styrene (12)	120	121$[M+H]^+$	103$[M+H-H_2O]^+$; 93[Phenolate]$^+$
C. Standard hydroxy CA			
Cinnamic acid	148	149$[M+H]^+$	121$[M+H-CO]^+$; 103$[M+H-HCOOH]^+$; 77 $[C_6H_5]^+$; 65$[C_5H_5]$
p-Coumaric acid	164	165$[M+H]^+$	147[p-hydroxy cinnamoyl]$^+$; 119$[M+H-HCOOH]^+$; 77$[C_6H_5]^+$
p-OMe-CA	178	179 $[M+H]^+$	161$[M+H-H_2O]^+$;133$[M+H-HCOOH]^+$;77$[C_6H_5]^+$; 118$[M+H-(HCOOH+CH_3)]^+$;109$[M+H-SC]^+$; 103$[M+H-(CO_2+OCH_3]^+$
Caffeic acid	180	179$[M-H]^-$	135$[M-H-CO_2]^-$; 89 $[M-H-(CO_2+HCOOH]^-$
Ferulic acid	194	193$[M-H]^-$	178$[M-H-CH_3]^-$; 149[M−H−CO2]−; 134$[M-H-(CO_2+CH_3)]^-$

SA, salicylic acid; P-phenol; p-CoBA, para-coumaroylbenzoic acid; P-tri-OHCi, phenyltrihydroxycinnamate; 2-CBA, 2-caffeoyl benzoic acid; 2-FBA, 2- feruloylbenzoic acid; 2-Tri-OH-CiBA, 2-trihydroycinnamoylbenzoic acid; *p*-CoBA MeEt, *para* -coumaroylbenzoic acid methyl ether; *p*-Co-MeEt, *para*-coumaric acid methyl ether; *p*-MeEt of Co-MeEs, *para*-methy ether of methylcoumarate; *p*-OMe-CA, *para*-methoxycinnamic acid.

originated by decarboxylation of SA either by the bacterium [16] or by collision induced dissociation of unreacted SA. Negative ESI-MS/MS spectra of carboxylic acids (Table-1) are known to eliminate carbon dioxide (CO_2) leading to the formation of $[M-H]^-$ and $[(M-H)-CO_2]$ anions [17,18]. Conjugate of phenol with trihydroxycinnamic acid (4) has also been identified as one of the bioconversion products of SA indicative of decarboxylation of SA to phenol.

The deprotonated $[M-H]^-$ ion at m/z 283, gave single intense fragment at m/z 163 in its product ion mass spectrum (Figure S1A in File S1, Table-1) which suggested that it is an esterified form of coumaric acid [17] with SA. The diagnostic fragmentation pattern of coumaric acid conjugate involved cleavage of molecule into intact p-hydroxy cinnamyl and benzoic acid fragments. Therefore, the product is a conjugate of SA with coumaric acid, probably 2-(p- hydroxycinnamoyl) benzoic acid (3) and not 2-(hydroxycinnamoyl) salicylate because the expected fragment at m/z 239 [17,18] due to elimination of CO_2 was not observed. Probably intramolecular hydrogen bonding between the ortho positioned hydroxyl and carboxylic acid group affects the decarboxylation [19]. Further, in the positive ionization mode, all the biotransformed compounds identified are ortho-CA esters of benzoic acid. These observations led to propose structure (3) for the molecule with deprotonated $[M-H]^-$ ion at m/z 283. Conjugate (3) appears as $[M+H-H_2O]$ at m/z 267 in positive full mass spectrum of the ethyl acetate extract and as methyl ether at m/z 299 (Figure 2C).

Identification of Metabolites Under Positive ESI-MS Conditions

Positive ESI-MS profile (Figure 2C) besides protonated $[M+H]^+$ ions, also displayed $[M+H-H_2O]^+/[M+H-2H_2O]^+$ peaks due to keto (site of protonation) and hyroxyl groups in the benzene ring (loss of water). In addition, it also showed clear fragmentation patterns and the occurrence of common ion fragments (Figure 2C). Product ion spectra were recorded in order to identify classes of compounds and to obtain structural information of the compounds. However, from the intensity of the signal it cannot be determined if the greater intensity is attributable to higher concentration of the compound in the fermentation mixture or greater ionization efficiency of compounds. It should be noted that

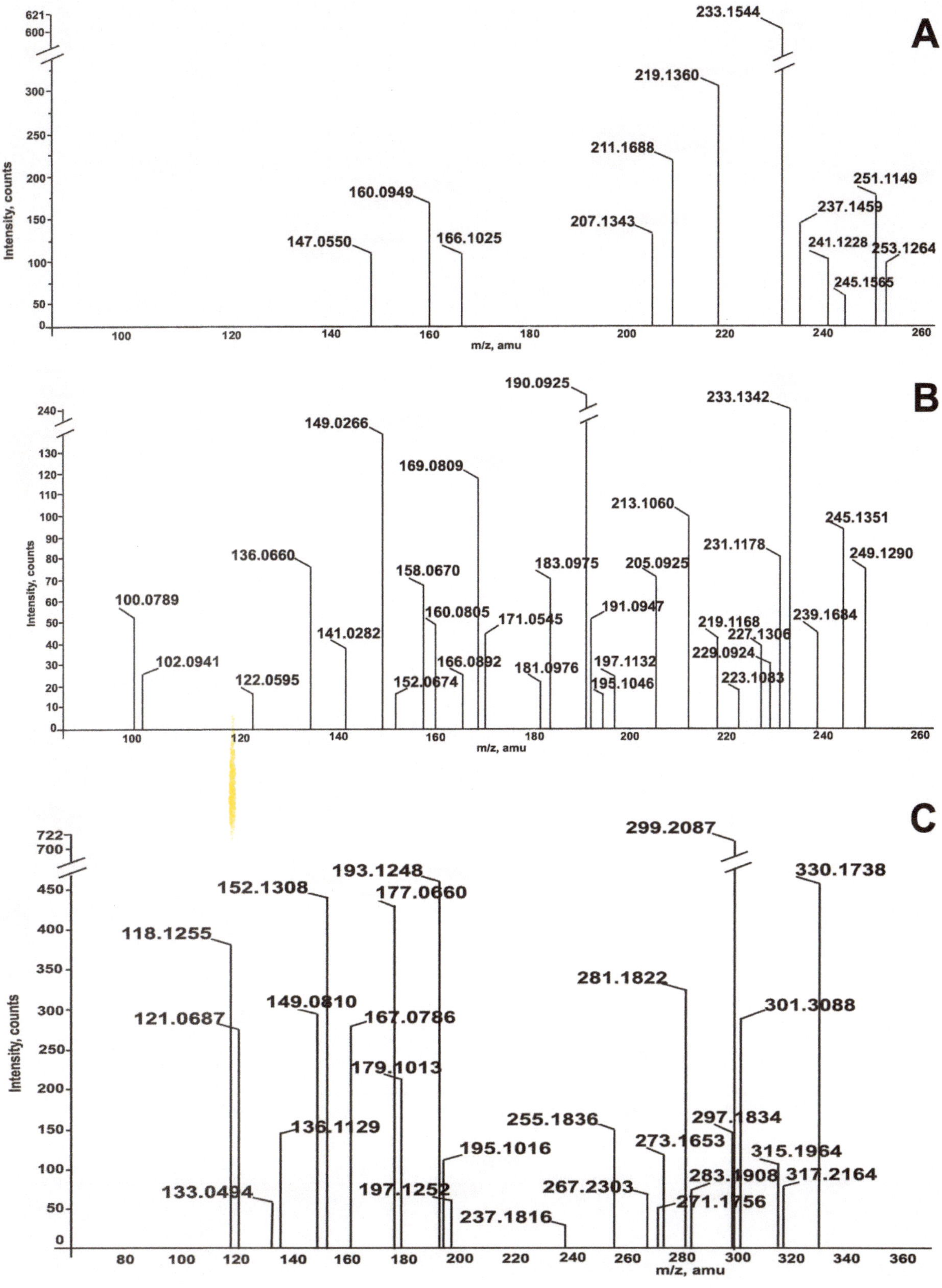

Figure 2. Negative ESI-MS profile of products of fermentation of *Moraxella* spp. MB1 in aqueous medium without salicylic acid (A). Positive ESI-MS profile of products of fermentation of *Moraxella* spp. MB1 without salicylic acid in biphasic medium (B). Positive ESI-MS profile of bioconversion products of salicylic acid (C).

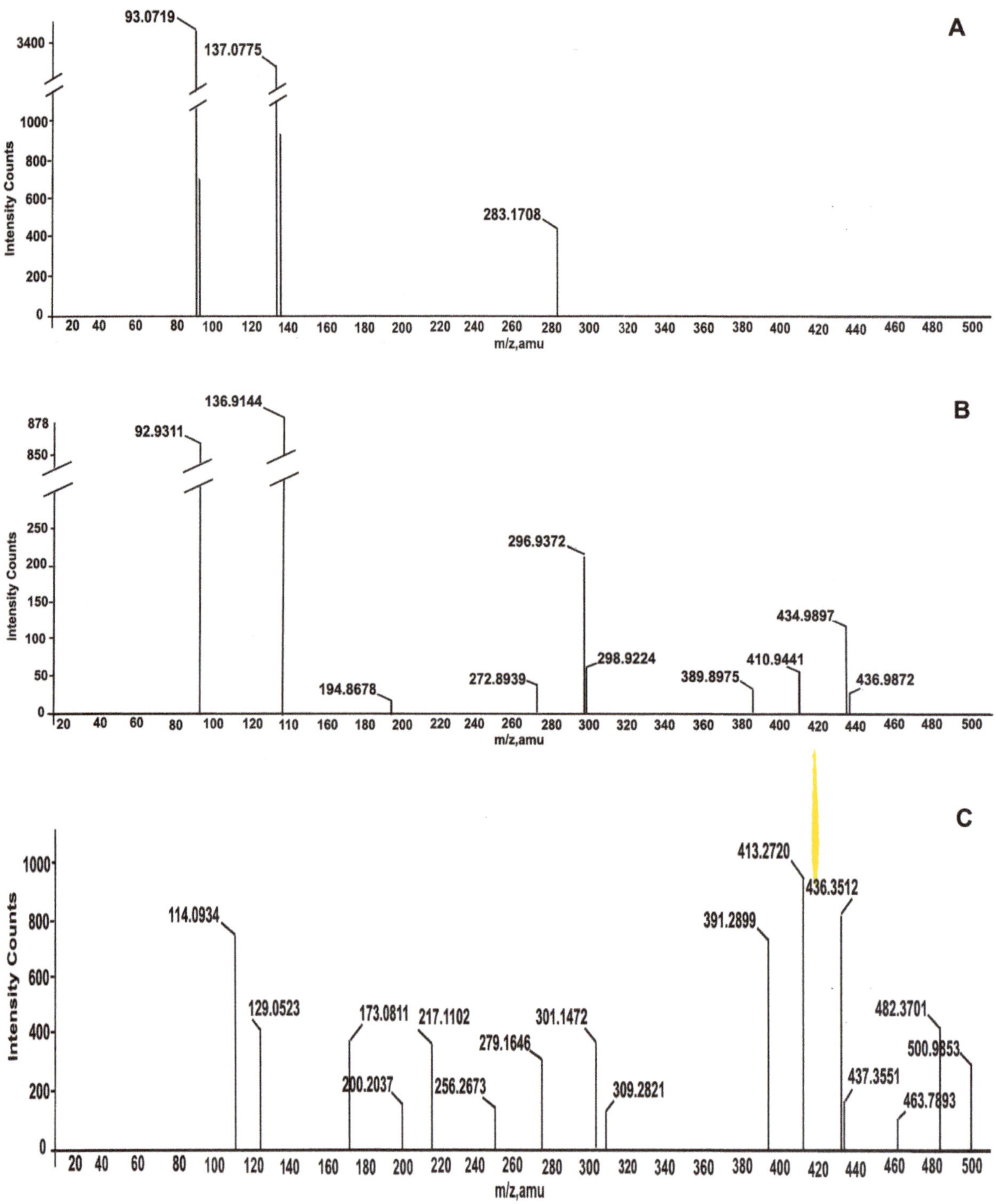

Figure 3. Negative ESI-MS profile of bioconversion products in methanol (A). Negative ESI-MS profile of bioconversion products in methanol+1% formic acid (B). Positive ESI-MS profile of ethyl acetate extract of symbiont *Bryopsis plumosa* (C).

ions observed are likely to be mixture of isomers [18]. As the experiment involves bioconversion of SA, esterification probably occurs at the ortho position. ESI-MS profile of the bioconversion mixture (Figure 2C) in the positive ionisation mode yielded several peaks, each signal corresponded to protonated $[M+H]^+$ molecular mass of individual molecular species. Each signal was subjected to product ion spectrum (MS^2). The following conjugates of phenol and hydroxyl cinnamic acid derivatives with salicylic acid were identified (Figure 4A, 3–8) based on the fragmentation observed in their tandem mass spectra (MS/MS) (Figure S2: B–F in File S1, Table-1).

Characterization of 2-caffeoyl Benzoic Acid (5)

Tandem Mass (MS^2) spectrum of the $[M+H]^+$ ion at m/z 301 (Figure S2B in File S1) produced the base fragment at m/z 283 by losing a water molecule $[M+H-H_2O]^+$ and a strong ion at m/z

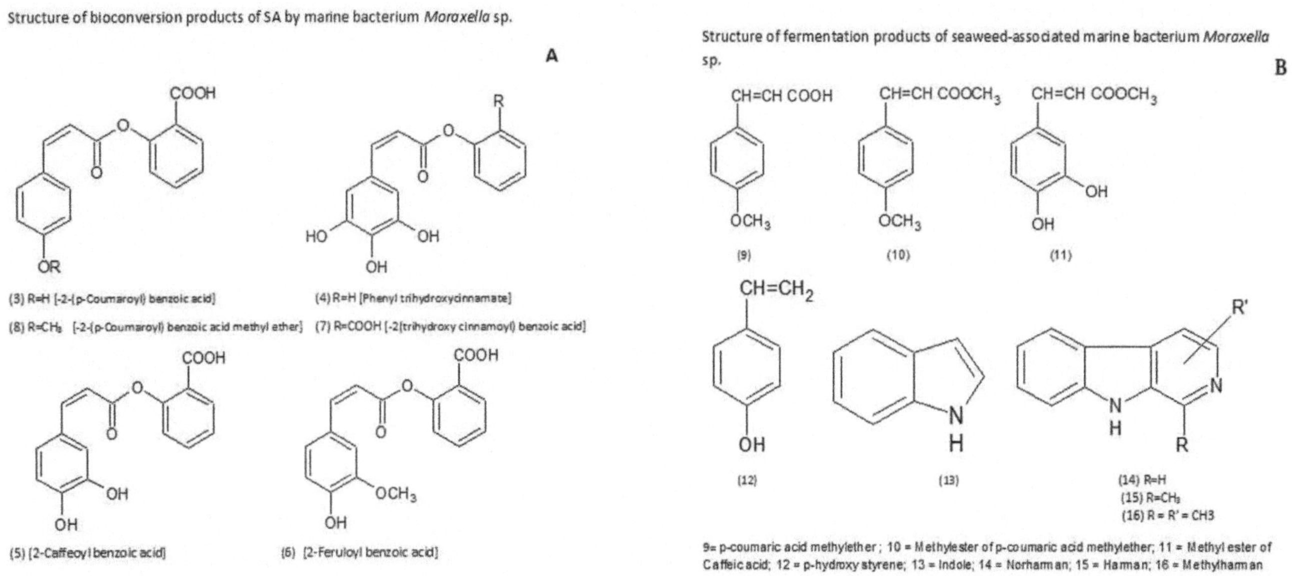

Figure 4. Structure of bioconversion products of salicylic acid by *Moraxella* spp. MB1 (A). Structure of fermentation products of seaweed associated *Moraxella* spp. MB1(B).

257 $[M+H-CO_2]^+$ with the elimination of carbon dioxide. Simultaneous elimination of these two neutral molecules resulted in an intense signal at m/z 239. Caffeoyl ion was evident at m/z 163 $[caffeoyl]^+$ whereas caffeic acid appeared as a small peak at m/z 181 $[caffeic\ acid+H]^+$. Based on foregoing data the signal at $[M+H]^+$301 was attributed to 2-caffeoyl benzoic acid (5).

Characterization of 2-feruloyl Benzoic Acid (6)

Collision induced dissociation (CID) spectrum of the molecular species with $[M+H]^+$ at m/z 315 (Figure S2C in File S1) exhibited base peak at m/z 177 due to loss of ferulic acid along with water which further loses CO_2 to yield fragment at m/z 133. The presence of ferulic acid was evident by a small peak at m/z 195. The spectrum also displayed small ions at m/z 297 and 253 due to the loss of H_2O and simultaneous elimination of H_2O and CO_2 respectively from the molecule. These data are in accordance with the structure of 2-feruloyl benzoic acid (6) for the molecule of $[M+H]^+$at m/z 315.

Characterization of Trihydroxycinnamates (4) and (7)

Targeted MS^2 experiments at m/z at 273 (Figure S2D in File S1) and 317 (Figure S2E in File S1), the $[M+H]^+$ ions of compounds 4 and 7 respectively produced a fragment at m/z 179 characteristic of trihydroxy cinnamoyl residue, and peak at m/z 197 due to the molecular ion of the corresponding acid. Elimination of a water and formic acid from compound 7 yielded the base fragment at m/z 299 $[M+H-H_2O]^+$ and a fragment at m/z 271 $[M+H-HCOOH]^+$ respectively. An abundant fragment was also observed at m/z 219 due to simultaneous loss of $3H_2O$ and a CO_2 from the parent molecule. Similarly, in the product ion spectrum of compound 4 consecutive loss of one and two H_2O molecules from the trihydroxy cinnamoyl moiety resulted in the base peak at m/z 161 and the fragment ion at m/z 143 respectively. Further, the conjugate 4 eliminates 2 and 3 water molecules to yield fragments at m/z 237 and 219 respectively. On the basis of foregoing data compound 4 was characterized as phenyl trihydroxycinnamate (4) a conjugate of trihydroxy cinnamic acid with phenol, and compound (7) as 2-(trihydroxycinnamoyl) benzoic acid (7) a conjugate of trihydroxy cinnamic acid with SA.

Characterization of Methyl Ether of 2-(p-coumaroyl) Benzoic Acid (8)

The MS^2 spectrum focused at m/z 299 yielded a fragment ion at m/z 161(Figure S2F in File S1) characteristic of p-methoxy cinnamoyl residue. This was further supported by the signals at m/z 281 and m/z 255 due to loss of water and CO_2. Simultaneous elimination of water and carbon dioxide resulted in the product ion at m/z 237. Loss of two water molecules along with carbon dioxide yielded a fragment at m/z 219. Loss of methanol gave a signal at m/z 267. This compound is not an artefact of experimental procedure adopted as ethyl acetate was used for extraction.

In addition to the signals due to conjugates and their dehydration products (m/z 283, 281, 267, 255) several other signals probably due to hydroxyl cinnamic acids (9–12) and other metabolites (13–16) produced by the strain MB1, were also observed in full scan mass spectrum (Figure 2C) in positive ion mode of the products of biotransformation (Table-1). In analogy with literature reports [20,21], the proposed fragmentation pathway involved in MS/MS spectra of these CA esters in positive as well as negative mode ionization is as depicted in (Figure 5A, 5B).

Decarboxylation of CA or its derivatives is expected to give styrene/styrene derivatives preferably when hydroxyl group is in 4-position (para). Positive ESI-MS profile of the crude ethyl acetate extract of bioconversion products indicated the presence of a compound with $[M+H]^+$ at m/z 121 whose tandem mass spectrum (Figure S2A in File S1) is well in agreement with the structure of p-hydroxy-styrene (12, Figure 4B) derived from p-coumaric acid.

Fermentation Products of *Moraxella* in Aqueous as well as Biphasic Media

The origin of hydroxyl cinnamic acid derivatives forming conjugates with SA was determined by cultivating *Moraxella* spp.,

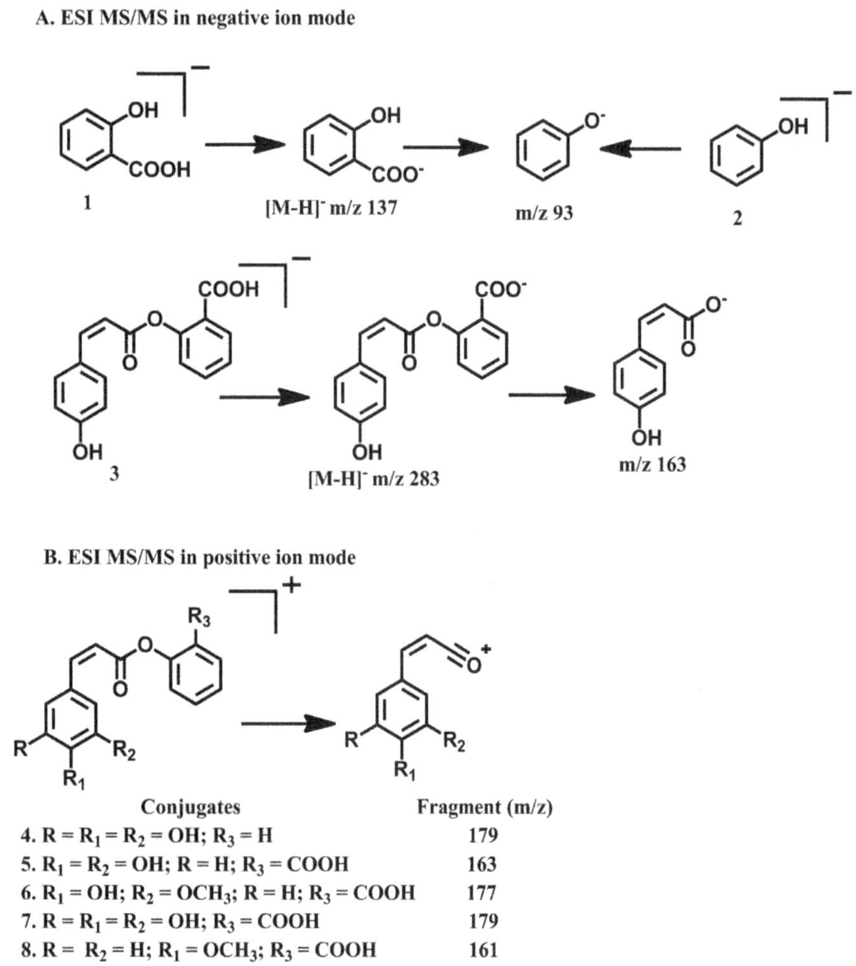

Figure 5. Proposed diagnostic mass spectral fragmentation of hydroxyl cinnamic acid esters. ESI-MS/MS in positive ionization mode (A); ESI-MS/MS in negative ionization mode (B).

strain MB1 under identical conditions but without the addition of SA in aqueous as well as biphasic medium. The metabolites were identified by ESI-MS (Figure 2A, 2B) and by TLC comparison with standards (Figure S3 in File S1). Analysis revealed that CA and its derivatives are in fact produced by the bacterium under aqueous as well as biphasic conditions as evident from the ESI-MS profile of the fermentation products of the bacterium (Figure 2A–C) which show signals for the presence of these compounds (MS^2 spectra have not been included here but the fragmentation observed are presented in Table-1). ESI-MS profile and MS^2 analysis (Figure S4 in File S1) was also indicative of the strain producing adenine (m/z 136), guanine (m/z 152), indole (m/z 118), β-carbolines (m/z 169, m/z 183 and m/z 197). The identification of these metabolites is based on the comparison of fragmentation observed in their tandem mass spectra with the literature reports [22,23].

Influence of Phenylalanine Addition to the Medium

Amino acid phenylalanine is a precursor of CA and it was found that there was enhancement in the production of phenolic acids on the addition of the amino acid to the cultivating media. This was evident from TLC analysis (Figure S3 in File S1).

ESI-MS Analysis of Ethyl Acetate Extract of *Bryopsis plumosa*

To rule out the possibility of CA derivatives arising out of symbiotic association with the alga, *Bryopsis plumosa*, ESI-MS profile (Figure 3C) (Supriya Tilvi, Dissertation, 2005) of the ethyl acetate extract of the seaweed was examined. No signal for the protonated CA or hydroxyl-cinnamic acid derivatives was observed.

It is evident from the results of the present investigation that condensation reaction with elimination of water has taken place between salicylic acid and the fermentation products of marine bacterium, *Moraxella* MB1, hydroxyl cinnamic acid derivatives. There are several reports of similar condensation reactions leading to the production of bio-based polymers of commercial importance from microorganisms by fermentation. Typical examples are the family of polyesters, polyhydroxyalkanoates (PHA) that can be produced by various strains of bacteria, and the simplest one polyhydroxybutyrate was discovered as a constituent of bacterium *Bacillus megaterium* [24]. More than 150 PHA monomers have been identified as the constituents of PHAs [25]. Such diversity allows production of polymers with a wide range of properties tailored for specific applications. Pullulan is a linear water soluble polysaccharide mainly consisting of maltotriose units connected by α-1, 6 glycosidic units. It was first obtained from the fermentation broth of *Aureoasidium pullulans*. This polysaccharide is used as food additive, flocculant, blood plasma substitute, and as adhesives [26]. Alginate, a 1,4 linked linear polysaccharide of β-*d*-mannuronic and α–*l*-guluronic acid, well known as product of brown seaweeds is also biosynthesised by some bacteria, mostly derived from the genus *Pseudomonas* and belonging to the RNA superfamily I, and a soil bacterium *Azotobacter vinelandii* [27]. It is widely used as a gelling agent in pharmaceutical and food applications. Besides bio-based polymers of microbial origin, other well known natural polymers include proteins, nucleic acids, and polysaccharides (starch, cellulose, chitin and chitosan).

Generally, chemical condensation/polymerization between the two reactants involving removal of water molecule is a reversible reaction if not performed under anhydrous conditions. In nature, formation of bio-based polymers, in biological systems, are examples of *in vivo* enzymatic condensation reaction which are not affected by the presence of water. This may be due to the fact that water is indispensable for the change in chemical transformation that made life possible.

This study may be of direct application in preparative organic synthesis particularly condensation reactions involving removal of water. Enzymatic syntheses are preferred to chemical syntheses as the latter involves use of chemicals as well as formation of by products which are harmful as pollutants and needs rigorous treatment prior to disposal.

Discussion

Bacteria belonging to genus *Moraxella* have been widely used for biotransformation. Gregg et al. [28] reports conversion of halo acetates to hydroxyl acetates catalyzed by fluoroacetate dehalogenase from *Moraxella* spp. B, while Spain et al. [29] described enzymatic substitution of nitro group by a hydroxyl group in p-nitrophenol degradation by a *Moraxella* strain. Yet another *Moraxella*, strain OA3, is known to degrade homogentisate, wherein, it is cleaved by a 1,2-dioxygenase yielding maleylacetoacetate [30]. There is a solitary report in the literature [15] on salicylate degradation via gentisate pathway by *Moraxella* strain VG45 isolated from oil field samples. *Moraxella* spp. MB1, strain of the present investigation, was reported earlier to decarboxylate nephrotoxin citrinin.

Results of the present investigation show that salicylic acid is not degraded by the seaweed associated marine bacterium *Moraxella* strain MB1 but undergoes esterification with hydroxycinnamic acid derivatives leading to the formation of conjugates of phenolic acids with salicylic acid (Figure 4A,Table-1). Hydroxycinnamic acids needed for condensation reaction are metabolites produced by *Moraxella* MB1 and do not arise from the symbiotic association of the strain with the seaweed, *Bryopsis plumosa* (Figure 3C). Though *Moraxella* spp. MB1 was reported earlier to decarboxylate nephrotoxin citrinin, in the present investigation it has been observed that in practically all the conjugates, except 4, the carboxyl group of salicylic acid moiety is intact probably because of intramolecular hydrogen bonding affects the decarboxylation. In decarboxylation of benzoic acid and CA and their hydroxyl derivatives by *Aerobacter aerogens* B-2614, it has been reported that enzyme activity requires a relatively unhindered 4-hydroxyl group on the aromatic ring and no decarboxylation is observed when the hydroxyl group is in ortho or para position when the reaction was carried out for 30 minutes [19]. The conjugates in the present study are not further metabolized as evident by the absence of signals in positive as well as negative ESI-MS profile of the bioconversion mixture (Figure 2C, Figures 3A and 3B) due to fission products of aromatic ring like 3-keto adipic acid (m/z 161), pyruvic acid (m/z 89), maleic/fumaric acids (m/z 117), succinic acid (m/z 119) and acetaldehyde (m/z 45). These products are expected when SA is degraded either via gentisic acid pathway or catechol pathway by scission of the benzene nucleus of catechol by 1,2/2,3 dioxygenase or of gentisate by gentisate 1,2 dioxygenase [15,31,32].

Proposed Metabolic Pathway of Salicylic Acid Bioconversion Products

The degradation of aromatic compounds generally proceeds in two stages: ring substitution that results in the formation of dihydroxylated benzene derivatives (catechol, protocatechuate, gentisate, homogentisate and homoprotocatechuate) [33] followed by the degradation of the dihydroxylated benzene derivatives with ring fission and subsequent reactions linked to central metabolism in the cell. Ring fission is catalysed by ring cleavage dioxygenase that use molecular oxygen to open the aromatic ring between the two hydroxyl groups (ortho cleavage, catalysed by intra diol dioxygenases) [34] or proximal to one of the two hydroxyl groups (meta cleavage, catalysed by extradiol dioxygenases) [35].

In analogy with literature [36] and the bioconversion metabolites identified, pathways involved in the metabolism of salicylic acid by marine bacterium *Moraxella* spp. MB1 have been envisaged as shown in Scheme-1(Figure 6). However, the various isoforms of the conjugates were not investigated in this communication because of the limitations of the technique.

Presumably, the bacterium synthesizes CA from phenylalanine by deamination, catalyzed by phenyl ammonia -lyase (PAL). PAL homologs are found in various plants, fungi, and yeasts but only in three prokaryotes *Streptomyces maritimus* [37], *Photorhadus luminescens* [38] and actinomycete *Saccharothrix espanaensis* [39]. Beginning with CA, a series of hydroxylation and methylation reactions catalyzed by enzymes hydoxylase and O-methyltransferase leads to the sequential synthesis of the common hydroxycinnamates [40]. Thus, hydroxylation of CA gives p-coumaric acid which on the introduction of second hydroxyl group gives caffeic acid and introduction of yet another hydroxyl group yields trihydroxycinnamic acid. Methylation of caffeic acid leads to the formation of ferulic acid while methylation of coumaric acid gives methyl ether of coumaric acid. These hydroxycinnamates are finally activated as hydroxycinnamate-CoA forms catalysed by hydroxycinnamate-CoA ligases. The activation of phenolic compounds as hydroxycinnamate-CoAs serve as building block molecules for further downstream modification which includes condensation and conjugation resulting in the biotransformation products of salicylic acid. This remains to be elucidated.

CA and its derivatives are secondary metabolites with antioxidants and antibacterial properties produced by plants in response to stressful conditions such as infections or wounding [41]. Fungi and yeast are also known to produce phenolic acids but as mentioned earlier PAL enzyme activity is rare in bacteria. CA possesses not only anti-tuberculosis properties [42] but is also known to inhibit the growth of several bacteria [43] in particular *Escherichia coli* O157:H7 [44]. It is also an active compound in medicinal plants with anti- *Helicobacter pylori* [45]. The bacterium of the present investigation besides CA derivatives, that forms conjugates with salicylic acid, also produces antimicrobial indole and β-carbolines, norharman, harman and its methyl derivative. All these compounds seem to play a protective role towards the host organisms and the environment.

Overall, the present study provides structural characterization by ESI-MS^n of hydroxycinnamates conjugates with SA as the metabolites of bioconversion of SA by marine bacterium *Moraxella* MB1. It also gives us information about the antimicrobial indole

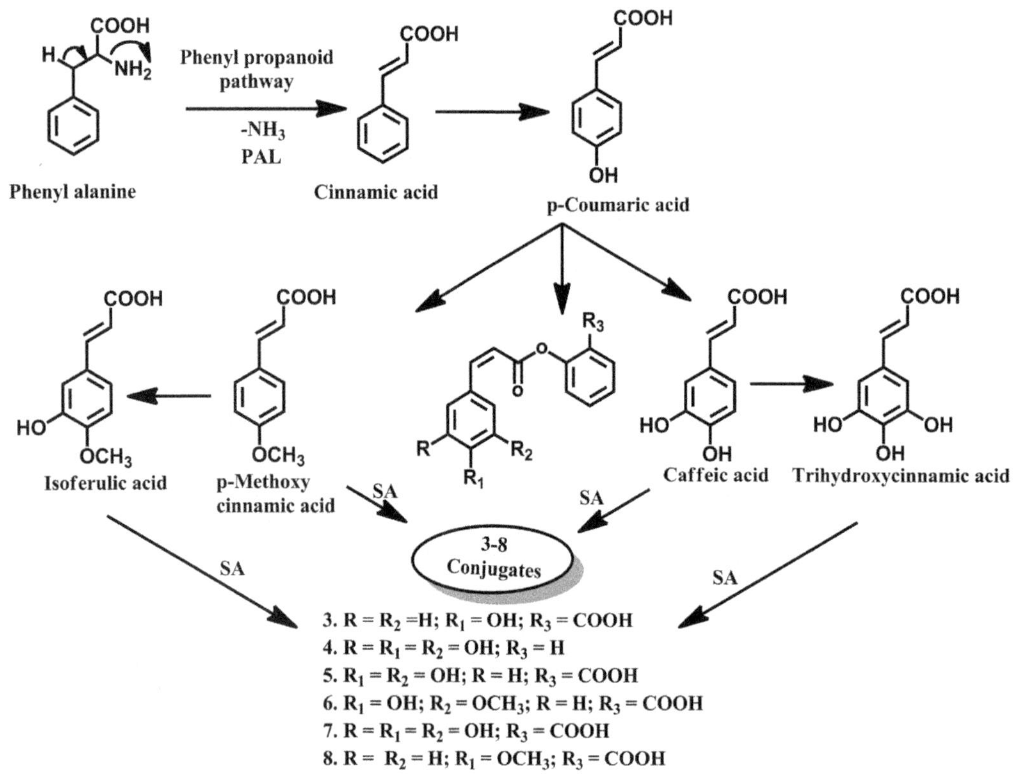

Figure 6. Scheme-1-Proposed metabolic pathway of formation of salicylic acid conjugates with hydroxycinnamic acid derivatives by *Moraxella* sp. MB1.

and β-carbolines as being the metabolites produced by the bacterium. These products are not only harmless to the environment but also play a protective role towards the host organisms. This study is of particular interest because SA makes its way into the sea, as pollutant, from different sources. Finally, we believe, this is a first report on the formation of conjugates of salicylic acid by a bacterium of genus *Moraxella*.

Materials and Methods

No animal work has been conducted. All other work performed has been carried out according to the Institutional rules and regulations. The bacterial strain used for the study was isolated as an endobiont from the seaweed *Bryopsis plumosa* collected from Malvan (16°15′N, 17°338 35′E), Maharashtra coast, India, by the procedure described earlier [16]. Being an Oceanographic Institute, we are permitted to collect marine samples other than the endangered/protected species and samples from sensitive areas. The seaweed collected for the study was a very small quantity and do not belong to the category of protected species.

Instrumentation

Nuclear Magnetic Resonance (NMR) spectra were recorded on a Bruker Avance AC 300 MHz instrument in deuterated chloroform ($CDCl_3$) containing tetramethyl silane (TMS) as the internal standard. Mass spectra (ESI-MS) were acquired, in positive (+) as well as negative (−) ionization mode, using a QTOF-XL MS/MS, Applied Biosystem equipped with the MDS Sciex Analyst Software. The following setting was used for recording mass spectra so as to obtain optimum fragmentation: nebulizer gas (N_2) 28 (arbitrary units); curtain gas (N2) 18 arbitrary units; ion spray voltage 5500 V for positive and −4500 V for negative mode, declustering potential (DP) 60.0; focusing potential (FP) 300 V; declustering potential (DP2) 15; and collision gas (CAD) 3 (arbitrary units). Full scan data acquisition was performed, scanning from m/z 0–400 in the profile mode and using a cycle time of 1 second. MS/MS were recorded at different collision energies in the range 10–40V. The identity of compounds was based on the fragmentation pattern observed and comparison with those of authentic standard cinnamic acid derivatives.

Thin Layer Chromatographic Analysis

TLC analysis were performed on precoated kiselgel $60F_{254}$ (Merck) plates developed in petroleum ether: ethyl acetate: acetic acid (75:25:1) along with the authentic standards. Spots were visualized either by spraying with 5% methanolic sulfuric acid/alcoholic ferric chloride followed by heating at 100°C or by UV (ultraviolet) illumination under a dual wavelength (254 nm/330 nm) UV lamp.

Bacterial Culture Isolation and Growth

The bacterial strain an endobiont from the seaweed *Bryopsis plumosa* collected from Malvan was maintained on nutrient agar slants which comprised of 5 g of peptic digest of animal tissue, 5 g of NaCl, 1.5 g of beef extract, 1.5 g of yeast extract and 20 g of agar dissolved in 1 litre of seawater-distilled water (1:1, v/v), having pH 7.5 and stored at 4°C. The culture was screened for solvent tolerance by the procedure described by Ogino et al. [46]. The culture, with deposition no. NIOCC/OSTB-MB1 has been deposited at the Bioorganic Chemistry Laboratory of the National Institute of Oceanography, Goa, India.

Identification of the Bacterium

Ethyl acetate solvent tolerant marine bacterium of the present investigation is a Gram negative, non-motile, cocco bacillus with cells as short rods, aerobic, oxidase and catalase positive and carbonic anhydrase negative. The culture did not produce acid from glucose and was found to be very sensitive to penicillin. On the basis of the above characteristics, the bacterium was identified as *Moraxella* spp. strain MB1. In addition, sequence analysis was carried out as described earlier [16]. Briefly, DNA was extracted from the stationary phase culture using a Bioron DNA isolation kit (Kit no. 501001). PCR amplification was performed in a total volume of 50 ml containing the appropriate reaction buffer and reagents and the universal primer 27 f (5′-GAGTTT-GATCCTGGCTCA-3′) corresponding to *Escherichia coli* 16S rDNA numbering. The PCR conditions were as follows: initial denaturation (2 min at 95°C), followed by 30 cycles of denaturation (1 min at 95°C), primer annealing (1 min at 52°C), and primer extension (1.5 min at 72°C). The PCR amplification product was purified using a Qiagen kit (Kit no. 28104). The recovered fragment was sequenced using ABI 3700 Sequencer and partial 16SrRNA sequence was established as GGATGTTAG-CGGCGGACGGGTGAGTACACGTGGGTAACCTGCCTG-TAAGACTGGGATAACTCCG. The sequence obtained was subjected to BLAST search for closest match in the database. This bacterium showed 98% homology with the deposited sequence of uncultured marine bacterium clone ISA-3133 16S rRNA gene, partial sequence (535 bp), with GenBank accession no. AY936933.

Growth of *Moraxella* spp. MB1 and Biotransformation of SA

The bacterium was grown in two flasks containing 100 ml of nutrient broth (Hi Media) at 28°C on a rotary shaker at 150 rpm whereas two uninoculated flasks with the same medium and under the same conditions served as control. Nutrient broth (Hi Media) comprised of 5 g of peptic digest of animal tissue, 5 g of NaCl, 1.5 g of beef extract, 1.5 g of yeast extract. After 8 hours, 200 mg of SA (Sigma) dissolved in 100 ml ethyl acetate was added aseptically to all the flasks and incubation continued for additional 40 hours. At the end of incubation period, the contents of the flasks were transferred to a separatory funnel, the organic layer separated from the biphasic cultivation system and the aqueous layer centrifuged (cell mass: 16 mg, dry weight) and the supernatant extracted thrice with ethyl acetate, the combined organic layer was washed with distilled water and dried over anhydrous sodium sulphate. Solvent from the extract was removed under vacuum on a rotavapor to yield a residue (43 mg/L, 21.5%) containing biotransformed products which were analysed spectrometrically using NMR (Figure 1B) and ESI-MS (Electrospray ionization mass spectra) data [Figure 2, 3; (Figure S1, S2 in File S1)].

Fermentation of *Moraxella* spp. MB1 in Aqueous and Biphasic Media in the Absence of SA

The above experiment was repeated in aqueous and biphasic media without the addition of SA. After 48 hours of fermentation the contents were centrifuged at 10,000 rpm for 5 minutes (cell mass : 28 mg/L) and the supernatant worked out as described above to obtain ethyl acetate extract (80 mg/L, 40%) which was subjected to ESI-MS (Figure 2A, 2B) analysis in methanol and TLC (Figure S3 in File S1) using standards as reference compounds.

Effect of Addition of Phenyl Alanine on the Fermentation of *Moraxella* spp

The marine bacterium, *Moraxella* MB1, was incubated at 28°C for 48 hours in nutrient broth (Hi Media) with the addition of phenylalanine (20 mg/L) a well known precursor of cinnamic acid (CA). At the end of the fermentation period the cells were harvested by centrifugation and the supernatant extracted with ethyl acetate and worked as mentioned earlier. The extract was spotted on TLC (Figure S3 in File S1) and the results (intensity of the spots) show that there is enhancement in the production of p-coumaric acid. Further evidenced by the fragmentation pattern observed using ESI-MS/MS conditions. Broth extract: 77 mg; cell mass: 26 mg).

Supporting Information

File S1 Combined Supporting Information File containing Figures S1–S4. Figure S1: Negative ESI-MS/MS spectra of deprotonated $[M-H]^-$ ions at m/z283 identified as 2-(hydroxycinnamoyl) benzoic acid (A); (m/z 137) unreacted salicylic acid; (m/z 93) phenol. **Figure S2:** Positive ESI-MS/MS spectra of protonated $[M+H]^+$ ion at m/z 121 identified as p-hydroxystyrene (A); (m/z 301) 2-(caffeoyl) benzoic acid (B); (m/z 315) 2-(feruloyl) benzoic acid (C); (m/z 273) phenyl trihydroxycinnamate (D); (m/z 317) 2-(trihydroxycinnamoyl) benzoic acid (E); m/z 299 methyl ether of p-coumaroyl benzoic acid (F). **Figure S3:** TLC of metabolites produced under different conditions. *Moraxella* sp. MB1 grown in the presence of Phenylalanine [(A): Ethyl acetate extract E A (1), Methanol extract Cell mass (1)]; Control[Ethyl acetate extract E A (2), Methanol extract Cell mass (2)]. (B) Ethyl acetate extracts of broth of *Moraxella* sp. MB1 grown in the presence of solvent EA (3) and Control EA (4). **Figure S4:** Positive ESI-MS/MS of indole (m/z 118) and β-carbolines, norharman (m/z 169), harman (m/z 183) and methyl harman (m/z 197).

Acknowledgments

The authors are grateful to Director, National Institute of Oceanography (NIO-CSIR), for constant encouragement. We wish to thank Nicholas Piramal Research Centre (NPRC) for providing 16S rRNA data and Dr Supriya Tilvi for recording of spectral data. Dr. Solimabi Wahidullah wishes to thank CSIR for the award of Emeritus Scientist. This manuscript has NIO contribution no. 5502.

Author Contributions

Conceived and designed the experiments: SW DNN PD. Performed the experiments: DNN PD. Analyzed the data: SW DNN. Contributed reagents/materials/analysis tools: SW PD. Wrote the paper: SW DNN PD.

References

1. Xue-Qin T, Lu GN, Dang Z, Yang C, Yi XY (2007) A phenanthrene degrading strain *Sphingomonas* sp. GY2B isolated from contaminated soils. Process Biochem 42: 401–408.
2. Jia Y, Yin H, Ye JS, Peng H, He BY, et al. (2008) Characteristics and pathways of naphthalene degradation by *Pseudomonas* sp. N7. Huang Jing Ke Xue 29: 756–762.
3. Gaille C, Kast P, Haas D (2002) Salicylate biosynthesis in *Pseudomonas aeruginosa*. J Biol Chem 277: 21768–21775.
4. Cohen SP, Levy SB, Foulds J, Rosner JL (1993) Salicylate induction of antibiotic resistance in *Escherichia coli*: activation of the mar-operon and a mar-independent pathway. J Bacteriol 175: 7856–7862.

5. Henschel KP, Wenzel A, Diedrich M, Fliedner NA (1997) Environmental hazard assessment of pharmaceuticals. Regul Toxicol Pharmacol 25: 220–225.
6. Crouse BA, Ghoshdastidar AJ, Tong AZ (2012) The presence of acidic and neutral drugs in treated sewage effluents and receiving waters in the Cornwallis and Annapolis River water sheds and the Mill Cove Sewage Treatment Plant in Nova Scotia, Canada. Environ Res 112: 92–99.
7. Shamsuzzamann KM, Barnsley EA (1974) The regulation of naphthalene oxygenase in *Pseudomonas*. J Gen Microbiol 83: 165–170.
8. Manohar S, Karegoudar TB (1995) Degradation of naphthalene by a *Pseudomonas* strain NGK1. Indian J Expt Biol 33: 353–356.
9. Suemori A, Nakajima K, Kuraneand R, Nakamura Y (1995) o-m-and p-hydroxybenzoate degradative pathways in *Rhodococcus erythropolis*. FEMS Microbiol Lett 125: 31–36.
10. Buswell JA, Paterson A, Salkinoja-Salonen MS (1980) Hydroxylation of salicylic acid to gentisate by a bacterial enzyme. FEMS Microbiol Lett 8: 135–137.
11. Haribabu BA, Kamath V, Vaidyanathan CS (1984) Degradation of substituted benzoic acids by a *Micrococcus* species. FEMS Microbiol Lett 21: 197–200.
12. Grund E, Knorr C, Eichenlaub R (1990) Catabolism of benzoate and monohydroxylated benzoates by *Amycolatopsis* and *Streptomyces* spp. Appl Environ Microbiol 56: 1459–1464.
13. Suzuki K, Katagiri M (1981) Mechanism of salicylate hydroxylase catalysed decarboxylation. Biochim Biophys Acta 657: 530–534.
14. Goetz FE, Harmuth LJ (1992) Gentisate pathway in *Salmonella typhimurium*: metabolism of m-hydroxybenzoate and gentisate. FEMS Microbiol Lett 76: 45–49.
15. Mamta R, Prakash D, Sobti RC, Jain RK (1996) Plasmid mediated degradation of o-phthalate and salicylate by a *Moraxella* sp. Biochem Biophys Res Com 220: 377–381.
16. Prabha Devi, Naik CG, Rodrigues C (2006) Biotransformation of citrinin to decarboxycitrinin using an organic solvent tolerant marine bacterium *Moraxella* sp. MB1. Mar Biotechnol 8: 129–138.
17. Hossain M, Dilip K, Brunton N, Martin-Diana A, Barry-Ryan C (2010) Characterization of phenolic composition in Lamiaceae Spices by LC-ESI-MS/MS. J Agric Food Chem 58: 10576–10581.
18. McIntyre C, Jardine D, McRae C (2001) Electrospray mass spectrometry of aquatic fulvic acids. Rapid Commun Mass Spectrom 15: 1974–1975.
19. Finkle BJ, Lewis JC, Corse JW, Lundin RE (1962) Enzyme reactions with phenolic compounds : Formation of hydroxystyrenes through the decarboxylation of 4- hydroxycinnamic acids by *Aerobacter*. J Biol Chem 237: 2926–2932.
20. Fang N, Yu S, Prior RL (2002) LC/MS/MS characterization of Phenolic constituents in Dried Plums. J Agric Food Chem 50: 3579–3585.
21. Parveen I, Threadgill MD, Hauck B, Donnison I, Winters A (2011) Isolation, identification and quantitation of hydroxycinnamic acid conjugates, potential platform chemicals, in the leaves and stems of *Miscanthus* x *giganteus* using LC-ESI-MS^n. Phytochemistry 72: 2376–2384.
22. Cardoso AM, Correia AJF (1999) Fragmentation reaction of Molecular ion and di-cations of indoleamines. Eur J Mass Spectrom 5: 11–18.
23. Crotti AEM, Gates PJ, Lopes JLC (2010) Electrospray MS-based characterization of β- carbolines- mutagenic constituents of thermally processed meat. Mol Nutr Food Res 54: 433–439.
24. Lemoigne M (1923) Production d'acide β-oxybutyrique par certainesbact'eriesdu groupe du Bacillus subtilis. CR. Hebd Seances Acad Sci 176: 1761–1763.
25. Steinbüchel A, Valentin HE (1995) Diversity of bacterial polyhydroxyalkanoic acids. FEMS Microbiol Lett 128: 219–228.
26. Cheng KC, Demicri A, Catchmark JM (2011) Pullulan: biosynthesis, production, and application Appl Microbiol Biotechnol 92: 29–44.
27. DragerKI, Skjak-Braek G, Smidsrød O (1997) Alginate based new materials. Int J Biol Macromol 21: 47–55.
28. Gregg K, Cooper CL, Schafer DJ, Sharpe H, Beard CE,et al. (1994) Detoxification of the plant toxin fluoroacetate by a genetically modified rumen bacterium. Nat Biotechnol 12: 1361–1365.
29. Spain JC, Wyss O, Gibson DT (1979) Enzymatic oxidation of p-nitrophenol. Biochem Biophys Res Commun 88: 634–641.
30. Crawford RL (1976) Degradation of homogentisate by strains of *Bacillus* and *Moraxella*. Can J Microbiol 22: 276–280.
31. Karegoudar TB, Kim C-K (2000) Microbial degradation of Monohydroxybenzoic acid. J Microbiol 38: 53–61.
32. Ishiyama D, Vujaklija D, Davies J (2004) Novel Pathway of Salicylate Degradation *by Streptomyces sp.* Strain WA46. Appl Environ Microbiol 70: 1297–1306.
33. Jimenez JI, Minambres B, Garcia JL, Diaz E (2002) Genomic analysis of the aromatic catabolic pathways from *Pseudomonas putida* KJ2440. Environ Microbiol 4: 824–841.
34. Travkin VM, Jadan AP, Briganti F, Scozzafava A, Golovleva LA (1997) Characterization of an intradiol dioxygenase involved in the biodegradation of the chlorophenoxy herbicides 2,4- D and 2,4,5-T. FEBS Lett 407: 69–72.
35. Stolz A, Knackmuss HJ (1993) Degradation of 2,4 dihydroxybenzoate by *Pseudomonas* sp. BN 9. FEMS Microbiol Lett 108: 219–229.
36. Chalabaev S, Turlin E, Bay S, Ganneau C, Brito-Fravallo E, et al. (2008) Cinnamic acid, an Auto inducer of its own Biosynthesis, is Processed via *Hca* Enzymes in *Photorhabdus luminescens*. Appl Environ Microbiol 74: 1717–1725.
37. Xiang L, Moore BS (2002) Inactivation, complementation and heterologus expression of *encP*, a novel antibacterial phenylalanine ammonia lyase gene. J Biol Chem 277: 32505–32509.
38. Williams JS, Thomas M, Clarke DJ (2005) The gene *stl*A encodes a phenylalanine lyase that is involved in the production of stilbene antibiotic in *Photorhabdus luminescens* TT01. Microbiology 151: 2543–2550.
39. Berner M, Krug D, Bihlmaier C, Vente A, Muller R, Bechthold A (2006) Genes and Enzymes involved in Caffeic acid Biosynthesis in the Actinomycete *Saccharothrix espanaensis*. J Bacteriol 188: 2666–2673.
40. Hakkinen S (2000) Flavonols and Phenolics acids in Berries and Berry Products. Dissertation, Kuopio University Publications.
41. Korkina LG (2007) Phenylpropanoids as naturally occurring antioxidants: from plant defense to human health. Cell Mol Biol (Noisy-le- grand) 53: 15–25.
42. Barnes CC, Smalley MK, Manfredi KP, Kindscher K, Loring H, et al. (2003) Characterization of an anti-tuberculosis resin glycoside from the prairie medicinal plant *Ipomoea leptophylla*. J Nat Prod 66: 1457–1462.
43. Burt S (2004) Essential oils: their antibacterial properties and potential applications in foods- a review. Int J Food Microbiol 94: 223–253.
44. Wells JE, Berry ED, Varel VH (2005) Effects of common forage phenolic acids on *Escherichia* coli O157:H7 viability in bovine faeces. Appl Environ Microbiol 71: 7974–7979.
45. Bae EA, Han MJ, Kim NJ, Kim DH (1998) Anti- *Helicobacter pylori* activity of herbal medicines. Biol Pharm Bull 21: 990–992.
46. Ogino H, Miyamoto K, Ishikawa H (1994) Organic solvent–tolerant bacterium which secretes organic solvent stable lipolytic enzymes. Appl Environ Microbiol 60: 3884–3886.

Can Particulate Air Sampling Predict Microbial Load in Operating Theatres for Arthroplasty?

Maria Luisa Cristina*, Anna Maria Spagnolo, Marina Sartini, Donatella Panatto, Roberto Gasparini, Paolo Orlando, Gianluca Ottria, Fernanda Perdelli

Department of Health Sciences, University of Genoa, Genoa, Italy

Abstract

Several studies have proposed that the microbiological quality of the air in operating theatres be indirectly evaluated by means of particle counting, a technique derived from industrial clean-room technology standards, using airborne particle concentration as an index of microbial contamination. However, the relationship between particle counting and microbiological sampling has rarely been evaluated and demonstrated in operating theatres. The aim of the present study was to determine whether particle counting could predict microbiological contamination of the air in an operating theatre during 95 surgical arthroplasty procedures. This investigation was carried out over a period of three months in 2010 in an orthopedic operating theatre devoted exclusively to prosthetic surgery. During each procedure, the bacterial contamination of the air was determined by means of active sampling; at the same time, airborne particulate contamination was assessed throughout the entire procedure. On considering the total number of surgical operations, the mean value of the total bacterial load in the center of the operating theatre proved to be 35 CFU/m^3; the mean particle count was 4,194,569 no./m^3 for particles of diameter $\geq$0.5 μm and 13,519 no./m^3 for particles of diameter $\geq$5 μm. No significant differences emerged between the median values of the airborne microbial load recorded during the two types of procedure monitored. Particulates with a diameter of $\geq$0.5 μm were detected in statistically higher concentrations ($p<0.001$) during knee-replacement procedures. By contrast, particulates with a diameter of $\geq$5 μm displayed a statistically higher concentration during hip-replacement procedures ($p<0.05$). The results did not reveal any statistically significant correlation between microbial loads and particle counts for either of the particle diameters considered ($\geq$0.5 μm and $\geq$5 μm). Consequently, microbiological monitoring remains the most suitable method of evaluating the quality of air in operating theatres.

Editor: Herman Tse, The University of Hong Kong, Hong Kong

Funding: The authors have no support or funding to report.

Competing Interests: The authors have declared that no competing interests exist.

* E-mail: cristinaml@unige.it

Introduction

All surgical procedures carry a risk of post-operative infection, which can be particularly serious in orthopedic surgery such as joint replacement [1]. The prevention of surgical site infection (SSI) after orthopedic implant surgery is a hot topic for politicians, hospital administrators and clinicians, given the enormous amount of resources these infections consume in terms of extra costs of medications, reoperations, and prolonged hospitalization [2]. Prevention involves identifying and controlling potential sources of microbial contamination.

Factors causing surgical site infections are multivarious [3], including surgical characteristics (e.g. type of procedure, duration of the operation etc), the appropriateness of staff behaviors (limited number of personnel and restricted movements) and microbial contamination of the air (especially in clean surgery) [4], [5]. In turn, the latter factor can be, at least partially, determined by factors linked to the surgical characteristics and staff behaviors.

Several studies have proposed that the microbiological quality of the air in operating theatres be indirectly evaluated by means of particle counting, a technique derived from industrial clean-room technology standards [6], using airborne particle concentration as an index of microbial contamination [6], [7]. One of the sources of airborne particulates in operating theatres may be the surgical smoke produced by particular instruments or techniques, such as the ultrasonic scalpel, electrocautery etc, during surgical procedures.

However, the relationship between particle counting and microbiological sampling has rarely been evaluated and demonstrated in operating theatres. Seal and Clark [8] compared airborne particle counting in eight size-ranges with the numbers of bacteria-carrying particles, in ultra-clean and turbulently ventilated operating theatres. They found that the number of particles in the 5–7 μm size-range correlated significantly with microbiological contamination. Landrin et al [6] compared these two methods in empty operating theatres equipped with conventional ventilation through HEPA filters. No correlation was observed between the two methods (Spearman correlation coefficient = 0.06, P = 0.6). These authors concluded that there was no reason to replace microbiological sampling with particle counting for the routine evaluation of microbiological contamination in conventionally ventilated operating theatres.

Another study that investigated whether the density of airborne particulates could predict the density of viable airborne bacteria was conducted by Stocks et al [9] during 22 arthroplasty procedures in operating rooms with a conventional ventilation system (turbulent air flow). They found a correlation between the presence of larger particles (>5 μm) and microbial contamination, which was attributed to the capability of larger particles to carry bacteria. Thus, given the inhomogeneity of the results obtained, the literature regarding the relationship between airborne particulates and airborne microbes remains controversial.

The aim of the present study was to determine whether particle counting could predict microbiological contamination of the air in an operating theatre during 95 surgical arthroplasty procedures (hip and knee). In addition, we examined the possible correlations among the microbial load, particle counting and other variables: i.e. the frequency of use of instruments that generate particulates, door-opening rates, the number of persons present in the operating theatre and the duration of the surgical procedure.

Methods

This investigation was carried out over a period of three months in 2010 in an orthopedic operating theatre devoted exclusively to prosthetic surgery and situated within a hospital facility in the north-west of Italy. About 1,200 hip- and knee-replacement procedures (ICD9-CM: 81.51–81.54) are performed per year in this operating theatre, which has a surface area of 35.51 m^2 and is part of a complex that provides adequate space for reception, anesthesia, surgery, recovery and observation.

We monitored 95 surgical arthroplasty procedures: 59 hip replacements (ICD9-CM 81.51) and 36 knee replacements (ICD9-CM 81.54). During each procedure, the bacterial contamination of the air was determined by means of active sampling; at the same time, airborne particulate contamination was assessed throughout the entire procedure. Moreover, particle counts were recorded during the use of instruments able to produce surgical smoke. Figure 1 shows the floor plan of the operating theater and the positions of the air and particulate samplers.

During each operation, detailed information on the surgical procedure was collected, including the length of the procedure, the number of people present in the theatre, the door-opening rate per procedure, and the duration of use of instruments that produce surgical smoke.

At the beginning of the study, when the operating theatre was at rest, we recorded the number of air exchanges provided by the air-conditioning system and checked for any microbial contamination of the airflow from the inlet ports of the system. The operating theatre has a turbulent-flow ventilation system equipped with HEPA filters, which are 99.97% efficient in removing airborne particles of 0.3 μm or larger; the filters are replaced every 6 months and maintenance work on the system is carried out periodically. The operating room is under positive pressure in relation to the adjacent rooms (≥5 Pa); air temperature and humidity are set respectively at 17°C ±1°C and 50% ±5%.

Determining Bacterial Contamination of the Air from the Inlet Ports and in the Centre of the Room

To determine the total airborne bacterial load, we used an SAS SUPER 100 (PBI International®) impactor equipped with RODAC plates (Ø = 55 mm) containing γ-irradiated TSA (Tryptone Soy Agar) culture medium (Biotest Italia s.r.l.). In order to sample the air in the centre of the room, the instrument was positioned in the immediate vicinity of the operating table, at a height of 1.5 m. During each procedure, a 1000 L volume of air was aspirated by means of a multi-aspiration modality; the impactor was switched on by remote control just as the skin was incised, and was switched off on completion of suturing.

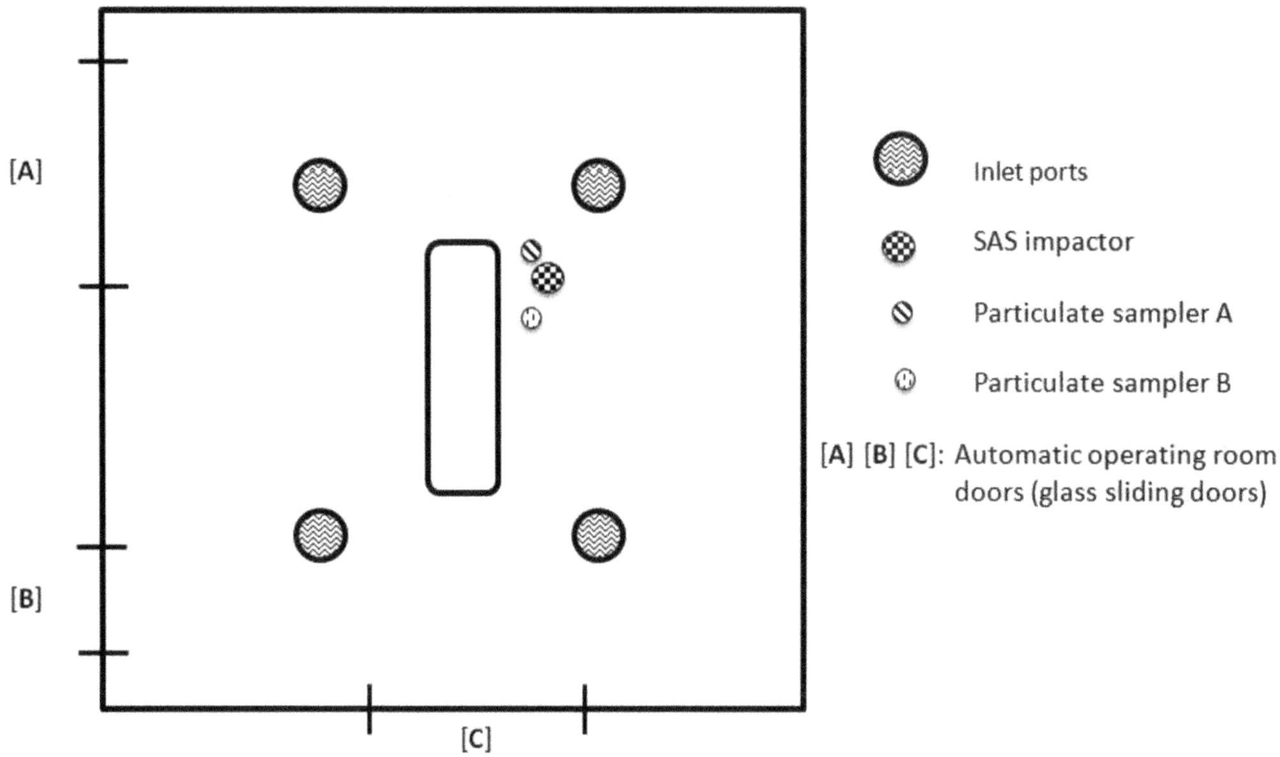

Figure 1. Floor plan of the operating theater, showing the positions of the samplers.

To determine the bacterial load of the airflow from the inlet ports of the ventilation system in at-rest conditions, the sampler was positioned in proximity to the flow-regulators. The total volume of air sampled was 1000 L.

Plates were incubated at 37°C for 48 h before the total aerobic bacterial count was measured. Microbiological results are expressed as CFU (Colony Forming Units)/m^3.

Determining Contamination by Airborne Particulates

Airborne particulates (particles ≥0.5 μm and particles ≥5 μm in size) were counted by means of two light-scattering particle analyzers (Met One, Pacific Scientific Instruments, Grants Pass, OR, USA) at a flow rate of 2.8 L/min., and expressed as numbers of particles per m^3. Both apparatuses were placed in the same position as the SAS impactor; one (Sampler A) was activated by remote control in parallel with the impactor, i.e. for the entire duration of the procedure, while the other (Sampler B), again activated by remote control, recorded the particle count both on each occasion when surgical smoke-producing instruments were used and when such instruments were not used.

Determining the Number of Efficacious Air Exchanges

The efficacy of the air-conditioning system was assessed in at-rest conditions by measuring the decay of the concentration of tracer gas by means of a portable GA301 meter (Eco-CONTROL, Milan) connected to a computer for the collection and analysis of data, as described by Sartini et al [10].

Statistical Analysis

Statistical analysis was carried out by means of STATA SE9™ software (Stata Corp LP - USA). The results were analyzed in terms of descriptive statistics, and the relationships between data were examined by using the non-parametric Mann-Whitney-Wilcoxon ranksum test. The rho Spearman rank correlation was used to assess the degree of association among particulate diameter ≥5 μm, particulate diameter ≥0.5 μm, bacterial load, procedure duration, percentage frequency of ultrasonic scalpel use, number of persons present and door-opening rate.

Moreover, the sensitivity and specificity of particle counting in discriminating between microbiological counting values were evaluated by means of receiver-operating characteristic (ROC) analysis.

Ethics Statement

We didn't need ethics approval because the study was carried out as part of routine control tests that we conduct in the operating theatres of the Hospital. As is the case of all studies conducted in the hospital environment, the General Management of the hospital approved the study protocol. The General Management is responsible for ensuring the ethical aspects of all activities of the hospital. Furthermore, the entire study was organised in accordance with a protocol agreed upon with the operating theatre teams. On entering the hospital, all patients sign an informed consent form regarding treatments in the hospital and the conditions of those treatments. Finally, the research was carried out in full respect of the Italian law on the privacy (Decreto legislativo 30 giugno 2003, n. 196).

Results

With regard to the characteristics of the air-conditioning system, 20 efficacious air exchanges were carried out per hour. The microbial load of the airflow through the inlet ports proved to be <1 CFU/m^3.

On considering the total number of surgical operations, the mean value of the total bacterial load in the center of the operating theatre proved to be 35 CFU/m^3; the mean particle count measured by Sampler A was 4,194,569 no./m^3 for particles of diameter ≥0.5 μm and 13,519 no./m^3 for particles of diameter ≥5 μm (Table 1).

Table 1 also reports the data on the microbial load and on the counts of particles of the 2 diameters (≥0.5 and ≥5 μm), subdivided by type of procedure. No significant differences emerged between the median values of the airborne microbial load (hip replacement median value: 35 CFU/m^3; knee replacement median value: 40 CFU/m^3) recorded during the two types of procedure monitored ($p = 0.4558$).

Particulates with a diameter of ≥0.5 μm were detected in statistically higher concentrations ($z = -6.013$; $p = 0.0000$) during knee-replacement procedures (median value 6,468,327 no./m^3). By contrast, particulates with a diameter of ≥5 μm displayed a statistically higher concentration ($z = 2.013$, $p = 0.0441$) during hip-replacement procedures (median value 14,520 no./m^3).

Table 2 reports the data on the duration of the procedures, the number of persons present in the room and door-opening rates. The difference in the median duration of the two types of procedure (hip-replacement : 40 min; knee-replacement: 45 min) proved to be statistically significant ($z = -3.569$; $p = 0.0004$).

Table 1. Mean values, standard deviations (SD), minimum and maximum values, median and quartiles (Q1–Q3) of the total bacterial load (CFU/m^3) and counts of particles (Sampler A) of diameter ≥0.5 μm and diameter ≥5 μm (no./m^3) in all the procedures monitored and in each of the two types of procedure.

	Total airborne bacterial load (CFU/m^3)		
	Hip replacement	**Knee replacement**	**All procedures**
Mean ± SD	35±16	34±11	35±15
Min-max	10–70	20–45	10–70
Median	35	40	40
Q1–Q3	25–40	22–45	25–45
	Count of airborne particles of diameter ≥0.5 μm (no./m^3)		
	Hip replacement	**Knee replacement**	**All procedures**
Mean ± SD	2,538,425±1,480,054	6,908,804±3,269,100	4,194,569±3,142,263
Min-max	600,353–5,164,279	2,925,281–11,219,627	600,353–11,219,627
Median	2,418,086	6,468,327	3,127,466
Q1–Q3	1,294,603–3,586,316	3,760,520–9,683,440	1,630,903–5,164,279
	Count of airborne particles of diameter ≥5 μm (no./m^3)		
	Hip replacement	**Knee replacement**	**All procedures**
Mean ± SD	13,915±3,592	12,868±3,488	13,519±3,571
Min-max	5,879–19,756	8,352–17,764	5,879–19,756
Median	14,520	12,207	14,520
Q1–Q3	12,657–16,608	9,926–16,046	11,500–16,608

Table 2. Mean, standard deviation (SD), median and range of the values of procedure length (min), number of persons present in theatre, and door-opening rate (no./min) for all surgical procedures and for hip and knee replacements.

All surgical procedures		
	Mean±SD	**Median (range)**
Operation length (min)	45±10	40 (35–70)
N° persons present	8±1	7 (6–10)
Door-opening rate (no./min)	0.20±0.03	0.20 (0.12–0.26)
Hip replacement (ICD9-CM 81.51)		
Operation length (min)	41±6	40 (35–50)
N° persons present	8±1	8 (6–10)
Door-opening rate (no./min)	0.20±0.03	0.20 (0.12–0.26)
Knee replacement (ICD9-CM 81.54)		
Operation length (min)	50±12	45 (40–70)
N° persons present	6±1	6 (6–7)
Door-opening rate (no./min)	0.20±0.04	0.18 (0.16–0.25)

During each operation, doors were opened a mean of 0.20 times per minute, with no statistically significant difference emerging between the two types of procedure ($p = 0.2334$).

All operating theatre staff wore nonwoven fabric suits; surgeons also wore highly effective isolation helmets, and instrument-handlers wore semi-integral masks, while the anesthetist and circulating nurses wore surgical masks and hair covering.

Evaluation of the surgical instruments able to release surgical smoke revealed that the ultrasonic scalpel was the instrument most frequently used in both types of procedure. Figure 2 reports the duration of the use of the ultrasonic scalpel as a percentage of procedure time (calculated as minutes of use/procedure duration x 100), subdivided by type of procedure. In this regard, statistically significant differences ($z = -3.145$; $p = 0.0017$) emerged between the two types of procedure.

Airborne particulate concentration as a function of the use of the ultrasonic scalpel was evaluated by means of Sampler B. This evaluation revealed that the median concentration of particulates with a diameter of ≥0.5 µm was significantly higher ($z = -4{-}432$; $p = 0.000$) when this instrument was used (8,014,873 no./m^3) than when it was not used (2,958,536 no./m^3). A similar pattern was seen in the concentration of ≥5 µm diameter particles (13,965 no./m^3 vs. 12,878 no./m^3), though the difference was not statistically significant ($p = 0.2019$).

A highly significant correlation emerged between the percentage use of the ultrasonic scalpel and the duration of the procedures ($rho = 0.8254$ $p < 0.001$). Table 3 reports of the Spearman correlation coefficients (rho) and significances (p) of the mean concentrations of particulates (≥0.5 µm and ≥5 µm) recorded by Sampler A in relation to procedure duration (min), percentage use of the ultrasonic scalpel, number of persons present and door-opening rate, both for the total number of procedures and for each procedure type (hip and knee replacement).

The results did not reveal any statistically significant correlation between microbial loads and particle counts for either of the particle diameters considered (≥0.5 µm and ≥5 µm).

A statistically significant correlation ($rho = 0.2344$ $p < 0.05$) emerged between the number of persons present in the operating theatre and the airborne microbial load during both types of procedure.

On the basis of ROC analysis, no particulate count value for the diameters of ≥0.5 and ≥5 µm can predict a microbiological count higher than 20 CFU/m^3 (ROC area = 0.4661 and 0.1080, respectively). The choice of this cut-off was determined by the fact that the national standard reference value for operating theatres with turbulent-flow ventilation (180 CFU/m^3) has never been reached; the most suitable cut-off value for our data proved to be the Italian and British standard value indicated for operating theatres in which ultraclean surgery is performed, despite the fact that laminar-flow ventilation is recommended.

Discussion

During surgical procedures, dust particles, textile fibers, skin scales and respiratory aerosols loaded with viable microorganisms are released from the surgical team and patient into the surrounding air of the operating room [11]. Moreover, the use of some particular instruments during surgical procedures increases the dispersion of particulates in the air. The size of such particles closely depends on the particular instruments used. Indeed, various investigations have revealed that the electrocautery releases the smallest particles, with a mean aerodynamic size of 0.07 µm, whereas laser tissue coagulation produces larger particles (0.31 µm) and the largest particles are generated by the ultrasonic scalpel (0.35–6.5 µm) [12].

The aim of the present study was to determine whether the concentration of the airborne particulates could predict microbial contamination during surgical operations. Particle counting could offer several advantages over microbiological sampling. Indeed, microbiological sampling is time-consuming and the results may only be available after several days [6]. Another critical point concerning microbiological sampling is the absence of uniform standardized sampling procedures, which prevents data collected in different countries from being compared. In addition, a range of instruments are used in volumetric air-sampling, which hampers the correlation of results on account of their variability [3].

Conversely, particle counting is less demanding and yields immediate results. However, the information obtained is indirect [6]. This method enables results to be compared with reference values applicable to classes of maximum particle concentration. Most industrialized countries have set their own standards, based on measuring the presence of particles of varying sizes and number; many of these standards have been amended to conform to the International Standards Organization (ISO) 14644 [3], [13].

Some similar studies [6], [8], [9] have yielded results which conflict with one another, thus leaving the question unresolved.

With regard to microbial contamination in operating theatres, it is well known that the number of airborne bacteria depends on several factors, such as the number of people present in the room and their activities and behavior.

Few countries have set bacterial threshold limits in conventionally-ventilated operating theatres, although most recommend 20 air exchanges per hour, so as not to exceed values of 50–150 colony forming units (CFU)/m^3 of air during surgical operations.

In some countries, the concentration should not exceed 180 CFU/m^3 for an average 5-min period during activity [3].

In France [14], the microbiological limits are more restrictive than in Italy [15] and the UK, with values of ≤20 CFU/m^3 and ≤5 CFU/m^3 being indicated for turbulent and unidirectional airflows, respectively [16].

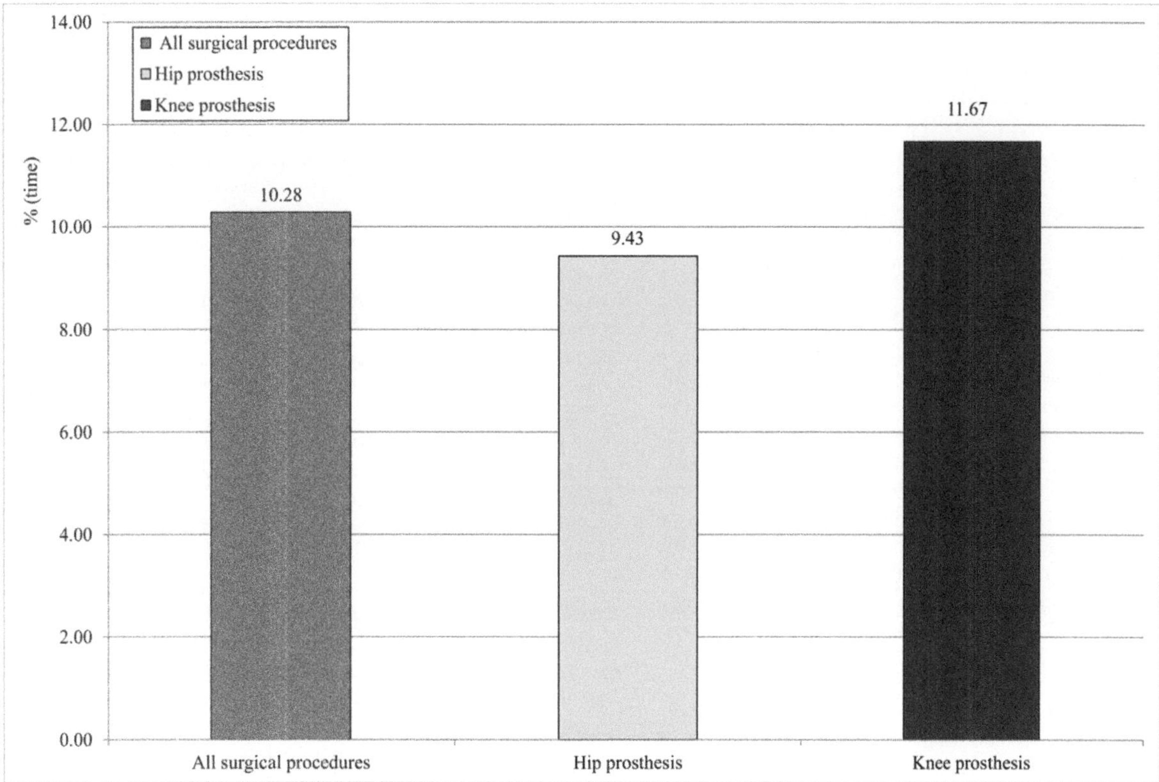

Figure 2. Duration of use of the ultrasonic scalpel as a percentage of operating time, subdivided by type of procedure.

In the present study, the mean values of the airborne microbial load (35 CFU/m^3) proved to be below the standard values for conventionally-ventilated [15] operating theatres in Italy. However, they exceeded the values proposed in some countries, such as France, and the generally accepted level of airborne microbial contamination during arthroplasty (10 CFU/m^3) [10]. The values that we recorded prove to be higher than those reported in other studies of the same types of surgical procedure [9], [17].

Our analysis of the quality of the air provided by the ventilation system revealed the microbial load of <1 CFU/m^3; it can

Table 3. Spearman's correlation coefficients (Rho) and significances (p) between the mean concentration of particulate measured by Sampler A (≥5 and ≥0.5 μm) and data on procedure duration (min), ultrasonic scalpel use, number of persons present in theatre and door-opening rates, referred to all procedures and broken down by procedure type (hip and knee replacement).

	Mean concentration of particulate ≥5 μm	**Mean concentration of particulate ≥0.5 μm**
Procedure length (min)°°	rho = 0.1706; p = 0.0984 (*)	rho = 0.6865; **p<0.001** (*)
	rho = 0.2418; p = 0.0650 (°)	rho = 0.1888; p = 0.1521 (°)
	rho = 0.1343; p = 0.4350 (§)	rho = 0.7622; **p<0.001**(§)
% use of ultrasonic scalpel	rho = 0.1460; p = 0.1581 (*)	rho = 0.7082; **p<0.001** (*)
	rho = 0.2347; p = 0.0736 (°)	rho = 0.8612; **p<0.001** (°)
	rho = 0.5881; **p<0.001** (§)	rho = 0.7933; **p<0.001** (§)
N° persons present	rho = 0.2854; **p<0.01** (*)	rho = 0.1624; p = 0.1158 (*)
	rho = 0.4723; **p<0.001** (°)	rho = 0.2104; p = 0.1097 (°)
	rho = 0.3199; p = 0.0571 (§)	rho = 0.1464; p = 0.3943 (§)
Door-opening rate (no./min)	rho = 0.1184; p = 0.2530 (*)	rho = 0.0378; p = 0.7159 (*)
	rho = 0.2418; p = 0.0650 (°)	rho = 0.0645; p = 0.6275 (°)
	rho = 0.6895; **p<0.001** (§)	rho = 0.1389; p = 0.4192 (§)

°° skin - skin.
(*) all procedures.
(°) hip replacement.
(§) knee replacement.

therefore be supposed that the contamination detected was produced during surgical activity.

With regard to particle counting, the results obtained revealed mean values that exceeded the European ISO 14644 Standard limits for ISO 7 clean-rooms ($352 \times 10^3 \geq 0.5$ µm particles/m^3 and $2{,}930 \geq 5$ µm particles/m^3) [13]. Moreover, the concentration values recorded in the present study, with regard to both sizes of particulates, differed from those reported by other similar studies conducted on conventionally ventilated operating theatres. However, it should be pointed out that these studies also differed considerably among themselves, again with regard to both sizes of particulates [9], [18].

It may reasonably be supposed that the utilization of surgical instruments that produce surgical smoke contributed to the high concentration of particulates. This supposition is supported by the fact that, with regard to particulates with a diameter of ≥ 0.5 µm, we found a significantly higher concentration ($p<0.001$) when the ultrasonic scalpel was being used than when it was not. Moreover, a statistically significant correlation emerged between the percentage use of this instrument and the mean concentration of ≥ 0.5 µm particles, and between procedure duration and the mean concentration of ≥ 0.5 µm particles. On the basis of these findings, it may be hypothesized that, as the duration of the procedure increases, so also does the use of the ultrasonic scalpel, i.e. the device responsible for the production of these particles.

For what concerns the heavier particles (≥ 5 µm), a correlation with the percentage use of the ultrasonic scalpel emerged only with regard to knee-replacement procedures. This could be explained by the fact that, as a proportion of the total particulates released by the ultrasonic scalpel, which is known to produce particulates in the 0.35–6.5 µm range, particulates with a diameter of ≥ 0.5 µm may prevail; by this token, the heavier particles could account for a smaller proportion of the total, becoming more detectable only after prolonged use of the device.

With regard to the correlation between ≥ 5 µm particles and the number of persons present in the room, the results obtained could be explained in terms of the fact that such particulates also includes the skin debris, the dimensions of which vary from 2.5 to 20 µm [18], and which are constantly being released by both theatre staff and patients. Moreover, the number of persons present influences the amount of movement taking place in the room, which, as already pointed out in other papers [18], tends to raise any dust that has already settled. Likewise, this can explain the correlation between microbial concentration and the number of persons present in the room.

In conclusion, this study showed that the use of surgical instruments that produce surgical smoke, such as the ultrasonic scalpel, can markedly contribute to the production of airborne particulates. However, neither fraction of these particulates (≥ 0.5 µm or ≥ 5 µm) displayed any correlation with the microbial load in either of the types of procedure considered.

However, one of the limitations of the present research is that only two particulate fractions were studied. We cannot therefore rule out the possibility that there may be a correlation between the number of particles/m^3 and CFU/m^3 if particulate fractions in narrower size-ranges are considered, as has been reported by Stocks et al. [9]. Moreover, further investigations should be carried out in operating theaters equipped with laminar-flow ventilation systems, an issue which has not yet been sufficiently addressed in studies of this kind.

Until these issues have been settled, microbiological monitoring remains the most suitable method of evaluating the quality of air in operating theatres.

Acknowledgments

The authors thank Dr. Bernard Patrick for translating the manuscript into English.

Author Contributions

Conceived and designed the experiments: MLC PO FP. Performed the experiments: GO DP. Analyzed the data: MS. Contributed reagents/materials/analysis tools: PO RG DP. Wrote the paper: AMS MLC.

References

1. Al-Waked R (2010) Effect of Ventilation Strategies on Infection Control inside Operating Theatres. Eng Appl Comp Fluid 4: 1–16.
2. Andersson AE, Bergh I, Karlsson J, Eriksson BI, Nilsson K (2012) Traffic flow in the operating room: An explorative and descriptive study on air quality during orthopedic trauma implant surgery. Am J Infect Control 40: 750–755.
3. Dharan S, Pittet D (2002) Environmental controls in operating theatres. J Hosp Infect 51: 79–84.
4. Weaving P, Cox F, Milton S (2008) Infection prevention and control in the operating theatre: reducing the risk of surgical site infections (SSIs). J Perioper Pract 18: 199–204.
5. Gould D (2001) Clean surgical wounds: prevention of infection. Nurs Stand 15: 45–52.
6. Landrin A, Bissery A, Kac G (2005) Monitoring air sampling in operating theatres: can particle counting replace microbiological sampling? J Hosp Infect 61: 27–29.
7. Wan GH, Chung FF, Tang CS (2011) Long-term surveillance of air quality in medical center operating rooms. Am J Infect Control 39: 302–308.
8. Seal DV, Clark RP (1990) Electronic particle counting for evaluating the quality of air in operating theatres: a potential basis for standards. J Appl Bacteriol 68: 225–230.
9. Stocks GW, Self SD, Thompson B, Adame XA, O'Connor DP (2010) Predicting bacterial populations based on airborne particulates: a study performed in nonlaminar flow operating rooms during joint arthroplasty surgery. Am J Infect Control 38: 199–204.
10. Sartini M, Ottria G, Dallera M, Spagnolo AM, Cristina ML (2006) Nitrous oxide pollution on operating theatres in relation to the type of leakage and the number of efficacious air exchanges per hour. J Prev Med Hyg 47: 155–159.
11. Diab-Elschahawi M, Berger J, Blacky A, Kimberger O, Oguz R, et al. (2011) Impact of different-sized laminar air flow versus no laminar air flow on bacterial counts in the operating room during orthopedic surgery. Am J Infect Control 39: e25–9.
12. Brüske-Hohlfeld I, Preissler G, Jauch KW, Pitz M, Nowak D, et al. (2008) Surgical smoke and ultrafine particles. J Occup Med Toxicol 3: 31.
13. EN ISO 14644-1 (ISO/TC 209) (1999) Cleanrooms and associated controlled environments, Part 1: Classification of air cleanliness International Organization for Standardization Technical Committee.
14. NF S 90–351 (2003) Norme - Etablissements de santé - Salles propres et environnements apparentés - Exigences relatives pour la maîtrise de la contamination aéroportée. Paris: Agence française de normalisation (AFNOR).
15. Istituto Superiore per la Prevenzione e Sicurezza sul Lavoro (ISPESL) (2010) Linee guida sugli standard di sicurezza e di igiene del lavoro nel reparto operatorio. Available: http://www.ispesl.it/linee_guida/Comparto_o_Settore/ISPESL-LG-SaleOperatorie.pdf. Accessed 2012 Jul 18.
16. Cristina ML, Spagnolo AM, Orlando P (2012) Prevention of Surgical Site Infections and Features of the Operating Theatre In: Nova Sciences Publishers, editors. Advances in Medicine and Biology vol. 34 63–92.
17. Orlando P, Cristina ML, Sartini M, Orlando M, Cordara R, et al. (1999) Risk of microbial contamination in operating room [Il rischio da contaminazione microbica in sala operatoria.] L'Igiene Moderna 112: 1253–1262.
18. Scaltriti S, Cencetti S, Rovesti S, Marchesi I, Bargellini A, et al. (2007) Risk factors for particulate and microbial contamination of air in operating theatres. J Hosp Infect 66: 320–326.

Peroxidase-Like Catalytic Activity of Ag_3PO_4 Nanocrystals Prepared by a Colloidal Route

Yuanjun Liu[1], Guoxing Zhu[2,3]*, Jing Yang[2], Aihua Yuan[1]*, Xiaoping Shen[2]

1 School of Environmental and Chemical Engineering, Jiangsu University of Science and Technology, Zhenjiang, China, **2** School of Chemistry and Chemical Engineering, Jiangsu University, Zhenjiang, China, **3** State Key Laboratory of Coordination Chemistry, Nanjing University, Nanjing, P. R. China

Abstract

Nearly monodispersed Ag_3PO_4 nanocrystals with size of 10 nm were prepared through a colloidal chemical route. It was proven that the synthesized Ag_3PO_4 nanoparticles have intrinsic peroxidase-like catalytic activity. They can quickly catalyze oxidation of the peroxidase substrate 3, 3, 5, 5-tetramethylbenzidine (TMB) in the presence of H_2O_2, producing a blue color. The catalysis reaction follows Michaelis-Menten kinetics. The calculated kinetic parameters indicate a high catalytic activity and the strong affinity of Ag_3PO_4 nanocrystals to the substrate (TMB). These results suggest the potential applications of Ag_3PO_4 nanocrystals in fields such as biotechnology, environmental chemistry, and medicine.

Editor: Jie Zheng, University of Akron, United States of America

Funding: Financial support was provided by the National Natural Science Foundation of China (No. 51203069, 51102117, 51072071), China Postdoctoral Science Foundation (2012T50439), Jiangsu Natural Science Foundation of China (No. BK2012276), and Cultivating Project of Young Academic Leader from Jiangsu University. The funders had no role in study design, data collection and analysis, decision to publish, or preparation of the manuscript.

Competing Interests: The authors have declared that no competing interests exist.

* Email: zhuguoxing@ujs.edu.cn (GZ); aihuayuan@just.edu.cn (AY)

Introduction

Owing to its excellent photocatalytic properties and broad range of applications such as in water-splitting, photocatalytic reactions, silver phosphate has got extensive study and has become a well studied material [1]. Especially, partly owing to the highly dispersive Ag *s*-Ag *s* bands without localized *d* states [2], Ag_3PO_4 semiconductor exhibits extremely high photooxidative ability for O_2 evolution from water as well as organic dye decomposition under visible light irradiation [3–6]. A much higher quantum efficiency (up to 90%) than the previously reported values at wavelengths longer than 420 nm was also achieved with it [1].

Up to now, various methods have been proposed to further enhance and optimize the photoelectric and photocatalytic properties of Ag_3PO_4 *via* microstructure control or forming composites with other components to improve its stability, bandgap structure and surface area [7–12]. Although extensive studies have been made for the photocatalytic applications of various Ag_3PO_4 micro-/nanoparticles and their composites, the application of Ag_3PO_4 in biological systems, for example used as biocatalyst, has rarely been studied, while the presence of phosphorus in biological systems is well known.

Recently, it was found that Fe_3O_4 nanoparticles have intrinsic enzyme-like activity similar to peroxidases found in nature, though Fe_3O_4 are usually thought to be biological and chemical inert [13]. After that, several kinds of micro/nanoparticles with smaller size or special structure were prepared for developing enzyme mimics, including the ferromagnetic nanoparticles with peroxidase-like activity [14–23], ceria oxide nanoparticles [24–27], and V_2O_5 nanowires [28], carbon-based nanomaterials [29–36] and so on [37–42]. In contrast to natural enzymes, nanoparticles-based enzyme mimics own prominent advantages. First, they have greater resistance to extremes of pH and temperature, while natural enzymes are usually sensitive to the external conditions and also easily lose their activity. Secondly, nanoparticles-based mimic enzymes have higher stability, while natural enzymes can be digested by proteases. Thirdly, with the extensive development of nanoscience and nanotechnology in the past three decades, the preparation and surface modification of various nanoobjects can be easily carried out, while the synthesis and purification of natural enzymes are still time-consuming, expensive, and also difficult [14].

Exploitation of new functions of known nanomaterials is one of the most attractive aspects in nanoscience [37]. Inspired by the above pioneering research, we investigated the peroxidase-like activity of Ag_3PO_4 nanocrystals, considering that some Ag-based metal alloy nanoparticles own intrinsic peroxidase-like activity. Ag_3PO_4 nanoparticles with smaller size were obtained via a simple colloidal route. It was found that the obtained Ag_3PO_4 nanoparticles show their ability to catalyze peroxidatic reactions in aqueous media. The kinetic parameters were also tested and compared. The reaction catalyzed by these Ag_3PO_4 nanoparticles followed a Michaelis-Menten kinetic behavior with an excellent catalytic activity, making it a promising mimic of peroxidase. The new application of Ag_3PO_4 as peroxidase mimic will add new content to this interesting material.

Results and Discussion

A colloidal route was employed for the preparation of Ag_3PO_4 nanoparticles because it can produce Ag_3PO_4 nanoparticles with smaller size [6]. The preparation was carried out at room temperature with H_3PO_4 and $AgNO_3$ as raw materials, while toluene and oleylamine were used as solvent and surfactant. The crystal phase of the obtained Ag_3PO_4 nanoparticles was first determined by X-ray diffraction (XRD). Fig. 1a shows the corresponding XRD pattern, which can be easily indexed to cubic Ag_3PO_4 with JCPDS No. 06-0505. The relatively strong peaks at 21.3, 30.1, 33.6, 36.9, 53.1, 55.3, and 57.6° corresponds to the (110), (200), (210), (211), (222), (320), and (321) crystal planes of cubic Ag_3PO_4, respectively. No diffraction peak from Ag with zero-valent state is observed in the pattern. This reveals pure Ag_3PO_4 is obtained with this simple route. It should be noted that the XRD pattern shows relatively broad peak, indicating the smaller size of Ag_3PO_4 nanocrystals according to Scherrer formula. Fig. 1b shows a typical transmission electron microscopy (TEM) image of the obtained Ag_3PO_4 nanocrystals, from which spherical particles with small size are observed. The Ag_3PO_4 particles show relatively uniform size. The average diameter is about 10 nm.

Before property investigation, the obtained Ag_3PO_4 nanocrystals were firstly treated through usual ligand exchange route to transfer it to being hydrophilic state. Peroxidase-like activity of the Ag_3PO_4 nanocrystals was evaluated in the catalysis oxidation of a commonly used peroxidase substrate, 3, 3′, 5, 5′-tetramethylbenzidine sulfate (TMB), in the presence of H_2O_2. TMB is colorless and can be oxidized slowly by H_2O_2 (Fig. 2). The often observed oxidation products are two colored products [43]. The first product is a blue charge-transfer complex of diamine, which are formed in rapid equilibrium with the radical cation. Its maximal absorption wavelength locates at ~370 and ~652 nm. Another product is a yellow diimine, which is generated by further oxidation of the diamine with excess H_2O_2 or strong acidic condition. The diimine product is stable in acidic conditions with maximal absorption wavelength of 450 nm. The first-step reaction with the formation of blue diamine is often used as a model process to evaluate activity of peroxidases.

As can be seen in Figure 3, our preliminary experiment shows that Ag_3PO_4 nanocrystals can catalyze the oxidation of TMB by H_2O_2 in NaAc buffer producing a blue solution (inset of Figure 3), suggesting the formation of charge-transfer complex of diamine. The typical absorbance peak of this oxidation product of TMB is at 652 nm. The reaction system will turn to be yellow if it was overnight placed, which is due to the formation of diimine. Also, it was found that Ag_3PO_4 nanocrystals or H_2O_2 alone did not produce significant color change (inset of Figure 3). These results confirm that Ag_3PO_4 nanocrystals behave with peroxidase-like activity toward TMB.

To investigate the effect of pH values of buffer solution on catalytic properties, we performed the catalytic experiments in NaAc buffer with different pH values. Relative activity was analyzed based on the absorption at 652 nm. Fig. 4 shows the relative activity of the Ag_3PO_4 nanocrystals with reaction time of 30 min at room temperature. It was found that the catalytic activity of Ag_3PO_4 nanocrystals is significantly affected by pH values. Only very lower catalytic activity was demonstrated when the pH value of buffer exceeds 4.5. We then selected the buffer with pH of 4 for the subsequent study due to the consideration of the possible disability of Ag_3PO_4 nanocrystals in buffer with strong acidity. With the buffer of pH = 4, a contrast measure was conducted in the absence of Ag_3PO_4 nanocrystals, which give very low absorbency at 652 nm.

As shown in Fig. 5, the catalytic activity of Ag_3PO_4 nanocrystals is also H_2O_2 concentration dependent. With the increasing of H_2O_2 concentration, the peroxidase-like catalytic activity increases at first. When the concentration of H_2O_2 reaches about 2.2 mmol/L (that is 7.3 times that of TMB), the catalytic activity of the Ag_3PO_4 nanocrystals achieves its highest point. However, further increasing the H_2O_2 concentration causes a lower absorbance at 652 nm, which implies low catalytic activity at higher H_2O_2 concentration. In fact, many nanoparticle-based enzyme mimics show this kind hump-shaped relationship between H_2O_2 concentration and the reaction activity including the highly studied enzyme mimic material Fe_3O_4 [13,44–46]. This phenomenon is also similar to that observed with horseradish peroxidase [13,44]. It is reasonable that the reaction activity increases at first with the increase of H_2O_2 concentration, since more oxidant is involved in the reaction system. With high concentration of H_2O_2 in the reaction system, it is usually proposed that the H_2O_2 moleculars would cap on the surface of catalyst, inhibiting the attachment of substance to the surface of catalyst, and so weakening the catalytic activity. Thus, a hump-shaped relationship is obtained. While, this H_2O_2 concentration dependent catalytic activity is different from that of CuO [47], Au nanoparticles [38], Ag nanoparticles [48]. In those cases, the reaction actitvity increases monotonously with H_2O_2 concentration till a saturation state is obtained.

For biomolecular enzymes, the catalytic active center is usually the coordination unsaturated metal sites under the capping of protein networks. For nanoparticles, the surface atoms place in similar situation–coordination unsaturation under the capping of surfactant moleculars. Thus, it is possible that they may share some common points in catalytic process, although the catalysis mechanism of inorganic catalysts and enzymes are usually different. At present stage, the Michaelis-Menten model is widely used for the study of nanoparticle-based enzyme mimetics [13,26,49–52]. Therefore, in our study, the Michaelis-Menten

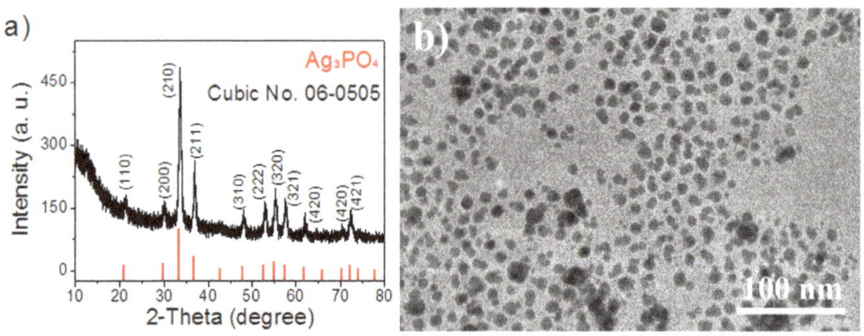

Figure 1. The prepared Ag_3PO_4 nanoparticles. a) XRD pattern and b) TEM image, the standard pattern of cubic Ag_3PO_4 with JCPDS No. 06-0505 is also shown for comparison.

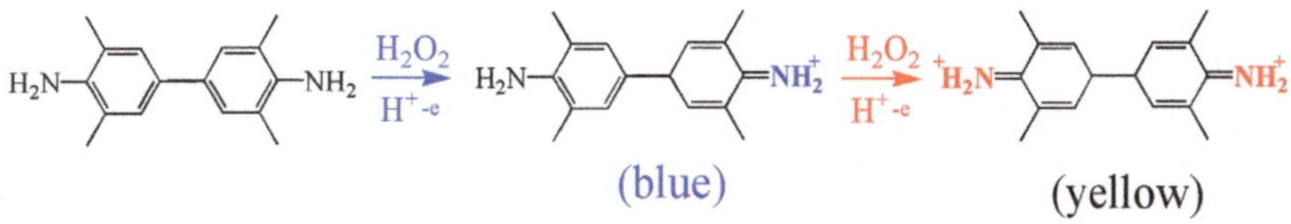

Figure 2. The oxidation reaction of TMB.

model was also selected to understand the peroxidase-like catalytic activity of the Ag_3PO_4 nanocrystals.

For further analyzing the catalytic kinetic parameters, the catalytic activity of Ag_3PO_4 nanocrystals was studied by the Michaelis-Menten model with TMB as substrate. The apparent steady-state kinetic parameters for the reaction were determined at 25°C with pH = 4 buffer. Fig. 6a shows the TMB concentration dependent catalytic activity. The absorbance increases with the increasing of TMB concentration, especially at higher TMB concentration range. Absorbance data were then back-calculated to concentrations by the Beer-Lambert Law using a molar absorption coefficient of 39000 M^{-1} cm^{-1} for TMB-derived oxidation product [53].

Kinetic parameters were then calculated based on the Michaelis–Menten equation, which

$$V_o = \frac{V_{max}[S]}{[S]+K_m} \qquad (1)$$

describes the relationship between the rate of substrate conversion by an enzyme catalyst and the concentration of the substrate. In this equation, V_o is the conversion rate, V_{max} is the maximum conversion rate, $[S]$ is the substrate concentration, and K_m is the Michaelis constant which denotes the affinity of the enzyme. According to (1), a corresponding plot of $1/V_o \sim 1/[S]$ is shown in Fig. 6b, which reveals that the reaction catalyzed by Ag_3PO_4 nanocrystals follows Michaelis-Menten kinetics in a certain range of substrate concentrations.

The corresponding values of K_m and V_{max} calculated from double reciprocal plots with TMB substrate are 0.327 mM and 2.01×10^{-5} mM s^{-1} (Fig. 6b). By comparing the apparent kinetic parameters, the K_m value of Ag_3PO_4 nanocrystals with TMB substrate is lower than that of reported horserahish peroxidase (HRP, 0.434) tested in buffer solution with pH 3.5 at 40°C [13], suggesting that Ag_3PO_4 nanocrystals have a higher affinity for TMB than HRP. The V_{max} value is similar to that of typical nanomaterial-based enzyme mimetics, Fe_3O_4 nanoparticles. The values of K_m and V_{max} calculated with H_2O_2 substrate are 0.216 mM and 1.27×10^{-5} mM s^{-1} (inset of Fig. 5). The K_m value obtained here with H_2O_2 substrate is lower than that of Fe_3O_4 nanoparticles and HRP [13,19] reported under similar test conditions (buffer solution with pH 3.6–4.6, TMB concentration of 0.2–0.8 mM, temperature of 25°C), suggesting that the Ag_3PO_4 nanocrystals exhibit strong affinity towards TMB and H_2O_2. The strong affinity would be a reason for the excellent peroxidase-like activity.

Fig. 7 shows the time-dependent catalytic activity of four similar reaction systems with different amounts of Ag_3PO_4 nanocrystals. The absorbance of the Ag_3PO_4 catalyzed system is much higher than the one without Ag_3PO_4 catalyst. Among the three Ag_3PO_4

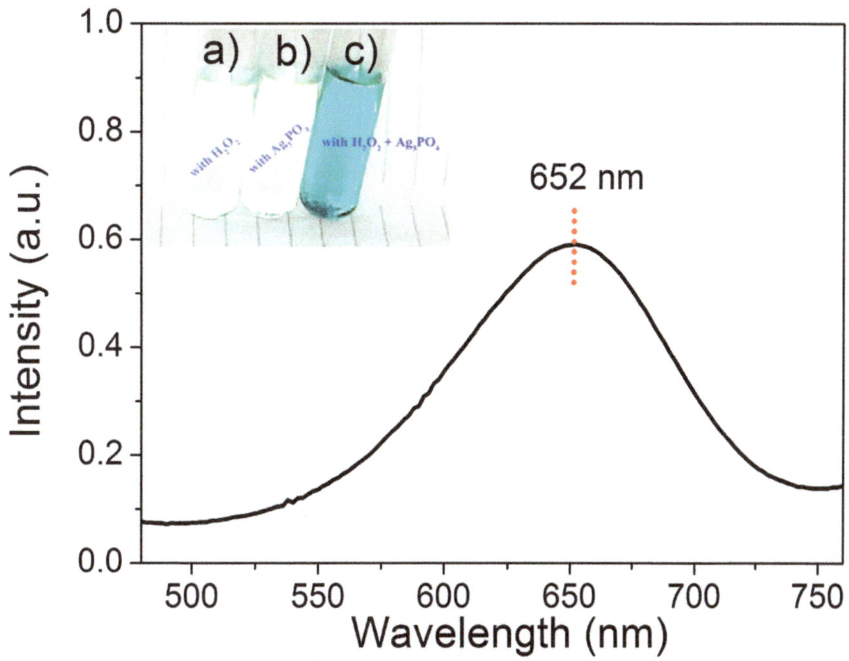

Figure 3. UV-vis spectrum of the reaction system with $H_2O_2 + Ag_3PO_4$ nanocrystals. The inset shows images of oxidation color reaction of TMB in NaAc buffer with a) H_2O_2, b) Ag_3PO_4 nanocrystals and c) H_2O_2 + Ag_3PO_4 nanocrystals. Reaction conditions: 0.3 mM of TMB, 2 mg/mL of Ag_3PO_4 (if with), 3.6 mM of H_2O_2 (if with) in 5 mL of NaAc buffer with pH 4.0. The reaction proceeded at 25°C with time of 30 min.

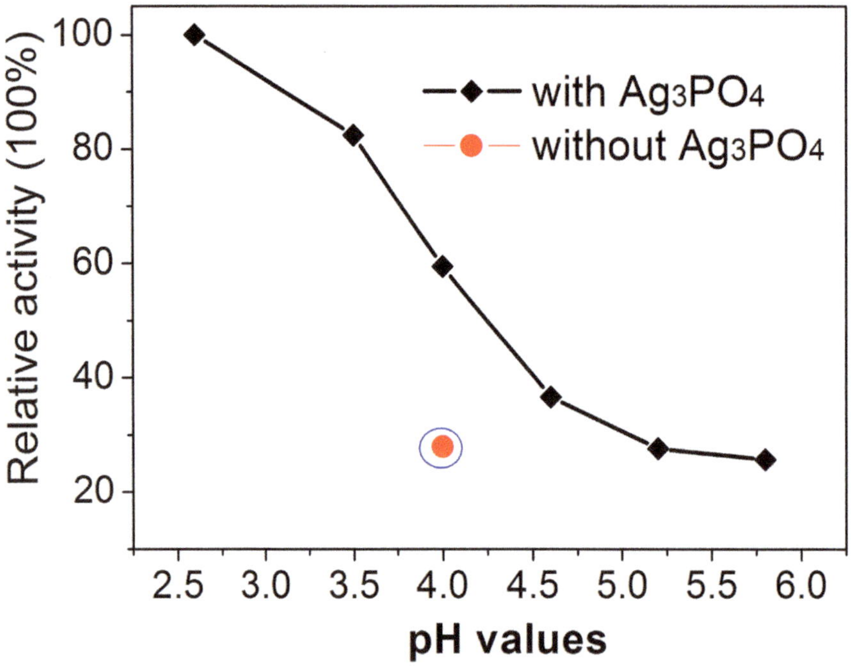

Figure 4. pH value-dependent peroxidase-like catalytic activity of Ag_3PO_4 nanocrystals. Reaction conditions: 0.3 mM of TMB, 2 mg/mL of Ag_3PO_4, 3.6 mM of H_2O_2 in 5 mL of NaAc buffer with different pH values. The reaction proceeded at 25°C with time of 30 min.

catalyzed systems, the more catalyst is involved, the higher absorbency is shown. Control experiment to examine the time-dependent absorption spectrum of Ag_3PO_4 nanocrystals dispersed in buffer solution (pH = 4) at different concentrations (1, 2, 4 mg/mL) gives no obvious change of absorbency with time, indicating that the increasing absorbency in reaction system is related to the oxidation of TMB, but not originated from the increased concentration of Ag_3PO_4 nanocrystals (Fig. 7c). These results show that higher reaction rates are obtained with high concentration of Ag_3PO_4 catalyst. With 4 mg/mL of Ag_3PO_4 nanocrys-

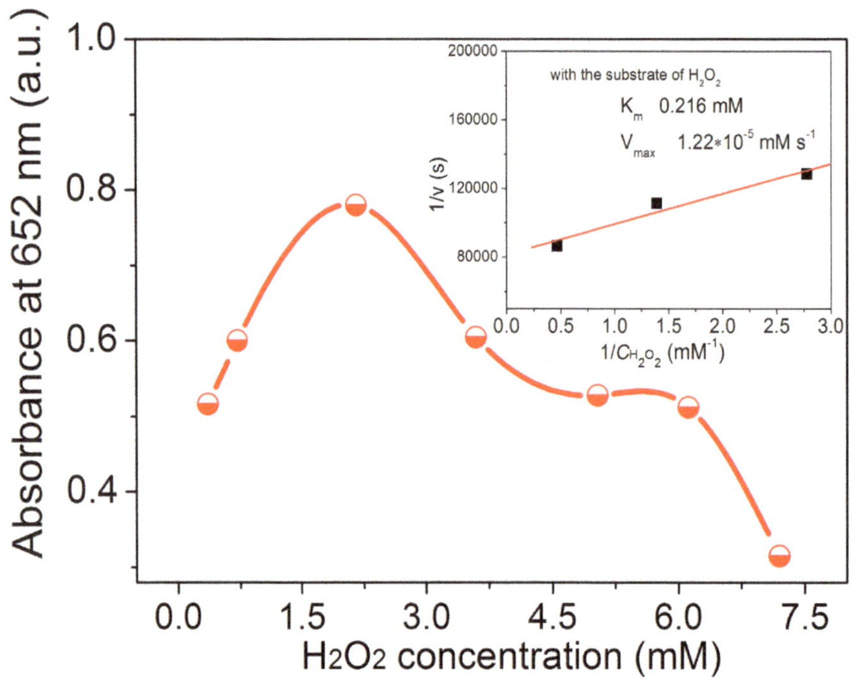

Figure 5. H_2O_2 concentration dependent peroxidase-like catalytic activity of Ag_3PO_4 nanocrystals. Reaction conditions: 0.3 mM of TMB, 2 mg/mL of Ag_3PO_4 in 5 mL of NaAc buffer with pH 4.0. The concentration of H_2O_2 varies in the range of 0.36–7.2 mM. The reaction proceeded at 25°C with time of 30 min.

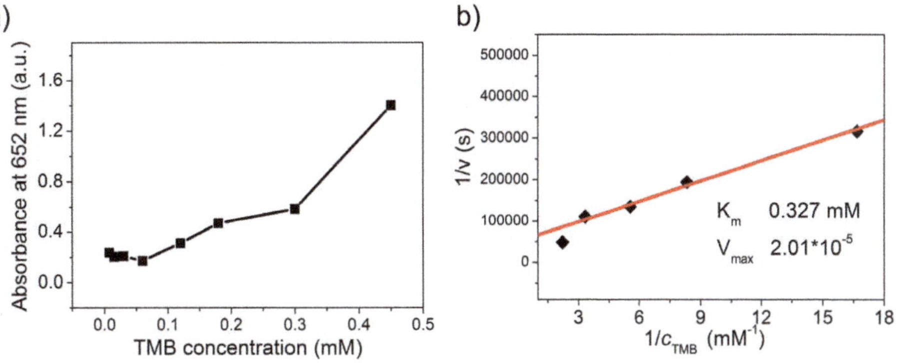

Figure 6. Steady-state kinetic assay of the Ag_3PO_4 nanocrystals. a) The concentration of H_2O_2 was 5 mM and the TMB concentration was varied. b) Double reciprocal plots of $1/V \sim 1/C_{TMB}$. Reaction conditions: 2 mg/mL of Ag_3PO_4 in 5 mL of NaAc buffer with pH 4.0 at 25°C.

tals, it seems that the reaction arrives equilibrium at about 50 minutes. The influence of temperature (10°C, 25°C, 35°C) on the peroxidase-like catalytic activity is also investigated with 20 mg of Ag_3PO_4 nanocrystals. As shown in Fig. 7b, the Ag_3PO_4 nanocrystals show the higher peroxidase-like catalytic activity at temperature of 25°C, although in the initial 15 min, the system has a relatively higher reaction rate at 35°C. The temperature dependent catalytic activity is similar to that of natural enzyme or Fe_3O_4 nanoparticles, which have a preferred temperature.

The use of Ag_3PO_4 nanocrystals as catalysts for electron-transfer reactions has been rarely investigated, and the in-depth catalytic mechanism is also not clear at present stage. It is proposed Ag_3PO_4 nanocrystals plays a role of transferring electrons to hydrogen peroxide, causing them to decompose. It is assumed that the oxygen-oxygen bond of H_2O_2 will rapidly broken by the catalytic action of Ag_3PO_4 nanocrystals to give OH radicals. The OH radicals stabilize at the surface of the Ag_3PO_4 nanocrystals, and react with TMB.

In summary, colloidal Ag_3PO_4 nanocrystals with smaller size were prepared. The peroxides-like catalytic activity of these Ag_3PO_4 nanocrystals were systematically investigated. The results show that they have a higher activity at acid environment. The catalytic activity was also dependent on H_2O_2 concentration, temperature, and catalsy amount. Kinetic analysis indicates that the catalysis reaction is in accord with typical Michaelis-Menten kinetics. The apparent kinetic parameters suggest the higher affinity of Ag_3PO_4 nanocrystals than that of horserahish peroxidase. Our research gives new content to well-known Ag_3PO_4 material and provides a new nanomaterial-based peroxide enzyme mimitics, which would found applications in medical diagnostics and biochemistry.

Materials and Methods

Synthesis of Ag_3PO_4 nanocrystals: Ag_3PO_4 nanocrystals were synthesized with reported methods with minor adjustments [6]. In brief, 8.5 g of $AgNO_3$ and 32 mL of oleylamine were dispersed in 150 mL of toluene and stirred for about 2 h at room temperature. After $AgNO_3$ was fully dissolved, an ethanol solution containing 50 mL of ethanol, 2 mL of H_2O, 2.84 mL of H_3PO_4 was added into the above solution. The solution tuned into yellow colloid quickly. After reaction for 30 mins at room temperature, Ag_3PO_4 nanocrystals were precipitated by adding ethanol, and washed several times with toluene and ethanol. The dark-yellow precipitate was dried in an oven.

Characterization of Ag_3PO_4 nanocrystals: The phase structure of the as-synthesized products were characterized using X-ray diffraction (XRD, Bruker D8 ADVANCE) with Cu-Kα radiation (λ = 1.5406 Å) at a scanning rate of 6° min^{-1}. The morphology and size of the products were examined by a transmission electron microscope (TEM, JEOL JEM-2100) with an accelerating voltage of 200 kV. The Ag_3PO_4 product dispersed in ethanol was dropped onto a holey copper grid covered with an amorphous carbon film for the TEM examination.

Surface modification for Ag_3PO_4 nanocrystals: Before property investigation, the obtained Ag_3PO_4 nanocrystals were firstly treated through usual ligand exchange route [54] to transfer it to being hydrophilic state. Briefly, about 50 mg of Ag_3PO_4 nanocrystals were dispersed into the mixture of hexane (35 mL),

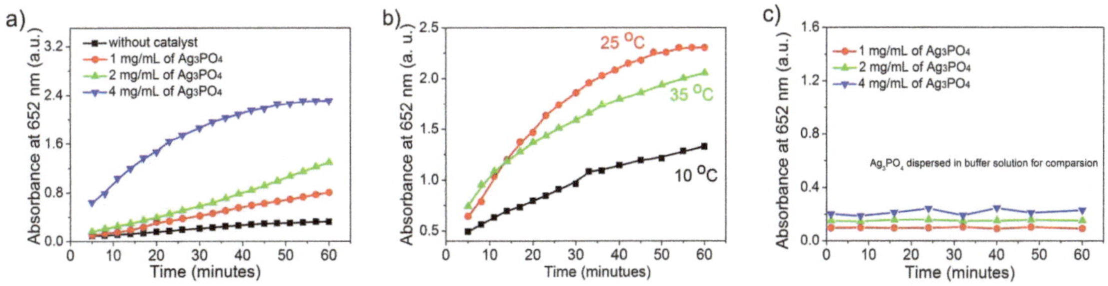

Figure 7. Time-dependent catalytic activity with a) different amounts of Ag_3PO_4 nanocrystals at 25°C and b) at different temperature with 4 mg/mL of Ag_3PO_4 nanocrystals. Reaction conditions: 0.3 mM of TMB, 0–4 mg/mL of Ag_3PO_4, 3.6 mM of H_2O_2 in 5 mL of NaAc buffer with pH of 4. c) Control experiment to show the time-dependent absorption spectra of Ag_3PO_4 nanocrystals dispersed in buffer solution (pH = 4) at different concentrations (1, 2, 4 mg/mL).

distilled water (15 mL), and ethanol (30 mL) through magnetic stirring. Then, 6-amino caproic acid (0.13 g) and equivalent molar $NH_3 \cdot H_2O$ in 5 mL of distilled water was added into the above system. After that, the mixture was heated to 70°C and kept at that temperature for 4 h. The nanocrystals were then collected by centrifugation and washed with water. Through this process, the hydrophobic Ag_3PO_4 nanocrystals were transformed into hydrophilic state, which can be dispersed in water.

Peroxidase-like catalytic activity of Ag_3PO_4 nanocrystals: The peroxidase-like activity of freshly treated Ag_3PO_4 nanocrystals was determined by measuring the formation of a blue charge-transfer complex of diamine from TMB at 652 nm ($\varepsilon = 39000\ M^{-1}\ cm^{-1}$). The TMB oxidation activity measurement, unless otherwise specified, was conducted in sodium acetate buffer (pH 4.0) in the presence of Ag_3PO_4 nanocrystals (2 mg mL^{-1}) with 0.3 mM of TMB and 3.6 mM of H_2O_2. The reaction proceeded at 25°C with time of 30 min.

pH Measurements: The activity of the Ag_3PO_4 nanocrystals at different pH values was performed using the same conditions as above, except different buffer compositions (with different concentration ratios of HAc to NaAc) for different pH values were employed. The reaction was carried out with 2 mg mL^{-1} of Ag_3PO_4 nanocrystals to which TMB (0.3 mM) and H_2O_2 (3.6 mM) were added. The pH of the different buffers was adjusted by using a pH meter.

Determination of kinetic parameters: The steady-state kinetics were performed by varying one of the concentrations of Ag_3PO_4 nanocrystals (0–4 mg mL^{-1}), H_2O_2 (0.35–7 mM), or TMB (0–0.45 mM) at a time. The reaction was carried out in acetate buffer (pH 4.0) for 30 min and monitored by measuring the absorbency at 652 nm. The kinetic curves were adjusted according to the Michaelis-Menten model.

Author Contributions

Conceived and designed the experiments: YJL. Performed the experiments: YJL GXZ JY. Analyzed the data: AHY XPS. Contributed reagents/materials/analysis tools: GXZ. Contributed to the writing of the manuscript: YJL.

References

1. Katsumata H, Taniguchia M, Kanecoa S, Suzuki T (2013) Photocatalytic Degradation of Bisphenol A by Ag_3PO_4 under Visible Light. Catal Commun 34: 30–34.
2. Umezawa N, Shuxin O, Ye J (2011) Theoretical Study of High Photocatalytic Performance of Ag_3PO_4. Phys Rev B 83: 035202.
3. Yi Z, Ye J, Kikugawa N, Kako T, Ouyang S, et al. (2010) An Orthophosphate Semiconductor with Photooxidation Properties under Visible-light Irradiation. Nat Mater 9: 559–564.
4. Bi Y, Ouyang S, Umezawa N, Cao J, Ye J (2011) Facet Effect of Single-Crystalline Ag_3PO_4 Sub-microcrystals on Photocatalytic Properties. J Am Chem Soc 133: 6490–6492.
5. Bi Y, Hu H, Ouyang S, Lu G, Cao J, et al. (2012) Photocatalytic and Photoelectric Properties of Cubic Ag_3PO_4 Sub-microcrystals with Sharp Corners and Edges. Chem Commun 48: 3748–3750.
6. Dinh CT, Nguyen TD, Kleitz F, Do TO (2011) Large-scale Synthesis of Uniform Silver Orthophosphate Colloidal Nanocrystals Exhibiting high Visible Light Photocatalytic Activity. Chem Commun 47: 7797–7799.
7. Liang Q-H, Ma W-J, Shi Y, Li Z, Yang X-M (2012) Hierarchical Ag_3PO_4 Porous Microcubes with Enhanced Photocatalytic Properties Synthesized with the Assistance of Trisodium Citrate. CrystEngComm 14: 2966–2973.
8. Li G-P, Mao L-Q (2012) Magnetically Separable Fe_3O_4-Ag_3PO_4 Sub-micrometre Composite: Facile Synthesis, High Visible Light-driven Photocatalytic Efficiency, and Good Recyclability. RSC Adv 2: 5108–5111.
9. Liu L, Liu J-C, Sun DD (2012) Graphene Oxide Enwrapped Ag_3PO_4 Composite: Towards a Highly Efficient and Stable Visible-Light-Induced Photocatalyst for Water Purification. Catal Sci & Techn 2: 2525–2532.
10. Zhang H-C, Huang H, Ming H, Li H-T, Zhang L-L, et al. (2012) Carbon Quantum dots/Ag_3PO_4 Complex Photocatalysts with Enhanced Photocatalytic Activity and Stability under Visible Light. J Mater Chem 22: 10501–10506.
11. Zhang L-L, Zhang H-C, Huang H, Liu Y, Kang Z-H (2012) Ag_3PO_4/SnO_2 Semiconductor Nanocomposites with Enhanced Photocatalytic Activity and Stability. New J Chem 36: 1541–1544.
12. Bi Y-P, Ouyang S-X, Cao J-Y, Ye J-H (2011) Facile Synthesis of Rhombic Dodecahedral AgX/Ag_3PO_4 (X = Cl, Br, I) Heterocrystals with Enhanced Photocatalytic Properties and Stabilities. Phys Chem Chem Phys 13: 10071–10075.
13. Gao L-Z, Zhuang J, Nie L, Zhang J-B, Zhang Y, et al. (2007) Intrinsic Peroxidase-like Activity of Ferromagnetic Nanoparticles. Nat Nanotechnol 2: 577–583.
14. Wei H, Wang E (2008) Fe_3O_4 Magnetic Nanoparticles as Peroxidase Mimetics and their Applications in H_2O_2 and Glucose Detection. Anal Chem 80: 2250–2254.
15. Yu F, Huang Y, Cole AJ, Yang VC (2009) The Artificial Peroxidase Activity of Magnetic Iron Oxide Nanoparticles and its Application to Glucose Detection. Biomaterials 30: 4716–4722.
16. Wang N, Zhu L, Wang M, Wang D, Tang H (2010) Sono-assisted Preparation of Highly-efficient Peroxidase-like Fe_3O_4 Magnetic Nanoparticles for Catalytic Removal of Organic Pollutants with H_2O_2. Ultrason Sonochem 17: 526–533.
17. Zuo X, Peng C, Huang Q, Song S, Wang L, et al. (2009) Design of a Carbon nanotube/Magnetic Nanoparticle-based Peroxidase-like Nanocomplex and its Application for Highly Efficient Catalytic Oxidation of Phenols. Nano Res 2: 617–623.
18. He S, Shi W, Zhang X, Li J, Huang Y (2010) β-Cyclodextrins-based Inclusion Complexes of $CoFe_2O_4$ Magnetic Nanoparticles as Catalyst for the Luminol Chemiluminescence System and their Applications in Hydrogen Peroxide Detection. Talanta 82: 377–383.
19. Zhang XQ, Gong SW, Zhang Y, Yang T, Wang CY, et al. (2010) Prussian Blue Modified Iron Oxide Magnetic Nanoparticles and Their High Peroxidase-like Activity. J Mater Chem 20: 5110–5116.
20. Liu S, Lu F, Xing R, Zhu JJ (2011) Structural Effects of Fe_3O_4 Nanocrystals on Peroxidase-Like Activity. Chem Eur J 17: 620–625.
21. Niu H, Zhang D, Zhang S, Zhang X, Meng Z, et al. (2011) Humic Acid Coated Fe_3O_4 Magnetic Nanoparticles as Highly Efficient Fenton-like Catalyst for Complete Mineralization of Sulfathiazole. J Hazard Mater 190: 559–565.
22. Park KS, Kim MI, Cho DY, Park HG (2011) Label-Free Colorimetric Detection of Nucleic Acids Based on Target-Induced Shielding Against the Peroxidase-Mimicking Activity of Magnetic Nanoparticles. Small 7: 1521–1525.
23. Shi W, Zhang X, He S, Huang Y (2011) $CoFe_2O_4$ Magnetic Nanoparticles as a Peroxidase Mimic Mediated Chemiluminescence for Hydrogen Peroxide and Glucose. Chem Commun 47: 10785–10787.
24. Korsvik C, Patil S, Seal S, Self WT (2007) Superoxide Dismutase Mimetic Properties Exhibited by Vacancy Engineered Ceria Nanoparticles. Chem Commun 1056–1058.
25. Heckert EG, Karakoti AS, Seal S, Self WT (2008) The Role of Cerium Redox State in the SOD Mimetic Activity of Nanoceria. Biomaterials 29: 2705–2709.
26. Asati A, Santra S, Kaittanis C, Nath S, Perez JM (2009) Oxidase-Like Activity of Polymer-Coated Cerium Oxide Nanoparticles. Angew Chem Int Ed 48: 2308–2312.
27. Pirmohamed T, Dowding JM, Singh S, Wasserman B, Heckert E, et al. (2010) Nanoceria Exhibit Redox State-dependent Catalase Mimetic Activity. Chem Commun 46: 2736–2738.
28. Andre R, Natalio F, Humanes M, Leppin J, Heinze K, et al. (2011) V_2O_5 Nanowires with an Intrinsic Peroxidase-Like Activity. Adv Funct Mater 21: 501–509.
29. Song Y, Qu K, Zhao C, Ren J, Qu X (2010) Graphene Oxide: Intrinsic Peroxidase Catalytic Activity and its Application to Glucose Detection. Adv Mater 22: 2206–2210.
30. Song Y, Wang X, Zhao C, Qu K, Ren J, et al. (2010) Label-Free Colorimetric Detection of Single Nucleotide Polymorphism by Using Single-Walled Carbon Nanotube Intrinsic Peroxidase-Like Activity. Chem Eur J 16: 3617–3621.
31. Qu F, Li T, Yang M (2011) Colorimetric Platform for Visual Detection of Cancer Biomarker Based on Intrinsic Peroxidase Activity of Graphene Oxide. Biosens Bioelectron 26: 3927–3931.
32. Cui R, Han Z, Zhu JJ (2011) Helical Carbon Nanotubes: Intrinsic Peroxidase Catalytic Activity and Its Application for Biocatalysis and Biosensing. Chem Eur J 17: 9377–9384.
33. Song Y, Chen Y, Feng L, Qu X (2011) Selective and Quantitative Cancer Cell Detection Using Target-directed Functionalized Graphene and its Synergetic Peroxidase-like Activity. Chem Commun 47: 4436–4438.
34. Shi W, Wang Q, Long Y, Cheng Z, Chen S, et al. (2011) Carbon Nanodots as Peroxidase Mimetics and their Applications to Glucose Detection. Chem Commun 47: 6695–6697.
35. Wang X, Qu K, Xu B, Ren J, Qu X (2011) Multicolor Luminescent Carbon Nanoparticles: Synthesis, Supramolecular Assembly with Porphyrin, Intrinsic Peroxidase-like Catalytic Activity and Applications. Nano Res 4: 908–920.
36. Song Y, Wei W, Qu X (2011) Colorimetric Biosensing Using Smart Materials. Adv Mater 23: 4215–4236.

37. He W, Wu X, Liu J, Hu X, Zhang K, et al. (2010) Design of AgM Bimetallic Alloy Nanostructures (M = Au, Pd, Pt) with Tunable Morphology and Peroxidase-like Activity. Chem Mater 22: 2988–2994.
38. Jv Y, Li B, Cao R (2010) Positively-charged Gold Nanoparticles as Peroxidiase Mimic and their Application in Hydrogen Peroxide and Glucose Detection. Chem Commun 46: 8017–8019.
39. Fan J, Yin JJ, Ning B, Wu X, Hu Y, et al. (2011) Direct Evidence for Catalase and Peroxidase Activities of Ferritin–platinum nanoparticles. Biomaterials 32: 1611–1618.
40. He W, Liu Y, Yuan J, Yin JJ, Wu X, et al. (2011) Au@ Pt nanostructures as Oxidase and Peroxidase Mimetics for Use in Immunoassays. Biomaterials 32: 1139–1147.
41. Liu J, Hu X, Hou S, Wen T, Liu W, et al. (2011) Screening of Inhibitors for Oxidase Mimics of Au@ Pt Nanorods by Catalytic Oxidation of OPD. Chem Commun47: 10981–10983.
42. Wang X-X, Wu Q, Shan Z, Huang QM (2011) BSA-stabilized Au Clusters as Peroxidase Mimetics for use in Xanthine Detection. Biosens Bioelectron 26: 3614–3619.
43. Josephy PD, Eling TE, Mason RP (1982) The Horseradish Peroxidase-catalyzed Oxidation of 3,5,3′,5′-Tetramethylbenzidine: Free Radical and Charge-transfer Complex Intermediates. J Biol Chem 257: 3669–3675.
44. Tian J-Q, Liu S, Luo YL, Sun X-P (2012) Glucose Oxidation Using Au-containing Bimetallic and Trimetallic Nanoparticles. Catal Sci Technol 2: 432–436.
45. Ma M, Zhang Y, Gu N (2011) Peroxidase-like Catalytic Activity of Cubic Pt Nanocrystals. Colloids and Surfaces A: Physicochem. Eng. Aspects 373: 6–10.
46. Dong Y-L, Zhang H-G, Rahman ZU, Su L, Chen XJ, et al. (2012) Graphene oxide-Fe_3O_4 Magnetic Nanocomposites with Peroxidase-like Activity for Colorimetric Detection of Glucose. Nanoscale 4: 3969–3976.
47. Chen W, Chen J, Feng YB, Hong L, Chen QY, et al. (2012) Peroxidase-like Activity of Water-soluble cupric Oxide Nanoparticles and its Analytical Application for Detection of Hydrogen Peroxide and Glucose. Analyst 137: 1706–1712.
48. Jiang H, Chen Z, Cao H, Huang Y (2012) Peroxidase-like Activity of Chitosan Stabilized Silver Nanoparticles for Visual and Colorimetric Detection of Glucose. Analyst 137: 5560–5564.
49. Liu J-B, Hua X, Hou S, Wen T, Liu W-Q, et al. (2012) Au@Pt Core/shell Nanorods with Peroxidase- and Ascorbate Oxidase-like Activities for Improved Detection of Glucose. Sensors and Actuators B166–167: 708–714.
50. Dutta AK, Maji SK, Srivastava DN, Mondal A, Biswas P, et al. (2012) Synthesis of FeS and FeSe Nanoparticles from a Single Source Precursor: A Study of Their Photocatalytic Activity, Peroxidase-Like Behavior, and Electrochemical Sensing of H_2O_2. ACS Appl. Mater. Interfaces 4: 1919–1927.
51. Sun X-L, Guo S-J, Chung C-S, Zhu W-L, Sun S-H (2013) A Sensitive H_2O_2 Assay Based on Dumbbell-like PtPd-Fe_3O_4 Nanoparticles. Adv Mater 25: 132–136.
52. Wang Q, Liu S-W, Sun H-Y, Lu Q-F (2014) Synthesis and Intrinsic Peroxidase-Like Activity of Sisal-Like Cobalt Oxide Architectures. Ind. Eng. Chem. Res. 53: 7917–7922.
53. Karaseva EI, Losev YP, Metelitsa DI (2002) Peroxidase-catalyzed Oxidation of 3, 3″, 5, 5″-Tetramethylbenzidine in the Presence of 2, 4-Dinitrosoresorcinol and Polydisulfide Derivatives of Resorcinol and 2, 4-Dinitrosoresorcinol. Russ J Bioorg Chem 28: 128–135.
54. Tian Q-W, Jiang F-R, Zou R-J, Liu Q, Chen Z-G, et al. (2011) Hydrophilic Cu_9S_5 Nanocrystals: A Photothermal Agent with a 25.7% Heat Conversion Efficiency for Photothermal Ablation of Cancer Cells in Vivo. ACS Nano 5: 9761–9771.

23

Supported Palladium Nanoparticles Synthesized by Living Plants as a Catalyst for Suzuki-Miyaura Reactions

Helen L. Parker[1], Elizabeth L. Rylott[2], Andrew J. Hunt[1], Jennifer R. Dodson[1], Andrew F. Taylor[2], Neil C. Bruce[2]*, James H. Clark[1]*

1 Green Chemistry Centre of Excellence, Department of Chemistry, University of York, York, United Kingdom, 2 Centre for Novel Agricultural Products, Department of Biology, University of York, York, United Kingdom

Abstract

The metal accumulating ability of plants has previously been used to capture metal contaminants from the environment; however, the full potential of this process is yet to be realized. Herein, the first use of living plants to recover palladium and produce catalytically active palladium nanoparticles is reported. This process eliminates the necessity for nanoparticle extraction from the plant and reduces the number of production steps compared to traditional catalyst palladium on carbon. These heterogeneous plant catalysts have demonstrated high catalytic activity in Suzuki coupling reactions between phenylboronic acid and a range of aryl halides containing iodo-, bromo- and chloro- moieties.

Editor: Andrew C. Marr, Queen's University Belfast, United Kingdom

Funding: Funding was received from EPSRC, grant code: EP/P505178/1. Funding was also received from "The G8 Research Councils Initiative on Multilateral Research Funding" (funds came from EPSRC) grant code: EP/K022482/1. The funders had no role in study design, data collection and analysis, decision to publish, or preparation of the manuscript.

Competing Interests: The authors have declared that no competing interests exist.

* E-mail: neil.bruce@york.ac.uk (NCB); james.clark@york.ac.uk (JHC)

Introduction

The use of plants to capture specific elements, particularly metals, from contaminated soils or aqueous waste has long been recognized as a potential method for environmental clean-up. [1,2] There has only been a limited number of studies into platinum group metals (PGMs) accumulation by plants, with most work focusing on the capture of metals such as, nickel, zinc and cadmium. [3] Whilst efforts to use plants as a tool for remediation of contaminated areas has demonstrated some promise, the full potential of this technology has not yet been realized.

Metallic nanoparticles (MNP) are receiving considerable interest from the scientific community due to the remarkable properties they present compared to the bulk of the same metal. [4] Previous work in the area of sustainable MNP production using plant extracts, algae and bacteria to form bio-MNPs has shown promise. [5–8] For example, Lloyd *et al* has synthesised palladium catalysts using bacteria such as, *Geobacter sulfurreducens* and *Escherichia coli* that show catalytic activity in a variety of reactions including Heck couplings. [9,10] Yang *et al* have also shown that viruses can be used to make bio-MNPs, where tobacco mosaic virus-templated palladium nanoparticles were used as effective catalysts for Suzuki reactions. [11] Unfortunately, there are certain limitations when using bacteria for MNP synthesis, for example, a reducing agent such as H_2 is often required in order for nanoparticle formation. [11,12] There is also the problem that bacteria may not be stable under certain reaction conditions, for instance, many bacteria will disintergrate at high (>60°C) temperatures. Sulphur is an essential constituent of microbial biomass and when cells disintegrate this sulphur will be released into the reaction consequently poisoning the catalyst [8].

It has been proposed that a possible route to synthesis of MNPs is through biological production by living plants. [13] Plants are known to uptake metal salts and are capable of reducing these metal ions to form crystalline nanostructures. [14] Such MNPs in plants have been described for Au, Ag and Cu. [15–17] After formation of MNPs, it is common that extraction from biomass is attempted by methods such as freeze-thawing, biomass incineration, and chemical leaching. However, some limitations of these activities are that they are laborious, energy intensive and can destroy the MNP structure thus interfering with the desired nano-material properties. [5,18] This paper demonstrates the first production of palladium nanoparticles (PdNPs) by living plants and, as an advantage over extraction methods outlined above, their direct use as effective catalysts for a range of Suzuki-Miyaura coupling reactions. Pd is arguably the most ubiquitous metal currently used in organic synthesis. Whilst some investigations into the potential bioaccumulation of palladium by plants have been carried out, [19] to the best of the authors knowledge no studies have investigated the fate of the palladium once it is accumulated or any potential applications of these materials.

Materials and Methods

Materials

Palladium on carbon 10% (Pd/C), palladium acetate ($Pd(OAc)_2$), potassium tetrachloropalladate (K_2PdCl_4) and all reagents and solvents used in this work were purchased from Sigma Aldrich and used as received.

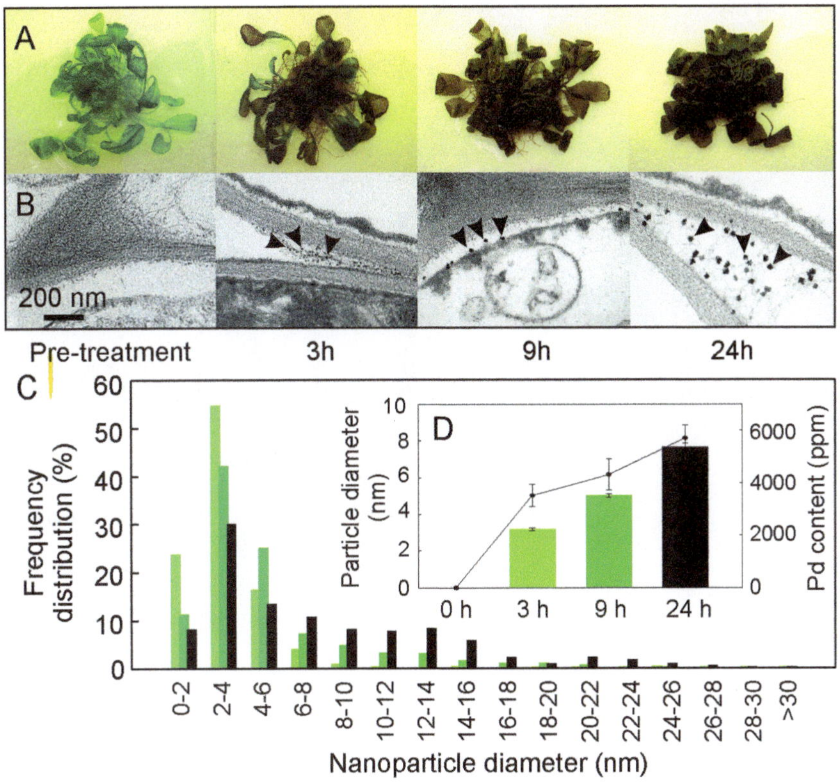

Figure 1. Pd uptake and PdNP formation in Arabidopsis: (A) appearance of 3-week-old plants 24 h after treatment with K_2PdCl_4. (B) TEM showing accumulation over time of PdNPs in cell wall corners. (C) Distribution of NP sizes in leaf tissue with time. (D) Mean NP diameter and Pd concentration with time.

Growth of *Arabidopsis* Containing Palladium

Wild-type Arabidopsis ecotypes Col-0 were obtained from the Nottingham Arabidopsis Stock Centre at the University of Nottingham. Seeds were imbibed for 4 days at 277 K on agar plates containing 20 mM sucrose and half strength Murashige and Skoog salts. Seedlings were grown in 16 h of light with 125 mmol m^{-2} s^{-1} white light. For liquid culture experiments, individual seedlings were transferred to 100-mL conical flasks containing 20 mL of half-strength Murashige and Skoog medium and 20 mM sucrose. Plants were grown under 20 mmol m^{-2} s^{-1} light on a rotary shaker at 130 rpm. The medium was then replaced with 10 mM potassium tetrachloropalladate(II). The plants remained in the palladium solution for 24 hours. Samples were taken for TEM periodically to 24 hours, then remaining tissue was washed, dried and ground to a powder.

Transmission Electron Microscopy

Plant leaf tissues were fixed in 2.5% (v/v) glutaraldehyde, 4% formaldehyde (v/v) in 50 mM phosphate buffer for 3.5 hours a secondary fix of 1% osmium tetroxide. Samples were then dehydrated through a 25–100% acetone series and infiltrated with Spurrs resin (Agar Scientific) (25, 50 and 75%) with overnight polymerisation at 343 K. Sections were mounted on 400 mesh thin-bar Athene grids (Agar Scientific), stained with saturated uranyl acetate and Reynolds lead citrate and viewed using a Tecnai 12 Bio Twin TEM operating at 120 kV.

Measuring Nanoparticle Size

Nanoparticle cross-sectional diameter was measured from TEM images of the plant. A minimum of 100 (but often >250) nanoparticles were measured for each material.

X-Ray Photoelectron Spectroscopy

XPS spectra were recorded on a Kratos Axis Ultra DLD photoelectron spectrometer using a hemispherical photoelectron analyser with a monochromatic AlKα X-ray source (75–150 W) and analyser pass energies of 160 eV (for survey scans) and 40 eV (for detailed scans). Samples were mounted using double-sided tape. Binding energies were referenced to the C 1 s binding energy 285 eV. Prior to analysis samples were degassed overnight at ultrahigh vacuum ($<5\times10^{-10}$ Torr). CasaXPS software was used to carry out analysis of the spectra.

Formation of the Catalyst

Two catalysts were prepared from the plant material. To produce the catalysts the dried plant powder was pyrolised using the Netzsch STA 409 under N_2 at 1 K min^{-1} to 573 K (300°C) (Pd-P-300) and 1073 K (800°C) (Pd-P-800) respectively.

Testing Catalytic Activity

For the reaction of aryl halide with phenylboronic acid typical reaction conditions were used: [20] Into a 25 ml round bottom flask the following reagents were measured; aryl halide (0.6 mmol), phenylboronic acid (0.7 mmol), sodium bicarbonate (2 mmol) and tetrabutylammonium bromide (TBAB) (1 mmol). Ethanol (2 ml) and water (1 ml) was then added and the flask heated with stirring to 328 K. Once the flask had heated to the required temperature

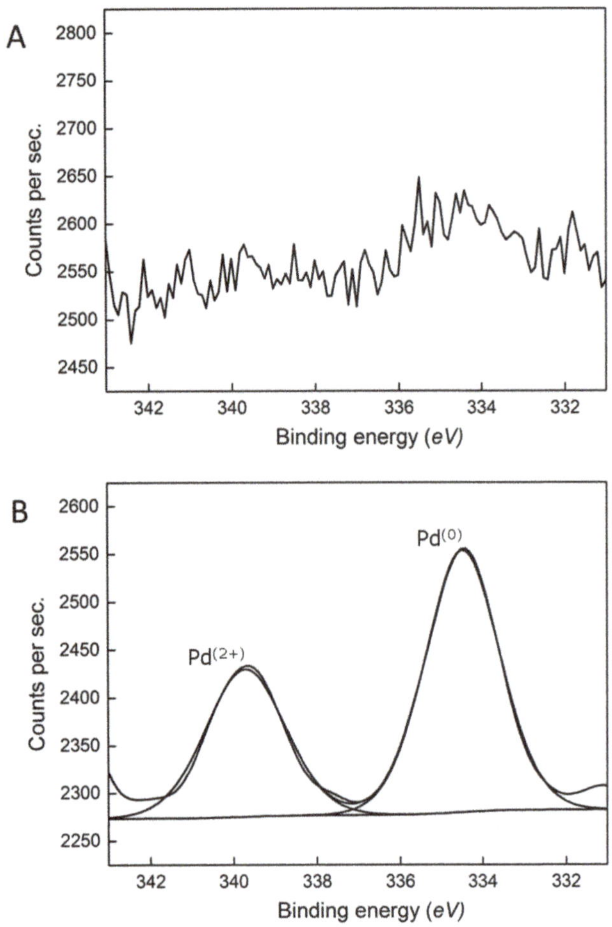

Figure 2. XPS spectra showing: a) undosed control plant, b) plant dosed with palladium for 24 h.

Pd catalyst was added, equivalent amount of catalyst was added to achieve a 0.12 mol % Pd concentration (based on aryl halide concentration). For control experiments no catalyst was added, also a reaction was run with plant material not containing palladium, 8 mg of material was used. The reaction was allowed to proceed for 20 hours at 328 K. Reaction was monitored by GC-FID using diethyl succinate as a standard. After 20 hours the reaction was allowed to cool. The cooled reaction mixture was filtered to remove the catalyst, undissolved sodium bicarbonate and TBAB. The filtered solution was purified by flash column chromatography using pentane as the solvent. Purified, isolated products were characterised by ^{1}HNMR and ^{13}CNMR obtained using a JOEL JNM-ECS400 NMR operating at 400 MHz and 100 MHz respectively, 1024 scans were taken for each sample.

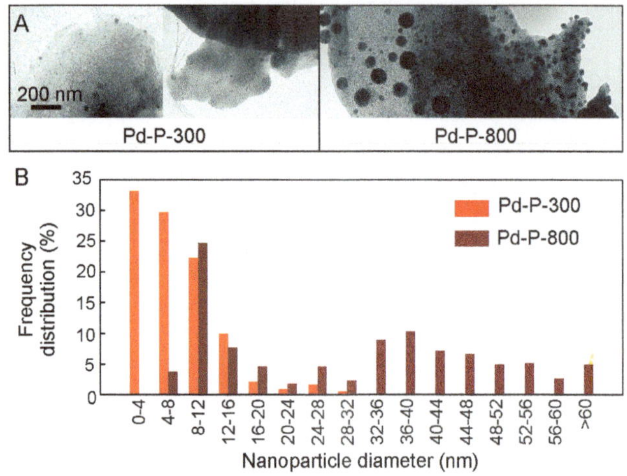

Figure 4. (A) TEM showing PdNPs in Pd-P-300 and Pd-P-800. (B) Distribution of PdNP sizes in the plant catalysts.

Results and Discussion

The preparation and characterisation of Pd plant materials utilized techniques adapted from previously published work by Sharma *et al.* for the synthesis of gold nanoparticles. [21] For the work presented here, three-week-old, liquid culture-grown *Arabidopsis thaliana* (L.) (Arabidopsis) ecotype Col0 were dosed with an aqueous solution of K_2PdCl_4 over a period of 24 hours. Liquid culture was chosen over a soil-based system in order to easily control accurate dosing of metals. This method also allowed the plant material to readily be washed clean from the growth substrate after dosing, thus reducing contamination with non-plant material. After 24 hours from the time of dosing with palladium the plants appeared brown and wilted (Figure 1(a)), and were removed from the growth medium, washed, dried and ground to a powder. Transmission electron micrograph (TEM) images of sections of the plants revealed that within three hours of dosing, the plants had successfully formed well-dispersed, spherical MNPs with a mean profile diameter of 3 nm (Figure 1 (b)–(d)). Increasing concentrations and diameters of NPs were observed over time, with larger particles up to 32 nm formed by the end of the incubation (Figure 1 (c)–(d)). MNPs were distributed evenly adaxially to abaxially across the leaf (Figure S1), predominantly concentrated in the wall and apoplast regions of cell junctions. This distribution of MNPs across the leaf has not been reported before under the conditions used here. When the plants were grown in liquid culture, they were partially submerged, and the levels of cuticular wax on leaf surfaces appeared to be compromised. Together, these factors may have contributed to an increase in the passive uptake of metal ions through the leaf membranes. Inductively coupled plasma (ICP) analysis of palladium concentration in the plants over time showed the plants reached maximum palladium concentration after 18 hours of exposure (Figure 1 (d)).

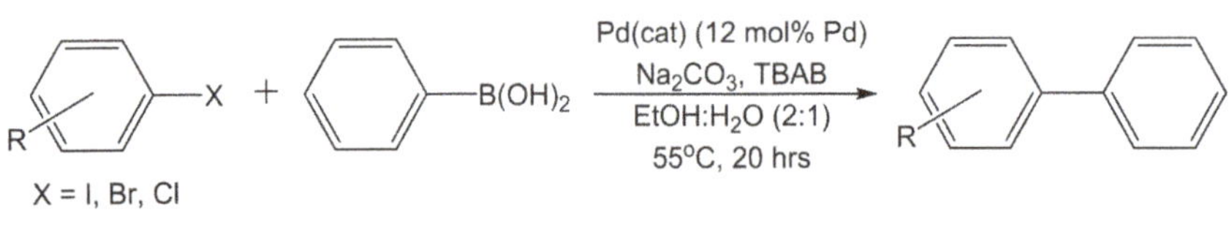

Figure 3. Suzuki-Miyaura C-C coupling reaction of arylhalide with phenylboronic acid.

Table 1. Suzuki-Miyaura coupling of aryl halides with phenylboronic acid under Pd-P-300 catalysis.[a]

Entry	R	X	Yield (%)[b]
1	H	I	>99 (91)[c] (>99)[d]
2	Cl	I	98
3	4-C(O)CH3	I	>99
4	3-NO2	I	93
5	4-C(O)CH3	Br	98
6	2-C(O)CH3	Br	94
7	4-NO2	Br	79
8	2-NO2	Br	>99
9	4-CN	Br	93 (60)[c] (78)[d]
10	3-CN	Br	>99
11	4-CN	Cl	81 (9)[c] (35)[d]

[a]*Reaction conditions:* aryl halide (0.6 mmol), phenylboronic acid (0.7 mmol), TBAB (1 mmol), Na_2CO_3 (2 mmol), Pd-P-300 (12 mol % Pd) in EtOH:H_2O (2 ml:1 ml), NMR characterisation of products available in supporting information.
[b]Isolated yields after column chromatography with *n*-pentane.
[c]Isolated yield for Palladium on Carbon 10%.
[d]Isolated yield for palladium acetate.

Analysis using X-ray Photoelectron Spectroscopy (XPS) on the dried Pd-plant material confirmed that the observed MNPs were palladium deposits. The spectra showed two chemically distinct spin-orbit pairs in the Pd 3d levels, centred at (I) 334.5 eV and (II) 339.7 eV binding energies, these peaks were not present in the control plant material (Figure 2). Peak (I) corresponds to fully reduced $Pd^{(0)}$ nanoparticles, whilst peak (II) is due to unreduced $Pd^{(2+)}$ ions. [22] The mechanism for metal reduction inside living plants is not yet understood but binding to carboxyl, amino and sulfhydryl groups are likely to be important in binding prior to reduction. A number of bacteria can reduce metals using enzymes such as the arsenate reductases and mercury reductases. [23,24] However, similar enzymes have so far not been found in any plant species and the reduction of $Pd^{(2+)}$ to $Pd^{(0)}$ and formation of MNPs under the experimental conditions used here, could be a chemical, rather than biological, process.

Palladium catalyzed Suzuki-Miyaura C-C bond forming reactions are important methods for the synthesis of pharmaceutical intermediates and other high value molecules and were chosen to determine the catalytic activity of the plant catalysts (Figure 3). [25] In order to provide an effective support for the PdNPs the plant material was carbonized under nitrogen with a heating rate of 1 K min^{-1}. This heating caused the diminution of heteroatoms such as oxygen and hydrogen as volatile gaseous emissions, resulting in the formation of a stable, carbon support that could withstand the reaction conditions. To determine the effect of carbonization temperature on PdNP state, and catalytic activity, the plant material was heated to 300°C or 800°C to produce materials named Pd-P-300 and Pd-P-800 respectively. During carbonization thermal gravimetric-infrared (TGIR) analysis of the material showed a 45% mass loss between 100–300°C, attributed to the loss of water and carbon dioxide (Figure S2). These catalytic materials contained a palladium concentration of 15%, determined by ICP, with the remaining 85% of the material made up predominantly of carbon and oxygen. The XPS analysis of Pd-P-300 showed the appearance of PdO after carbonisation of the material (Figure S3 and Table S1). The presence of PdO could be attributed to interaction between the $Pd^{(2+)}$ and oxygen that is given off during heating. Table S1 also shows the XPS specta for palladium species on Pd/C, results showed that there are no $Pd^{(0)}$ species present, only $Pd^{(2+)}$ and PdO, this may result in a difference in reactivity of Pd/C and Pd-P-300.

Analysis using TEM revealed highly dispersed, spherical palladium NPs in both catalytic materials (Figure 4). Further examination of Pd-P-300 revealed the mean cross-sectional diameter and frequency distribution of PdNPs was broadly similar to the pre-carbonized samples. However, for Pd-P-800, a marked shift in NP size was observed, with a 400% increase in mean diameter and wider frequency distribution. This indicates that between 300°C and 800°C PdNPs have diffused through the carbon support to form larger aggregations. This observation is consistent with previous observations by others, including Joo *et al.* and Jiang *et al.* [26,27].

The driving force for NP aggregation comes from the higher chemical potential and increased surface free energy of metal atoms in small NPs in comparison with bulk metal surfaces. [28] This often leads to the deactivation of heterogeneous or supported NP catalysts and therefore may affect the activity of Pd-P-800. Initial reactions carried out using Pd-P-800 catalysts were unsuccessful resulting in no product likely due to the large PdNPs present and no further work was carried out with this material. Control reactions also exhibited no conversion without catalyst and when dried and carbonized plant material (containing no palladium) was used.

To test the activity of Pd-P-300 initially various temperatures, bases (organic or inorganic), solvents and mixtures of solvent were screened in order to determine the optimal reaction conditions (Table S2). The reaction of iodobenzene (1 mol eqv) with phenylboronic acid (1.2 mol eqv) in the presence of Na_2CO_3 (3 mol eqv) and tetrabuthylammonium bromide (TBAB) (2 mol eqv) in a solvent mixture of EtOH-H_2O (2 ml:1 ml) at 55°C resulted in an isolated yield of >99% of the coupling product (Table 1, entry 1), and these reaction conditions were chosen for further experiments. TBAB was used in this reaction as a phase transfer agent and a stabiliser for the PdNPs, [29] previous work has shown that the presence of TBAB in Suzuki reactions facilitates the reaction [30].

Using this catalytic system Pd-P-300 was successfully applied to a range of other Suzuki-Miyaura reactions (Table 1). Yields were consistently high for the different aryliodides tested (Table 1, entries 1–4). The Pd-P-300 was also successful at catalyzing reactions involving a wide range of arylbromides (entries 5–11) and even a reaction with an arylchloride (entry 12) which are notoriously more difficult to react than the iodo-containing reagents. [31] This activity is excellent when compared against other non-conventional palladium catalysts. [11,20,32] As an example, Suzuki reactions carried out by Søbjerg *et al.*, under the same conditions, using bio-$Pd^{(0)}$ formed using the gram-negative proteobacteria *Cupriavidus nectar* failed to show any reaction for the bromoarene tested (coupling using chloroarenes was not attempted). [20] In comparison to commercially available catalyst Pd on carbon 10% (Pd/C) and $Pd(OAc)_2$, the reactivity of Pd-P-300 was superior in the case of all the reactions tested which is highly promising for the future of this material (Table 1, entry 1, 9 and 11).

Conclusion

In summary, this work has demonstrated the ability of Arabidopsis to produce PdNPs in a relatively simple way without the requirement for toxic chemicals or energy intensive processes.

These PdNPs show excellent catalytic activity across a range of Suzuki-Miyaura coupling reactions involving I, Br and Cl leaving groups, also producing higher yields than commercial Pd catalyst.

This method of plant-mediated MNP synthesis may represent a greener alternative to many other techniques. There is still further work to undertake, namely testing plant species more suited to in-field application, soil-based systems and the use of waste streams containing Pd to act as feedstocks.

Supporting Information

Figure S1 Mean nanoparticle diameter in cells from adaxial to abaxial leaf surface. TEM sections from 3-week-old, liquid culture grown Arabidopsis plants were treated with 10 mM potassium tetrachloropalladate and the mean palladium nanoparticle profile areas measured across the leaf over time.

Figure S2 (A) Thermal gravimetric analysis of plant material pyrolisis to 300°C (B) DSC signal showing significant mass loss (C) Infra-red spectrum of emission at (B).

Figure S3 Pd XPS spectra of Pd-P-300.

Figure S4 Biphenyl ^{1}H NMR and ^{13}C NMR.

Figure S5 4-Chlorobiphenyl ^{1}H NMR and ^{13}C NMR.

Figure S6 4-Acetylbiphenyl ^{1}H NMR and ^{13}C NMR.

Figure S7 2-Acetylbiphenyl ^{1}H NMR and ^{13}C NMR.

Figure S8 4-Nitrobiphenyl ^{1}H NMR and ^{13}C NMR.

Figure S9 2-Nitrobiphenyl ^{1}H NMR and ^{13}C NMR.

Figure S10 4-Cyanobiphenyl ^{1}H NMR and ^{13}C NMR.

Figure S11 3-Cyanobiphenyl ^{1}H NMR and ^{13}C NMR.

Table S1 Composition palladium present in carbonised palladium plant (Pd-P-300) from XPS spectra.

Table S2 Screening conditions to find optimal conditions for Suzuki reactions.

Table S3 ^{1}H NMR and ^{13}C NMR data for isolated products.

Acknowledgments

The authors would like to thank Ms M. Stark for assistance with the TEM and SEM analysis and Dr B. Johnson for the XPS analysis.

Author Contributions

Conceived and designed the experiments: HLP ELR AJH JRD AFT NCB JHC. Performed the experiments: HLP ELR. Analyzed the data: HLP ELR AJH. Wrote the paper: HLP ELR AJH JRD AFT NCB JHC.

References

1. Salt DE, Blaylcock M, Kumer NPBA, Dushenkov V, Ensley BD, et al. (1995) Pytoremediation: a novel strategy for the removal of toxic metals from the environment using plants. Nat Biotechnol 13 468–474.
2. Brooks RR (1998a) Plants that Hyperaccumulate Heavy Metals: International, New York, NY. 1–14 p.
3. Yang XE, Long XX, Ye HB, He ZL, Calvert DV, et al. (2004) Cadmium tolerance and hyperaccumulation in a new Zn-hyperaccumulating plant species (*Sedum alfredii* Hance). Plant Soil 259: 181–189.
4. Astruc D, Lu F, Aranzaes JR (2005) Nanoparticles as Recyclable Catalysts: The Frontier between Homogeneous and Heterogeneous Catalysis. Angew Chem Int Ed 44: 7852–7872.
5. Iravani S (2011) Green synthesis of metal nanoparticles using plants. Green Chem 13: 2638.
6. Bennett JA, Creamer NJ, Deplanche K, Macaskie LE, Shannon IJ, et al. (2010) Palladium supported on bacterial biomass as a novel heterogeneous catalyst: A comparison of Pd/Al_2O_3 and bio-Pd in the hydrogenation of 2-pentyne. Chem Eng Sci 65: 282–290.
7. Singaravelu G, Arockiamary JS, Ganesh Kumar V, Govindaraju K (2007) A novel extracellular synthesis of monodisperse gold nanoparticles using marine alga, Sargassum wightii Greville. Colloids and Surfaces B: Biointerfaces 57: 97–101.
8. De Corte S, Hennebel T, De Gusseme B, Verschuere S, Boon N (2012) Bio-palladium: from metal recovery to catalytic applications. Microbial Biotechnology 5: 5–17.
9. Coker VS, Bennett JA, Telling ND, Henkel T, Charnock JM, et al. (2010) Microbial Engineering of Nanoheterostructures: Biological Synthesis of a Magnetically Recoverable Palladium Nanocatalyst ACS Nano 4: 2577–2584.
10. Foulkes JM, Malone KJ, Coker VS, Turner NJ, Lloyd JR (2011) Engineering a Biometallic Whole Cell Catalyst for Enantioselective Deracemization Reactions. ACS Catalysis 1: 1589–1594.
11. Yang C, Manocchi AK, Lee B, Yi H (2011) Viral-templated palladium nanocatalysts for Suzuki coupling reaction. J Mater Chem 21: 187–194.
12. Dodson J, Parker H (2013) Mining municipal waste: prospective for elemental recovery. In: Hunt A, editor. Element Recovery and Sustainability. London: Royal Society of Chemistry. 220–258.
13. Haverkamp RG, Marshall AT (2009) The mechanism of metal nanoparticle formation in plants: limits on accumulation. J Nanopart Res 11: 1453–1463.
14. Anderson CWN, Bhatti SM, Gardea-Torresdey J, Parsons J (2013) In vivo effects of copper and silver on synthesis of gold nanoparticles inside living plants. ACS Sustainable Chem Eng 1: 640–648.
15. Anderson CWN, Brooks RR, Stewart RB (1998) Harvesting a crop of gold in plants. Nature 395: 553–554.
16. Harris AT, Bali R (2008) On the formation and extent of uptake of silver nanoparticles by live plants. J Nanopart Res 10: 691–695.
17. Manceau A, Nagy KL, Marcus MA, Lanson M, Geoffroy N, et al. (2008) Formation of metallic copper nanoparticles at the soil-root interface. Environ Sci Technol 42: 1766–1772.
18. Losfeld G, Escande V, La Blache PV, L'Huillier L, Grison C (2012) Design and performance of supported Lewis acid catalysts derived from metal contaminated biomass for Friedel-Crafts alkylation and acylation. Catal Today 189: 111–116.
19. Nemutandani T, Dutertre D, Chimuka L, Cukrowska E, Tutu H (2006) The potential of Berkheya coddii for phytoextraction of nickel, platinum, and palladium contaminated sites. Toxicol Environ Chem 88: 175–185.
20. Sobjerg LS, Gauthier D, Lindhardt AT, Bunge M, Finster K, et al. (2009) Bio-supported palladium nanoparticles as a catalyst for Suzuki-Miyaura and Mizoroki-Heck reactions. Green Chem 11: 2041–2046.
21. Sharma NC, Sahi SV, Nath S, Parsons JG, Gardea-Torresdey JL, et al. (2007) Synthesis of plant-mediated gold nanoparticles and catalytic role of biomatrix-embedded nanomaterials. Environ Sci Technol 41: 5137–5142.
22. Mandal S, Roy D, Chaudhari RV, Sastry M (2004) Pt and Pd nanoparticles immobilized on amine-functionalized zeolite: excellent catalysts for hydrogenation and Heck reactions. Chem Mater 16: 3714–3724.
23. Oremland RS, Stolz JF (2003) The Ecology of Arsenic. Science 300: 939–944.
24. Fox B, Walsh CT (1982) Mercuric reductase: homology to glutathione reductase and lipoamide dehydrogenase. Iodoacetamide alkylation and sequence of the active site peptide. J Biol Chem 257: 2498–2503.
25. Torborg C, Beller M (2009) Recent applications of Palladium-Catalyzed coupling reaction in the pharmaceutical, agrochemical and fine chemical industries. Adv Synth Catal 351: 3027–3043.
26. Joo SH, Park JY, Tsung C-K, Yamada Y, Yang P, et al. (2009) Thermally stable Pt/mesoporous silica core-shell nanocatalysts for high-temperature reactions. Nature Materials 8: 126–131.
27. Jiang L, Hsu A, Chu D, Chen R (2009) Size-dependent activity of palladium nanoparticles for oxygen electroreduction in alkaline solutions. J Electrochem Soc 156: B643–B649.
28. Datye AK, Xu Q, Kharas KC, McCarty JM (2006) Particle size distributions in heterogeneous catalysts: What do they tell us about the sintering mechanism? Catal Today 111: 59–67.

29. Adak L, Chattopadhyay K, Ranu BC (2009) Palladium Nanoparticle-Catalyzed C−N Bond Formation. A Highly Regio- and Stereoselective Allylic Amination by Allyl Acetates. J Org Chem 74: 3982–3985.
30. Badone D, Baroni M, Cardamone R, Ielmini A, Guzzi U (1997) Highly Efficient Palladium-Catalyzed Boronic Acid Coupling Reactions in Water: Scope and Limitations. J Org Chem 62: 7170–7173.
31. Wolfe JP, Singer RA, Yang BH, Buchwald SL (1999) Highly active palladium catalysts for suzuki coupling reactions. J Am Chem Soc 121: 9550–9561.
32. Karami K, Ghasemi M, Naeini NH (2013) Palladium nanoparticles supported on polymer: An efficient and reusable heterogeneous catalyst for the Suzuki cross-coupling reactions and aerobic oxidation of alcohols. Catal Commun 38: 10–15.

From Arylamine *N*-Acetyltransferase to Folate-Dependent Acetyl CoA Hydrolase: Impact of Folic Acid on the Activity of (HUMAN)NAT1 and Its Homologue (MOUSE)NAT2

Nicola Laurieri[1,2], Julien Dairou[3], James E. Egleton[1,2], Lesley A. Stanley[4], Angela J. Russell[1,2], Jean-Marie Dupret[3], Edith Sim[1,5]*, Fernando Rodrigues-Lima[3]*

1 Department of Pharmacology, University of Oxford, Oxford, United Kingdom, **2** Department of Chemistry, Chemistry Research Laboratory, University of Oxford, Oxford, United Kingdom, **3** Université Paris Diderot, Sorbonne Paris Cité, Unit of Functional and Adaptive Biology, Paris, France, **4** Consultant in Investigative Toxicology, Linlithgow, West Lothian, United Kingdom, **5** Faculty of Science, Engineering and Computing, Kingston University, Kingston on Thames, United Kingdom

Abstract

Acetyl Coenzyme A-dependent *N*-, *O*- and *N,O*-acetylation of aromatic amines and hydrazines by arylamine *N*-acetyltransferases is well characterised. Here, we describe experiments demonstrating that human arylamine *N*-acetyltransferase Type 1 and its murine homologue (Type 2) can also catalyse the direct hydrolysis of acetyl Coenzyme A in the presence of folate. This folate-dependent activity is exclusive to these two isoforms; no acetyl Coenzyme A hydrolysis was found when murine arylamine *N*-acetyltransferase Type 1 or recombinant bacterial arylamine *N*-acetyltransferases were incubated with folate. Proton nuclear magnetic resonance spectroscopy allowed chemical modifications occurring during the catalytic reaction to be analysed in real time, revealing that the disappearance of acetyl CH_3 from acetyl Coenzyme A occurred concomitantly with the appearance of a CH_3 peak corresponding to that of free acetate and suggesting that folate is not acetylated during the reaction. We propose that folate is a cofactor for this reaction and suggest it as an endogenous function of this widespread enzyme. Furthermore, *in silico* docking of folate within the active site of human arylamine *N*-acetyltransferase Type 1 suggests that folate may bind at the enzyme's active site, and facilitate acetyl Coenzyme A hydrolysis. The evidence presented in this paper adds to our growing understanding of the endogenous roles of human arylamine *N*-acetyltransferase Type 1 and its mouse homologue and expands the catalytic repertoire of these enzymes, demonstrating that they are by no means just xenobiotic metabolising enzymes but probably also play an important role in cellular metabolism. These data, together with the characterisation of a naphthoquinone inhibitor of folate-dependent acetyl Coenzyme A hydrolysis by human arylamine *N*-acetyltransferase Type 1/murine arylamine *N*-acetyltransferase Type 2, open up a range of future avenues of exploration, both for elucidating the developmental role of these enzymes and for improving chemotherapeutic approaches to pathological conditions including estrogen receptor-positive breast cancer.

Editor: Peiwen Fei, University of Hawaii Cancer Center, United States of America

Funding: This work was supported by Cancer Research United Kingdom through a Cancer Medicinal Chemistry Studentship (NL) and an Oxford Cancer Research Centre Prize D.Phil. Studentship (C38302/A12450, JEE). ARR was supported by a Research Councils United Kingdom fellowship. The funders had no role in study design, data collection and analysis, decision to publish, or preparation of the manuscript. No other external funding was received.

Competing Interests: The authors have declared that no competing interests exist.

* E-mail: fernando.rodrigues-lima@univ-paris-diderot.fr (FR-L); E.Sim@kingston.ac.uk (ES)

Introduction

Arylamine *N*-acetyltransferases (NATs) are well characterised xenobiotic-metabolising enzymes which catalyse the acetyl Coenzyme A (AcCoA)-dependent *N*-, *O*- and *N,O*-acetylation of aromatic amines and hydrazines. They have also been shown to play important endogenous roles as well as having potential as novel targets for pharmacological intervention [1,2].

The human genome contains two adjacent NAT genes and a pseudogene (*NATP*). Both functional genes encode enzymes which catalyse the transfer of an acetyl group from AcCoA to an arylamine resulting in the formation of an amide bond, although their individual substrate specificities differ [2,3]. One of these enzymes, arylamine *N*-acetyltransferase Type 2 ((HUMAN)NAT2), is responsible for the acetylation of a range of drugs and xenobiotics and came to prominence because of its role in the polymorphic metabolism of the anti-tubercular agent isoniazid. Specific combinations of single nucleotide polymorphisms (SNPs) in the *(HUMAN)NAT2* gene are associated with slow acetylation of a number of therapeutic agents as well as industrial chemicals such as arylamines.

The second functional human NAT gene encodes human arylamine *N*-acetyltransferase Type 1 ((HUMAN)NAT1) and has also been shown to be polymorphic, although the consequences of the individual SNPs at this locus have been called into question because many of them are outside the coding region [4]. The substrate specificity of (HUMAN)NAT1 differs from that of

(HUMAN)NAT2 in that its preferred model substrate is the arylamine *para*-aminobenzoic acid (pABA) rather than the arylhydrazine compounds preferred by (HUMAN)NAT2.

Determination of the crystal structures of (HUMAN)NAT1 and (HUMAN)NAT2 has allowed their substrate specificities to be explained. These studies have shown that the catalytic triad of Cys69, His107 and Asp122 initially identified in the prokaryotic NAT of *Salmonella typhimurium* ((SALTY)NAT) [5] has the same orientation in both human NAT enzymes in which Cys is located at position 68 and His at 106 [6]. The architecture of the active sites of (HUMAN)NAT1 and (HUMAN)NAT2 reinforced earlier evidence identifying the importance of residues 125–129 in determining substrate specificity [7] and substantiated the importance of the *C*-terminus in controlling activity, as originally demonstrated by Sinclair and Sim [8]. These observations help to explain why, whilst (HUMAN)NAT1 metabolises a spectrum of substrates which overlaps that of (HUMAN)NAT2, its specificity profile is unique.

While (HUMAN)NAT1 is structurally similar to (HUMAN)-NAT2, it is biologically distinct in that it is expressed during embryonic development (as early as the four-cell stage), in placenta during trimester one and throughout pregnancy [9–11], and in stem cells. The enzymatic profile of NAT1 expression in different human foetal organs [12] illustrates its widespread tissue distribution, which differs from the much more restricted distribution of (HUMAN)NAT2, and many strands of evidence suggest a link between (HUMAN)NAT1 and folate [13–15].

The potential role of (HUMAN)NAT1 in folate clearance in the embryo and placenta is of particular interest because of the fact that folic acid supplementation is effective in the prevention of neural tube defects (NTDs) [16]. Rare loss-of-function mutations in *(HUMAN)NAT1* are associated with reduced risk of spina bifida, and *(HUMAN)NAT1* genotype has therefore been identified as a risk factor for this condition [17]. The chromosomal region marked by *(HUMAN)NAT1* and *(HUMAN)NAT2* has also been shown to influence the incidence of sporadic and teratogen-induced orofacial clefting [18]. Moreover, folate supplementation during pregnancy seems to reduce the incidence of oral clefts in newborns [19].

Like humans, mice have two functional *NAT* genes [20–22]. Historically, the characterisation of NAT polymorphisms in the mouse preceded the discovery of polymorphism in (HUMAN)-NAT1; the polymorphic murine NAT is therefore designated (MOUSE)NAT2 by analogy with (HUMAN)NAT2 [23], despite the fact that it shares the wide tissue distribution, pattern of expression during embryonic development [24] particularly in foetal neural tissues [25] and substrate specificity of (HUMAN)-NAT1 [20].

Both biochemical and transgenic evidence suggest a link between NAT and folate (Figure 1). In humans, intracellular folate levels in erythrocytes appear to be inversely correlated with cellular pABA *N*-acetylation activity [26], and folate and its analogues have been identified as inhibitors of *in vitro* arylamine *N*-acetylation measured with several sources of enzymes including pure recombinant (HUMAN)NAT1 [14,26]. Furthermore, both (HUMAN)NAT1 and (MOUSE)NAT2 are able to acetylate the folate catabolite *para*-aminobenzoyl-1-glutamate (pABAglu) [20] as well as pABA (Figure 1).

Over the last decade or so, the development of mouse models has greatly enhanced our understanding of the biological role of (MOUSE)NAT2 (and, by analogy, (HUMAN)NAT1). Observations in knockout mouse lines have suggested that (HUMAN)-NAT1 and its homologue (MOUSE)NAT2 may play a role in folate catabolism [27]. Targeted insertion of a *LacZ* transgene into the (MOUSE)NAT2 locus provided a functional knockout which also acts as a reporter model for expression of the *(MOUSE)Nat2* gene [28]. The resulting NAT2 null mice appear phenotypically normal, but their livers, kidneys and urine contain no detectable *N*-acetyl-pABAglu and they exhibit biochemical abnormalities in 5-methyl tetrahydrofolate and *S*-adenosylmethionine homeostasis [29,30]. Whole-mount histochemical staining indicates that the transgene is expressed in the closing neural tube of the developing embryo in this line. The closure process occurs normally in the majority of transgenic embryos, but an increased incidence of posterior NTDs and occasional ocular defects is detected when there is a genetic imbalance between the parents [29,31].

Mice engineered with constructs designed to drive overexpression of (HUMAN)NAT1 exhibit poor survival and the progeny of the few surviving founders only exhibit small increases in hepatic pABA acetylation [32,33], suggesting that (HUMAN)NAT1 overexpression during development could be harmful, possibly due to disruption of folate homeostasis. Moreover, the overexpression of (HUMAN)NAT1 in mice provokes NTDs usually associated with folate deficiency [32,34,35] and has complex effects on the incidence and severity of orofacial clefting [33].

Folate-dependent methylation appears to play a role in the epigenetic regulation of (MOUSE)NAT2 expression: specifically, tissue-specific CpG methylation of the promoter region increases in a strongly cell cycle-associated fashion after functional deletion of the gene or following dietary folate supplements [36], consistent with the aberrant expression of (HUMAN)NAT1 in certain developmental defects and cancers. In humans, promoter hypomethylation associated with increased *(HUMAN)NAT1* mRNA expression has been demonstrated in infiltrating ductal carcinomas of the breast [37].

The overexpression of (HUMAN)NAT1 may be of pathological significance. Proteomic analysis has identified (HUMAN)NAT1 as one of the predominant proteins overexpressed in invasive ductal and lobular breast cancers; moreover, cellular proliferation and the acquisition of etoposide resistance was observed when (HUMAN)NAT1 overexpression was induced in non-cancerous human breast cell lines [38]. In addition, transcriptional profiling has identified (HUMAN)NAT1 as a potential prognostic biomarker in male breast cancer [39].

Studies using human breast tumour-derived cell lines have indicated a 50-fold range of (HUMAN)NAT1 activities among estrogen-receptor (ER)-positive cell lines, the highest activity being observed in the ZR-75-1 line [3,40]. These observations led various investigators to evaluate small molecule libraries with a view to identifying (HUMAN)NAT1 inhibitors with chemotherapeutic potential [1,41–45]. Particular attention has been paid to the small molecule naphthoquinone **1** because of its striking selectivity for (HUMAN)NAT1/(MOUSE)NAT2 (Figure 1A) and its diagnostic/prognostic potential in ER-positive breast tumours [44,45].

The aim of the present study was to determine whether (HUMAN)NAT1 and its murine homologue can catalyse the direct hydrolysis of acetyl Coenzyme A in the presence of folate. We characterised enzyme selectivity using a broad panel of eukaryotic and prokaryotic NAT enzymes, and describe here experiments demonstrating that (HUMAN)NAT1 and its homologue (MOUSE)NAT2, but not prokaryotic NATs, can catalyse this reaction even in the absence of any arylamine substrate. We also investigated the catalytic mechanism and final products of the reaction. Folate was not acetylated during the reaction; instead, the acetate released from AcCoA entered the bulk solvent. We propose that folate is a cofactor for the hydrolysis of AcCoA by

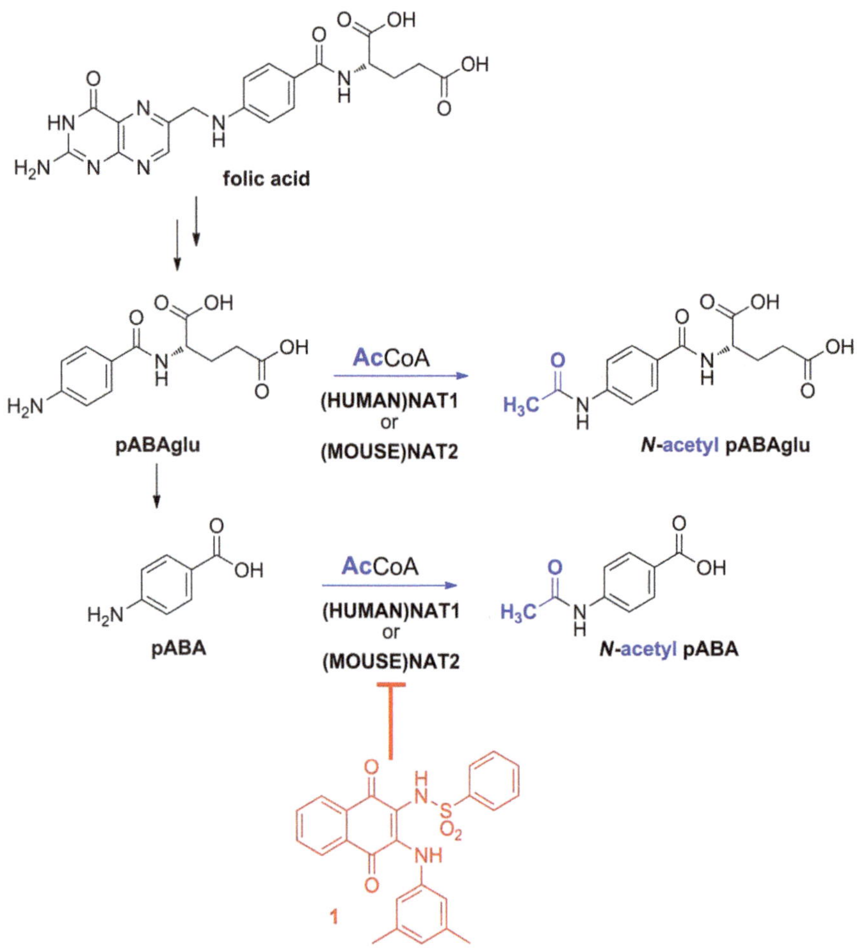

Figure 1. Potential role of (HUMAN)NAT1 and (MOUSE)NAT2 in folate degradation. One endogenous role of (HUMAN)NAT1 and (MOUSE)NAT2 appears to be related to the breakdown of folate through the acetylation of the folate catabolites pABAglu and pABA. Small molecule inhibitors such as naphthoquinone **1** (shown in red) are valuable tools which can be used to enhance our understanding of this proposed role.

(HUMAN)NAT1 and its mouse equivalent (MOUSE)NAT2 and suggest this reaction as an endogenous function of this widespread enzyme, differing from that of (HUMAN)NAT2, where the presence of an added arylamine is essential to permit appreciable rates of AcCoA hydrolysis to occur.

Materials and Methods

Chemicals

All chemicals were purchased from Sigma-Aldrich unless otherwise stated. Molecular biology reagents were obtained from Promega (Southampton, UK). Competent cells were purchased from Promega and Invitrogen (Carlsbad, USA).

Expression of Pure Recombinant NAT Enzymes

Recombinant NAT enzymes were prepared as described previously [5,20,46–53]. Human ((HUMAN)NAT1), murine ((MOUSE)NAT1 and (MOUSE)NAT2) and prokaryotic NATS (from *Salmonella typhimurium*, *Mycobacterium marinum*, *Mycobacterium smegmatis*, *Pseudomonas aeruginosa* and *Mycobacterium tuberculosis* ((SALTY)NAT, (MYCMR)NAT, (MYCSM)NAT, (PSEAE)NAT) and (MYCTU)NAT)) were evaluated.

Detection of Catalytic AcCoA Hydrolysis by HPLC Analysis

The activity of pure recombinant (HUMAN)NAT1 protein was measured using reverse-phase HPLC as previously described by Grant *et al.* [54]. Enzyme samples were first incubated with folate (500 μM, final 5% DMSO (v/v)) in assay buffer (25 mM Tris-HCl, pH 7.5) at 37°C for 5 min. The reaction was started by adding AcCoA to a final concentration of 400 μM (final volume 100 μL) and allowed to proceed at 37°C for various periods of time (up to 60 min). The reaction was quenched by adding 100 μL of ice-cold aqueous perchlorate (15% w/v), precipitated proteins were removed by centrifugation for 5 min at 12,000 *g* and 20 μL of the supernatant was injected onto a C18 reverse-phase HPLC column.

Determination of NAT Activity using Colorimetric Assays

Two methods which have previously been demonstrated to give equivalent results [55] were used to determine NAT activity; one quantified the direct disappearance of the arylamine over time, the other estimated the formation of the Coenzyme A (CoA) product. Assays were conducted in triplicate and the rates and kinetic parameters quoted represent the mean of three independent measurements ± standard deviation unless otherwise stated. In some experiments, the ability of *n*-propionyl CoA to replace AcCoA was investigated using the same methods.

Determination of NAT activity in terms of substrate acetylation. A modification of the method of Andres *et al.* [56] was used to determine the rates of enzymatic reactions with arylamine substrates. Samples of (MOUSE)NAT2 (5 ng), with pABA as the arylamine substrate (150 μM), were pre-incubated at 37°C in assay buffer (20 mM Tris-HCl, pH 8.0 containing 1 mM DTT) for 5 min in a 96-well plate (Corning). The reaction was initiated by adding pre-warmed AcCoA to a final concentration of 400 μM (final volume 45 μL), allowed to proceed for the appropriate time at 37°C and quenched by adding an equal volume of 20% (w/v) aqueous trichloroacetic acid solution at 0°C. The colorimetric reagent *para*-dimethylaminobenzaldehyde (DMAB) (5% (w/v) in 90% aqueous CH_3CN solution, 150 μL) was added and after 15 min incubation at room temperature the absorbance was measured at 450 nm in a 96-well plate reader (Tecan Sunrise). The remaining arylamine concentration was calculated by reference to a standard curve. Folate could be detected using a similar method because its pteridine moiety possesses a primary amino group which can react with DMAB to form a coloured product detectable at 450 nm. The reactivity of folic acid with DMAB is lower than that of pABA but sufficient concentration-dependent linearity was observed to permit this reaction to be used to detect free folate and evaluate the potential for folate acetylation.

Determination of NAT activity in terms of AcCoA hydrolysis. The formation of CoA was determined spectrophotometrically using the colorimetric agent 5,5′-dithio-bis(2-nitrobenzoic acid) (Ellman's reagent, DTNB) as described previously [55], with some minor modifications. This method is applicable to a wide range of arylhydrazines and arylhydroxylamines as well as arylamines; for the purposes of this study, the appropriate substrate was chosen according to Westwood *et al.* [57] depending upon the enzyme under investigation. Following preincubation of a reaction mixture containing the relevant enzyme with its arylamine substrate in assay buffer (20 mM Tris-HCl, pH 8.0) for 5 min at 37°C, pre-warmed AcCoA in assay buffer was added to 400 μM final concentration, starting the reaction (final volume 100 μL), which was allowed to proceed at 37°C. After the appropriate incubation time, simultaneous quenching and colour development was achieved by addition of DTNB (5 mM in 100 mM Tris-HCl, 6.4 M guanidine.HCl, pH 7.3; 25 μL). The absorbance was read immediately at 405 nm and the rate of reaction was determined by reference to a standard curve. The same method was used to examine the hydrolysis of *n*-propionyl CoA.

Use of NMR to Determine Folate-dependent AcCoA Hydrolysis by (MOUSE)NAT2

Prior to the commencement of enzymatic experiments, ^1H-NMR spectra of all the reagents and potential products of the assay were recorded using a Bruker Avance AVIII 700 spectrometer at the following concentrations: AcCoA (400 μM), folate (150 μM), CoA (400 μM), and acetic acid (AcOH) (400 μM), all in phosphate-buffered saline solution in D_2O (PBS-D_2O; pD 7.4) (Figures S1, S2, S3). Each peak was assigned to the corresponding proton consistent with previous full assignments [58–60]. Most of the chemical shifts of AcCoA and CoA protons were similar, but the chemical shifts of protons 6″ and 9″ differed and were chosen as markers of these two molecules in a real-time ^1H-NMR assay. The concentration of folate used in the assay (150 μM) permitted recognition of all spectral bands and the possible AcOH product could be identified from a singlet at 1.97 ppm. The characteristic resonances corresponding to all reagents and products were found to be distinguishable in the spectra of the assay mixture at the concentrations stated; the range from 3.2 to 1.6 ppm was chosen as the most adequate region to follow the enzymatic AcCoA hydrolysis by (MOUSE)NAT2 in the presence of folate (Figure S4).

For the determination of folate-dependent AcCoA hydrolysis activity, the reaction mixture comprised purified recombinant (MOUSE)NAT2 (25 μg), folate (150 μM) and AcCoA (400 μM) in PBS-D_2O at 37°C. Following the collection of basal data the reaction was initiated by adding AcCoA to a final concentration of 400 μM and monitored by real-time ^1H-NMR using a Bruker Avance AVII 500 MHz spectrometer. The field was locked by external referencing to the relevant deuteron resonance (D_2O). The integration values were corrected appropriately and referenced to a common internal standard (DMSO; 5% (v/v)).

Structural Docking Simulations

All images showing protein structures were generated using PyMOL software [61]. The ground state conformation of folate was predicted using the molecular editor Avogadro. Analysis of the possible interactions between folate and the active site of (HUMAN)NAT1 (pdb: 2PQT) [6] were performed using the licensed software GOLD [62]. A docking site was defined as a region of 10 Å within the active pocket of the enzyme and the generated solutions were ranked by their GOLD Score Fitness functions.

Modelling was also used to compare the shape and electron density of folate with that of CoA. The conformation of CoA used in these studies was that extracted from the co-crystal structure of CoA with (HUMAN)NAT2 [6]. Modelling was carried out using the licensed software FORGE (Cresset Group, Litlington, Cambridgeshire, UK). CoA from its co-crystal structure with (HUMAN)NAT2 was used as the reference molecule, and folate was used as the test ligand. Modelling was carried out comparing the two molecules both in free space and subsequently using (HUMAN)NAT2 as an excluded volume.

Results and Discussion

(HUMAN)NAT1 and (MOUSE)NAT2, but not other NATs, can act as Folate-dependent Acetyl CoA Hydrolases

A truncated form of (HUMAN)NAT1 with a shortened *C*-terminus is known to be able to catalyse the hydrolysis of AcCoA without transfer of the acetyl moiety to the arylamine substrate [8]. Such hydrolysis has also been demonstrated with truncated forms of the NAT enzyme from *S. typhimurium* [63], but has not, to date, been observed in routine enzymatic assays using full length mammalian NAT isoforms.

We initially used HPLC analysis to determine whether pure recombinant (HUMAN)NAT1 could act as an AcCoA hydrolase in the presence of folate alone and found that, over a 30 minute reaction period, the peak corresponding to AcCoA gradually reduced in area while the area of the CoA peak increased (Figure 2). The disappearance of the peak corresponding to AcCoA and appearance of the CoA peak was clear from the sequential HPLC traces; however, no change in the folate peak was observed, indicating that the role of folate in this reaction was not as a substrate. This behaviour was characteristic of (HUMAN)NAT1 as opposed to (HUMAN)NAT2, and was also observed in the presence of other folate analogues or moieties, such as methotrexate, pterin, and pteroic acid, but with lower efficiency (results not shown).

Following the discovery that (HUMAN)NAT1 can hydrolyse AcCoA in the presence of folate, the ability of its homologue, (MOUSE)NAT2, to act as a folate-dependent AcCoA hydrolase

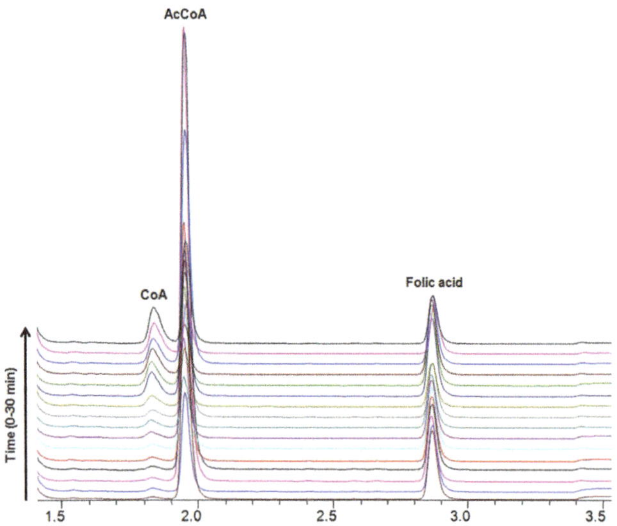

Figure 2. (HUMAN)NAT1 hydrolyses AcCoA in the presence of folate. The time course of AcCoA hydrolysis by (HUMAN)NAT1 in the presence of folate was monitored. The reaction mixture was prepared and incubated as described. Aliquots were taken every 2 minutes over a 30 minute time course and analysed by HPLC.

was also investigated. Single time point analytical assays were performed with (MOUSE)NAT2 in order to evaluate the possible fate of the reaction participants. In preliminary experiments to determine whether (MOUSE)NAT2 can hydrolyse AcCoA in the presence of folate, reactions were carried out in the presence of: pABA and AcCoA; pABA, folate (in equimolar concentrations) and AcCoA; or folate and AcCoA in the absence of pABA (Figure 3A). The results obtained confirmed the ability of (MOUSE)NAT2 to acetylate the arylamine pABA (150 μM) in the presence of AcCoA and demonstrated that the addition of 150 μM folate to the reaction did not affect this activity, whether measured in terms of substrate acetylation or release of CoA. The third set of conditions demonstrated that (MOUSE)NAT2 could catalyse the time-dependent generation of CoA from AcCoA in the presence of folate but that no acetylation of folate occurred, at least under the single time-point conditions used. These results suggest a possible role of folate as a cofactor for AcCoA hydrolysis by both (HUMAN)NAT1 and (MOUSE)NAT2, rather than an acetylation substrate. The ability of (MOUSE)NAT2, but not (MOUSE)NAT1, to hydrolyse AcCoA in the presence of folate corresponds with the ability of (HUMAN)NAT1 to act as a folate-dependent AcCoA hydrolase, in contrast with (HUMAN)NAT2.

The time course of the AcCoA hydrolysis reaction catalysed by (MOUSE)NAT2 in the presence of folate indicated that CoA release approximated to linearity over 25 minutes (Figure 3B). No hydrolysis of AcCoA was detected in the absence of (MOUSE)-NAT2, confirming that hydrolysis of AcCoA in the presence of folate was enzyme-dependent under the conditions used, nor was AcCoA hydrolysis observed with (MOUSE)NAT2 alone, consistent with conventional negative controls used in *N*-acetylation assays with DTNB as a colorimetric quenching agent [55].

The AcCoA hydrolysing capacity of a range of other murine and prokaryotic NATs was evaluated by means of similar enzymatic assays performed in the presence of: 5-aminosalicylic acid (5AS; 150 μM) as the arylamine substrate; both 5AS and folate (each at 150 μM); or folate (150 μM) alone (Figure 3C). No inhibition of 5AS acetylation activity was observed for any NAT isoform in the presence of folate. The only isoform capable of hydrolysing AcCoA in the presence of folate alone without an acceptor arylamine substrate being added to the reaction mixture was (MOUSE)NAT2. The rate of this reaction was estimated to be 360 $\mu M.sec^{-1}.mg^{-1}$. Another acyl donor substrate of (MOUSE)-NAT2, *n*-propionyl CoA, was hydrolysed by the enzyme in the presence of folate with a slightly lower activity (290 $\mu M.sec^{-1}.mg^{-1}$) compared to AcCoA.

In order to estimate the folate concentration dependence of AcCoA hydrolysis by (HUMAN)NAT1 and (MOUSE)NAT2, increasing concentrations of folate were tested with each enzyme. The hydrolytic activities measured were consistent with Michaelis-Menten kinetics. Apparent dissociation constants for folate (assumed to be acting as a cofactor) were estimated to be 75 μM for (HUMAN)NAT1 and 71 μM for (MOUSE)NAT2. The corresponding V_{max} values were 583 $\mu M.sec^{-1}.mg^{-1}$ and 613 $\mu M.sec^{-1}.mg^{-1}$, respectively (Figure 3D).

In order to investigate whether the enzyme selectivity of folate-dependent AcCoA hydrolysis is related to specific interactions between the AcCoA substrate and the NAT enzyme, the structures of several CoA-complexed NAT enzymes were compared. This comparison indicated that the binding site for CoA (analogous to AcCoA) involves the *C*-terminus but that the AcCoA binding site differs from one NAT enzyme to another. In addition, the existence of a loop between domains two and three appears to be of great importance with regard to CoA binding specificity [2]: shortening the third domain of (SALTY)NAT, either completely or partially, leads to increased AcCoA hydrolysis and inability to acetylate arylamines following formation of the acetyl-enzyme intermediate [63]. In previous studies, a truncated version of (HUMAN)NAT1 comprising only the first 204 residues was found to hydrolyse AcCoA without concomitant acetylation of the arylamine substrate [8]. Furthermore, Wang *et al.* [64] demonstrated that the acetylated derivative of Syrian hamster NAT Type 2 could be hydrolysed in the presence of arylamines whose weak nucleophilicity delays attack on the acetyl-Cys68 thioester bond. In the absence of an arylamine substrate, hydrolysis of the rapidly formed acetylated intermediate was only detected (at a velocity of 31–67 nM/sec) when high concentrations (4–8 μM) of enzyme were used [65]. Moreover, under these conditions the half-life of the acetylated enzyme intermediate was around 1 min, confirming that the acetyl-enzyme intermediate was stable and suggesting that spontaneous deacetylation was rate-limiting [65,66]. Enzyme concentrations in the (MOUSE)NAT2 reactions presented in this study were always ≤3 nM and negative control assays confirmed that the production of free CoA by (MOUSE)NAT2 as a side product of the reaction without the need of folate was close to or below the limit of detection.

It is interesting that the exclusive function of both (HUMAN)-NAT1 and (MOUSE)NAT2, which have identical *C*-termini, with respect to AcCoA hydrolysis in the presence of folate corresponds with the ability of these, but not other NAT isoforms, to acetylate the bulkier arylamine pABAglu [20]. In the human isoforms, this is thought to be a consequence of the unique presence of three key residues which have bulky side chains (Phe125, Arg127 and Tyr129) in the active site of (HUMAN)NAT1 compared with that of (HUMAN)NAT2, which has serines at these positions [6], making the substrate binding pocket of (HUMAN)NAT1 smaller than that of (HUMAN)NAT2 and conferring this substrate selectivity on (HUMAN)NAT1 compared with (HUMAN)NAT2.

Fate of the Acetyl Group Removed during NAT-mediated AcCoA Hydrolysis

The finding that (HUMAN)NAT1 and (MOUSE)NAT2 could hydrolyse AcCoA in the presence of folate but without an

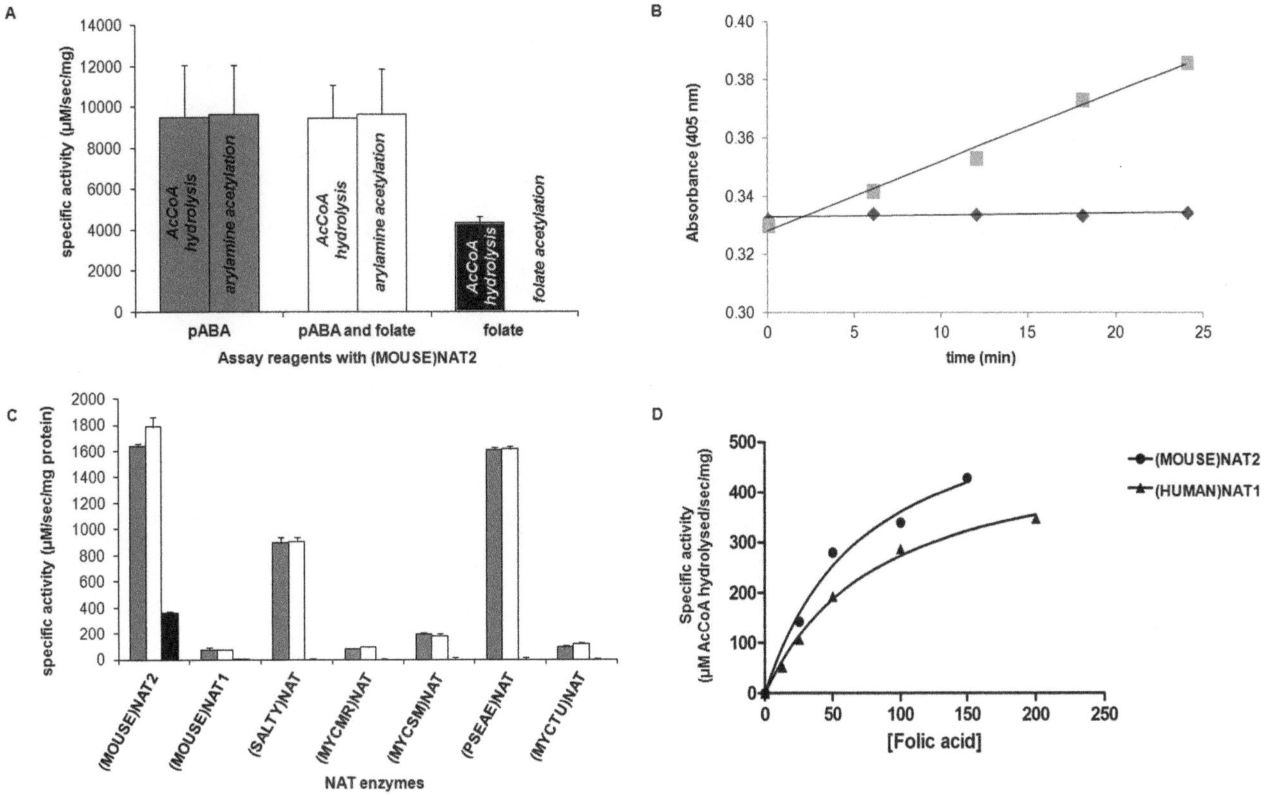

Figure 3. AcCoA hydrolysis by (MOUSE)NAT2 in the presence of folate. (A) Specific pABA (150 µM) acetylation activity by (MOUSE)NAT2 (5 ng) was followed in the absence (grey columns) or in the presence (white columns) of an equimolar concentration of folate. The concentration of AcCoA was 400 µM in all assays. Specific AcCoA (400 µM) hydrolysis activity by (MOUSE)NAT2 (5 ng) was also followed in the presence of folate (150 µM), but in the absence of an arylamine substrate (black columns). For each set of reagents, both primary amine acetylation and AcCoA hydrolysis methods were used in parallel, yielding equivalent results. No acetylation of folate was detected in a corresponding acetylation assay. (B) Time course of assay containing AcCoA (400 µM) and folate (150 µM), with 5 ng (MOUSE)NAT2 (squares); negative control: assay containing AcCoA (400 µM) and folate (150 µM) only, without (MOUSE)NAT2 (diamonds). (C) Activity of mammalian and bacterial NATs as AcCoA hydrolases in the presence of folate. Different NAT enzymes were tested in AcCoA hydrolysis assays in the presence of either the arylamine substrate 5AS (150 µM, grey columns), both 5AS and folate (each at 150 µM, white columns) or folate alone (150 µM, black columns). (D) Initial velocities of AcCoA hydrolysis by (HUMAN)NAT1 (triangles) and (MOUSE)NAT2 (circles) in the presence of different concentrations of folate. A Michaelis-Menten curve was fitted using the formula: $y = (V_{max} \times [folate])/(K_m + [folate])$.

arylamine substrate raised the question of whether the acetyl group is released directly into the bulk solvent (Figure 4, Pathway (a)) or forms an unstable *N*-acetylfolate intermediate which immediately decomposes, releasing free AcOH (Figure 4, Pathway (b)).

Preliminary HPLC studies did not reveal any *N*-acetylated folate species after AcCoA hydrolysis by (HUMAN)NAT1, nor was AcOH detectable using this analytical technique; however, the biochemical techniques used with (MOUSE)NAT2 were also unable to provide information about the intermediate steps of the catalytic mechanism. In order to resolve this issue, ^1H-NMR analysis was performed; (MOUSE)NAT2 was used in this experiment because sufficiently large amounts of pure recombinant protein (25 µg) were available to permit ^1H-NMR analysis. This allowed chemical modifications occurring during the catalytic reaction to be analysed in real time, revealing that the disappearance of acetyl CH_3 from AcCoA (at 2.41 ppm) occurred concomitantly with the appearance of a CH_3 peak (at 1.97 ppm) corresponding to that of free AcOH (Figure 5A). This was consistent with the hypothesis that (MOUSE)NAT2 could catalyse direct hydrolysis of AcCoA in the presence of folate, and that the acetyl group of AcCoA was transferred to the bulk solvent as AcOH. In contrast, no change in the chemical shifts of folate protons was observed, suggesting that the chemical structure of folate was not modified during the reaction; formation of an *N*-acetylated intermediate of folate was therefore considered unlikely. However, the formation of a transient *N*-acetylfolate intermediate could not be excluded at this stage because of the possibility that it could be hydrolysed in D_2O too rapidly to be detected on the NMR timescale.

In order to visualise the kinetics of folate-dependent AcCoA hydrolysis by (MOUSE)NAT2, reaction progress curves were drawn using the integration values of characteristic resonances arising from AcCoA, CoA and AcOH over time in relation to those of DMSO (5% (v/v)) as an internal standard. The peaks corresponding to the acetyl CH_3 groups of AcCoA and AcOH, and the methylene groups in the α position relative to sulfur in both AcCoA and CoA (numbered as 6″ and 9″ in Figures S2–S3), were also taken into account. The resulting curves indicated non-linear progression of the reaction, both in terms of loss of starting reagent and formation of products, consistent with Michaelis-Menten kinetics. The disappearance of the initial AcCoA was apparently complete after ~23 min and the formation of both products reached a maximum after the same period of time. When all the integrations were converted into concentration values and

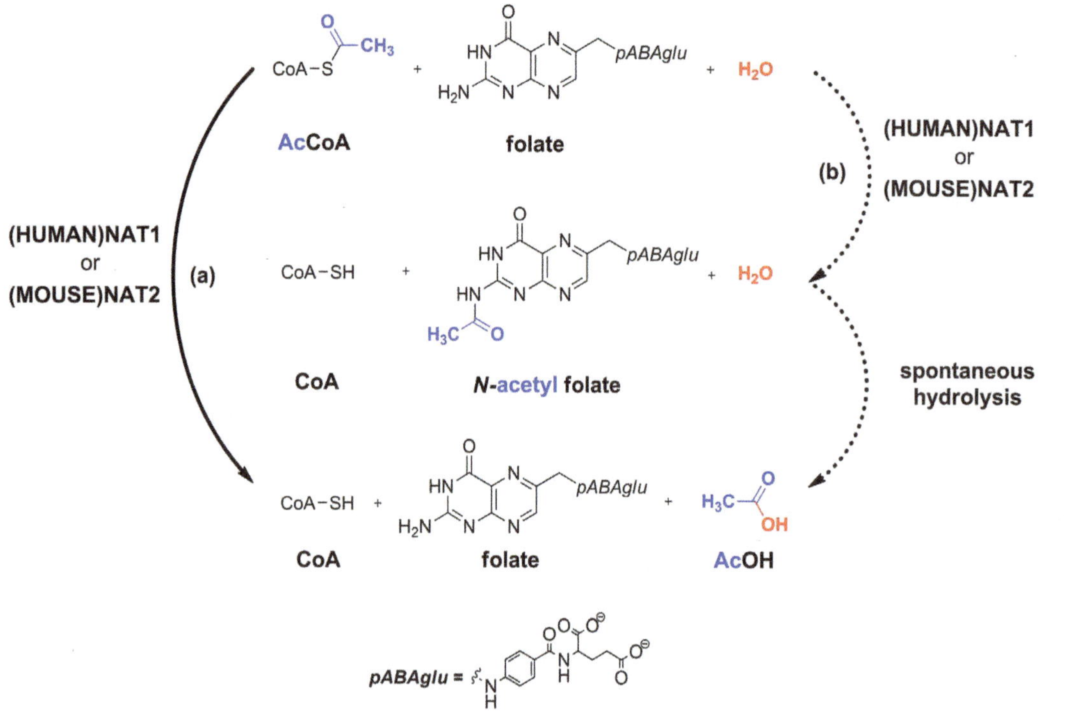

Figure 4. Two hypotheses for the mechanism of the AcCoA hydrolysis reaction catalysed by (MOUSE)NAT2 in the presence of folate. Pathway (a) corresponds to hydrolysis of AcCoA with direct release of AcOH into the bulk solvent; Pathway (b) corresponds to the formation of an unstable *N*-acetylfolate intermediate which immediately decomposes, releasing free AcOH.

plotted against time (Figure 5B), the initial reaction velocity was calculated to be 21.6 ± 2.1 μM.min^{-1}.

One possibility is that the primary amine moiety of folate is initially acetylated to form an *N*-acetylfolate intermediate which is subsequently hydrolysed; however, we did not detect such an intermediate by any method used, including ^{1}H-NMR. This may be because no such intermediate exists; alternatively, if it does, it may be turned over too rapidly for detection on the NMR timescale and/or may never accumulate to a high enough level to be detected by ^{1}H-NMR. Given that the time for running a ^{1}H-NMR spectrum is around 4 sec, the initial rate of AcCoA hydrolysis by (MOUSE)NAT2 suggests an estimate of ~1.4 μM

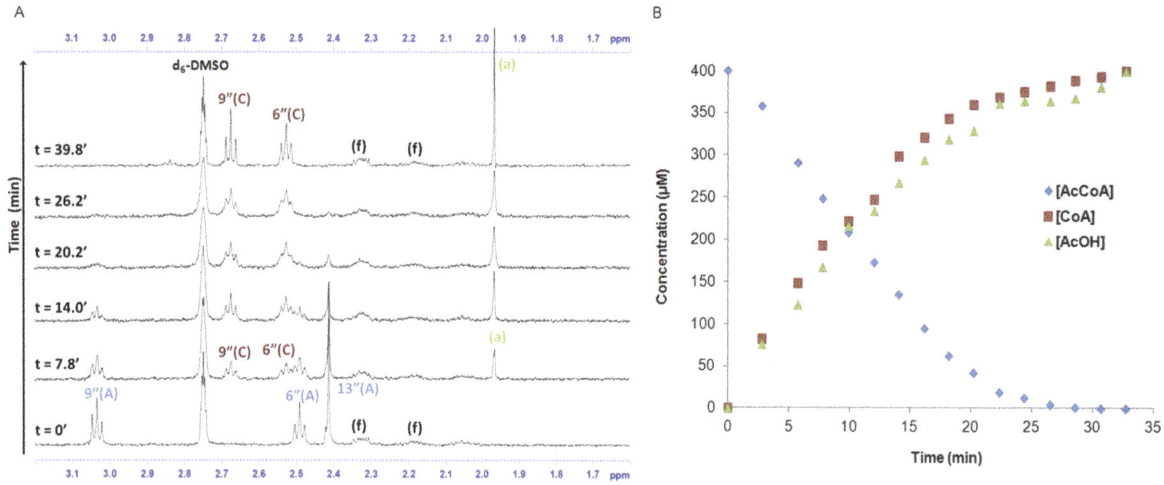

Figure 5. Real-time ^{1}H-NMR analysis of (MOUSE)NAT2 activity as a folate-dependent AcCoA hydrolase performed in PBS-D_2O. (A) Purified recombinant (MOUSE)NAT2 (25 μg) was mixed with folate (150 μM) and AcCoA (400 μM) in buffer PBS-D_2O (pD 7.4) and incubated at 37°C. NMR data were collected before adding the enzyme (t = 0 min), and subsequently 5 min after enzyme addition. A selected region of the spectra (3.2 to 1.6 ppm) is shown, the signals being labelled as follows: folate (f), AcCoA (A), CoA (C), AcOH (a). (B) Enzymatic reaction progress curves of folate-dependent AcCoA hydrolytic activity by (MOUSE)NAT2. Folate-dependent AcCoA hydrolysis by (MOUSE)NAT2 was monitored by following the characteristic peaks of both reagent and products *versus* d_6-DMSO as an internal standard with appropriate correction. Diamonds: AcCoA; Squares: CoA; Triangles: AcOH.

for the concentration of the possible *N*-acetylfolate intermediate, corresponding to ~1% of the initial amount of folate. This lies at the very limit of detection by ^{1}H-NMR and is certainly beyond the limit of quantification by this method using standard techniques [67]. These data therefore confirm that CoA and AcOH were produced during (MOUSE)NAT2-mediated catalysis in the presence of folate; however, it was not possible to determine whether the reaction proceeded *via* direct hydrolysis of the AcCoA with release of the acetyl group as free AcOH to the bulk solvent or whether a transient intermediate was formed. Since the primary amine of folic acid would be predicted to be less nucleophilic than the primary amine of pABA, due to enhanced resonance stabilisation of the nitrogen lone pair by the pterdine moiety, folate is less likely to act as a nucleophile in the hydrolysis of acetyl-Cys68 than pABA. Taking all this evidence together, therefore, we consider that (MOUSE)NAT2, and by analogy (HUMAN)NAT1, are likely to act as AcCoA hydrolases in the presence of folate without forming an *N*-acetylated folate intermediate.

Inhibition of AcCoA Hydrolysis by a Selective Naphthoquinone Inhibitor of (HUMAN)NAT1 and (MOUSE)NAT2

In order to characterise the folate-dependent hydrolysis of AcCoA further, we made use of a small molecule inhibitor (for background information on small molecule inhibitors of NATs, see [2]). Naphthoquinone **1** (Figure 6A) is a selective competitive inhibitor of *N*-acetylation by (HUMAN)NAT1 and (MOUSE)-NAT2 [44], and Scatchard analysis indicates that it binds to a single site, specifically the active site, on (MOUSE)NAT2. In addition, **1** shows the remarkable quality of changing colour from red to blue on binding specifically to (HUMAN)NAT1 or (MOUSE)NAT2, whereas it does not change colour on binding to other isoforms, in particular (HUMAN)NAT2 [45].

The availability of large amounts of stable pure recombinant (MOUSE)NAT2 allowed us to investigate the ability of naphthoquinone **1** to inhibit (MOUSE)NAT2-dependent AcCoA hydrolysis in three assay formats: arylamine *N*-acetylation in the absence of folate, arylamine *N*-acetylation in the presence of folate, and the folate-dependent AcCoA hydrolysis reaction (Figure 6B). As demonstrated previously, naphthoquinone **1** is a potent inhibitor of (MOUSE)NAT2 in the arylamine *N*-acetylation reaction; however, the extent of inhibition was reduced by at least 20% in the presence of added folate. In contrast, folate-dependent AcCoA hydrolysis by (MOUSE)NAT2 was totally inhibited by naphthoquinone **1**.

In order to verify that inhibition of folate-dependent AcCoA hydrolysis by naphthoquinone **1** was not an artefact of the enzymatic assay, its concentration dependence was characterised (Figure 6C). This analysis indicated an IC_{50} of 0.14 μM ($p<0.01$). This was roughly seven-fold lower than the IC_{50} of naphthoquinone **1** for arylamine *N*-acetylation (0.99 μM) [44].

In silico Docking of Folate within the Active Site of (HUMAN)NAT1

The mechanism of AcCoA hydrolytic catalysis depends both on the binding site and the sequence in which AcCoA and folate bind to the enzyme. Folate-dependent AcCoA hydrolysis corresponds to the first step of the conventional ping-pong bi-bi mechanism of NAT catalysis in the presence of an arylamine substrate [2]. However, in an arylamine *N*-acetylation reaction the aromatic primary amine is subsequently believed to attack the thioester functional group of the acetyl-Cys68 intermediate, resulting in transfer of the *N*-acetyl group to the aromatic amine [64], whereas the proposed second step in an AcCoA hydrolysis reaction by (MOUSE)NAT2 in the presence of folate is deacetylation of the acetyl-Cys68 by a molecule of water to form AcOH.

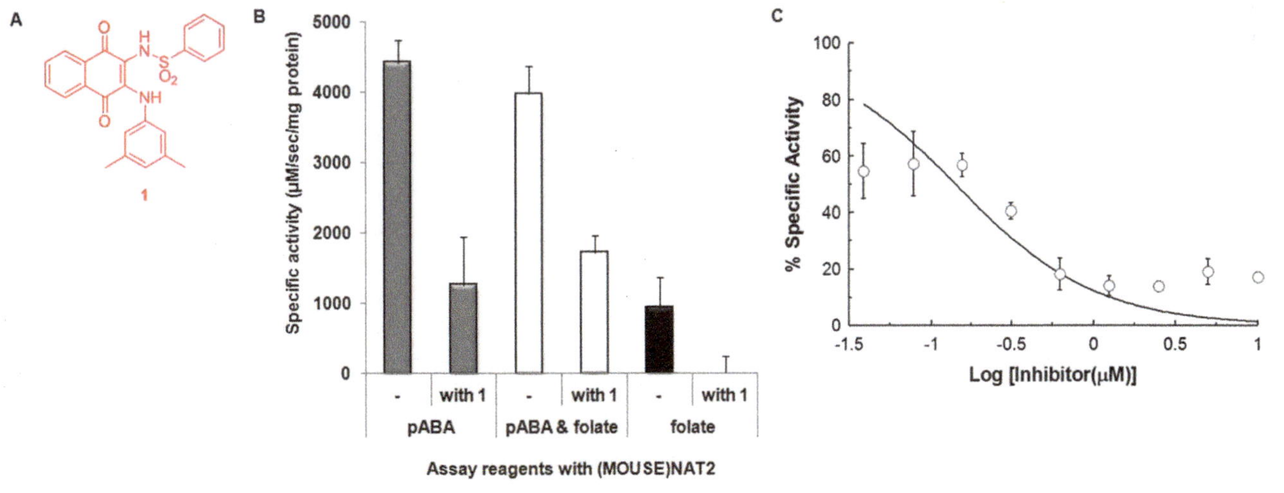

Figure 6. Inhibition of (MOUSE)NAT2-dependent AcCoA hydrolysis by the small molecule inhibitor naphthoquinone 1. (A) Structure of naphthoquinone **1**, a selective inhibitor of (HUMAN)NAT1. (B) Screening of different enzymatic assays with (MOUSE)NAT2 in the presence of naphthoquinone **1**. AcCoA hydrolysis assays were carried out in order to monitor the inhibitory potency of naphthoquinone **1** at 10 μM. AcCoA was present at 400 μM in all assays. Specific pABA (150 μM) acetylation activities by (MOUSE)NAT2 (5 ng) were performed in the absence (grey columns) or presence (white columns) of an equimolar concentration of folate. Specific AcCoA hydrolysis activity was also detected in the presence of folate (150 μM) in a DTNB assay without an arylamine substrate (black columns). The percentage inhibition was calculated in terms of the reduction in (MOUSE)NAT2 specific activity measured against a control in which DMSO (5% (v/v)) alone was added to the assay. (C) Naphthoquinone **1** as an inhibitor of the folate-dependent AcCoA hydrolytic activity of (MOUSE)NAT2. AcCoA hydrolysis assays were performed using (MOUSE)NAT2 (510 ng) and AcCoA (400 μM) in the presence of folate (150 μM) and serial dilutions of naphthoquinone **1**. Data points represent the mean ± standard deviation of triplicate assays. Percentage inhibition was determined as the ratio of specific activity with the inhibitor related to specific activity without inhibitor (100%); 100% activity was 198 $\mu M.sec^{-1}.mg^{-1}$. The regression equation used was: $y = 100/[1+10^{\wedge}(x\text{-}\log(IC_{50})]$.

This hypothetical mechanism requires a water molecule to gain access to the active site acetyl-Cys68 in order to release the acetyl group and contrasts with the situation pertaining in the case of the arylamine *N*-acetylation reaction in which, following the acetylation of the catalytic cysteine, only highly hydrophobic molecules such as arylamines and arylhydrazines can access the active site pouch [6,57,68].

The observation that no appreciable AcCoA hydrolysis occurs in control reactions with (MOUSE)NAT2 in the absence of an arylamine substrate (or folate) is consistent with this hypothesis [55]. Conversely, the observed hydrolysis of acetyl-Cys68 in the presence of folate, despite the hydrophobicity of the active site pocket, suggests that folate may modify the hydrophobicity of the acetylated enzyme's active site, thereby allowing water to enter the active site pocket and cleave acetyl-Cys68, possibly as a consequence of altered folding induced by the binding of folate.

We undertook modelling studies in an effort to predict the nature of the interactions between (HUMAN)NAT1 and folate in the case that it binds to the catalytic active site, which is likely because the known (HUMAN)NAT1 specific substrates, pABAglu and pABA, are chemical moieties of folate (Figure 1A). Modelling using GOLD software suggests that the γ-COOH of folate could form hydrogen bonds with the backbone carbonyl of Ile106 and the thiolate moiety of Cys68, and the α-COOH of folate could interact with the backbone carbonyl of Gly124 and the imidazole side chain of His107. Additionally, the pteridine carbonyl of folate could potentially interact with the guanidine moiety of Arg127 and the primary amine of folate with the backbone carbonyl of Lys188 and the hydroxyl groups of Thr96 and Tyr129 (Figure 7).

Comparing the conformation of CoA as it is bound to the active site of (HUMAN)NAT2, using an available co-crystal structure (pdb code:2PFR), against free folate suggests marked similarities between the shape and electron density of the two species (Figure 8A). Repeatable simulations show that regions of positive charge, negative charge and neutral regions for the two ligands align well in space. Furthermore, when (HUMAN)NAT2 was used as an excluded volume in this modelling, the same conclusions could also be reliably drawn (Figure 8B).

It should be noted that these FORGE experiments involved a comparison between the conformation of CoA as bound to (HUMAN)NAT2 and that of free folate; thus the simulations pertain to free folate, not the predicted conformation of hypothetical enzyme-bound conformations of folate. We used (HUMAN)NAT2 as an excluded volume to ensure that folate and CoA were excluded from the same space, permitting a reliable comparison between the shapes of the two molecules. This does not necessarily suggest that folate is binding to the active site of (HUMAN)NAT2 in this way; interestingly, when folate is docked into the active site of (HUMAN)NAT2 it does give a good predicted fit, but this does not necessarily imply binding of folate to (HUMAN)NAT2 and even if folate does bind to (HUMAN)-NAT2, it may not induce the conformational changes required for AcCoA hydrolysis.

It is not possible to state whether our prediction regarding the orientation of folate within the active site of (HUMAN)NAT1 is an accurate representation of the real-life situation since no structures of (HUMAN)NAT1 or other NAT enzymes complexed with folate are yet available. Docking studies based on the structure of an acetylated form of (HUMAN)NAT1 would be particularly informative in this context. Biochemical studies have yielded discordant results concerning the mode of inhibition of eukaryotic NATs by folate; however, the data presented in this paper suggest a possible explanation in that folate-dependent inhibition of arylamine *N*-acetylation may be due to concomitant depletion of AcCoA *via* hydrolysis. In particular, some previous folate inhibitor studies with human NATs [26], which measured NAT activity in terms of arylamine *N*-acetylation, may not have taken into account the possibility of direct hydrolysis of AcCoA.

A representative crystal structure of a folate-bound NAT enzyme is required for a complete definition of the folate binding site of (HUMAN)NAT1, which is needed before a hypothesis for the catalytic mechanism for AcCoA hydrolysis by (HUMAN)-

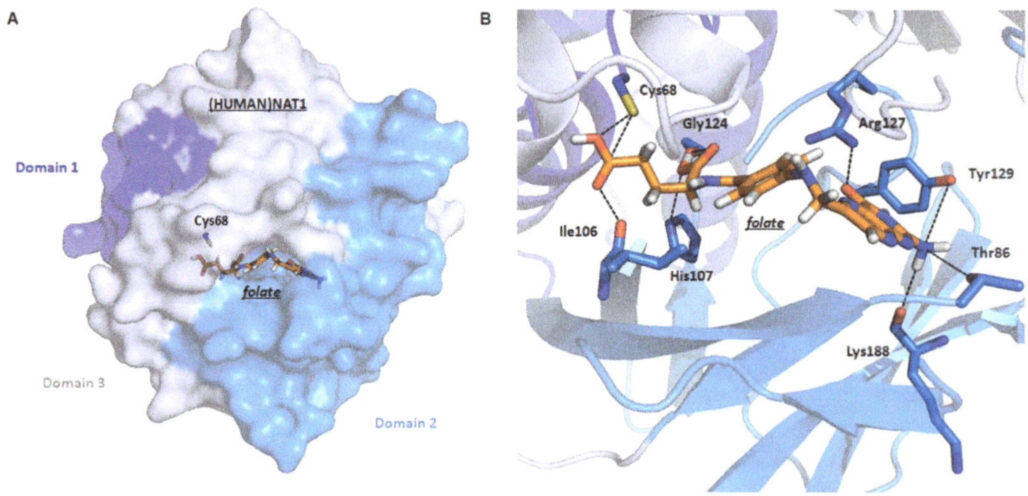

Figure 7. *In silico* docking of folate into the active site of (HUMAN)NAT1. (A) Folic acid was docked into the active site of (HUMAN)NAT1 using the program GOLD [62]; the highest scoring solution is shown. The structure of (HUMAN)NAT1 (pdb code: 2PQT) [6] is shown in surface format with the three domains labelled and coloured in different shades of blue. The domains are numbered from the amino terminus with Domain 3 corresponding to the *C*-terminus of the protein. Folic acid is labelled with carbon atoms in orange, nitrogen in blue, oxygen in red, and hydrogen in white. (B) Maximised view of the (HUMAN)NAT1 active site with folic acid docked. The structure of (HUMAN)NAT1 is shown in cartoon format with the three domains coloured as above. The side chains of the key residues involved in predicted folic acid binding within the catalytic pocket are labelled with nitrogen in blue, oxygen in red, and sulfur in yellow. Folic acid is labelled with carbon atoms in orange, nitrogen in blue, oxygen in red, and hydrogen in white.

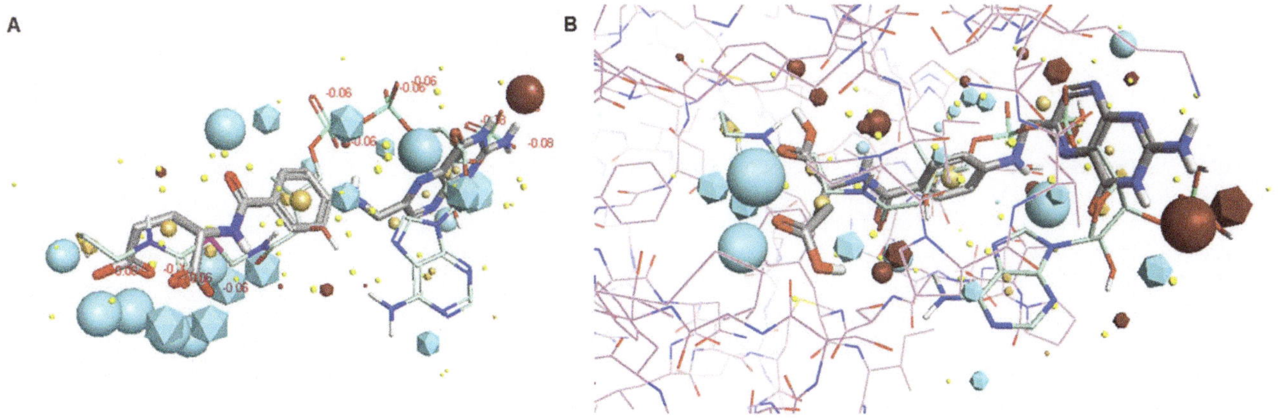

Figure 8. Structural similarity between folate and AcCoA. (A) Predicted alignment between folic acid and CoA in free space using the software program FORGE, based on the conformation of CoA as it is bound in the co-crystal structure with (HUMAN)NAT2 (pdb code: 2PFR). Folic acid is shown in thick sticks, with carbons in grey, and CoA is shown in thin sticks, with carbon atoms in green. Nitrogen atoms are shown in blue, oxygens in red, and hydrogens in white. Regions of negative charge are shown in cyan, regions of positive charge are shown in dark red, and lipophilic regions are shown in yellow. Icosahedra represent the electronic fields of CoA and spheres represent those of folic acid. (B) Predicted alignment between folic acid and CoA using (HUMAN)NAT2 as an excluded volume. Folic acid, CoA and their respective electron density distributions are represented as above. The (HUMAN)NAT2 structure is represented in thin sticks with the carbon atoms in lilac.

NAT1 can be proposed. Although folate has a much higher molecular weight (441.4 g.mol^{-1}) than selective (HUMAN)NAT1 arylamine substrates (<266.1 g.mol^{-1} (pABAglu)) [6], our docking simulations have highlighted potentially favourable complementary binding partners for the polar functional groups of folate in the hydrophobic catalytic pocket of (HUMAN)NAT1 *via* hydrogen bond interactions. It is possible that the accommodation of a large molecule such as folate could induce further opening of the (HUMAN)NAT1 active site and mediate the access of a molecule of water to the cavity. Furthermore, modelling studies suggest that the electron density distribution of folate may resemble that of CoA, allowing it to bind to the active sites of human NATs in a similar manner.

Crystallographic structural data are therefore required to clarify the mechanism by which folate selectively facilitates AcCoA hydrolysis by (HUMAN)NAT1. More broadly, our data raise questions about the possible role that certain NAT enzymes could play at the interface between AcCoA homeostasis and folate metabolism.

Folic acid has been reported to act as a free radical scavenger [69,70], and it is interesting to hypothesise a link between the discovery that (HUMAN)NAT1 and (MOUSE)NAT2 can act as AcCoA hydrolases in the presence of an appropriate amount of folic acid and a potential role for (HUMAN)NAT1/(MOUSE)-NAT2 and folate in the regulation of AcCoA/CoA homeostasis in cells, especially under conditions of oxidative stress. The availability of naphthoquinone **1** as a selective inhibitor of folate-dependent AcCoA hydrolysis by (HUMAN)NAT1/(MOUSE)NAT2 adds to the available tool kit for the further investigation of such possibilities. Indeed, (HUMAN)NAT1 has been postulated as a novel biomarker in ER-positive breast cancer [40] and the development of analogues of the naphthoquinone inhibitor **1** for potential diagnostic and chemotherapeutic applications in breast cancer treatment is a current area of interest [71].

Conclusions

The experiments described in this paper demonstrate that (HUMAN)NAT1 and its homologue (MOUSE)NAT2 can catalyse the direct hydrolysis of acetyl Coenzyme A in the presence of folate. This folate-dependent activity is exclusive to these two isoforms; no AcCoA hydrolysis was found when (HUMAN)NAT2, (MOUSE)NAT1 or recombinant bacterial NATs were incubated with folate. Proton NMR allowed chemical modifications occurring during the catalytic reaction to be analysed in real time; the disappearance of acetyl CH_3 from AcCoA occurred concomitantly with the appearance of a CH_3 peak corresponding to that of free AcOH, suggesting that folate is not acetylated during the reaction. We propose that folate is a cofactor for this reaction and suggest it as an endogenous function of this widespread enzyme. Furthermore, *in silico* docking of folate within the active site of (HUMAN)NAT1 suggests that folate may bind at the enzyme's active site and facilitate AcCoA hydrolysis.

The evidence presented in this paper both adds to our growing understanding of the endogenous roles of (HUMAN)NAT1 and (MOUSE)NAT2 and expands the catalytic repertoire of these enzymes, demonstrating that they are by no means just xenobiotic metabolising enzymes but probably play an important role in cellular metabolism. Our data highlight the important question of whether, and how, the NAT enzymes play a role at the interface between AcCoA homeostasis and folate metabolism [2]. Furthermore, the characterisation of naphthoquinone **1** as an inhibitor of folate-dependent AcCoA hydrolysis reinforces the value of this compound as a powerful chemical tool which is likely to open up a range of future avenues of exploration *via* selective modulation of (HUMAN)NAT1 and (MOUSE)NAT2 function, both for improved understanding of the developmental role of these enzymes and for better chemotherapeutic approaches to ER-positive breast cancer.

Supporting Information

Figure S1 Structure and proton NMR spectra of folic acid. 500 MHz ^{1}H-NMR spectrum of 400 μM AcCoA in D_2O, pD 7.4, 25°C (room temperature). Fully resolved spectral assignments were made on the basis of previous studies [72,73].

Figure S2 Structure and proton NMR spectra of AcCoA. 500 MHz ^{1}H-NMR spectrum of 400 μM AcCoA in D_2O, pD 7.4, 25°C (room temperature). Fully resolved spectral assignments were made on the basis of previous studies [58,59]. The spectral bands of proton 2′, 3′ and 4′ were assumed to be hidden by the D_2O signal. Traces of Tris.HCl buffer from AcCoA stock solutions were observed.

Figure S3 Structure and proton NMR spectra of CoA. 700 MHz spectrum of 400 μM CoA in D_2O, pD 7.4, 25°C (room temperature). Fully resolved spectral assignments were made on the basis of previous studies [58–60].

Figure S4 NMR spectra of reagents and possible products present in an AcCoA hydrolysis assay with (MOUSE)NAT2. NMR spectra were recorded with a Bruker AVC 500 spectrometer at 500 MHz or, for CoA, an AV700 spectrometer at 700 MHz at the optimal concentrations used in the AcCoA enzymatic assay: 400 μM for AcOH, CoA, AcCoA, and 150 μM for folate. The blue square highlights the range of chemical shifts chosen to follow the enzymatic assay over time.

Acknowledgments

We would like to thank Dr Tim Claridge (Chemistry Research Laboratory, University of Oxford) and Helen Storr (Department of Pharmacology, University of Oxford) for their help with the NMR experiments. We are grateful to Larissa Wakefield, Valerie Cornish and Hilary Long for helpful discussions.

Author Contributions

Conceived and designed the experiments: NL JD ES FR-L. Performed the experiments: NL JD JEE. Analyzed the data: NL JD JEE LAS AJR J-MD ES FR-L. Contributed reagents/materials/analysis tools: JD J-MD ES FR-L. Wrote the paper: NL JEE LAS AJR ES.

References

1. Butcher NJ, Minchin RF (2012) Arylamine *N*-acetyltransferase 1: a novel drug target in cancer development. Pharmacol Rev 64: 147–165.
2. Sim E, Fakis G, Laurieri N, Boukouvala S (2012) Arylamine *N*-acetyltransferases–from drug metabolism and pharmacogenetics to identification of novel targets for pharmacological intervention. Adv Pharmacol 63: 169–205.
3. Sim E, Walters K, Boukouvala S (2008) Arylamine *N*-acetyltransferases: from structure to function. Drug Metab Rev 40: 479–510.
4. Hein DW (2009) *N*-acetyltransferase SNPs: emerging concepts serve as a paradigm for understanding complexities of personalized medicine. Expert Opin Drug Metab Toxicol 5: 353–366.
5. Sinclair JC, Sandy J, Delgoda R, Sim E, Noble ME (2000) Structure of arylamine *N*-acetyltransferase reveals a catalytic triad. Nat Struct Biol 7: 560–564.
6. Wu H, Dombrovsky L, Tempel W, Martin F, Loppnau P, et al. (2007) Structural basis of substrate-binding specificity of human arylamine *N*-acetyltransferases. J Biol Chem 282: 30189–30197.
7. Goodfellow GH, Dupret JM, Grant DM (2000) Identification of amino acids imparting acceptor substrate selectivity to human arylamine acetyltransferases NAT1 and NAT2. Biochem J 348: 159–166.
8. Sinclair J, Sim E (1997) A fragment consisting of the first 204 amino-terminal amino acids of human arylamine *N*-acetyltransferase one (NAT1) and the first transacetylation step of catalysis. Biochem Pharmacol 53: 11–16.
9. Derewlany LO, Knie B, Koren G (1994) Arylamine *N*-acetyltransferase activity of the human placenta. J Pharmacol Exp Ther 269: 756–760.
10. Derewlany LO, Knie B, Koren G (1994) Human placental transfer and metabolism of *p*-aminobenzoic acid. J Pharmacol Exp Ther 269: 761–765.
11. Smelt VA, Upton A, Adjaye J, Payton MA, Boukouvala S, et al. (2000) Expression of arylamine *N*-acetyltransferases in pre-term placentas and in human pre-implantation embryos. Hum Mol Genet 9: 1101–1107.
12. Pacifici GM, Bencini C, Rane A (1986) Acetyltransferase in humans: development and tissue distribution. Pharmacology 32: 283–291.
13. Minchin RF (1995) Acetylation of *p*-aminobenzoylglutamate, a folic acid catabolite, by recombinant human arylamine *N*-acetyltransferase and U937 cells. Biochem J 307: 1–3.
14. Ward A, Summers MJ, Sim E (1995) Purification of recombinant human *N*-acetyltransferase type 1 (NAT1) expressed in *E. coli* and characterization of its potential role in folate metabolism. Biochem Pharmacol 49: 1759–1767.
15. Minchin RF, Hanna PE, Dupret JM, Wagner CR, Rodrigues-Lima F, et al. (2007) Arylamine *N*-acetyltransferase I. Int J Biochem Cell Biol 39: 1999–2005.
16. Murphy M, Seagroatt V, Hey K, O'Donnell M, Godden M, et al. (1996) Neural tube defects 1974–94–down but not out. Arch Dis Child Fetal Neonatal Ed 75: F133–134.
17. Jensen LE, Hoess K, Mitchell LE, Whitehead AS (2006) Loss of function polymorphisms in NAT1 protect against spina bifida. Hum Genet 120: 52–57.
18. Erickson RP (2010) Genes, environment, and orofacial clefting: *N*-acetyltransferase and folic acid. J Craniofac Surg 21: 1384–1387.
19. Czeizel AE, Timar L, Sarkozi A (1999) Dose-dependent effect of folic acid on the prevention of orofacial clefts. Pediatrics 104: e66.
20. Kawamura A, Westwood I, Wakefield L, Long H, Zhang N, et al. (2008) Mouse *N*-acetyltransferase type 2, the homologue of human *N*-acetyltransferase type 1. Biochem Pharmacol 75: 1550–1560.
21. Boukouvala S, Price N, Sim E (2002) Identification and functional characterization of novel polymorphisms associated with the genes for arylamine *N*-acetyltransferases in mice. Pharmacogenetics 12: 385–394.
22. Martell KJ, Levy GN, Weber WW (1992) Cloned mouse *N*-acetyltransferases: enzymatic properties of expressed *Nat-1* and *Nat-2* gene products. Mol Pharmacol 42: 265–272.
23. Hein DW, Boukouvala S, Grant DM, Minchin RF, Sim E (2008) Changes in consensus arylamine *N*-acetyltransferase gene nomenclature. Pharmacogenet Genomics 18: 367–368.
24. Mitchell MK, Futscher BW, McQueen CA (1999) Developmental expression of *N*-acetyltransferases in C57Bl/6 mice. Drug Metab Dispos 27: 261–264.
25. Stanley LA, Copp AJ, Pope J, Rolls S, Smelt V, et al. (1998) Immunochemical detection of arylamine *N*-acetyltransferase during mouse embryonic development and in adult mouse brain. Teratology 58: 174–182.
26. Ward A, Hickman D, Gordon JW, Sim E (1992) Arylamine *N*-acetyltransferase in human red blood cells. Biochem Pharmacol 44: 1099–1104.
27. Payton M, Smelt V, Upton A, Sim E (1999) A method for genotyping murine arylamine *N*-acetyltransferase type 2 (NAT2): a gene expressed in preimplantation embryonic stem cells encoding an enzyme acetylating the folate catabolite *p*-aminobenzoylglutamate. Biochem Pharmacol 58: 779–785.
28. Cornish VA, Pinter K, Boukouvala S, Johnson N, Labrousse C, et al. (2003) Generation and analysis of mice with a targeted disruption of the arylamine *N*-acetyltransferase Type 2 gene. Pharmacogenomics J 3: 169–177.
29. Wakefield L, Cornish V, Long H, Griffiths WJ, Sim E (2007) Deletion of a xenobiotic metabolizing gene in mice affects folate metabolism. Biochem Biophys Res Commun 364: 556–560.
30. Witham KL, Butcher NJ, Sugamori KS, Brenneman D, Grant DM, et al. (2013) 5-Methyl-tetrahydrofolate and the *S*-adenosylmethionine cycle in C57BL/6J Mouse Tissues: Gender differences and effects of arylamine *N*-acetyltransferase-1 deletion. PLoS One 8: e77923.
31. Wakefield L, Long H, Lack N, Sim E (2007) Ocular defects associated with a null mutation in the mouse arylamine *N*-acetyltransferase 2 gene. Mamm Genome 18: 270–276.
32. Cao W, Chau B, Hunter R, Strnatka D, McQueen CA, et al. (2005) Only low levels of exogenous *N*-acetyltransferase can be achieved in transgenic mice. Pharmacogenomics J 5: 255–261.
33. Erickson RP, Cao W, Acuna DK, Strnatka DW, Hunter RJ, et al. (2008) Confirmation of the role of *N*-acetyltransferase 2 in teratogen-induced cleft palate using transgenics and knockouts. Mol Reprod Dev 75: 1071–1076.
34. Sim E, Pinter K, Mushtaq A, Upton A, Sandy J, et al. (2003) Arylamine *N*-acetyltransferases: a pharmacogenomic approach to drug metabolism and endogenous function. Biochem Soc Trans 31: 615–619.
35. Sim E, Westwood I, Fullam E (2007) Arylamine *N*-acetyltransferases. Expert Opin Drug Metab Toxicol 3: 169–184.
36. Wakefield L, Boukouvala S, Sim E (2010) Characterisation of CpG methylation in the upstream control region of mouse *Nat2*: evidence for a gene-environment interaction in a polymorphic gene implicated in folate metabolism. Gene 452: 16–21.
37. Kim SJ, Kang HS, Chang HL, Jung YC, Sim HB, et al. (2008) Promoter hypomethylation of the *N*-acetyltransferase 1 gene in breast cancer. Oncol Rep 19: 663–668.
38. Adam PJ, Berry J, Loader JA, Tyson KL, Craggs G, et al. (2003) Arylamine *N*-acetyltransferase-1 is highly expressed in breast cancers and conveys enhanced growth and resistance to etoposide *in vitro*. Mol Cancer Res 1: 826–835.
39. Johansson I, Nilsson C, Berglund P, Lauss M, Ringner M, et al. (2012) Gene expression profiling of primary male breast cancers reveals two unique subgroups and identifies *N*-acetyltransferase-1 (NAT1) as a novel prognostic biomarker. Breast Cancer Res 14: R31.
40. Wakefield L, Robinson J, Long H, Ibbitt JC, Cooke S, et al. (2008) Arylamine *N*-acetyltransferase 1 expression in breast cancer cell lines: a potential marker in estrogen receptor-positive tumors. Genes Chromosomes Cancer 47: 118–126.
41. Russell AJ, Westwood IM, Crawford MH, Robinson J, Kawamura A, et al. (2009) Selective small molecule inhibitors of the potential breast cancer marker,

human arylamine *N*-acetyltransferase 1, and its murine homologue, mouse arylamine *N*-acetyltransferase 2. Bioorg Med Chem 17: 905–918.

42. Ballester PJ, Westwood I, Laurieri N, Sim E, Richards WG (2010) Prospective virtual screening with Ultrafast Shape Recognition: the identification of novel inhibitors of arylamine *N*-acetyltransferases. J R Soc Interface 7: 335–342.
43. Tiang JM, Butcher NJ, Minchin RF (2010) Small molecule inhibition of arylamine *N*-acetyltransferase Type I inhibits proliferation and invasiveness of MDA-MB-231 breast cancer cells. Biochem Biophys Res Commun 393: 95–100.
44. Laurieri N, Crawford MH, Kawamura A, Westwood IM, Robinson J, et al. (2010) Small molecule colorimetric probes for specific detection of human arylamine *N*-acetyltransferase 1, a potential breast cancer biomarker. J Am Chem Soc 132: 3238–3239.
45. Laurieri N, Egleton JE, Varney A, Thinnes CC, Quevedo CE, et al. (2013) A novel color change mechanism for breast cancer biomarker detection: naphthoquinones as specific ligands of human arylamine *N*-acetyltransferase 1. PLoS One 8: e70600.
46. Dairou J, Atmane N, Dupret JM, Rodrigues-Lima F (2003) Reversible inhibition of the human xenobiotic-metabolizing enzyme arylamine *N*-acetyltransferase 1 by *S*-nitrosothiols. Biochem Biophys Res Commun 307: 1059–1065.
47. Westwood IM, Holton SJ, Rodrigues-Lima F, Dupret JM, Bhakta S, et al. (2005) Expression, purification, characterization and structure of *Pseudomonas aeruginosa* arylamine *N*-acetyltransferase. Biochem J 385: 605–612.
48. Sandy J, Mushtaq A, Kawamura A, Sinclair J, Sim E, et al. (2002) The structure of arylamine *N*-acetyltransferase from *Mycobacterium smegmatis*–an enzyme which inactivates the anti-tubercular drug, isoniazid. J Mol Biol 318: 1071–1083.
49. Fullam E, Kawamura A, Wilkinson H, Abuhammad A, Westwood I, et al. (2009) Comparison of the arylamine *N*-acetyltransferase from *Mycobacterium marinum* and *Mycobacterium tuberculosis*. Protein J 28: 281–293.
50. Abuhammad A, Lack N, Schweichler J, Staunton D, Sim RB, et al. (2011) Improvement of the expression and purification of *Mycobacterium tuberculosis* arylamine *N*-acetyltransferase (TBNAT) a potential target for novel anti-tubercular agents. Protein Expr Purif.
51. Sandy J, Holton S, Fullam E, Sim E, Noble M (2005) Binding of the anti-tubercular drug isoniazid to the arylamine *N*-acetyltransferase protein from *Mycobacterium smegmatis*. Protein Sci 14: 775–782.
52. Fullam E, Westwood IM, Anderton MC, Lowe ED, Sim E, et al. (2008) Divergence of cofactor recognition across evolution: Coenzyme A binding in a prokaryotic arylamine *N*-acetyltransferase. J Mol Biol 375: 178–191.
53. Wang H, Vath GM, Kawamura A, Bates CA, Sim E, et al. (2005) Over-expression, purification, and characterization of recombinant human arylamine *N*-acetyltransferase 1. Protein J 24: 65–77.
54. Grant DM, Blum M, Beer M, Meyer UA (1991) Monomorphic and polymorphic human arylamine *N*-acetyltransferases: a comparison of liver isozymes and expressed products of two cloned genes. Mol Pharmacol 39: 184–191.
55. Brooke EW, Davies SG, Mulvaney AW, Pompeo F, Sim E, et al. (2003) An approach to identifying novel substrates of bacterial arylamine *N*-acetyltransferases. Bioorg Med Chem 11: 1227–1234.
56. Andres HH, Klem AJ, Szabo SM, Weber WW (1985) New spectrophotometric and radiochemical assays for acetyl-CoA: arylamine *N*-acetyltransferase applicable to a variety of arylamines. Anal Biochem 145: 367–375.
57. Westwood IM, Kawamura A, Fullam E, Russell AJ, Davies SG, et al. (2006) Structure and mechanism of arylamine *N*-acetyltransferases. Curr Top Med Chem 6: 1641–1654.
58. Patel SS, Walt DR (1987) Substrate specificity of acetyl Coenzyme A synthetase. J Biol Chem 262: 7132–7134.
59. Wu WJ, Tonge PJ, Raleigh DP (1998) Stereospecific H-1 and C-13 NMR assignments of crotonyl CoA and hexadienoyl CoA: Conformational analysis and comparison with protein-CoA complexes. Journal of the American Chemical Society 120: 9988–9994.
60. Rossi C, Donati A, Sansoni MR (1993) Folic-Acid - Solution Structure and NMR Strategy for Conformational-Analysis. Spectrosc Lett 26: 1603–1611.
61. DeLano WL (2002) PyMOL: An open-source molecular graphics tool. DeLano Scientific, San Carlos, California, USA; http://www.ccp4.ac.uk/newsletters/newsletter40/11_pymol.pdf.
62. Verdonk ML, Cole JC, Hartshorn MJ, Murray CW, Taylor RD (2003) Improved protein-ligand docking using GOLD. Proteins 52: 609–623.
63. Mushtaq A, Payton M, Sim E (2002) The COOH terminus of arylamine *N*-acetyltransferase from Salmonella typhimurium controls enzymic activity. J Biol Chem 277: 12175–12181.
64. Wang H, Liu L, Hanna PE, Wagner CR (2005) Catalytic mechanism of hamster arylamine *N*-acetyltransferase 2. Biochemistry 44: 11295–11306.
65. Wang H, Vath GM, Gleason KJ, Hanna PE, Wagner CR (2004) Probing the mechanism of hamster arylamine *N*-acetyltransferase 2 acetylation by active site modification, site-directed mutagenesis, and pre-steady state and steady state kinetic studies. Biochemistry 43: 8234–8246.
66. Jencks WP, Gresser M, Valenzuela MS, Huneeus FC (1972) Acetyl Coenzyme A: arylamine acetyltransferase. Measurement of the steady state concentration of the acetyl-enzyme intermediate. J Biol Chem 247: 3756–3760.
67. Claridge TDW, Davies SG, Polywka MEC, Roberts PM, Russell AJ, et al. (2008) "Pure by NMR"? Organic Letters 10: 5433–5436.
68. Abuhammad AM, Lowe ED, Fullam E, Noble M, Garman EF, et al. (2010) Probing the architecture of the *Mycobacterium marinum* arylamine *N*-acetyltransferase active site. Protein Cell 1: 384–392.
69. Matte C, Mackedanz V, Stefanello FM, Scherer EB, Andreazza AC, et al. (2009) Chronic hyperhomocysteinemia alters antioxidant defenses and increases DNA damage in brain and blood of rats: protective effect of folic acid. Neurochem Int 54: 7–13.
70. Moat SJ, Bonham JR, Cragg RA, Powers HJ (2000) Elevated plasma homocysteine elicits an increase in antioxidant enzyme activity. Free Radic Res 32: 171–179.
71. Egleton JE, Thinnes CC, Seden PT, Laurieri N, Lee SP, et al. (2014) Structure-activity relationships and colorimetric properties of specific probes for the putative cancer biomarker human arylamine *N*-acetyltransferase 1. Bioorg Med Chem In press.
72. Lam Y-F, Kotowycz G (1972) Self association of folic acid in aqueous solution by Proton Magnetic Resonance. Canadian Journal of Chemistry 50: 2357–2363.
73. Rossi C, Donati A, Sansoni MR (1993) Folic acid: Solution structure and NMR strategy for conformational analysis. Spectrosc Lett 26: 1603–1611.

25

Improved Lanthipeptide Detection and Prediction for antiSMASH

Kai Blin[1,2¤a], Daniyal Kazempour[1], Wolfgang Wohlleben[1,2], Tilmann Weber[1,2*¤b]

1 Microbiology/Biotechnology, Interfaculty Institute of Microbiology and Infection Medicine, Eberhard-Karls-University of Tübingen, Tübingen, Germany, **2** German Centre for Infection Research, Tübingen, Germany

Abstract

Lanthipeptides are a class of ribosomally synthesised and post-translationally modified peptide (RiPP) natural products from the bacterial secondary metabolism. Their name is derived from the characteristic lanthionine or methyl-lanthionine residues contained in the processed peptide. Lanthipeptides that possess an antibacterial activity are called lantibiotics. Whereas multiple tools exist to identify lanthipeptide gene clusters from genomic data, no programs are available to predict the post-translational modifications of lanthipeptides, such as the proteolytic cleavage of the leader peptide part or tailoring modifications based on the analysis of the gene cluster sequence. antiSMASH is a software pipeline for the identification of secondary metabolite biosynthetic clusters from genomic input and the prediction of products produced by the identified clusters. Here we present a novel antiSMASH module using a rule-based approach to combine signature motifs for biosynthetic enzymes and lanthipeptide-specific cleavage site motifs to identify lanthipeptide clusters in genomic data, assign the specific lanthipeptide class, predict prepeptide cleavage, tailoring reactions, and the processed molecular weight of the mature peptide products.

Editor: Eugene A. Permyakov, Russian Academy of Sciences, Institute for Biological Instrumentation, Russian Federation

Funding: Funding was provided by the German Ministry of Education and Research (BMBF) (Grant 0315585A to TW), German Centre for Infection Research (DZIF) (8000-402-2 to TW and WW), LAPTOP (Grant 245066 to WW). Funding for open access charge: Deutsche Forschungsgemeinschaft (DFG) and Open Access Publishing Fund of Tübingen University. The funders had no role in study design, data collection and analysis, decision to publish, or preparation of the manuscript.

Competing Interests: The authors have declared that no competing interests exist.

* E-mail: tilmann.weber@biotech.uni-tuebingen.de

¤a Current address: Max Planck Institute for Biology of Ageing, Cologne, Germany.
¤b Current address: The Novo Nordisk Foundation Center for Biosustainability, Technical University of Denmark, Hørsholm, Denmark

Introduction

Lanthipeptides

Lanthipeptides are polycyclic peptides named after the thioether-linked amino acids lanthionine and (2 *S*,3 *S*,6 *R*)-3-methyllanthionine contained in the mature peptide. Formerly called lantibiotics from "lanthionine-containing antibiotics", the new name lanthipeptide was proposed to also include non-antibiotic peptides of the same biosynthetic origin [1]. The best-known lanthipeptide is Nisin, which has been used as a food preservative since the 1940 s. Still a hot spot of natural products research, other lanthipeptides, for example NVB302, entered phase 1 clinical trials in 2011 as a treatment against *Clostridium difficile* infections [2]. Lanthipeptides are ribosomally synthesised and post-translationally modified peptides (RiPPs). The extensive post-translational modifications enhance the stability of the mature peptide against proteolysis and temperature stress. Lanthipeptides are encoded on the genome as a precursor peptide containing a leader and a core peptide part. The leader peptide, which is removed at a very late stage of the biosynthesis by a cluster-encoded or host-encoded protease, serves as a docking site for the modifying enzymes acting on the core peptide [3]. The lanthionine (Lan) and methyllanthionine (MeLan) residues are introduced in a two-step reaction. First, serine (Ser) and threonine (Thr) residues are dehydrated to dehydroalanine (Dha) and dehydrobutyrine (Dhb), respectively, usually with an intermediate phosphorylation step. In the second step, a Michael-type addition by cysteine (Cys) residues onto the dehydro amino acids then yields the thioether cross-links.

Depending on the biosynthetic enzymes installing the thioether cross-links, lanthipeptides are divided into different classes [4]. Currently, four lanthipeptide classes are known. In class I lanthipeptides, the dehydration is catalysed by a dedicated dehydratase commonly called LanB. Cyclisation is carried out by a cyclase called LanC. In specific gene clusters, the generic enzyme names might be replaced by a more specific name: for example in the nisin gene cluster, the LanB-type dehydratase is called NisB and the LanC-type cyclase is called NisC. For the remaining class II, III and IV lanthipeptides, both dehydration and cyclisation are catalysed by a single bi-functional enzyme. A class II LanM enzyme carries an N-terminal dehydratase domain with little sequence similarity to other characterised enzymes. The C-terminal cyclisation domain is similar to the LanC enzymes from class I lanthipeptide cyclases. The bi-functional enzymes for class III (LanKC) and IV (LanL) have a common N-terminal phospho- serine/phosphothreonine lyase domain and a central kinase domain. The C-terminal cyclisation domain in class III enzymes, while similar to the cyclisation domains from the other classes, lacks three zinc- bindinding residues that are conserved in the other classes. In class IV, those residues are present. In

Table 1. Lanthipeptide-related HMM profiles and scores.

Name	Description	Cutoff	File
LANC_like	LanC-like lantibiotics biosynthesis protein	17	LANC_like.hmm
DUF4135	Lantibiotic-associated domain	150	PF13575.hmm
Lant_dehyd_N	Lantibiotic dehydratase, N-terminus	20	Lant_dehyd_N.hmm
Lant_dehyd_C	Lantibiotic dehydratase, C-terminus	20	Lant_dehyd_C.hmm
Flavoprotein	Lantibiotic aminovinyl flavoprotein	20	PF0241.hmm
Trp_halogenase	Tryptophan halogenase	20	PF04820.hmm
p450	P450 oxygenase	60	PF00067.hmm
Pkinase	Protein kinase domain	30	PF00069.hmm
adh_short	Short-chain dehydrogenase	100	PF00106.hmm
adh_short_C2	Short-chain dehydrogenase, C-terminus	100	PF13561.hmm
Antimicr18	Lantibiotic antimicrobial peptide 18	20	Antimicrobial18.hmm
Gallidermin	Gallidermin	20	Gallidermin.hmm
L_biotic_A	Lantibiotic, type A	20	L_biotic_typeA.hmm
TIGR03731	Lantibiotic, gallidermin/nisin family	18	TIGR03731.hmm
leader_d	Lantibiotic leader lacticin 481 group	20	LE-LAC481.hmm
leader_eh	Lantibiotic leader mersacidin cinnamycin group	20	LE-MER+2PEP.hmm
leader_abc	Lantibiotic leader LanBC modified	20	LE-LanBC.hmm
mature_d	Lantibiotic peptide lacticin 481 group	20	MA-LAC481.hmm
mature_ab	Lantibiotic peptide nisin epidermin group	20	MA-NIS+EPI.hmm
mature_a	Lantibiotic peptide nisin group	20	MA-NIS.hmm
mature_b	Lantibiotic peptide epidermin group	20	MA-EPI.hmm
mature_ha	Lantibiotic peptide two component alpha	20	MA-2PEPA.hmm
mature_h_beta	Lantibiotic peptide two component beta	20	MA-2PEPB.hmm
lacticin_l	lantibiotic leader lacticin 481 group (Dufour et al)	20	LE-DUF.hmm
lacticin_mat	lantibiotic peptide lacticin 481 group (Dufour et al)	20	MA-DUF.hmm
LD_lanti_pre	FxLD family lanthipeptide	20	TIGR04363.hmm
strep_PEQAXS	*Streptomyces* PEQAXS motif lanthipeptide	20	strep_PEQAXS.hmm

addition to the introduction of Lan and MeLan, a number of further post-translational modifications may occur if the appropriate tailoring enzymes are present in the gene cluster. Among the modifications found in lanthipeptides is the formation of *S*-[(*Z*)-2-aminovinyl]-D-cysteine (AviCys) or *S*-[(*Z*)-2-aminovinyl]-(3 *S*)-3-methyl-D-cysteine (AviMeCys) [5]. The formation of AviCys and AviMeCys is catalysed by an enzyme of the family of homo-oligomeric flavin-containing cysteine decarboxylases.The enzyme with the generic designation LanD catalyses the oxidative decarboxylation of a C-terminal cysteine residue to a reactive thio–enol intermediate, which then cyclises with a Dha or Dhb residue, respectively, yielding AviCys or AviMeCys. An example would be the AviCys residue in epidermin introduced by EpiD [6]. Another post-translational modification is the chlorination of tryptophan residues catalysed by a flavin-dependent tryptophan halogenase designated LanH. This kind of reaction has been observed in the chlorination of tryptophan by MibH in microbisporicin biosynthesis [7]. If the cluster contains a cytochrome P450 oxygenase designated LanO, amino acids in the modified precursor peptide can be hydroxylated, as observed in the hydroxylation of proline in microbisporicin biosynthesis [7]. If the N-terminal amino acid is Dha and an oxidoreductase is present in the cluster, the N-terminal amino acid can be converted to lactate, as observed in the epicidin 280 cluster [8].

antiSMASH

antiSMASH, the antibiotics and secondary metabolite analysis shell, is a software pipeline for the automated identification of secondary metabolite biosynthesis clusters in whole genome sequenes of bacteria and fungi [9,10]. In antiSMASH, the user, who does not need specialized bioinformatics training, can upload microbial genome sequences which then are mined for secondary metabolite biosynthetic pathways in a fully automated manner. In the initial release of antiSMASH, the prediction of the products of the biosynthetic pathways was only possible for non-ribosomal peptide synthase (NRPS) and polyketide synthase (PKS) gene clusters [9]. In 2013, we released antibiotics and secondary metabolite analysis shell (antiSMASH) 2.0 [10]. In the new release, the architecture of the software was redesigned, now making it possible to add new predictors as self-contained plug-ins. antiSMASH is available as a web service at http://antismash.secondarymetabolites.org and can also be downloaded to run standalone. It is released under the GNU Affero Public License version 3, an OSI-approved Open Source license. Here we present the implementation of a lanthipeptide-specific analysis module for antiSMASH 2. With this module, lanthipeptide biosynthetic pathways can be automatically detected, classified and their products predicted. In addition, the module can identify putative protease cleavage sites for the most abundant class I and class II

Table 2. Class I cleavage site motif sequences.

Name	Position	Sequence
mutacin_1140	22...41	FAFDTTDTTIVASNDDPDTR
mutacin_Ny266	22...41	FTFDTTDTIVAESNDDPDTR
salivaricin_D	6...23	FNLDLVEVSK-SNTGASAR
nisin_U	6...24	FNLDLIKISK-ENNSGASPR
nisin_A	6...23	FNLDLVSVSKK-DSGASPR
nisin_Z	6...23	FNLDLLSVSKK-DSGASPR
nisin_Q	6...23	FNLDLVSVSKT-DSGASTR
gallidermin	11...30	FDLDVKVNAKESNDSGAEPR
epidermin	11...30	FNLDVKVNAKESNDSGAEPR
entianin	7...24	FDLDVVKVSKQ-DSKITPQ
Pep5	8...26	FDLEIKKETSQNTD-ELEPQ
epicidin_280	8...26	FDLEIKKDNME-NNNELEPQ
epilancin_K7	6...24	FDLNLNKGVETQK-SDLSPQ
geobacillin_I	7...23	FDLDIVVK-KQ-DDVVQPN
streptin	8...23	FDLDLKTNKK-D-TATPY
microbisporicin	17...33	LDLDLSIGVEE-ITAGPA

Table 3. Class II cleavage site motif sequences.

Name	Position	Sequence
mutacin_II	19...52	EL-TILGG
variacin	15...47	ELDAILGG
salivaricin_A	22...51	ELMEVAGG
butyrivibriocin	16...48	ELEQILGG
streptococcin_A_FF22	20...51	ELDNLLGG
lichenicidin_A1	30...74	EQHSIAGG
lichenicidin_A2	27...72	ELKALVGG
thermophilin_1277	18...66	ELEMLIGG
lacticin_A1	16...59	FDEDVFGA
lacticin_A2	26...65	EGDESHGG
nukacin_KQ_131	23...57	ELNEVLGA
macedocin	18...51	ELDQIIGA
salivaricin_B	24...56	ELDNVLGA
haloduracin_A1	33...69	ILAGVNGA
haloduracin_A2	28...65	ELSSLAGS
cytolysin	12...68	EMEAIQGS
plantaricin_W	24...59	NLLNVNGA
cinnamycin	52...59	IAATEAFA
mersacidin	41...48	QMDKLVGA
actagardine	35...64	EDRTIYAA
michiganin_A	36...66	RRVVSPYM

lanthipeptides and predict the molecular masses of the final products. The module is shipped with the antiSMASH 2.2 release and is also running on the public web server.

Design, Implementation and Validation

Secondary metabolite clusters in antiSMASH are identified using Hidden Markov Models (HMMs) of protein motifs for key biosynthetic enzymes. Which profiles are required to be identified for a specific secondary metabolite type is described by a rules file containing one rule-set per cluster type. Rule-sets can be simple hits against a single profile, AND and OR combinations of multiple profiles, or a selection of more complex rules, e.g. requiring a match against a minimum of n hits of a set of profiles. New secondary metabolite types can be included by adding new profile HMMs and extending the rules file. Once the cluster detection has identified a secondary metabolite cluster of a certain type, specific analysis modules can be run to generate a more detailed analysis of the pathway and the prediction of the product of a given cluster. Specific analysis modules are written as self-contained plug-ins that are loaded from the user's PYTHONPATH at run-time.

Identification of Lanthipeptide Biosynthetic Gene Clusters

To make more detailed cluster information available to the downstream specific analysis module, the cluster detection rules have been extended to include domain-specific Pfam [11] HMMs for the N-terminal domain (PFAM: PF13575) of class II LanM enzymes, the central kinase domain (PFAM: PF00069) of class III and IV enzymes, LanD-type flavin-dependent decarboxylases (PFAM: PF02441), LanH-type flavin-dependent halogenases (PFAM: PF04820), LanO-type cytochrome P450 oxygenases (PFAM: PF00067) and EciO-type short chain dehydrogenases (PFAM: PF00106, PF13561) (see Table 1 for details). Cutoff-values were determined empirically by analysing known sequences with the HMM profiles and manually adjusted to gain optimal sensitivity while keeping false positive hits low.

Prediction of the Lanthipeptide Class

Lanthipeptide classes are assigned by determining the domains present in the biosynthetic enzymes. Characteristic for class I lanthipeptides is the separate LanB enzyme containing the dehydratase domain, so the class prediction checks for a hit against the Lant_dehyd_N (PFAM: PF04737) or Lant_dehyd_C (PFAM: PF04738) domains. The dehydratase domain of class II LanM-type enzymes is characteristic as well, so if the cluster contains this dehydratase domain (PFAM: PF13575), the lanthipeptide will be considered class II. Class III LanKC-type and class IV LanL-type enzymes are identified via the central kinase domain (PFAM: PF00069). To differentiate between class III and IV enzymes, the algorithm checks if the conserved zinc binding sites in the C-terminal cyclase domain are absent (class III) or present (class IV).

Cleavage Site prediction

In the final step in lanthipeptide biosynthesis, a protease cleaves the leader peptide part off the modified precursor peptide to yield the mature peptide. Depending on the class of the lanthipeptide, the cleavage site motives vary widely. In order to predict the cleavage site, we have created a manually curated set of HMMs, one for lanthipeptide classes I and II each (Tables 2, 3). Profiles for the HMMer 2.3.2 software [12] were generated using the command hmmbuild profile.hmm alignment.fa; hmmcalibrate profile.hmm. As the method depends on the size of the seed sequence data set, we decided not to include cleavage site predictions for class III (only six seed sequences available) and class IV (no experimentally verified sequences available) lanthipeptides. Once more seed sequences become available for these two classes,

Table 4. Stability of the prediction motifs.

Class	# of Sequences	# Found	% Found	# Correct	% Correct
I	16	16	1.00	16	1.00
II	21	21	1.00	17	0.81

adding cleavage site predictions using the same method will be straightforward.

Monoisotopic Mass, Molecular Weight and Alternative Weights

Once the cleavage site is predicted, both the monoisotopic mass and the average molecular weight are calculated. For the calculation of these numbers it is assumed that all Ser and Thr residues are dehydrated to Dha and Dhb respectively. As a lack of dehydration is frequently observed but the mechanism behind this has not been elucidated, we also calculate alternative weights under the assumption that one up to n Ser or Thr are not dehydrated, where n is the number of Ser and Thr residues in the core peptide subtracted by the number of Cys residues in the core peptide. This upper bound is set to account for the observation that all Cys residues tend to participate in Lan or MeLan bridges with Dha or Dhb residues.

Predicting Tailoring Reactions

Tailoring reactions are not performed by the core biosynthetic enzymes that perform the dehydration and cyclisation but instead by additional enzymes also encoded on the cluster.

AviCys and AviMeCys formation. The unusual amino acids AviCys and AviMeCys are formed by oxidative decarboxylation of the C-terminal Cys residue. The resulting thio–enol intermediate cyclises with a Dha or Dhb side-chain respectively. This reaction is catalysed by a LanD-type flavin-dependent decarboxylase, identified by a hit against the PFAM PF02441 profile with a score ≥20. The formation of AviCys or AviMeCys reduces the predicted peptide weight by 46 Da.

Halogenation. LanH-type halogenases chlorinate an amino acid side chain, increasing the predicted peptide weight by 34 Da. These enzymes are identified by a hit against the PFAM PF04820 profile with a score ≥20.

Hydroxylation. LanO-type cytochrome P450 oxygenases catalyse the regiospecific oxidation of non-activated hydrocarbons. The enzymes are identified by a hit against PFAM PF00067 with a score ≥60. The hydroxylation increases the predicted peptide weight by 16 Da.

Lactate formation. EciO-type short-chain dehydrogenases, identified by a hit against PFAM PF00106 or PFAM PF13561 with a score ≥100, catalyse the final step of the conversion of the N-terminal Dha residue to lactate. This increases the predicted peptide weight by 2 Da.

Predicting the number of Lan and MeLan bridges. To predict the number of Lan and MeLan bridges, a simple heuristic is applied using the formula.

$$|b| = \min(|S|+|T|,|C|) - v,$$

$$v = \begin{cases} 1 & \text{if AviCys or AviMeCys residue is present} \\ 0 & \text{otherwise} \end{cases}$$

where |b| is the number of bridges, and |S|, |T|, |C| is the number of amino acids Ser, Thr, and Cys in the core peptide.

Validation and Benchmarking

To validate the robustness of the cleavage site HMM-profiles, we used *n-fold cross validation.* For a seed alignment of size n, we built n different profiles by including $n - 1$ sequences, and then checked if a cleavage site was predictable and correct for the left out sequence. A cleavage site was predictable if the profile produced a hit with a score above the threshold. A cleavage site was considered correct if the prediction matched the ungapped seed sequence not used for building the profile. A cleavage site was predicted in all test sequences for both class I and II lanthipeptides. Cleavage sites were correct in all class I test inputs and in 81% of the class II test inputs (Table 4). To benchmark the overall performance of the prediction, we ran a number of lanthipeptide

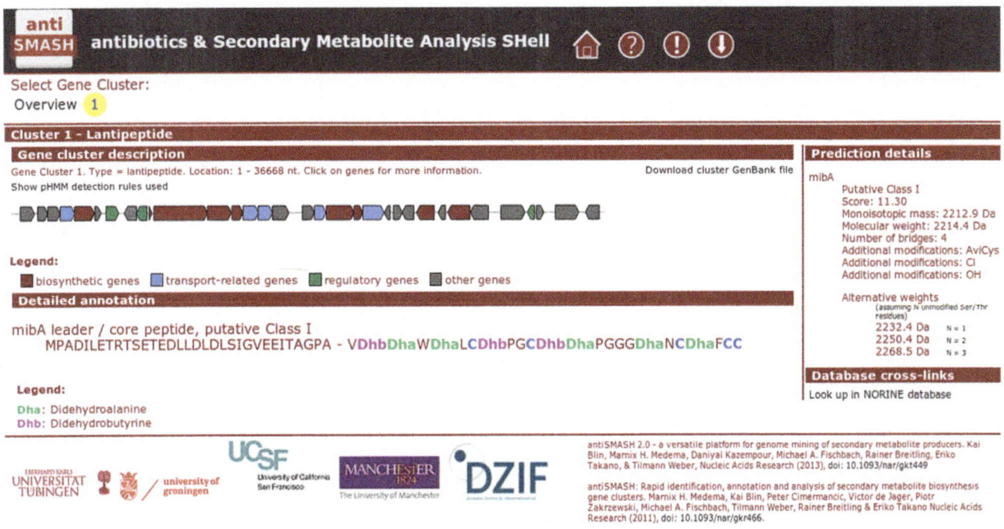

Figure 1. Example lanthipeptide output antiSMASH 2.2 output for the microbisporicin [5] gene cluster, showing the predicted leader/core peptide split and the predicted tailoring reactions and weights in the sidebar.

Table 5. Benchmark results.

Substance	Class	Predicted Mass (Da)	Actual Mass (Da)	Number of bridges	Source
Salivaricin D	I	3466.7	3467.5	4	[19]
Nisin U	I	3029.6	3029.0	5	[20]
Nisin A	I	3353.9	3354.5	5	[21]
Nisin Z	I	3330.9	3331.5	5	[21]
Nisin Q	I	3326.9	3327.3	5	[21]
Gallidermin	I	2164.0	2164.0	4	[6]
Epidermin	I	2164.0	2164.0	4	[6]
Entianin	I	3346.7	3346.0	5	[22]
Pep5	I	3487.1	3488.0	3	[6]
Epicidin 280	I	3135.6	3135.0	3	[8]
Geobacillin I	I	3261.5	3265.0	7	[23]
Streptin 1′	I	2441.9	2442.0	3	[24]
Microbisporicin A2	I	2232.4	2232.0	5	[25]
Mutacin II	II	3243.5	3244.0	3	[26]
Salivaricin A2	II	2366.6	2368.0	3	[27]
Streptococcin A-FF22	II	2796.1	2795.0	3	[28]
Lichencidin A1	II	3250.7	3251.0	4	[18]
Lichencidin A2	II	3632.8†	3021.0	4	[18]
Thermophilin 1277	II	3395.9‡	3428.0	2‡	[16]
Lacticin 3147 A1	II	3322.6	3322.3	4	[29]
Lacticin 3147 A2	II	2843.2	2847.5	3	[29]
Salivaricin B	II	2733.1	2740.0	3	[30]
Cinnamycin	II	2043.2	2041.0	3	[31]
Mersacidin	II	2399.0†	1826.3	3	[32]
Actagardine	II	1856.2	1860.5	4	[33]
Michiganin A	II	2145.5	2145.0	4*	[34]
Planosporicin	I	2193.3	2194.0	5	[13]
Epilancin 15X	I	3171.8	3171.7	3	[14]

†N-terminal removal of six amino acids not predicted.
‡Contains a disulphite bridge.
* Not shown experimentally.

biosynthetic gene clusters through antiSMASH. We checked if the gene cluster was identified, the precursor peptide was detected, and finally the peptide mass was predicted correctly.

Among the clusters run for benchmarking, we included the planosporicin and epilancin 15X clusters. The cleavage sites of both of these lanthipeptides were not part of the seed alignments. For planosporicin, the algorithm predicted a mass of 2193.3 Da (actual mass 2194 Da [13]) and correctly predicted 5 lanthionine bridges. For epilancin 15X, the algorithm predicted a mass of 3171.8 Da (actual mass 3171.7 Da [14]) and correctly predicted 3 lanthionine bridges and the N-terminal lactate.

Results and Discussion

Only few tools are currently available that allow the automated identification of RiPPs. Apart from antiSMASH, there is BAGEL, recently released in version 3 [15]. BAGEL targets a large number of different ribosomally synthesised peptides. For lanthipeptides, BAGEL only predicts the leader peptide and the class, but does not attempt to predict tailoring reactions, number of Lan and MeLan bridges or the molecular weight.

The detection of enzymes responsible for tailoring reactions is central in the prediction of the mature peptide mass. antiSMASH can predict the presence of AviCys residues which are found, for example, in epidermin and microbisporicin. Halogenations and hydroxylations, which occur in microbisporicin biosynthesis are also detected. A remaining issue in mass prediction is that not all Ser and Thr residues are dehydrated in all the peptides, resulting in mass predictions that are 18 Da too low per undehydrated amino acid. antiSMASH assists in detecting the presence of undehydrated residues by providing alternative mass predictions for lanthipeptides that carry more Ser and Thr than Cys residues. Once the tailoring reactions have been predicted, the final step is the prediction of the number of Lan and MeLan bridges. The naïve heuristic of counting the Cys and Ser/Thr residues and then using the smaller number fails if the mature peptide contains an AviCys or AviMeCys residue and needs to be adjusted accordingly.

The prediction details of the lanthipeptide module are included in the standard antiSMASH output. For the HTML output (Figure 1), lanthipeptide class and leader/core peptide split predicted are shown in the "detailed annotation" section of the cluster page. The score of the class prediction, predicted monoisotopic mass and molecular weights, the number of bridges and the identified additional modifications are shown in the "prediction details" sidebar.

To test the accuracy of the antiSMASH lanthipeptide predictions, a benchmark dataset was analyzed and the results evaluated:

The determination of the lanthipeptide class based on the biosynthetic enzymes in the cluster is straightforward, and antiSMASH performed this tasks flawlessly on the benchmark data set (Table 5). Using the advanced heuristic, antiSMASH was able to correctly predict the number or bridges in almost all lanthipeptides of the benchmark data-set. The heuristic only fails if two Cys residues form a disulphide bridge, a rare occurrence observed in e.g. thermophilin 1277 [16]. Unfortunately, the enzyme catalysing the formation of the disulphide bridge is not present in the gene cluster and thus cannot be used to predict a disulphide bridge formation [16]. Predicting the core peptide sequence is more difficult. The cleavage site motif of class II lanthipeptides (Table 3) is relatively uniform, largely consisting in two amino acids with small side chains that are preceded by alternating hydrophobic and hydrophilic residues. In fact, the four class II cleavage sites incorrectly predicted during validation (Table 4) all differed from this pattern and contained a site that more closely matches the motif upstream of the actual cleavage site. Some class II core peptides like mersacidin [17] or lichenicidin A2 [18] lose an additional six amino acids at the N-terminus after the proteolytic cleavage [18], so it seems likely that the predicted cleavage sites may be accurate and some additional enzyme catalyses the N-terminal modifications. Class I leader peptides also carry a short motif of alternating hydrophobic and hydrophilic amino acids, usually called the FNLD motif. The spacer between this motif and the actual cleavage site varies. At position −2 in front of the cleavage site, many leader peptides carry a proline residue (Table 2). During validation, all cleavage sites were predicted correctly (Table 4). Due to the strong signal of the FNLD motif, class I prediction (stability 100%) is even more robust than the class II prediction (stability 81%) with the shorter motif. As a proof of concept, we used the gene clusters of the recently published planosporicin [13] and the lactate-containing epilancin 15X [14]. Both precursor peptides contain cleavage sites that are distinct from all the sequences included in the class I seed alignment. For both lanthipeptides the algorithm is able to correctly predict the mass, number of Lan/MeLan bridges and tailoring modifications.

Conclusions

With the algorithm described in this paper, antiSMASH gains extensive lanthipeptide-specific predictive capabilities. antiSMASH is the only software currently available that will predict lanthipeptide class, core peptide cleavage, tailoring reactions, number of Lan and MeLan bridges, and the molecular weight of the mature peptide product. These informations obtained by analyzing the genome sequence of the producers can give crucial hints for the identification of the lanthipeptide compounds in chemical analytics.

Author Contributions

Conceived and designed the experiments: KB TW. Performed the experiments: KB DK. Analyzed the data: KB DK. Wrote the paper: KB WW TW.

References

1. Knerr PJ, van der Donk WA (2012) Discovery, Biosynthesis, and Engineering of Lantipeptides. Annual Review of Biochemistry 81: 1–27.
2. Crowther GS, Baines SD, Todhunter SL, Freeman J, Chilton CH, et al. (2013) Evaluation of NVB302 versus vancomycin activity in an in vitro human gut model of *Clostridium difficile* infection. Journal of Antimicrobial Chemotherapy 68: 168–176.
3. Plat A, Kluskens LD, Kuipers A, Rink R, Moll GN (2011) Requirements of the engineered leader peptide of nisin for inducing modification, export, and cleavage. Applied and Environmental Microbiology 77: 604–611.
4. Arnison PG, Bibb MJ, Bierbaum G, Bowers AA, Bugni TS, et al. (2013) Ribosomally synthesized and post-translationally modified peptide natural products: overview and recommendations for a universal nomenclature. Natural Product Reports 30: 108–160.
5. Sit CS, Yoganathan S, Vederas JC (2011) Biosynthesis of Aminovinyl-Cysteine-Containing Peptides and Its Application in the Production of Potential Drug Candidates. Accounts of Chemical Research 44: 261–268.
6. Kellner R, Jung G, Hörner T, Zähner H, Schnell N, et al. (1988) Gallidermin: a new lanthionine-containing polypeptide antibiotic. European Journal of Biochemistry 177: 53–59.
7. Foulston LC, Bibb MJ (2010) Microbisporicin gene cluster reveals unusual features of lantibiotic biosynthesis in actinomycetes. Proceedings of the National Academy of Sciences 107: 13461–13466.
8. Heidrich C, Pag U, Josten M, Metzger J, Jack RW, et al. (1998) Isolation, characterization, and heterologous expression of the novel lantibiotic epicidin 280 and analysis of its biosynthetic gene cluster. Applied and Environmental Microbiology 64: 3140–3146.
9. Medema MH, Blin K, Cimermancic P, de Jager V, Zakrzewski P, et al. (2011) antiSMASH: rapid identification, annotation and analysis of secondary metabolite biosynthesis gene clusters in bacterial and fungal genome sequences. Nucleic Acids Research 39: W339–W346.
10. Blin K, Medema MH, Kazempour D, Fischbach MA, Breitling R, et al. (2013) antiSMASH 2.0– a versatile platform for genome mining of secondary metabolite producers. Nucleic Acids Research 41: W204–W212.
11. Punta M, Coggill PC, Eberhardt RY, Mistry J, Tate J, et al. (2012) The Pfam protein families database. Nucleic Acids Research 40: D290–D301.
12. Durbin R, Eddy SR, Krogh A, Mitchison G (1998) Biological sequence analysis: probabilistic models of proteins and nucleic acids: Cambridge University Press.
13. Sherwood EJ, Bibb MJ (2013) The antibiotic planosporicin coordinates its own production in the actinomycete *Planomonospora alba*. Proceedings of the National Academy of Sciences 110: E2500–E2509.
14. Velasquez JE, Zhang X, van der Donk WA (2011) Biosynthesis of the antimicrobial peptide epilancin 15X and its N-terminal lactate. Chemistry & Biology 18: 857–867.
15. van Heel AJ, de Jong A, Montalban-Lopez M, Kok J, Kuipers OP (2013) BAGEL3: automated identification of genes encoding bacteriocins and (non-) bactericidal posttranslationally modified peptides. Nucleic Acids Research 41: W448–W453.
16. Kabuki T, Uenishi H, Seto Y, Yoshioka T, Nakajima H (2009) A unique lantibiotic, thermophilin 1277, containing a disulfide bridge and two thioether bridges. Journal of Applied Microbiology 106: 853–862.
17. Altena K, Guder A, Cramer C, Bierbaum G (2000) Biosynthesis of the lantibiotic mersacidin: organization of a type B lantibiotic gene cluster. Applied and Environmental Microbiology 66: 2565–2571.
18. Dischinger J, Josten M, Szekat C, Sahl H-G, Bierbaum G (2009) Production of the novel two-peptide lantibiotic lichenicidin by *Bacillus licheniformis* DSM 13. PLoS ONE 4: e6788.
19. Birri DJ, Brede DA, Nes IF (2012) Salivaricin D, a novel intrinsically trypsin-resistant lantibiotic from *Streptococcus salivarius* 5M6c isolated from a healthy infant. Applied and Environmental Microbiology 78: 402–410.
20. Wirawan RE, Klesse NA, Jack RW, Tagg JR (2006) Molecular and Genetic Characterization of a Novel Nisin Variant Produced by *Streptococcus uberis*. Applied and Environmental Microbiology 72: 1148–1156.
21. Zendo T, Fukao M, Ueda K, Higuchi T, Nakayama J, et al. (2003) Identification of the lantibiotic nisin Q, a new natural nisin variant produced by *Lactococcus lactis* 61–14 isolated from a river in Japan. Bioscience, Biotechnology, and Biochemistry 67: 1616–1619.
22. Fuchs SW, Jaskolla TW, Bochmann S, Kötter P, Wichelhaus T, et al. (2011) Entianin, a novel subtilin-like lantibiotic from *Bacillus subtilis* subsp. *spizizenii* DSM 15029T with high antimicrobial activity. Applied and Environmental Microbiology 77: 1698–1707.
23. Garg N, Tang W, Goto Y, Nair SK, van der Donk WA (2012) Lantibiotics from *Geobacillus thermodenitrificans*. Proceedings of the National Academy of Sciences 109: 5241–5246.

24. Wescombe PA, Tagg JR (2003) Purification and characterization of streptin, a type A1 lantibiotic produced by *Streptococcus pyogenes*. Applied and Environmental Microbiology 69: 2737–2747.
25. Castiglione F, Lazzarini A, Carrano L, Corti E, Ciciliato I, et al. (2008) Determining the Structure and Mode of Action of Microbisporicin, a Potent Lantibiotic Active Against Multiresistant Pathogens. Chemistry & Biology 15: 22–31.
26. Novak J, Caufield PW, Miller EJ (1994) Isolation and biochemical characterization of a novel lantibiotic mutacin from *Streptococcus mutans*. Journal of Bacteriology 176: 4316–4320.
27. Wescombe PA, Upton M, Dierksen KP, Ragland NL, Sivabalan S, et al. (2006) Production of the lantibiotic salivaricin A and its variants by oral streptococci and use of a specific induction assay to detect their presence in human saliva. Applied and Environmental Microbiology 72: 1459–1466.
28. Hynes WL, Ferretti J, Tagg J (1993) Cloning of the gene encoding Streptococcin A-FF22, a novel lantibiotic produced by *Streptococcus pyogenes*, and determination of its nucleotide sequence. Applied and Environmental Microbiology 59: 1969–1971.
29. Ryan MP, Jack RW, Josten M, Sahl H-G, Jung G, et al. (1999) Extensive Post-translational Modification, Including Serine to D-Alanine Conversion, in the Two-component Lantibiotic, Lacticin 3147. Journal of Biological Chemistry 274: 37544–37550.
30. Hyink O, Wescombe PA, Upton M, Ragland N, Burton JP, et al. (2007) Salivaricin A2 and the novel lantibiotic salivaricin B are encoded at adjacent loci on a 190-kilobase transmissible megaplasmid in the oral probiotic strain *Streptococcus salivarius* K12. Applied and Environmental Microbiology 73: 1107–1113.
31. Widdick D, Dodd H, Barraille P, White J, Stein T, et al. (2003) Cloning and engineering of the cinnamycin biosynthetic gene cluster from *Streptomyces cinnamoneus cinnamoneus* DSM 40005. Proceedings of the National Academy of Sciences 100: 4316–4321.
32. Herzner AM, Dischinger J, Szekat C, Josten M, Schmitz S, et al. (2011) Expression of the Lantibiotic Mersacidin in *Bacillus amyloliquefaciens* FZB42. PLoS ONE 6: e22389.
33. Boakes S, Cortes J, Appleyard AN, Rudd BA, Dawson MJ (2009) Organization of the genes encoding the biosynthesis of actagardine and engineering of a variant generation system. Molecular Microbiology 72: 1126–1136.
34. Holtsmark I, Mantzilas D, Eijsink V, Brurberg M (2006) Purification, characterization, and gene sequence of michiganin A, an actagardine-like lantibiotic produced by the tomato pathogen *Clavibacter michiganensis* subsp. *michiganensis*. Applied and Environmental Microbiology 72: 5814–5821.

Permissions

All chapters in this book were first published in PLOS ONE, by The Public Library of Science; hereby published with permission under the Creative Commons Attribution License or equivalent. Every chapter published in this book has been scrutinized by our experts. Their significance has been extensively debated. The topics covered herein carry significant findings which will fuel the growth of the discipline. They may even be implemented as practical applications or may be referred to as a beginning point for another development.

The contributors of this book come from diverse backgrounds, making this book a truly international effort. This book will bring forth new frontiers with its revolutionizing research information and detailed analysis of the nascent developments around the world.

We would like to thank all the contributing authors for lending their expertise to make the book truly unique. They have played a crucial role in the development of this book. Without their invaluable contributions this book wouldn't have been possible. They have made vital efforts to compile up to date information on the varied aspects of this subject to make this book a valuable addition to the collection of many professionals and students.

This book was conceptualized with the vision of imparting up-to-date information and advanced data in this field. To ensure the same, a matchless editorial board was set up. Every individual on the board went through rigorous rounds of assessment to prove their worth. After which they invested a large part of their time researching and compiling the most relevant data for our readers.

The editorial board has been involved in producing this book since its inception. They have spent rigorous hours researching and exploring the diverse topics which have resulted in the successful publishing of this book. They have passed on their knowledge of decades through this book. To expedite this challenging task, the publisher supported the team at every step. A small team of assistant editors was also appointed to further simplify the editing procedure and attain best results for the readers.

Apart from the editorial board, the designing team has also invested a significant amount of their time in understanding the subject and creating the most relevant covers. They scrutinized every image to scout for the most suitable representation of the subject and create an appropriate cover for the book.

The publishing team has been an ardent support to the editorial, designing and production team. Their endless efforts to recruit the best for this project, has resulted in the accomplishment of this book. They are a veteran in the field of academics and their pool of knowledge is as vast as their experience in printing. Their expertise and guidance has proved useful at every step. Their uncompromising quality standards have made this book an exceptional effort. Their encouragement from time to time has been an inspiration for everyone.

The publisher and the editorial board hope that this book will prove to be a valuable piece of knowledge for researchers, students, practitioners and scholars across the globe.

List of Contributors

Zhaojuan Zheng, Cuiqing Ma, Chao Gao, Fengsong Li, Jiayang Qin, Haiwei Zhang, Kai Wang
State Key Laboratory of Microbial Technology, Shandong University, Jinan, People's Republic of China

Ping Xu
MOE Key Laboratory of Microbial Metabolism and School of Life Sciences and Biotechnology, Shanghai Jiao Tong University, Shanghai, People's Republic of China
State Key Laboratory of Microbial Technology, Shandong University, Jinan, People's Republic of China

Teng Bao, Xian Zhang, Zhiming Rao, Xiaojing Zhao, Taowei Yang
The Key Laboratory of Industrial Biotechnology of Ministry of Education, School of Biotechnology, Jiangnan University, Wuxi, Jiangsu, People's Republic of China

Rongzhen Zhang
School of Biotechnology, Jiangnan University, Wuxi, Jiangsu, People's Republic of China

Zhenghong Xu
School of Medicine and Pharmaceuticals, Jiangnan University, Wuxi, Jiangsu,
People's Republic of China

Shangtian Yang
Department of Chemical Engineering, Ohio State University, Columbus, Ohio, United States of America

Inna Biela, Naomi Tidten-Luksch, Florian Immekus, Serghei Glinca, Tran Xuan Phong Nguyen, Hans-Dieter Gerber, Andreas Heine, Gerhard Klebe, Klaus Reuter
Institut für Pharmazeutische Chemie, Philipps-Universität Marburg, Marburg, Germany

Siaw Cheng Lau, Hamid Reza Fard Masoumi
Department of Chemistry, Faculty of Science, Universiti Putra Malaysia, Serdang, Selangor, Malaysia

Hong Ngee Lim
Institute of Advanced Technology, University Putra Malaysia, Serdang, Selangor, Malaysia
Department of Chemistry, Faculty of Science, Universiti Putra Malaysia, Serdang, Selangor, Malaysia

Mahiran Basri
Department of Chemistry, Faculty of Science, Universiti Putra Malaysia, Serdang, Selangor, Malaysia
Laboratory of Molecular Biomedicine, Institute of Bioscience, Universiti Putra Malaysia, Serdang, Selangor, Malaysia

Asilah Ahmad Tajudin
Faculty of Biotechnology and Biomolecular Sciences, Universiti Putra Malaysia, Serdang, Selangor, Malaysia

Nay Ming Huang, Alagarsamy Pandikumar
Low Dimensional Materials Research Centre, Department of Physics, Faculty of Science, University of Malaya, Kuala Lumpur, Malaysia

Chi Hua Chia
School of Applied Physics, Faculty of Science and Technology, Universiti Kebangsaan Malaysia, Bangi, Selangor, Malaysia

Yoshito Andou
Eco-Town Collaborative Research and Development Center for the Environment and Recycling, Kyushu Institute of Technology, Kitakyushu-shi, Fukuoka, Japan

Balázs Könnyű
Department of Plant Taxonomy and Ecology, Institute of Biology, Eötvös Loránd University, Budapest, Hungary

Tamás Czárán
Theoretical Biology and Ecology Research Group, Institute of Biology, Eötvös Loránd University, Budapest, Hungary

Jun-Hui Zhang, Yu-Yan Jiang, Ying Lin, Yu-Fei Sun, Sui-Ping Zheng, Shuang-Yan Han
Guangdong Key Laboratory of Fermentation and Enzyme Engineering, School of Bioscience and Bioengineering, South China University of Technology, Guangzhou, P. R. China

Binbin Sheng, Min Lv, Haiwei Zhang, Tong Qin, Chao Gao, Cuiqing Ma, Ping Xu
State Key Laboratory of Microbial Technology, Shandong University, Jinan, People's Republic of China

Zhaojuan Zheng
College of Chemical Engineering, Nanjing Forestry University, Nanjing, People's Republic of China

Shuguang Yuan
Laboratory of Molecular Plant Physiology, Institute of Botany and Microbiology, KU Leuven, Heverlee, Belgium
Laboratory for Biomolecular Modelling, Department of Chemistry, Division of Biochemistry, Molecular and Structural Biology, KU Leuven, Heverlee, Belgium

Katrien Le Roy, Willem Lammens, Wim Van den Ende
Laboratory of Molecular Plant Physiology, Institute of Botany and Microbiology, KU Leuven, Heverlee, Belgium

Tom Venken, Marc De Maeyer
Laboratory for Biomolecular Modelling, Department of Chemistry, Division of Biochemistry, Molecular and Structural Biology, KU Leuven, Heverlee, Belgium

Martin R. Challand, James Spencer
School of Cellular and Molecular Medicine, University of Bristol Medical Sciences Building, Bristol, United Kingdom

Enrico Salvadori, Christopher W. M. Kay
Institute of Structural and Molecular Biology, University College London, London, United Kingdom
London Centre for Nanotechnology, University College London, London, United Kingdom

Rebecca C. Driesener
Chemistry, University of Southampton, Highfield, Southampton, United Kingdom

Peter L. Roach
Institute for Life Sciences, University of Southampton, Highfield, Southampton, United Kingdom
Chemistry, University of Southampton, Highfield, Southampton, United Kingdom

Daniel Krell, Mawuelikem Assoku
Molecular and Population Genetics Laboratory, Wellcome Trust Centre for Human Genetics, University of Oxford, Oxford, United Kingdom
Department of Medical Oncology, University College London Hospitals, London, United Kingdom

Malcolm Galloway
Department of Cellular Pathology, The Royal Free Hospital, London, United Kingdom

Paul Mulholland
Department of Medical Oncology, University College London Hospitals, London, United Kingdom

Ian Tomlinson, Chiara Bardella
Molecular and Population Genetics Laboratory, Wellcome Trust Centre for Human Genetics, University of Oxford, Oxford, United Kingdom

Gustavo Pierdominici-Sottile, Rodrigo Cossio Pérez, Juliana Palma
Departamento de Ciencia y Tecnologı´a, Universidad Nacional de Quilmes, Bernal, Argentina

Johan F. Galindo
Quantum and Computational Chemistry Group, Departamento de Química, Universidad Nacional de Colombia, Bogotá, Colombia

Chao Gao, Jianhua Qiu, Cuiqing Ma
State Key Laboratory of Microbial Technology, Shandong University, Jinan, People's Republic of China

Ping Xu
State Key Laboratory of Microbial Metabolism, Shanghai Jiao Tong University, Shanghai, People's Republic of China
State Key Laboratory of Microbial Technology, Shandong University, Jinan, People's Republic of China

Zhi Guan, Yan-Li Chen, Yi Yuan, Jian Song, Da-Cheng Yang, Yang Xue, Yan-Hong He
School of Chemistry and Chemical Engineering, Southwest University, Chongqing, P. R. China

Elisha Krieg, Haim Weissman, Boris Rybtchinski
Department of Organic Chemistry, Weizmann Institute of Science, Rehovot, Israel

Shira Albeck
Israel Structural Proteomics Center (ISPC), Weizmann Institute of Science, Rehovot, Israel

Eyal Shimoni
Department of Chemical Research Support, Weizmann Institute of Science, Rehovot, Israel

Irene Rodríguez, Elisa Longás, Miguel de Vega, Margarita Salas
Instituto de Biología Molecular "Eladio Viñuela" (Centro Superior de Investigaciones Científicas), Centro de Biología Molecular "Severo Ochoa" (Centro Superior de Investigaciones Cientı´ficas-Universidad Autónoma de Madrid), Cantoblanco, Madrid, Spain

Pier Luigi Buttigieg, Frank Oliver Glöckner
Microbial Genomics and Bioinformatics Group, Max Planck Institute for Marine Microbiology, Bremen, Germany
Jacobs University gGmbH, Bremen, Germany

Wolfgang Hankeln, Renzo Kottmann, Pelin Yilmaz
Microbial Genomics and Bioinformatics Group, Max Planck Institute for Marine Microbiology, Bremen, Germany

Ivaylo Kostadinov
Departament de Biologia Marina i Oceanografia, Institut de Ciències del Mar, Barcelona, Spain

Melissa Beth Duhaime
Department of Ecology and Evolutionary Biology, University of Michigan, Ann Arbor, Michigan, United States of America

Solimabi Wahidullah, Deepak N. Naik, Prabha Devi
Bioorganic Chemistry Lab, Chemical Oceanography Division, CSIR- National Institute of Oceanography, Dona Paula, Goa, India

Maria Luisa Cristina, Anna Maria Spagnolo, Marina Sartini, Donatella Panatto, Roberto Gasparini, Paolo Orlando, Gianluca Ottria, Fernanda Perdelli
Department of Health Sciences, University of Genoa, Genoa, Italy

Yuanjun Liu, Aihua Yuan
School of Environmental and Chemical Engineering, Jiangsu University of Science and Technology, Zhenjiang, China

Guoxing Zhu
School of Chemistry and Chemical Engineering, Jiangsu University, Zhenjiang, China
State Key Laboratory of Coordination Chemistry, Nanjing University, Nanjing, P. R. China

Jing Yang, Xiaoping Shen
State Key Laboratory of Coordination Chemistry, Nanjing University, Nanjing, P. R. China

Nicola Laurieri, Angela J. Russell, James E. Egleton
Department of Pharmacology, University of Oxford, Oxford, United Kingdom
Department of Chemistry, Chemistry Research Laboratory, University of Oxford, Oxford, United Kingdom

Julien Dairou, Fernando Rodrigues-Lima, Marie Dupret
Universite´ Paris Diderot, Sorbonne Paris Cité, Unit of Functional and Adaptive Biology, Paris, France

Lesley A. Stanley
Consultant in Investigative Toxicology, Linlithgow, West Lothian, United Kingdom

Jean-Edith Sim
Department of Pharmacology, University of Oxford, Oxford, United Kingdom
Faculty of Science, Engineering and Computing, Kingston University, Kingston on Thames, United Kingdom

Kai Blin, Wolfgang Wohlleben, Tilmann Weber
Microbiology/Biotechnology, Interfaculty Institute of Microbiology and Infection Medicine, Eberhard-Karls-University of Tübingen, Tübingen, Germany
German Centre for Infection Research, Tübingen, Germany

Daniyal Kazempour
Microbiology/Biotechnology, Interfaculty Institute of Microbiology and Infection Medicine, Eberhard-Karls-University of Tübingen, Tübingen, Germany

Index